AF356168

6.69x9.61_P5-1N6_M   9783725808434
2175_3172242
84351ELV8B

microRNAs in Health and Disease

microRNAs in Health and Disease

Editors

Milena Rizzo
Elena Levantini

Basel • Beijing • Wuhan • Barcelona • Belgrade • Novi Sad • Cluj • Manchester

Editors
Milena Rizzo
Institute of Clinical
Physiology (IFC), National
Research Council (CNR)
Pisa
Italy

Elena Levantini
Institute of Biomedical
Technologies (ITB), National
Research Council (CNR)
Pisa
Italy

Editorial Office
MDPI
St. Alban-Anlage 66
4052 Basel, Switzerland

This is a reprint of articles from the Special Issue published online in the open access journal *Biomedicines* (ISSN 2227-9059) (available at: https://www.mdpi.com/journal/biomedicines/special_issues/microRNA_Disease).

For citation purposes, cite each article independently as indicated on the article page online and as indicated below:

Lastname, A.A.; Lastname, B.B. Article Title. *Journal Name* **Year**, *Volume Number*, Page Range.

ISBN 978-3-7258-0843-4 (Hbk)
ISBN 978-3-7258-0844-1 (PDF)
doi.org/10.3390/books978-3-7258-0844-1

Contents

About the Editors

Milena Rizzo

Milena Rizzo holds a Master of Science degree in Biology (2000) and a Ph.D. in Molecular Biotechnology (2004). She is the head of the Non-coding RNA Unit at the Institute of Clinical Physiology, part of the National Research Council (CNR) in Pisa, Italy. As a researcher with an extensive molecular and cellular oncology background, she has major experience in the field of cancer research, with a particular focus on non-coding RNAs. Her research interests are centered on cancer biology, specifically the regulatory functions of non-coding RNA, primarily miRNAs, in the onset and progression of cancer, mechanisms of drug resistance and the role of extracellular miRNAs in cell-to-cell communication. Dr. Rizzo has been an integral part of numerous national projects, and has spearheaded initiatives aimed at investigating the role of miRNAs in drug resistance in metastatic prostate cancer and new therapeutic approaches for highly hypoxic glioblastoma. With a publication record of 44 papers, Dr. Rizzo's contributions to the scientific community extend beyond research. She has served as reviewer for several international journals and has played an editorial role for Special Issues. Her expertise is also recognized in her capacity as a reviewer for national and international public funding agencies.

Elena Levantini

Elena Levantini, PhD, currently leads the Molecular Oncology lab at the Institute of Biomedical Technologies (ITB) of the National Research Council (CNR) of Italy in Pisa. Supported by the Italian AIRC Foundation for Cancer Research and various international funding agencies, her work is committed to translating benchwork into clinical applications for pulmonary tumorigenesis. Her current research investigates transcriptional dysregulation in solid tumors at a high-resolution and spatial transcriptomic level, examining interactions with the tumor microenvironment to discover novel therapies for cancers driven by EGFR and KRAS mutations. Educated in hematopoiesis and leukemia at the University of Pisa and Harvard Medical School, Dr. Levantini's focus shifted to oncology. She has elucidated the critical role of CEBPa in tumorigenesis and identified BMI1 as a significant factor in epithelial hyperproliferation. Her work has contributed to the advancement of clinical trials for a novel compound targeting BMI1, reflecting her commitment to developing therapeutic options for lung cancer patients.

Dr. Levantini's received the Cancer Research Scholar Award and the Nolan Miller Discovery Award for Lung Cancer Research, among others. She is a member of numerous oncology journal editorial boards and a reviewer for global funding agencies. A passionate advocate for science education, she engages with diverse audiences, including schools and correctional facilities, emphasizing the importance of public scientific literacy.

 *biomedicines*

Editorial

miRNAs: From Master Regulators of Gene Expression to Biomarkers Involved in Intercellular Communication

Elena Levantini [1] and Milena Rizzo [2,*]

[1] Institute of Biomedical Technologies, CNR, Via Moruzzi 1, 56124 Pisa, Italy; elena.levantini@itb.cnr.it
[2] Institute of Clinical Physiology, CNR, Via Moruzzi 1, 56124 Pisa, Italy
* Correspondence: milena.rizzo@cnr.it

Citation: Levantini, E.; Rizzo, M. miRNAs: From Master Regulators of Gene Expression to Biomarkers Involved in Intercellular Communication. *Biomedicines* **2024**, *12*, 721. https://doi.org/10.3390/biomedicines12040721

Received: 12 March 2024
Accepted: 20 March 2024
Published: 25 March 2024

1. Introduction

MicroRNAs (miRNAs) are non-coding RNAs that act as master regulators of gene expression, fine-tuning the activity of thousands of genes in our cells, by modulating gene expression at the post-transcriptional level. They play a fundamental role in the regulation of almost all physiological processes, and their alteration can contribute to the development of several diseases.

This Special Issue collects some of the latest advances in the field, focusing on three main aspects: (i) the involvement of miRNAs in human disease, (ii) the role of extracellular miRNAs (ex-miRNAs) in intercellular communication and (iii) the use of ex-miRNAs as biomarkers to diagnose and monitor diseases in different body fluids. Moreover, this Special Issue also covers some of the "less conventional" aspects of miRNA research, such as the discovery of both small nucleolar (sno)-derived RNAs and transposable elements as novel players in the miRNA regulatory network. These findings reveal the diversity and complexity of the miRNA regulatory network, which is constantly evolving and expanding.

2. Extracellular miRNAs

In 2008, it was discovered that miRNAs are also present in the extracellular space (ex-miRNAs), creating a new dimension of gene regulation that crosses the boundaries of cells and tissues, thus opening up a new field of research with great potential for biomedical applications. Ex-miRNAs are released from cells and circulate in various biological fluids (blood, saliva, urine, milk, cerebrospinal fluid, etc.). For this reason, they are also called circulating miRNAs (c-miRNAs). They carry information about our health status and can be used as biomarkers for different diseases and disorders, thanks to their high stability and abundance in these fluids. Furthermore, the analysis of c-miRNAs has the clear advantage of being minimally invasive compared to tissue biopsy (e.g., [1,2]).

In this Special Issue, several papers explore the potential of c-miRNAs as biomarkers in different scenarios, such as pregnancy, brain damage, ovarian cancer, and breast milk. For instance, Thibeault et al. [3] identified some plasma miRNAs associated with the maternal body mass index in the first trimester of pregnancy, many of which were related to fatty acid and lipid metabolism according to an in silico analysis. Robles et al. [4] identified a set of c-miRNAs whose level changed in the serum of a mouse model of intracerebral hemorrhage (ICH), which could be evaluated as potential biomarkers of brain injury. Robotti et al. [5] reviewed how salivary miRNAs can act as potential biomarkers of ovarian cancer, emphasizing why saliva is a reliable and easy-to-manage source of biomarkers in tumor diagnosis. Kondracka et al. [6] reviewed the recent findings on human breast milk miRNAs. The authors highlighted their possible role in shaping the development of the infant's immune systems and how they are associated with some diseases in both infants and mothers, including breast neoplasms and neonatal jaundice. Related to this aspect, Pomar et al. [7] showed that changing the diet of diet-induced obese mice during lactation can reduce the abnormal level of miRNAs in their mammary glands, but not in their milk.

However, despite more than a decade of research, using c-miRNAs as biomarkers is not an easy task. There are still technical hurdles to overcome, such as the accuracy and reproducibility of their quantification [8,9]. Luckily, some technical advances have been made, especially with digital PCR (dPCR), which offers high sensitivity and specificity in detecting low-level miRNAs in human plasma. D'Alessandra et al. [10] confirmed the accuracy of dPCR in this context and also proposed a faster and simpler way to quantify circulating miRNAs with dPCR, without the need to extract them first.

Another aspect of ex-miRNAs, which is still debated, is their role in intercellular communication, either as free molecules or as encapsulated vesicles (such as exosomes or other types of extracellular vesicles). There are many indications that ex-miRNAs can act as signaling molecules in both an autocrine/paracrine and an endocrine manner, influencing physiological and pathological processes (e.g., [11–13]), including drug resistance (e.g., [14]). Zeng et al. [15] reviewed some examples of exosomal miRNAs that play a role in cell–cell communication in pathological states, highlighting their therapeutic potential. The authors focused on miRNAs secreted by mesenchymal stem cells and their potential application in bone regeneration and various diseases (such as cancer, Alzheimer's disease, spinal cord injury, ischemia) treatment. Barbosa et al. [16] showed that the secretome isolated from miR-124-silenced amyotrophic lateral sclerosis (ALS) motoneurons, if injected into the spine of ALS mice (at the early stage), was able to hamper the progression of the disease, improving locomotor behaviors and preventing motoneuronal dysfunction by reducing neurodegeneration.

3. miRNAs in Human Disease

Many pieces of evidence support the hypothesis that miRNAs are deregulated in all hallmarks of cancers. In this Special Issue, you can find some papers that show how miRNAs can act as both villains and heroes in different types of cancer. For example, Panella et al. [17] demonstrated how miR-22 can make triple-negative breast cancer cells more aggressive by promoting epithelial–mesenchymal transition (EMT). The authors also demonstrated how pharmacologically inhibiting miR-22 improved the survival of mice with breast tumors, highlighting the possible use of miRNAs as targets for therapeutic development [18]. In their review, Nguyen et al. [19] summarized how miR-29s can have opposite roles in different human cancers, acting as either oncogene or tumor suppressors. Chiantore et al. [20] defined a miRNA signature in actinic keratosis, a skin condition that can lead to cancer, that may affect the pathways involved in tumor development. De Almeida et al. [21] reported the most relevant findings on the epigenetic features of uterine leiomyosarcomas and endometrial stromal sarcoma, two rare and aggressive cancers. Among them, several miRNAs were found to be altered in these tumors. Finally, Galardi et al. [22] reviewed the most recent data on the role of circular RNAs (circRNAs) in pediatric cancer. CircRNAs are long (>200 bp) non-coding RNAs that regulate gene expression at various levels (by affecting the transcription efficiency, by interacting with proteins and modifying their activity, etc.) [23]. One of the most commonly reported functions of circRNAs, even for pediatric cancer, is trapping miRNAs (miRNA sponging). Recently, this function has been questioned, and the most likely conclusion is that for circRNAs to have a significant biological effect, they should either contain many miRNA binding sites or be highly expressed [23].

miRNAs are also involved in many other diseases and disorders. In this Special Issue, some papers show how miRNAs can affect the development and function in different organs and systems. For instance, Leavy et al. [24] reported that the impairment of normal brain maturation caused by early life brain injury (such as hypoxia) is accompanied by modification of miRNA expression, possibly driven by cMYC. In their review, Piquer-Gil et al. [25] focused on the importance of the balance between two key pathways (Wnt/β-catenin and Hippo pathways) in the progression of the pathological arrhythmogenic cardiomyopathy (ACM) phenotype, a form of heart disease that can cause sudden death. The authors, by using data from cancer research, hypothesized that non-coding

RNA (including miRNAs) could play a role in regulating these pathways in ACM. Finally, An et al. [26] investigated the association of specific miRNA polymorphisms with pregnancy loss.

4. "Non-Canonical" miRNAs and miRNA Regulation

In this Special Issue, authors discuss i) a neglected class of non-coding RNA with miRNA-like functions and ii) a new function of transposons in miRNA sponging.

The review by Coley et al. [27] focused on sno-derived RNAs (sdRNAs), small RNAs that originate from small nuclear RNAs (snoRNAs), which can act as miRNA-like molecules. This review summarized the recent literature on the role of sdRNAs in cancer gene regulation, including miRNAs derived from snoRNA transcripts. The authors highlighted the relevant role of this small non-coding RNA category, especially in cancer pathology, and urged the boosting of research in this novel field.

The paper by Esposito et al. [28] studied a possible new function of transposons (TEs), which are mobile genetic elements that can move around the genome and sometimes affect the expression of nearby genes. TEs, depending on where they jump, can insert functional domains such as miRNA binding sites transforming transcripts in miRNA sponges (see reference in the paper). This is part of a network of competitive endogenous RNAs (ceRNAs), where different RNA molecules compete for the same miRNAs and influence their availability and activity [29]. The authors investigated the role of TEs as miRNA sponges by analyzing in silico transcriptomic data of different cellular conditions in which a specific type of TE, called LINE L1, was more active. LINE L1, the most abundant and active TE in humans, can copy and paste itself to new locations in the genome. The authors found that, in a cellular condition where LINE L1 was overactive, genes that had many miRNA binding sites in common with LINE L1 were more upregulated compared to genes with fewer shared miRNA binding sites. This suggests that an increase in LINE L1 could increase specific miRNAs' sequestration, derepressing their targets. The authors proposed that this could be a new miRNA regulation modality that could impact both health and disease. However, they also acknowledged that this phenomenon needs to be experimentally validated and further investigated.

5. Conclusions

In conclusion, this Special Issue covers the current state of the art on miRNA research, providing novel insights on miRNA regulation and function that could advance our understanding and applications of these versatile biomolecules. We hope that this collection of papers will inspire further research and innovation in this exciting and promising field.

Author Contributions: E.L. and M.R. served as guest editors of the Special Issue "microRNAs in Health and Disease" and wrote the editorial. All authors have read and agreed to the published version of the manuscript.

Acknowledgments: E.L. acknowledges AIRC Investigator Grant 2021 ID 25734, the PNNR THE Spoke 1 Award, and private donations from the Gheraldeschi and the Pecoraro families.

Conflicts of Interest: The authors declare no conflicts of interest.

References

1. Suárez, B.; Solé, C.; Márquez, M.; Nanetti, F.; Lawrie, C.H. Circulating MicroRNAs as Cancer Biomarkers in Liquid Biopsies. *Adv. Exp. Med. Biol.* **2022**, *1385*, 23–73. [CrossRef] [PubMed]

2. Malhotra, S.; Miras, M.C.M.; Pappolla, A.; Montalban, X.; Comabella, M. Liquid Biopsy in Neurological Diseases. *Cells* **2023**, *12*, 1911. [CrossRef] [PubMed]

3. Thibeault, K.; Légaré, C.; Desgagné, V.; White, F.; Clément, A.-A.; Scott, M.S.; Jacques, P.-É.; Guérin, R.; Perron, P.; Hivert, M.-F.; et al. Maternal Body Mass Index Is Associated with Profile Variation in Circulating MicroRNAs at First Trimester of Pregnancy. *Biomedicines* **2022**, *10*, 1726. [CrossRef] [PubMed]

4. Robles, D.; Guo, D.-H.; Watson, N.; Asante, D.; Sukumari-Ramesh, S. Dysregulation of Serum MicroRNA after Intracerebral Hemorrhage in Aged Mice. *Biomedicines* **2023**, *11*, 822. [CrossRef] [PubMed]

5.   Robotti, M.; Scebba, F.; Angeloni, D. Circulating Biomarkers for Cancer Detection: Could Salivary MicroRNAs Be an Opportunity for Ovarian Cancer Diagnostics? *Biomedicines* **2023**, *11*, 652. [CrossRef] [PubMed]

6.   Kondracka, A.; Gil-Kulik, P.; Kondracki, B.; Frąszczak, K.; Oniszczuk, A.; Rybak-Krzyszkowska, M.; Staniczek, J.; Kwaśniewska, A.; Kocki, J. Occurrence, Role, and Challenges of MicroRNA in Human Breast Milk: A Scoping Review. *Biomedicines* **2023**, *11*, 248. [CrossRef]

7.   Pomar, C.A.; Castillo, P.; Palou, A.; Palou, M.; Picó, C. Dietary Improvement during Lactation Normalizes MiR-26a, MiR-222 and MiR-484 Levels in the Mammary Gland, but Not in Milk, of Diet-Induced Obese Rats. *Biomedicines* **2022**, *10*, 1292. [CrossRef]

8.   Takizawa, S.; Matsuzaki, J.; Ochiya, T. Circulating MicroRNAs: Challenges with Their Use as Liquid Biopsy Biomarkers. *Cancer Biomark.* **2022**, *35*, 1–9. [CrossRef]

9.   Lakkisto, P.; Dalgaard, L.T.; Belmonte, T.; Pinto-Sietsma, S.-J.; Devaux, Y.; de Gonzalo-Calvo, D. EU-CardioRNA COST Action CA17129 (https://cardiorna.eu/) Development of Circulating MicroRNA-Based Biomarkers for Medical Decision-Making: A Friendly Reminder of What Should NOT Be Done. *Crit. Rev. Clin. Lab. Sci.* **2023**, *60*, 141–152. [CrossRef] [PubMed]

10. D'Alessandra, Y.; Valerio, V.; Moschetta, D.; Massaiu, I.; Bozzi, M.; Conte, M.; Parisi, V.; Ciccarelli, M.; Leosco, D.; Myasoedova, V.A.; et al. Extraction-Free Absolute Quantification of Circulating MiRNAs by Chip-Based Digital PCR. *Biomedicines* **2022**, *10*, 1354. [CrossRef] [PubMed]

11. Mori, M.A.; Ludwig, R.G.; Garcia-Martin, R.; Brandão, B.B.; Kahn, C.R. Extracellular MiRNAs: From Biomarkers to Mediators of Physiology and Disease. *Cell Metab.* **2019**, *30*, 656–673. [CrossRef]

12. Lucero, R.; Zappulli, V.; Sammarco, A.; Murillo, O.D.; Cheah, P.S.; Srinivasan, S.; Tai, E.; Ting, D.T.; Wei, Z.; Roth, M.E.; et al. Glioma-Derived MiRNA-Containing Extracellular Vesicles Induce Angiogenesis by Reprogramming Brain Endothelial Cells. *Cell Rep.* **2020**, *30*, 2065–2074.e4. [CrossRef]

13. Xu, C.; Wang, P.; Guo, H.; Shao, C.; Liao, B.; Gong, S.; Zhou, Y.; Yang, B.; Jiang, H.; Zhang, G.; et al. MiR-146a-5p Deficiency in Extracellular Vesicles of Glioma-Associated Macrophages Promotes Epithelial-Mesenchymal Transition through the NF-ΚB Signaling Pathway. *Cell Death Discov.* **2023**, *9*, 206. [CrossRef]

14. Canovai, M.; Evangelista, M.; Mercatanti, A.; D'Aurizio, R.; Pitto, L.; Marrocolo, F.; Casieri, V.; Pellegrini, M.; Lionetti, V.; Bracarda, S.; et al. Secreted MiR-210-3p, MiR-183-5p and MiR-96-5p Reduce Sensitivity to Docetaxel in Prostate Cancer Cells. *Cell Death Discov.* **2023**, *9*, 445. [CrossRef] [PubMed]

15. Zeng, E.Z.; Chen, I.; Chen, X.; Yuan, X. Exosomal MicroRNAs as Novel Cell-Free Therapeutics in Tissue Engineering and Regenerative Medicine. *Biomedicines* **2022**, *10*, 2485. [CrossRef] [PubMed]

16. Barbosa, M.; Santos, M.; de Sousa, N.; Duarte-Silva, S.; Vaz, A.R.; Salgado, A.J.; Brites, D. Intrathecal Injection of the Secretome from ALS Motor Neurons Regulated for MiR-124 Expression Prevents Disease Outcomes in SOD1-G93A Mice. *Biomedicines* **2022**, *10*, 2120. [CrossRef] [PubMed]

17. Panella, R.; Cotton, C.A.; Maymi, V.A.; Best, S.; Berry, K.E.; Lee, S.; Batalini, F.; Vlachos, I.S.; Clohessy, J.G.; Kauppinen, S.; et al. Targeting of MicroRNA-22 Suppresses Tumor Spread in a Mouse Model of Triple-Negative Breast Cancer. *Biomedicines* **2023**, *11*, 1470. [CrossRef] [PubMed]

18. Levantini, E. Is MiR Therapeutic Targeting Still a MiRage? *Front. Biosci. (Landmark Ed)* **2021**, *26*, 680–692. [CrossRef] [PubMed]

19. Nguyen, T.T.P.; Suman, K.H.; Nguyen, T.B.; Nguyen, H.T.; Do, D.N. The Role of MiR-29s in Human Cancers-An Update. *Biomedicines* **2022**, *10*, 2121. [CrossRef] [PubMed]

20. Chiantore, M.V.; Iuliano, M.; Mongiovì, R.M.; Luzi, F.; Mangino, G.; Grimaldi, L.; Accardi, L.; Fiorucci, G.; Romeo, G.; Di Bonito, P. MicroRNAs Differentially Expressed in Actinic Keratosis and Healthy Skin Scrapings. *Biomedicines* **2023**, *11*, 1719. [CrossRef] [PubMed]

21. de Almeida, B.C.; Dos Anjos, L.G.; Dobroff, A.S.; Baracat, E.C.; Yang, Q.; Al-Hendy, A.; Carvalho, K.C. Epigenetic Features in Uterine Leiomyosarcoma and Endometrial Stromal Sarcomas: An Overview of the Literature. *Biomedicines* **2022**, *10*, 2567. [CrossRef]

22. Galardi, A.; Colletti, M.; Palma, A.; Di Giannatale, A. An Update on Circular RNA in Pediatric Cancers. *Biomedicines* **2022**, *11*, 36. [CrossRef]

23. Kristensen, L.S.; Jakobsen, T.; Hager, H.; Kjems, J. The Emerging Roles of CircRNAs in Cancer and Oncology. *Nat. Rev. Clin. Oncol.* **2022**, *19*, 188–206. [CrossRef]

24. Leavy, A.; Brennan, G.P.; Jimenez-Mateos, E.M. MicroRNA Profiling Shows a Time-Dependent Regulation within the First 2 Months Post-Birth and after Mild Neonatal Hypoxia in the Hippocampus from Mice. *Biomedicines* **2022**, *10*, 2740. [CrossRef]

25. Piquer-Gil, M.; Domenech-Dauder, S.; Sepúlveda-Gómez, M.; Machí-Camacho, C.; Braza-Boïls, A.; Zorio, E. Non Coding RNAs as Regulators of Wnt/β-Catenin and Hippo Pathways in Arrhythmogenic Cardiomyopathy. *Biomedicines* **2022**, *10*, 2619. [CrossRef] [PubMed]

26. An, H.-J.; Cho, S.-H.; Park, H.-S.; Kim, J.-H.; Kim, Y.-R.; Lee, W.-S.; Lee, J.-R.; Joo, S.-S.; Ahn, E.-H.; Kim, N.-K. Genetic Variations MiR-10aA>T, MiR-30cA>G, MiR-181aT>C, and MiR-499bA>G and the Risk of Recurrent Pregnancy Loss in Korean Women. *Biomedicines* **2022**, *10*, 2395. [CrossRef] [PubMed]

27. Coley, A.B.; DeMeis, J.D.; Chaudhary, N.Y.; Borchert, G.M. Small Nucleolar Derived RNAs as Regulators of Human Cancer. *Biomedicines* **2022**, *10*, 1819. [CrossRef]

28. Esposito, M.; Gualandi, N.; Spirito, G.; Ansaloni, F.; Gustincich, S.; Sanges, R. Transposons Acting as Competitive Endogenous RNAs: In-Silico Evidence from Datasets Characterised by L1 Overexpression. *Biomedicines* **2022**, *10*, 3279. [CrossRef]
29. Salmena, L.; Poliseno, L.; Tay, Y.; Kats, L.; Pandolfi, P.P. A CeRNA Hypothesis: The Rosetta Stone of a Hidden RNA Language? *Cell* **2011**, *146*, 353–358. [CrossRef] [PubMed]

biomedicines

Article

Maternal Body Mass Index Is Associated with Profile Variation in Circulating MicroRNAs at First Trimester of Pregnancy

Kathrine Thibeault [1], Cécilia Légaré [1], Véronique Desgagné [1,2], Frédérique White [3], Andrée-Anne Clément [1], Michelle S. Scott [1], Pierre-Étienne Jacques [3,4], Renée Guérin [1,2], Patrice Perron [4,5], Marie-France Hivert [5,6,7] and Luigi Bouchard [1,2,4,*]

1   Department of Biochemistry and Functional Genomics, Faculty of Medicine and Health Sciences (FMHS), Université de Sherbrooke, Sherbrooke, QC J1K 2R1, Canada; kathrine.thibeault@usherbrooke.ca (K.T.); cecilia.legare@usherbrooke.ca (C.L.); veronique.desgagne@usherbrooke.ca (V.D.); andree-anne.clement@usherbrooke.ca (A.-A.C.); michelle.scott@usherbrooke.ca (M.S.S.); renee.guerin2@usherbrooke.ca (R.G.)
2   Clinical Department of Laboratory Medicine, Centre Intégré Universitaire de Santé et de Services Sociaux (CIUSSS) du Saguenay-Lac-Saint-Jean—Hôpital de Chicoutimi, Saguenay, QC G7H 5H6, Canada
3   Département de Biologie, Faculté des Science, Université de Sherbrooke, Sherbrooke, QC J1K 2R1, Canada; frederique.white@usherbrooke.ca (F.W.); pierre-etienne.jacques@usherbrooke.ca (P.-É.J.)
4   Department of Medicine, FMHS, Université de Sherbrooke, Sherbrooke, QC J1K 2R1, Canada; patrice.perron@usherbrooke.ca
5   Centre de Recherche du Centre Hospitalier Universitaire de Sherbrooke (CR-CHUS), Université de Sherbrooke, Sherbrooke, QC J1H 5N4, Canada; mhivert@partners.org
6   Department of Population Medicine, Harvard Medical School, Harvard Pilgrim Health Care Institute, Boston, MA 02115, USA
7   Diabetes Unit, Massachusetts General Hospital, Boston, MA 02114, USA
*   Correspondence: luigi.bouchard@usherbrooke.ca; Tel.: +418-541-1234 (ext. 2475)

Citation: Thibeault, K.; Légaré, C.; Desgagné, V.; White, F.; Clément, A.-A.; Scott, M.S.; Jacques, P.-É.; Guérin, R.; Perron, P.; Hivert, M.-F.; et al. Maternal Body Mass Index Is Associated with Profile Variation in Circulating MicroRNAs at First Trimester of Pregnancy. *Biomedicines* **2022**, *10*, 1726. https://doi.org/10.3390/biomedicines10071726

Academic Editors: Milena Rizzo and Elena Levantini

Received: 2 June 2022
Accepted: 14 July 2022
Published: 18 July 2022

Publisher's Note: MDPI stays neutral with regard to jurisdictional claims in published maps and institutional affiliations.

Abstract: Many women enter pregnancy with overweight and obesity, which are associated with complications for both the expectant mother and her child. MicroRNAs (miRNAs) are short non-coding RNAs that regulate many biological processes, including energy metabolism. Our study aimed to identify first trimester plasmatic miRNAs associated with maternal body mass index (BMI) in early pregnancy. We sequenced a total of 658 plasma samples collected between the 4th and 16th week of pregnancy from two independent prospective birth cohorts (Gen3G and 3D). In each cohort, we assessed associations between early pregnancy maternal BMI and plasmatic miRNAs using DESeq2 R package, adjusting for sequencing run and lane, gestational age, maternal age at the first trimester of pregnancy and parity. A total of 38 miRNAs were associated (FDR q < 0.05) with BMI in the Gen3G cohort and were replicated (direction and magnitude of the fold change) in the 3D cohort, including 22 with a nominal *p*-value < 0.05. Some of these miRNAs were enriched in fatty acid metabolism-related pathways. We identified first trimester plasmatic miRNAs associated with maternal BMI. These miRNAs potentially regulate fatty acid metabolism-related pathways, supporting the hypothesis of their potential contribution to energy metabolism regulation in early pregnancy.

Keywords: microRNA; obesity; pregnancy; next-generation sequencing

1. Introduction

Overweight and obesity (OW/O) are worldwide health issues [1] increasingly affecting women of reproductive age [2,3] and are considered the most common health conditions in pregnancy [4]. The prevalence of obesity among women of reproductive age is 17.8% in Canada [5] and 39.7% in the United States [6]. Maternal OW/O can have serious short- and long-term consequences on the health of the mother and her child. OW/O in pregnancy are associated with an increased risk of gestational diabetes mellitus (GDM), gestational hypertension, preeclampsia (PE) and cesarean delivery [4]. Offspring exposed to maternal OW/O are more likely to suffer from macrosomia and neonatal hyperinsulinemic hypoglycemia [7]. Possibly through fetal metabolic programming (based on the Developmental Origin of Health and Disease-DOHaD), these offspring are also at increased risk of childhood obesity, type 2 diabetes, cardiovascular disease, asthma and neurodevelopmental disorders [8,9]. A better understanding of the pathophysiological mechanisms leading to pregnancy complications associated with maternal OW/O could lead to strategies to prevent to maternal morbidity and possibly lower their consequences on the exposed offspring.

MicroRNAs (miRNAs) are short non-coding, single-stranded, RNA molecules of 19 to 25 nucleotides. They regulate many biological processes by targeting messenger RNAs (mRNAs), leading to a decrease in protein synthesis [10]. Three miRNA clusters are predominantly expressed in the placenta: chromosome 19 miRNA (C19MC), chromosome 14 miRNA (C14MC) and miR-371-3 miRNA clusters. The abundance of miRNAs from C19MC gradually increases in maternity from conception until the end of pregnancy, whereas miRNAs from the C14MC cluster are more abundant at the beginning of pregnancy and their blood levels decrease throughout pregnancy [11,12]. These placental miRNAs may participate in the regulation of maternal physiology by intercellular communication through placental extracellular vesicle secretion in maternal circulation [13]. Indeed, miRNAs are suspected of having an important role in pregnancy, its maintenance as well as adaptation to its very specific physiologic needs [10,11]. Accordingly, dysregulation of the C19MC and C14MC has been associated with pregnancy complications, including PE, intrauterine growth restriction and insulin sensitivity regulation [13–17].

Until now, only a few studies have investigated miRNAs in pregnancy complicated by OW/O [18–21]. However, none of these studies were done by next-generation sequencing or in plasma in the first trimester of pregnancy. We hypothesized that maternal BMI at the first trimester of pregnancy is associated with variations in circulating levels of miRNAs. Therefore, our objectives were to identify plasmatic miRNAs associated with maternal BMI and the metabolic pathways they potentially regulate.

2. Materials and Methods

2.1. Discovery Cohort: Genetics of Glucose Regulation in Gestation and Growth (Gen3G) Cohort

We selected participants for the discovery step of the study from the Gen3G prospective pregnancy and birth cohort [22]. Briefly, we recruited women in the first trimester of pregnancy (between the 4th and the 16th week), and we followed them until delivery. An oral glucose tolerance test (OGTT—75 g) was performed between the 24th and 28th week of pregnancy for 854 women. For this study, our selection criteria were: women of European descent, 18 years old and older, not taking any medication that influences glycemia, free from pre-gestational diabetes, having singleton pregnancy as well as the availability of a plasma sample at the first trimester of pregnancy (500 µL), anthropometric measures (e.g., maternal BMI), follow-ups of the offspring at 3 and 5 years old, and genetics (Mother: Infinium MEGAEX Array, Illumina; Offspring: whole genome sequencing) and epigenetics (EPIC array) data. A total of 444 women fulfilled these criteria. Before the study began, consent was obtained from all participating women, and all protocols were approved by the Centre Hospitalier Universitaire de Sherbrooke (CHUS) ethics committee.

2.2. Anthropometric Measurements in Gen3G

BMI is used to classify individuals by dividing their weight (kg) by their height (m) squared. This measurement was taken in the first trimester of pregnancy (between the 4th and 16th week) [23], and a description of the BMI measurement has been published [22]. Briefly, weight was measured in kg on a calibrated electronic scale, and height was measured in meters with a wall stadiometer (without shoes). The BMI value was calculated from these data.

2.3. Replication Cohort: Design, Develop, Discover (3D) Cohort Study

A replication analysis was run on 226 participants selected from the 3D prospective birth cohort [24]. Briefly, the participants were recruited during their first trimester of pregnancy in nine different sites across the province of Québec, Canada. For this study, women of European descent with plasma samples (500 µL) at the 1st trimester of pregnancy and at least fasting and the 2 h post-OGTT glucose levels measured between the 24th and 28th week of pregnancy were included. Exclusion criteria included pre-existing diabetes, GDM diagnosed at the first trimester of pregnancy, chronic hypertension, or gestational hypertension diagnosed at the first trimester of pregnancy (Supplementary Figure S1). Biological specimens (e.g., plasma) and anthropometric measurements were collected (measured by research staff). BMI was measured between the 8th and 14th week of pregnancy during the first trimester of pregnancy. Women without BMI data were also excluded. All women gave their informed consent before the beginning of the study according to the Helsinki declaration, and the protocols have been accepted by the ethical committees.

2.4. RNA Extraction

We extracted total RNA with the MirVana PARIS kit (ThermoFisher Scientific, Waltham, United States, catalog #AM1556) from 500 µL of plasma collected between the 4th and 16th week of pregnancy following the standard protocol and eluted in 75 µL of nuclease-free water, in random order. We concentrated our RNA samples following the protocol established by Burgos et al. [25]. In brief, RNA was mixed with 35 µL of cold (4 °C) 7 M ammonium acetate solution (ThermoFisher Scientific, Waltham, United States, catalog #02002268) and mixed with 420 µL of chilled (−20 °C) absolute ethanol (Commercial Alcohols, ON, Canada; catalog #P006EAAN). RNA was precipitated overnight at −20 °C and centrifuged at 16,000× g at 4 °C for 30 min. The RNA pellet was washed twice with 200 µL of 80% ethanol and then centrifuged at 16,000× g, 4 °C for 5 min. The RNA pellet was dried for 30 min at room temperature and resuspended in 5 µL of nuclease-free water.

2.5. Library Preparation

We used the Truseq Small RNA Sample Prep kit (Illumina, BC, Canada, catalog #RS-200-0012) for library preparation. Concentrated RNA samples (5 µL) were randomly treated following the standard protocol adapted by Burgos et al. [25]. Briefly, half of the reagents were used for ligation of RNA at the 3′ and 5′ ends, reverse transcription, indexing (1–48, one index per sample) and PCR amplification (15 cycles), to maintain an optimal ratio between RNA and reagents. The libraries were purified by migration on a Novex polyacrylamide TBE Gel, 6% (ThermoFisher Scientific, Waltham, United States, catalog #EC265BOX) by selecting bands between 145–160 bp, eluted in 300 µL of nuclease-free water, and incubated overnight at room temperature and 500 RPM on an incubating microplate shaker (VWR, ON, Canada, catalog #12620-930). Then, the libraries were concentrated by precipitation following a standardized procedure (including a 30 min incubation of the precipitation mix at −80 °C), and the cDNA pellet was suspended in 25 µL of 10 mM Tris-HCl pH 8.5 buffer.

2.6. Library Quality Control and Sequencing

The libraries were sequenced at the McGill University and Génome Québec Innovation Centre (Montréal, Canada), either on a HiSeq 2500 or HiSeq 4000 platform (50 cycles, with 7 cycles indexing read) for the Gen3G samples. Twelve samples were extracted twice and sequenced on both the HiSeq 2500 and HiSeq 4000 platforms that we leveraged during our QC and normalization process to take into account potential technical and batch effects from the different sequencing platforms. Overall, Pearson correlation coefficients between miRNA levels measured on the 2 platforms for these 12 samples were ≥0.94. The data from both platforms were then combined for processing and statistical analysis but adjusted for run and lane as normally recommended [12]. Libraries were quantified by qPCR, equimolarly pooled (HiSeq 2500: 12 libraries with different indexes per lane at a molarity of 7 pM; HiSeq 4000: 20 libraries with different indexes per lane at a molarity of 10 pM), denatured and clustered on single-read Illumina flowcells (catalog # GD-401-3001 and catalog GD-410-1001) according to the manufacturer's standard protocol.

The 3D replication study samples were sequenced on the NovaSeq 6000 platform at the McGill University and Génome Québec Innovation Centre (Montréal, Canada). Libraries were prepared following the procedure applied to Gen3G samples and quantified using the Kapa Illumina GA with Revised Primers-SYBR Fast Universal kit (Kapa Biosystems, Wilmington, United States), normalized, equimolarly pooled (48 libraries with different indexes per lane at a molarity of 225 pM), denatured and clustered on an Illumina NovaSeq S1 flowcell following the Xp protocol from the manufacturer's recommendations. The run was performed for 100 cycles in single-end mode.

2.7. Bioinformatics Analysis

We first applied the extracellular RNA processing toolkit (exceRpt) pipeline (version 4.6) [26]. Briefly, exceRpt uses FASTX-Toolkit and FastQC to assess sequencing data quality after removing the adapters and poor quality (Phred score < 20 for 80% or more of the read) sequences. The remaining reads were then aligned to the human genome (GRCh37) and miRbase [27] (version 21) using STAR [28] (version 2.4.2a) alignment algorithm. After performing data visualization of the raw read counts, 9 outliers were excluded from the Gen3G cohort, and 3 outliers were excluded from the 3D cohort.

2.8. Statistical Analysis

Since all the participant characteristics (Gen3G and 3D cohorts) were not normally distributed based on a Shapiro-Wilk test, non-parametric Mann-Whitney U tests were applied to compare the two cohorts. Association between plasmatic miRNA levels and maternal BMI at the first trimester of pregnancy was assessed using the default parameters in DESeq2 R package [29]. The duplicate samples (n = 12) were combined using the collapseReplicates function from DESeq2 package. The statistical model was adjusted for sequencing run and lane, gestational age, maternal age at the first trimester of pregnancy, and parity. A sensitivity analysis was also performed by adding the fetal sex as a covariate to the analysis model. The results remain overall unchanged. MiRNAs were considered significantly associated with maternal BMI with a false discovery rate (FDR) adjusted q-value < 0.05, where the fold change represents the change in the miRNAs normalized read counts per unit of BMI. The EnhancedVolcano [30] package was used for the creation of the volcano plot. For replication, criteria were first the direction of the associations, the fold changes, and finally the nominal p-values (one-tailed). The statistical analyses were all done with R version 4.0.3 in R studio version 1.4.1103.

2.9. Biological Pathway Analysis

The potential biological function of the maternal BMI-associated miRNAs was evaluated using miRPath v.3 software from DIANA tools [31]. In brief, Kyoto Encyclopedia of Genes and Genomes (KEGG) metabolic pathway analysis was done on all the miRNAs that were significantly (q-values < 0.05) associated with maternal BMI at the first trimester of pregnancy and replicated for their fold change value in the 3D cohort. Tarbase 7.0 was selected for miRNA-mRNA interactions as it considers those validated experimentally. The pathway union parameter was used to merge the results with the Fisher exact test enrichment analysis method. The default settings in miRPath v.3 and FDR correction were applied, and the q-value threshold was set to 0.05.

3. Results

3.1. Participant's Characteristics

Table 1 presents the characteristic of the selected Gen3G and 3D cohorts' participants. Women from Gen3G were on average 28.48 ± 4.26 years old, had a mean BMI of

25.95 ± 5.98 kg/m^2 at the first trimester of pregnancy (mean gestational age: 9.63 ± 2.26 weeks), and had a mean parity of 0.69 ± 0.91 (53.56% nulliparous). On average, women from the 3D cohort were slightly older (30.60 ± 3.90 years old, $p < 0.001$) and had a greater gestational age in the 1st trimester visit (11.89 ± 1.52 weeks, $p < 0.001$) and a lower parity (0.47 ± 0.66, $p = 0.01$; 60.99% nulliparous), whereas their mean BMI (25.39 ± 5.91 kg/m^2, $p = 0.22$) was similar to that of the women from the Gen3G cohort.

Table 1. Maternal characteristics at the first trimester of pregnancy in Gen3G ($n = 435$) and 3D ($n = 223$) cohorts.

Characteristics	Gen3G Cohort		3D Cohort		Comparison between Cohorts (p-Value) *
	Mean ± SD	Range	Mean ± SD	Range	
BMI (kg/m^2)	25.95 ± 5.98	16.10–54.10	25.39 ± 5.91	16.80–48.50	0.22
Age (years)	28.48 ± 4.26	18–47	30.60 ± 3.90	20–42	<0.001
Gestational age (weeks)	9.63 ± 2.26	4.09–16.30	11.89 ± 1.52	7.71–16.43	<0.001
Parity(% nulliparous)	0.69 ± 0.91 (53.56)	0–6	0.47 ± 0.66 (60.99)	0–3	0.01

* Comparisons between Gen3G and 3D cohorts were performed using a Mann-Whitney U test. Abbreviations: 3D: Design, Develop, Discover birth cohort; BMI: body mass index; Gen3G: Genetics of Glucose regulation in Gestation and Growth birth cohort; SD: standard deviation.

3.2. Association between miRNA Levels and Maternal BMI at the First Trimester of Pregnancy

In Gen3G, 2170 miRNAs were expressed in the plasma of pregnant women. A miRNA was considered detected when it has >1 read per participant. A total of 61 miRNAs (FDR-adjusted q-values < 0.05) were associated with maternal BMI in the first trimester of pregnancy (Figure 1). The list of significant miRNAs with their mean normalized read counts, fold changes (FC), unadjusted p-values, and FDR-adjusted q-values are shown in Supplementary Table S1. Among these 61 miRNAs, higher BMI was associated with lower circulating levels for 48 miRNAs and with higher circulating levels for 13 miRNAs. Interestingly, 28 of these miRNAs (46%) are encoded by the C19MC, 1 miRNA by the C14MC, and 3 miRNAs by the miR-371-3 miRNAs cluster. Table 2 shows the top 10 miRNAs with the most significant associations with maternal BMI; for these 10 miRNAs, higher BMI was associated with lower circulating levels.

Table 2. Top 10 miRNAs most significantly associated with maternal BMI at first trimester of pregnancy in the Gen3G cohort.

miRNA	DESeq2 Normalized Read Count (Mean ± SD)	Fold Change *	Unadjusted p-Value	FDR-Adjusted q-Value
hsa-miR-1323 [a]	146.39 ± 230.60	0.957	2.79×10^{-10}	9.60×10^{-8}
hsa-miR-516b-5p [a]	101.50 ± 150.35	0.958	1.79×10^{-10}	9.60×10^{-8}
hsa-miR-371a-5p [b]	11.02 ± 17.09	0.941	2.13×10^{-10}	9.60×10^{-8}
hsa-miR-525-5p [a]	9.07 ± 15.31	0.952	1.63×10^{-9}	4.21×10^{-7}
hsa-miR-516a-5p [a]	31.25 ± 53.81	0.959	4.55×10^{-9}	7.83×10^{-7}
hsa-miR-524-5p [a]	8.64 ± 14.48	0.952	3.90×10^{-9}	7.83×10^{-7}
hsa-miR-518e-5p [a]	43.96 ± 62.52	0.962	8.92×10^{-9}	1.22×10^{-6}
hsa-miR-520a-3p [a]	86.75 ± 132.37	0.960	9.41×10^{-9}	1.22×10^{-6}
hsa-miR-518e-3p [a]	7.99 ± 14.22	0.956	2.02×10^{-8}	2.09×10^{-6}
hsa-miR-520d-5p [a]	5.92 ± 10.35	0.955	1.98×10^{-8}	2.09×10^{-6}

Model adjusted for sequencing run and lane, gestational age, maternal age at the first trimester of pregnancy, and parity. * Fold changes represent the change in miRNA abundance for each increase of one unit of maternal BMI at 1st trimester of pregnancy. [a] miRNAs from C19MC. [b] miRNAs from miR-371-3 miRNAs cluster. Abbreviations: FDR: false discovery rate; SD: standard deviation.

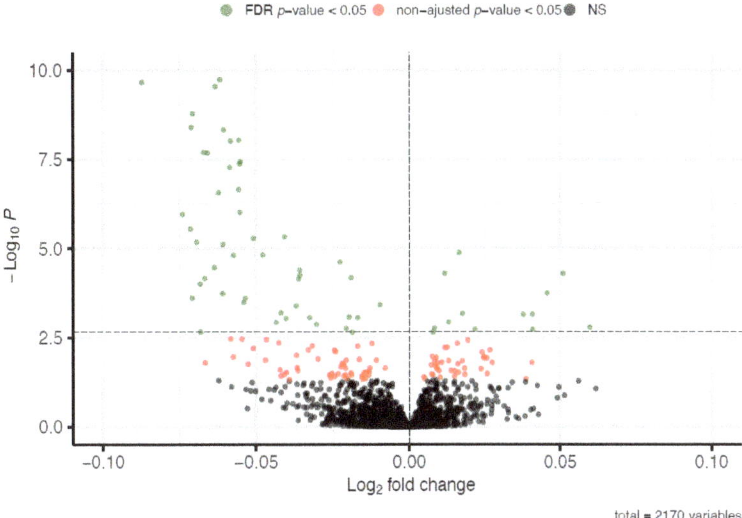

Figure 1. Plasmatic miRNAs associated with maternal BMI at 1st trimester of pregnancy. This volcano plot shows the miRNAs associated with maternal BMI. Each point represents a single miRNA. The vertical dotted line represents a log2 fold change of 0, and the horizontal dotted line represents the FDR-adjusted q-value threshold of <0.05. The model was adjusted for sequencing lane and run, maternal and gestational age at 1st trimester, and parity. The fold change represents the change in miRNA abundance for an increase of one unit of maternal BMI in 1st trimester of pregnancy. Abbreviations: FDR: false discovery rate; NS: not significant.

3.3. Replication of miRNAs Associated with Maternal BMI at the First Trimester of Pregnancy in the 3D Cohort

We conducted replication analyses for the 61 identified miRNAs (with q < 0.05) in 3D cohort (see Supplementary Table S1). Interestingly, we achieved full replication based on the strength of the associations (one tailed *p*-value < 0.05) as well as the direction and magnitude of the fold changes between Gen3G and 3D cohorts for 22 (36%) miRNAs (Table 3). Moreover, 8 (80%) out of the 10 most strongly associated miRNAs in Gen3G cohort were replicated in 3D (*p* < 0.05). Sixteen (16) additional miRNAs (total 38) had a similar fold change (direction and magnitude) between the Gen3G and 3D cohort without reaching our a priori selected statistical significance threshold (Supplementary Table S1). These miRNAs were included in the pathway analysis below.

Table 3. miRNAs significantly associated with maternal BMI at 1st trimester of pregnancy in Gen3G cohort and fully replicated in 3D cohort.

	Gen3G				3D		
miRNA	DESeq2 Normalized Read Count (Mean ± SD)	Fold Change *	Unadjusted *p*-Value	FDR-Adjusted q-Value	DESeq2 Normalized Read Count (Mean ± SD)	Fold Change *	Unadjusted *p*-Value
hsa-miR-1323 [a]	146.39 ± 230.60	0.957	2.79×10^{-10}	9.60×10^{-8}	676.45 ± 581.88	0.966	2.56×10^{-5}
hsa-miR-516b-5p [a]	101.50 ± 150.35	0.958	1.79×10^{-10}	9.60×10^{-8}	327.56 ± 247.26	0.966	2.92×10^{-5}
hsa-miR-525-5p [a]	9.07 ± 15.31	0.952	1.63×10^{-9}	4.21×10^{-7}	79.89 ± 71.16	0.982	0.04471
hsa-miR-516a-5p [a]	31.25 ± 53.81	0.959	4.55×10^{-9}	7.83×10^{-7}	150.92 ± 135.76	0.974	0.00141
hsa-miR-518e-5p [a]	43.96 ± 62.52	0.962	8.92×10^{-9}	1.22×10^{-6}	99.07 ± 88.52	0.981	0.02118
hsa-miR-520a-3p [a]	86.75 ± 132.37	0.960	9.41×10^{-9}	1.22×10^{-6}	220.91 ± 243.90	0.982	0.04009
hsa-miR-518e-3p [a]	7.99 ± 14.22	0.956	2.02×10^{-8}	2.09×10^{-6}	29.60 ± 25.75	0.971	0.00189
hsa-miR-512-3p [a]	287.04 ± 575.48	0.963	3.63×10^{-8}	3.41×10^{-6}	771.74 ± 982.11	0.982	0.02747

Table 3. *Cont.*

	Gen3G				3D		
miRNA	DESeq2 Normalized Read Count (Mean ± SD)	Fold Change *	Unadjusted *p*-Value	FDR-Adjusted *q*-Value	DESeq2 Normalized Read Count (Mean ± SD)	Fold Change *	Unadjusted *p*-Value
hsa-miR-1283 [a]	57.10 ± 87.82	0.962	4.17×10^{-8}	3.59×10^{-6}	203.52 ± 173.84	0.977	0.00635
hsa-miR-517a-3p [a]	20.19 ± 37.78	0.960	5.17×10^{-8}	4.11×10^{-6}	76.56 ± 73.88	0.980	0.01599
hsa-miR-526b-5p [a]	17.12 ± 25.90	0.962	2.19×10^{-7}	1.62×10^{-5}	39.62 ± 34.83	0.977	0.01056
hsa-miR-517-5p [a]	16.90 ± 29.75	0.962	9.56×10^{-7}	6.17×10^{-5}	70.67 ± 67.45	0.981	0.02191
hsa-miR-519c-3p [a]	11.08 ± 20.08	0.965	5.11×10^{-6}	0.00026	36.44 ± 32.83	0.966	0.00034
hsa-miR-519d-5p [a]	6.12 ± 9.51	0.961	1.53×10^{-5}	0.00063	10.75 ± 11.77	0.969	0.01205
hsa-miR-515-5p [a]	10.50 ± 19.12	0.967	1.50×10^{-5}	0.00063	26.70 ± 26.08	0.968	0.00076
hsa-miR-27b-3p	29,699.76 ± 20,347.78	1.008	4.98×10^{-5}	0.00172	55,214.27 ± 41,098.16	1.011	0.00110
hsa-miR-885-5p	13.50 ± 17.92	1.036	4.98×10^{-5}	0.00172	40.63 ± 78.98	1.034	0.00588
hsa-miR-520a-5p [a]	3.06 ± 5.66	0.963	0.00032	0.00830	27.63 ± 27.09	0.965	0.00136
hsa-miR-375	1938.28 ± 3939.32	0.975	0.00040	0.00986	2900.08 ± 2878.58	0.982	0.03889
hsa-miR-520d-3p [a]	9.80 ± 15.84	0.972	0.00091	0.01877	10.77 ± 13.37	0.966	0.00638
hsa-miR-592	1.64 ± 3.00	1.042	0.00162	0.03103	2.43 ± 4.89	1.058	0.00400
hsa-miR-21-5p	85,256.17 ± 61,108.55	1.005	0.00224	0.03811	82,095.57 ± 63,195.51	1.008	0.02738

Models adjusted for sequencing run and lane, gestational age, maternal age at the first trimester of pregnancy, and parity. * Fold changes represent the change in miRNA abundance for each increase of one unit of maternal BMI at 1st trimester of pregnancy. [a] miRNAs from C19MC. Abbreviations: 3D: Design, Develop, Discover birth cohort; FDR: false discovery rate; Gen3G: Genetics of Glucose regulation in Gestation and Growth birth cohort; SD: standard deviation.

3.4. Metabolic Pathway Analysis of miRNAs Associated with Maternal BMI at the First Trimester of Pregnancy

To assess the potential biological role of the 38 miRNAs associated with maternal BMI at the first trimester of pregnancy, we did a metabolic pathway analysis using miRPath v3 [32]. Seven KEGG pathways from the union pathway analysis were enriched with targets of these miRNAs. Figure 2 shows these targeted metabolic pathways and their FDR-adjusted q-value. Interestingly, the top 2 pathways were related to fatty acid metabolism ($p = 1 \times 10{-325}$, 6 miRNAs) and fatty acids biosynthesis ($p < 1 \times 10{-325}$, 4 miRNAs).

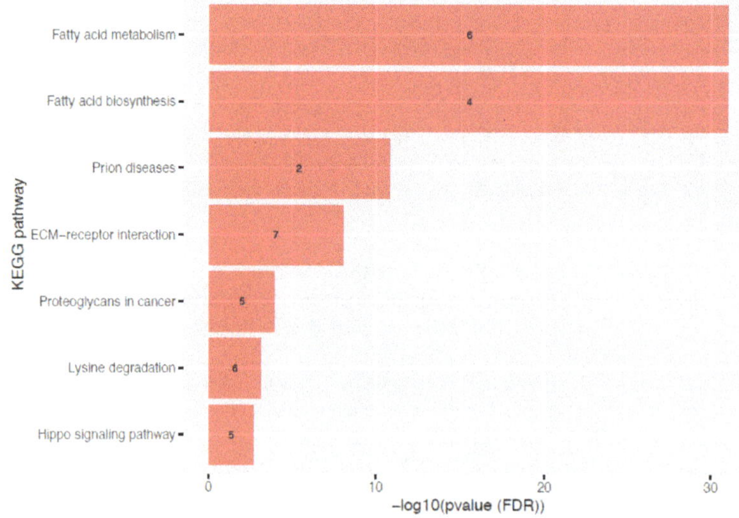

Figure 2. KEGG pathways targeted by miRNAs associated with maternal BMI at the first trimester of pregnancy. This bar graph is showing the KEGG pathways enriched with targets of miRNAs associated with maternal BMI. Each bar represents one pathway, ranked according to its level of significance (FDR adjusted q-value). The number of miRNAs involved in the pathway is shown directly in the bar. Abbreviations: ECM: extracellular matrix; FDR: false discovery rate; KEGG: Kyoto Encyclopedia of Genes and Genomes.

4. Discussion

Many women enter pregnancy with OW/O, which are associated with complications, such as GDM and pre-eclampsia. In the current study, we sequenced 658 plasma samples from two independent birth cohorts and identified over 20 miRNAs circulating in maternal plasma at first trimester that are associated with maternal BMI in early pregnancy. To the best of our knowledge, this is the largest study using next generation sequencing to identify miRNAs at the first trimester of pregnancy associated with maternal BMI.

Only a few studies have associated maternal OW/O in pregnancy with plasmatic miRNA profile dysregulation [15,16,18,19]. Interestingly, four of our top 10 miRNAs, all replicated in 3D, were previously associated with maternal obesity in cord blood [18]. Jing et al. reported that cord blood hsa-miR-1323, hsa-miR-516b-5p, hsa-miR-516a-5p, and hsa-miR-520a-3p levels were positively associated with maternal OW/O [18], which is contrary to what we have observed in the current study. This apparent discrepancy could be explained by the difference in sample origin (fetal whole cord blood vs. maternal plasma) and the timing during pregnancy (1st trimester vs. at delivery). These four miRNAs are all encoded in the C19MC, the expression of which is known to increase throughout pregnancy. Overall, our results suggest that the impact of OW/O might be dependent on the timing of collection and the tissue tested (maternal plasma vs. cord blood). In addition to these four miRNAs, our study also identified novel miRNAs associated with maternal BMI, of which many are also encoded by the C19MC. None of our miRNAs were reported in the other three studies [15,16,19].

One fascinating hypothesis is that the placenta secretes miRNAs into maternal blood contributing to feto-maternal communication and metabolic adaptation throughout pregnancy [32]. Indeed, more than 600 miRNAs dynamically expressed in the human placenta have been reported so far [10]. From these, 127 miRNAs are encoded into the C14MC, C19MC and miR-371-3 clusters. [10,11] Interestingly, 17 of the 22 (77%) replicated miRNAs are encoded by the C19MC. More broadly, 21 out of the 38 replicated miRNAs (55%) are from the C19MC, 1 from the C14MC, and 2 from the miR-371-3. C19MC is a large miRNA cluster specific to primates that encodes 46 miRNAs genes, 58 mature miRNAs and is maternally imprinted (paternally expressed) [10]. These miRNAs are also strongly expressed in trophoblasts, suggesting an important role in pregnancy and embryonic and fetal development [10]. In a previous study, we also identified miRNAs specific to pregnancy (pregnant vs. non-pregnant women) and varying between the 4th and the 16th week of pregnancy, using the same Gen3G microtranscriptomic dataset [12]. Accordingly, 5 more miRNAs (in addition to the 8 from the C19MC) of our replicated miRNAs were found upregulated in pregnancy in that previous study. This provides novel support of the roles of plasma miRNAs in pregnancy. Specifically, maternal BMI levels in pregnancy may influence the expression of miRNAs that are more abundant in pregnancy and may thus have a possible role in metabolic adaptation and development.

Maternal OW/O are also risk factors for GDM and PE and consequently the miRNAs we have identified could also contribute to the development of these complications. In another previous study from our group and using the same Gen3G microtranscriptomic dataset, we identified miRNAs in the first trimester of pregnancy associated with and predictive of insulin sensitivity in the second trimester of pregnancy. Among our replicated miRNAs, 15 were also associated with insulin sensitivity [17]. Similar analyses were also conducted to identify miRNAs associated and predictive of GDM still using this microtranscriptomics and Gen3G datasets. Among the replicated miRNAs, 16 were also associated with GDM [33]. Overall, 13 were associated with maternal obesity, insulin sensitivity and GDM. Moreover, higher circulating levels of serum hsa-miR-1323 has been associated with GDM [34], hsa-miR-517-5p and hsa-miR-520a-5p were reported to be downregulated in PE, which is consistent with our results with maternal BMI [16,35]. Finally, lower circulating levels of hsa-miR-526b-5p were detected in patients with metabolic syndrome – characterized by central excess adiposity—a direction of association that is concordant with our results [36]. Through their mechanism of action, the miRNAs that we found associated with maternal BMI might potentially have a role to play in the pathophysiology of how excess weight in early gestation may lead to pregnancy complications. However, further studies are needed to disentangle their specific roles and the causal relationship.

Our pathway analysis showed that the miRNAs associated with maternal BMI regulate a few metabolic pathways, of which lipid metabolism is of high interest for its biologic relevance. Overall, 10 miRNAs were linked to fatty acid biosynthesis (5 miRNAs), fatty acid metabolism (10 miRNAs), or fatty acid elongation (5 miRNAs) pathways. They were either up—(n = 5) or down—(n = 5) regulated in relation to maternal BMI. Obesity, inside and outside of pregnancy, is associated with dyslipidemia, among several other metabolic complications [37]. Early pregnancy is characterized by a lipogenic profile which is believed to favor energy storage that will be later used to meet the metabolic demand of both the mother and her growing fetus during pregnancy [38]. By targeting fatty acid biosynthesis and metabolism, our BMI-associated miRNAs could play an early role in the regulation of the lipid metabolism pathways in pregnancy or dysregulation in response to maternal OW/O. Further studies are also needed to understand the roles of plasma miRNAs in metabolic adaptation to pregnancy and their links with maternal OW/O.

Strengths and Limitations

To the best of our knowledge, this is the largest study investigating circulating miRNAs associated with maternal BMI in the first trimester of pregnancy using next-generation sequencing. Among the strengths, our study was conducted in two large, well-characterized prospective pregnancy and birth cohorts and on plasma samples collected early in pregnancy, from the 4th to 16th weeks. This allowed us to assess obesity-related miRNA dysregulation months before pregnancy complications, such as GDM and PE, develop. Finally, we applied next generation sequencing technology, which allowed us to fully profile and quantify plasma miRNAs. This technology is a robust and sensitive approach for miRNA quantification [39]. Finally, although our study was by design associative, the independent replication significantly improves the robustness of the results we are reporting.

Our study also has some limitations. First, BMI is an easy and convenient assessment of OW/O but remains a surrogate measure of fat mass. Nevertheless, BMI remains recommended by the WHO to assess OW/O. Also, our BMIs were measured at the time plasma samples were collected. Overall, BMI in the first trimester of pregnancy is considered an acceptable assessment of that before pregnancy as weight is relatively stable in the first weeks. Also, it seems clear that functional studies are needed to confirm the role of these miRNAs associated during pregnancy (metabolic adaptation, fetal development and in pregnancy complications.)

5. Conclusions

We have identified 22 plasma miRNAs associated with maternal BMI in the first trimester of pregnancy. Identified miRNAs were enriched in biological pathways related to fatty acids and lipids that are implicated in the pathophysiology of OW/O and mostly encoded by the C19MC, which is mainly expressed by the placenta. These results could provide new insights into the understanding of the effect of OW/O on the development of different complications in pregnancy such as GDM and PE.

Supplementary Materials: The following supporting information can be downloaded at: https://www.mdpi.com/article/10.3390/biomedicines10071726/s1, Figure S1: Selection of participants for the 3D cohort. This diagram is showing the different inclusion and exclusion criteria of our selected participants in the 3D cohort. Abbreviations: OGTT: oral glucose tolerance test, GDM: gestational diabetes mellitus, BMI: body mass index. Table S1: List of miRNAs significantly associated with maternal BMI at the first trimester of pregnancy in the Gen3G cohort and replication results obtained in the 3D cohort for miRNAs significantly associated with maternal BMI at 1st trimester of pregnancy in the Gen3G cohort.

Author Contributions: L.B. designed the study with the contribution of V.D., M.S.S., P.-É.J., P.P., M.-F.H. and R.G.; C.L. and V.D. performed data collection with K.T.; C.L., A.-A.C. and F.W. worked on bioinformatics analysis; K.T. performed statistical analyses; K.T. wrote the manuscript with the collaboration of V.D. and L.B.; P.P., M.-F.H., R.G. and L.B. supervised all steps of the study. All authors have read and agreed to the published version of the manuscript.

Funding: K.T., C.L. and A.-A.C. were supported by a doctoral research award from Fonds de la recherche du Québec en santé (FRQS). V.D. received funding from Diabète Québec. L.B., M.S.S. and P.É.J. are research scholars from the FRQS and a member of the CR-CHUS, a FRQS-funded Research Center. This study was supported by the CIHR (Grant #IGH-155183), the Fondation de ma vie of the CIUSSS du Saguenay–Lac-St-Jean—Hôpital Universitaire de Chicoutimi and Diabète Québec. The Gen3G birth cohort recruitment was supported by operating grants from the FRQS (Grant #20697), the CIHR (Grant #MOP 115071), Diabète Québec, and the Canadian Diabetes Association (CDA; Grant #OG-3-08-2622-JA).

Institutional Review Board Statement: The study was conducted in accordance with the Declaration of Helsinki and approved by the Ethics Committee of CIUSSS de l'Estrie-CHUS (protocol code #MP-31-2019-3059 and February 7 2019)" for studies involving humans.

Informed Consent Statement: Informed consent was obtained from all subjects involved in the study.

Data Availability Statement: The datasets used and/or analyzed during the current study are available from the corresponding author upon reasonable request.

Acknowledgments: We thank the McGill University and Génome Québec Innovation Centre (Montréal, Canada) staff for their work in libraries quality control and sequencing. We are also grateful to Calcul Québec (Montréal, Canada) and Compute Canada (Toronto, Canada) for their support in this research. Finally, we thank all the participants of the Gen3G and 3D cohorts.

Conflicts of Interest: The authors declare no conflict of interest.

References

1. Caballero, B. The Global Epidemic of Obesity: An Overview. *Epidemiol. Rev.* **2007**, *29*, 1–5. [CrossRef] [PubMed]
2. Valsamakis, G.; Kyriazi, E.; Mouslech, Z.; Siristatidis, C.; Mastorakos, G. Effect of Maternal Obesity on Pregnancy Outcomes and Long-Term Metabolic Consequences. *Hormones* **2015**, *14*, 345–357.e3. [CrossRef] [PubMed]
3. Tauqeer, Z.; Gomez, G.; Stanford, F.C. Obesity in Women: Insights for the Clinician. *J. Womens Health* **2018**, *27*, 444–457. [CrossRef] [PubMed]
4. Catalano, P.M.; Shankar, K. Obesity and Pregnancy: Mechanisms of Short Term and Long Term Adverse Consequences for Mother and Child. *BMJ Br. Med. J. Online* **2017**, *356*, j1. [CrossRef]
5. Berger, H.; Melamed, N.; Murray-Davis, B.; Hasan, H.; Mawjee, K.; Barrett, J.; McDonald, S.D.; Geary, M.; Ray, J.G. Prevalence of Pre-Pregnancy Diabetes, Obesity, and Hypertension in Canada. *J. Obstet. Gynaecol. Can.* **2019**, *41*, 1579–1588.e2. [CrossRef]
6. Overweight & Obesity Statistics | NIDDK. Available online: https://www.niddk.nih.gov/health-information/health-statistics/overweight-obesity (accessed on 21 March 2022).
7. Catalano, P.M.; McIntyre, H.D.; Cruickshank, J.K.; McCance, D.R.; Dyer, A.R.; Metzger, B.E.; Lowe, L.P.; Trimble, E.R.; Coustan, D.R.; Hadden, D.R.; et al. The Hyperglycemia and Adverse Pregnancy Outcome Study: Associations of GDM and Obesity with Pregnancy Outcomes. *Diabetes Care* **2012**, *35*, 780–786. [CrossRef]
8. Kelly, A.C.; Powell, T.L.; Jansson, T. Placental Function in Maternal Obesity. *Clin. Sci.* **2020**, *134*, 961–984. [CrossRef]
9. Godfrey, K.M.; Reynolds, R.M.; Prescott, S.L.; Nyirenda, M.; Jaddoe, V.W.V.; Eriksson, J.G.; Broekman, B.F.P. Influence of Maternal Obesity on the Long-Term Health of Offspring. *Lancet Diabetes Endocrinol.* **2017**, *5*, 53–64. [CrossRef]
10. Poirier, C.; Desgagné, V.; Guérin, R.; Bouchard, L. MicroRNAs in Pregnancy and Gestational Diabetes Mellitus: Emerging Role in Maternal Metabolic Regulation. *Curr. Diab. Rep.* **2017**, *17*, 35. [CrossRef]
11. Morales-Prieto, D.M.; Ospina-Prieto, S.; Chaiwangyen, W.; Schoenleben, M.; Markert, U.R. Pregnancy-Associated MiRNA-Clusters. *J. Reprod. Immunol.* **2013**, *97*, 51–61. [CrossRef]
12. Légaré, C.; Clément, A.-A.; Desgagné, V.; Thibeault, K.; White, F.; Guay, S.-P.; Arsenault, B.J.; Scott, M.S.; Jacques, P.-É.; Perron, P.; et al. Human Plasma Pregnancy-Associated MiRNAs and Their Temporal Variation within the First Trimester of Pregnancy. *Reprod. Biol. Endocrinol.* **2022**, *20*, 14. [CrossRef]
13. Morales-Prieto, D.M.; Favaro, R.R.; Markert, U.R. Placental MiRNAs in Feto-Maternal Communication Mediated by Extracellular Vesicles. *Placenta* **2020**, *102*, 27–33. [CrossRef] [PubMed]
14. Awamleh, Z.; Gloor, G.B.; Han, V.K.M. Placental MicroRNAs in Pregnancies with Early Onset Intrauterine Growth Restriction and Preeclampsia: Potential Impact on Gene Expression and Pathophysiology. *BMC Med. Genom.* **2019**, *12*, 91. [CrossRef]

15. Hromadnikova, I.; Kotlabova, K.; Ondrackova, M.; Pirkova, P.; Kestlerova, A.; Novotna, V.; Hympanova, L.; Krofta, L. Expression Profile of C19MC MicroRNAs in Placental Tissue in Pregnancy-Related Complications. *DNA Cell Biol.* **2015**, *34*, 437–457. [CrossRef]
16. Hromadnikova, I.; Dvorakova, L.; Kotlabova, K.; Krofta, L. The Prediction of Gestational Hypertension, Preeclampsia and Fetal Growth Restriction via the First Trimester Screening of Plasma Exosomal C19MC MicroRNAs. *Int. J. Mol. Sci.* **2019**, *20*, 2972. [CrossRef] [PubMed]
17. Légaré, C.; Desgagné, V.; Poirier, C.; Thibeault, K.; White, F.; Clément, A.-A.; Scott, M.S.; Jacques, P.-É.; Perron, P.; Guérin, R.; et al. First Trimester Plasma MicroRNAs Levels Predict Matsuda Index-Estimated Insulin Sensitivity between 24th and 29th Week of Pregnancy. *BMJ Open Diabetes Res. Care* **2022**, *10*, e002703. [CrossRef]
18. Jing, J.; Wang, Y.; Quan, Y.; Wang, Z.; Liu, Y.; Ding, Z. Maternal Obesity Alters C19MC MicroRNAs Expression Profile in Fetal Umbilical Cord Blood. *Nutr. Metab.* **2020**, *17*, 52. [CrossRef] [PubMed]
19. Tsamou, M.; Martens, D.S.; Winckelmans, E.; Madhloum, N.; Cox, B.; Gyselaers, W.; Nawrot, T.S.; Vrijens, K. Mother's Pre-Pregnancy BMI and Placental Candidate MiRNAs: Findings from the ENVIRONAGE Birth Cohort. *Sci. Rep.* **2017**, *7*, 5548. [CrossRef]
20. Enquobahrie, D.A.; Wander, P.L.; Tadesse, M.G.; Qiu, C.; Holzman, C.; Williams, M.A. Maternal Pre-Pregnancy Body Mass Index and Circulating MicroRNAs in Pregnancy. *Obes. Res. Clin. Pract.* **2017**, *11*, 464–474. [CrossRef]
21. Carreras-Badosa, G.; Bonmatí, A.; Ortega, F.-J.; Mercader, J.-M.; Guindo-Martínez, M.; Torrents, D.; Prats-Puig, A.; Martinez-Calcerrada, J.-M.; de Zegher, F.; Ibáñez, L.; et al. Dysregulation of Placental MiRNA in Maternal Obesity Is Associated with Pre- and Postnatal Growth. *J. Clin. Endocrinol. Metab.* **2017**, *102*, 2584–2594. [CrossRef]
22. Guillemette, L.; Allard, C.; Lacroix, M.; Patenaude, J.; Battista, M.-C.; Doyon, M.; Moreau, J.; Ménard, J.; Bouchard, L.; Ardilouze, J.-L.; et al. Genetics of Glucose Regulation in Gestation and Growth (Gen3G): A Prospective Prebirth Cohort of Mother–Child Pairs in Sherbrooke, Canada. *BMJ Open* **2016**, *6*, e010031. [CrossRef]
23. Krukowski, R.A.; West, D.S.; DiCarlo, M.; Shankar, K.; Cleves, M.A.; Saylors, M.E.; Andres, A. Are Early First Trimester Weights Valid Proxies for Preconception Weight? *BMC Pregnancy Childbirth* **2016**, *16*, 357. [CrossRef] [PubMed]
24. Fraser, W.D.; Shapiro, G.D.; Audibert, F.; Dubois, L.; Pasquier, J.; Julien, P.; Bérard, A.; Muckle, G.; Trasler, J.; Tremblay, R.E.; et al. 3D Cohort Study: The Integrated Research Network in Perinatology of Quebec and Eastern Ontario. *Paediatr. Perinat. Epidemiol.* **2016**, *30*, 623–632. [CrossRef] [PubMed]
25. Burgos, K.L.; Javaherian, A.; Bomprezzi, R.; Ghaffari, L.; Rhodes, S.; Courtright, A.; Tembe, W.; Kim, S.; Metpally, R.; Van Keuren-Jensen, K. Identification of Extracellular MiRNA in Human Cerebrospinal Fluid by Next-Generation Sequencing. *RNA* **2013**, *19*, 712–722. [CrossRef] [PubMed]
26. Rozowsky, J.; Kitchen, R.; Park, J.J.; Galeev, T.R.; Diao, J.; Warrell, J.; Thistlethwaite, W.; Subramanian, S.L.; Milosavljevic, A.; Gerstein, M. ExceRpt: A Comprehensive Analytic Platform for Extracellular RNA Profiling. *Cell Syst.* **2019**, *8*, 352–357.e3. [CrossRef] [PubMed]
27. Kozomara, A.; Griffiths-Jones, S. MiRBase: Annotating High Confidence MicroRNAs Using Deep Sequencing Data. *Nucleic Acids Res.* **2014**, *42*, D68–D73. [CrossRef]
28. Dobin, A.; Davis, C.A.; Schlesinger, F.; Drenkow, J.; Zaleski, C.; Jha, S.; Batut, P.; Chaisson, M.; Gingeras, T.R. STAR: Ultrafast Universal RNA-Seq Aligner. *Bioinformatics* **2013**, *29*, 15–21. [CrossRef]
29. Love, M.I.; Huber, W.; Anders, S. Moderated Estimation of Fold Change and Dispersion for RNA-Seq Data with DESeq2. *Genome Biol.* **2014**, *15*, 550. [CrossRef]
30. Blighe, K.; Rana, S.; Lewis, M. EnhancedVolcano: Publication-Ready Volcano Plots with Enhanced Colouring and Labeling. Available online: https://github.com/kevinblighe/EnhancedVolcano (accessed on 16 February 2022).
31. Vlachos, I.S.; Zagganas, K.; Paraskevopoulou, M.D.; Georgakilas, G.; Karagkouni, D.; Vergoulis, T.; Dalamagas, T.; Hatzigeorgiou, A.G. DIANA-MiRPath v3.0: Deciphering MicroRNA Function with Experimental Support. *Nucleic Acids Res.* **2015**, *43*, W460–W466. [CrossRef]
32. Luo, S.-S.; Ishibashi, O.; Ishikawa, G.; Ishikawa, T.; Katayama, A.; Mishima, T.; Takizawa, T.; Shigihara, T.; Goto, T.; Izumi, A.; et al. Human Villous Trophoblasts Express and Secrete Placenta-Specific MicroRNAs into Maternal Circulation via Exosomes. *Biol. Reprod.* **2009**, *81*, 717–729. [CrossRef]
33. Légaré, C.; Desgagné, V.; Thibeault, K.; White, F.; Clément, A.-A.; Poirier, C.; Luo, Z.-C.; Scott, M.; Jacques, P.-É.; Perron, P.; et al. First Trimester Plasma MicroRNA Levels Predict Risk of Developing Gestational Diabetes Mellitus. *Front. Endocrinol.* **2022**, *13*.
34. Liu, L.; Zhang, J.; Liu, Y. MicroRNA-1323 Serves as a Biomarker in Gestational Diabetes Mellitus and Aggravates High Glucose-Induced Inhibition of Trophoblast Cell Viability by Suppressing TP53INP1. *Exp. Ther. Med.* **2021**, *21*, 230. [CrossRef] [PubMed]
35. Fu, J.-Y.; Xiao, Y.-P.; Ren, C.-L.; Guo, Y.-W.; Qu, D.-H.; Zhang, J.-H.; Zhu, Y.-J. Up-Regulation of MiR-517-5p Inhibits ERK/MMP-2 Pathway: Potential Role in Preeclampsia. *Eur. Rev. Med. Pharmacol. Sci.* **2018**, *22*, 6599–6608. [CrossRef] [PubMed]
36. Liu, G.; Lei, Y.; Luo, S.; Huang, Z.; Chen, C.; Wang, K.; Yang, P.; Huang, X. MicroRNA Expression Profile and Identification of Novel MicroRNA Biomarkers for Metabolic Syndrome. *Bioengineered* **2021**, *12*, 3864–3872. [CrossRef] [PubMed]
37. Klop, B.; Elte, J.W.F.; Castro Cabezas, M. Dyslipidemia in Obesity: Mechanisms and Potential Targets. *Nutrients* **2013**, *5*, 1218–1240. [CrossRef]
38. Herrera, E. Lipid Metabolism in Pregnancy and Its Consequences in the Fetus and Newborn. *Endocrine* **2002**, *19*, 43–55. [CrossRef]
39. Tam, S.; de Borja, R.; Tsao, M.-S.; McPherson, J.D. Robust Global MicroRNA Expression Profiling Using Next-Generation Sequencing Technologies. *Lab. Investig. J. Tech. Methods Pathol.* **2014**, *94*, 350–358. [CrossRef]

 *biomedicines*

Article

Dysregulation of Serum MicroRNA after Intracerebral Hemorrhage in Aged Mice

Dominic Robles, De-Huang Guo, Noah Watson, Diana Asante and Sangeetha Sukumari-Ramesh *

Department of Pharmacology and Toxicology, Medical College of Georgia, Augusta University, 1120 15th Street, CB3618, 30912 Augusta, Georgia; dominic.s.robles@live.mercer.edu (D.R.); dehuangguo@yahoo.com (D.-H.G.); nowatson@augusta.edu (N.W.); dasante@augusta.edu (D.A.)
* Correspondence: sramesh@augusta.edu; Tel.: +706-446-3645; Fax: +706-721-2347

Abstract: Stroke is one of the most common diseases that leads to brain injury and mortality in patients, and intracerebral hemorrhage (ICH) is the most devastating subtype of stroke. Though the prevalence of ICH increases with aging, the effect of aging on the pathophysiology of ICH remains largely understudied. Moreover, there is no effective treatment for ICH. Recent studies have demonstrated the potential of circulating microRNAs as non-invasive diagnostic and prognostic biomarkers in various pathological conditions. While many studies have identified microRNAs that play roles in the pathophysiology of brain injury, few demonstrated their functions and roles after ICH. Given this significant knowledge gap, the present study aims to identify microRNAs that could serve as potential biomarkers of ICH in the elderly. To this end, sham or ICH was induced in aged C57BL/6 mice (18–24 months), and 24 h post-ICH, serum microRNAs were isolated, and expressions were analyzed. We identified 28 significantly dysregulated microRNAs between ICH and sham groups, suggesting their potential to serve as blood biomarkers of acute ICH. Among those microRNAs, based on the current literature, miR-124-3p, miR-137-5p, miR-138-5p, miR-219a-2-3p, miR-135a-5p, miR-541-5p, and miR-770-3p may serve as the most promising blood biomarker candidates of ICH, warranting further investigation.

Keywords: intracerebral hemorrhage; aging; microRNA

Citation: Robles, D.; Guo, D.-H.; Watson, N.; Asante, D.; Sukumari-Ramesh, S. Dysregulation of Serum MicroRNA after Intracerebral Hemorrhage in Aged Mice. *Biomedicines* **2023**, *11*, 822. https://doi.org/10.3390/biomedicines11030822

Academic Editors: Milena Rizzo and Elena Levantini

Received: 18 November 2022
Revised: 1 February 2023
Accepted: 6 February 2023
Published: 8 March 2023

1. Introduction

Stroke is one of the most severe health issues that plagues the healthcare system. Intracerebral hemorrhage (ICH) is the second most common type of stroke and has a higher risk of mortality and morbidity rates than other stroke types [1]. Notably, there is no effective treatment for ICH [2–5]. Therefore, preclinical and clinical research on this disease is essential. ICH arises in the form of blood vessel rupture in the brain, resulting in the accumulation of blood in the brain parenchyma and the development of hematoma [6]. ICH often causes severe brain damage that is categorized into primary and secondary brain injuries. The mass effect of the hematoma mostly contributes to primary brain damage, whereas the oxidative and inflammatory signaling pathways [7,8], induced by blood components such as thrombin, hemoglobin, hemin, and iron, are responsible for secondary brain damage [9,10]. In contrast to primary brain damage, secondary brain damage persists for a longer period of time, which could contribute to both acute and long-term neurological outcomes [11]. Hence, the molecular regulators of secondary brain damage are considered potential targets for therapeutic intervention [12]. However, a detailed mechanistic understanding of the molecular events underlying secondary brain injury after ICH is lacking [13]. This represents a significant gap in the literature and reflects on the lack of defined therapeutic targets.

MicroRNAs (miRNAs), short non-coding RNAs, comprise a group of regulatory molecules that modulate the expression of genes, which play critical roles in cellular processes such as inflammation and apoptosis [14,15]. Many studies have identified the changes in miRNA expression in ischemic stroke [16], while there remains a significant gap in our knowledge of their dysregulation in ICH, particularly in the elderly. Notably, circulating miRNAs undergo dysregulation in response to pathological conditions [17] and can be found in a remarkably stable form in serum or plasma [18]. Therefore, miRNAs could serve as non-invasive diagnostic and prognostic blood biomarkers. Specifically, diagnostic blood biomarkers may help distinguish ICH from ischemic stroke, while prognostic blood biomarkers may be able to predict mortality or poor outcomes after ICH.

Aging is characterized by the accumulation of degenerative processes. MiRNAs contribute to aging [19] and have regulatory roles in neurodegeneration [20,21]. Moreover, aging is listed as the most profound risk factor for cardiovascular and neurological diseases [22]. Notably, ICH incidence and mortality rates increase with aging [23–25], but the precise role of aging in the pathophysiology of ICH remains largely unknown. Therefore, it is highly required to characterize the molecular level changes that occur after ICH in aged subjects, as it may help develop novel strategies for the diagnosis and management of ICH. Though preclinical animal models of ICH are invaluable tools for studying disease pathophysiology, miRNA dysregulation post-ICH was mostly studied in young animal subjects [17]. Moreover, aging is associated with miRNA expression level changes in mice and humans [26–29]. Hence, the objective of this study is to identify circulating miRNAs that are dysregulated after ICH in aged mice, as it may help characterize the pathophysiology of ICH in the elderly.

2. Methods

2.1. ICH Induction

All animal studies were performed according to the protocols approved by the Institutional Animal Care and Use Committee, in accordance with the NIH and USDA guidelines. Intracerebral hemorrhage was induced in aged male C57BL/6 mice (18–24 months), (Jackson Laboratories, Bar Harbor, ME, USA), as previously reported by our laboratory [2,30–33]. Briefly, mice were anesthetized with isoflurane and positioned prone on a stereotaxic head frame (Stoelting, Wood Dale, IL, USA). Using a high-speed drill (Dremel, Racine, WI, USA), a burr hole (0.5 mm) was made 2.2 mm lateral to the bregma, and a small animal temperature controller (David Kopf Instruments, Los Angeles, CA, USA) was used to keep the body temperature at $37 \pm 0.5\ ^\circ\mathrm{C}$. Employing a Hamilton syringe (26-G), 0.04 U of bacterial type IV collagenase (Sigma, St. Louis, MO, USA) in 0.5 μL phosphate-buffered saline (phosphate buffered saline; pH 7.4 (PBS) was injected with the stereotaxic guidance 3.0 mm into the left striatum to induce ICH [2]. After removing the needle, bone wax was used to seal the burr hole and the incision was stapled. Sham mice underwent the same surgical procedure, but only PBS (0.5 μL) was injected, which served as the experimental control.

2.2. Neurobehavioral Analysis

Mice were analyzed for neurobehavioral deficits, as previously reported, using a 24-point scale [33–35], which estimates sensorimotor deficits. The neurobehavioral analysis consisted of six different tests: circling, climbing, beam walking, compulsory circling, bilateral grasp, and whisker response. Each test was graded from 0 (no impairment) to 4 (severe impairment) and the sum of the scores on all six tests established a composite neurological deficit score.

2.3. Serum Collection

Blood was collected from deeply anesthetized mice and allowed to clot, undisturbed, at room temperature. Then, the clot was removed by centrifugation at $1500\times g$ for 10 min in a refrigerated centrifuge. The supernatant or serum was collected and stored at $-80\ ^\circ\mathrm{C}$. Before miRNA isolation, the serum was thawed and centrifuged, and the supernatant was used for miRNA isolation.

2.4. miRNA Isolation

miRNA isolation procedure was performed using the miRNeasy Mini Kit (Qiagen, Hilden, Germany, catalogue. No: 217004), according to the manufacturer's instructions, with some modifications. Briefly, the TRIzol LS reagent was added to mouse serum (0.75 mL TRIzol per 0.25 mL serum). This was followed by the addition of chloroform (0.2 mL chloroform per 0.75 mL of TRIzol), and centrifugation at 12,000× g at 4 °C for phase separation. The aqueous phase was transferred to a new tube and 100% ethanol (1.5 volumes of the sample) was added and mixed thoroughly and transferred to the RNeasy Mini spin column to elute the miRNA, according to the manufacturer's instructions.

2.5. miRNA Sequencing

RNA quality and quantity were assessed by the Agilent 2100 bioanalyzer (Agilent Technologies, Santa Clara, CA, USA). Purified small RNA samples were processed for cDNA library preparations using the QIAseq miRNA Library kit (Qiagen, catalogue. No: 331502). Briefly, 15 ng of purified small RNA was ligated with a 3′ adaptor and 5′ adaptor, and converted to cDNA using RT primer with integrated unique molecular indices (UMI), to enable the quantification of individual miRNA molecules. The cDNA products were purified, enriched with PCR, and purified using QMN Beads (Qiagen, catalog. No: 331502) to create the final cDNA library. The prepared library was examined by a bioanalyzer and Qubit (Thermo Fisher, Waltham, MA, USA), to test the quality and quantity of the sequencing library, respectively. The libraries were pooled with the correspondingly identified bar codes for each sample and run on the NextSeq500 sequencing system using a 75-cycle paired-end protocol. BCL files generated by the NextSeq500 were converted to FASTQ files for downstream analysis. Reads that passed quality control with individual UMI counts were aligned to the murine reference miRNA sequences using a web-based tool, GeneGlobe Data Analysis Center of QIAGEN, which also performed differential expression analysis and generated a volcano plot, and a hierarchical clustering heatmap.

2.6. Statistical Analysis

Statistical analysis was performed using GraphPad Prism software and the student's t-test was used for two-group comparisons. $p < 0.05$ was taken as statistically significant.

3. Results

3.1. Serum microRNA Isolation and Analysis after ICH

ICH was induced in the striatum of aged, male C57BL/6 (18–24 months) mice, using the collagenase injection method. For miRNA analysis, we collected whole blood from mice on day 1 post-ICH, an acute time point, which exhibited profound neurodegeneration [36] and had significant predictive values in the patient prognosis [37]. Serum microRNA was then isolated, as described in the methods, and subjected to RNA sequencing using the Agilent 2100 bioanalyzer (Agilent Technologies). The serum miRNAs from sham animals served as the experimental controls and the schematic representation of the overall experimental design is depicted (Figure 1). The analysis of RNA sequencing data, using QIAGEN GeneGlobe Data Analysis Center, identified 1960 miRNAs, out of which 28 miRNAs exhibited a significant difference ($p < 0.05$) in their expression between ICH and sham (Table 1, Figures 2 and 3). Among those, the serum levels of 20 miRNAs were found to be significantly increased ($p < 0.05$) and the serum levels of 8 miRNAs were significantly decreased ($p < 0.05$) after ICH in comparison to sham (Table 1). Soon before collecting the blood samples for miRNA analysis, the animals were subjected to neurobehavioral analysis to confirm the ICH induction. Notably, ICH animals exhibited profound neurobehavioral deficits in comparison to sham ($p < 0.01$; Figure 4).

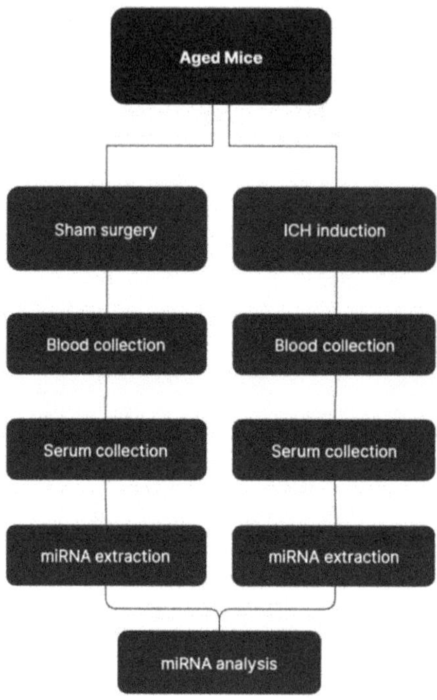

Figure 1. Schematic representation of the overall experimental design. Sham or ICH was induced in aged, male mice. Blood was collected at 24 h post-sham/ICH and serum miRNAs were extracted and subjected to microRNA sequencing.

Table 1. Serum microRNAs that exhibited differential expression between sham and ICH ($p < 0.05$).

MicroRNA	Fold-Change	p-Value
mmu-miR-122-5p	13.10	0.00000083
mmu-miR-122-3p	20.88	0.0000155
mmu-miR-9-3p	7.49	0.000124
mmu-miR-9-5p	10.83	0.000215
mmu-miR-137-3p	7.49	0.000884
mmu-miR-1298-5p	−6.62	0.00346
mmu-miR-219a-2-3p	19.81	0.00508
mmu-miR-384-5p	8.93	0.00520
mmu-miR-34b-3p	−4.03	0.00653
mmu-miR-124-3p	5.59	0.00803
mmu-miR-34c-5p	−5.7	0.01290
mmu-miR-34b-5p	−5.9	0.01691
mmu-miR-200b-3p	−2.88	0.01801
mmu-miR-135a-5p	4.83	0.01994
mmu-miR-148a-5p	7.5	0.02179
mmu-miR-133b-3p	2.23	0.02347

Table 1. *Cont.*

MicroRNA	Fold-Change	*p*-Value
mmu-miR-1199-5p	−3.87	0.02520
mmu-miR-133a-5p	2.51	0.02609
mmu-miR-451a	2.33	0.02788
mmu-miR-138-5p	3.06	0.02806
mmu-let-7g-5p	2.07	0.03187
mmu-miR-301a-3p	2.31	0.03210
mmu-miR-200c-5p	−8.98	0.03236
mmu-miR-770-3p	5.52	0.03542
mmu-miR-541-5p	2.85	0.03935
mmu-miR-194-5p	2.04	0.04143
mmu-miR-200c-3p	−2.44	0.04401
mmu-miR-216a-5p	4.09	0.04500

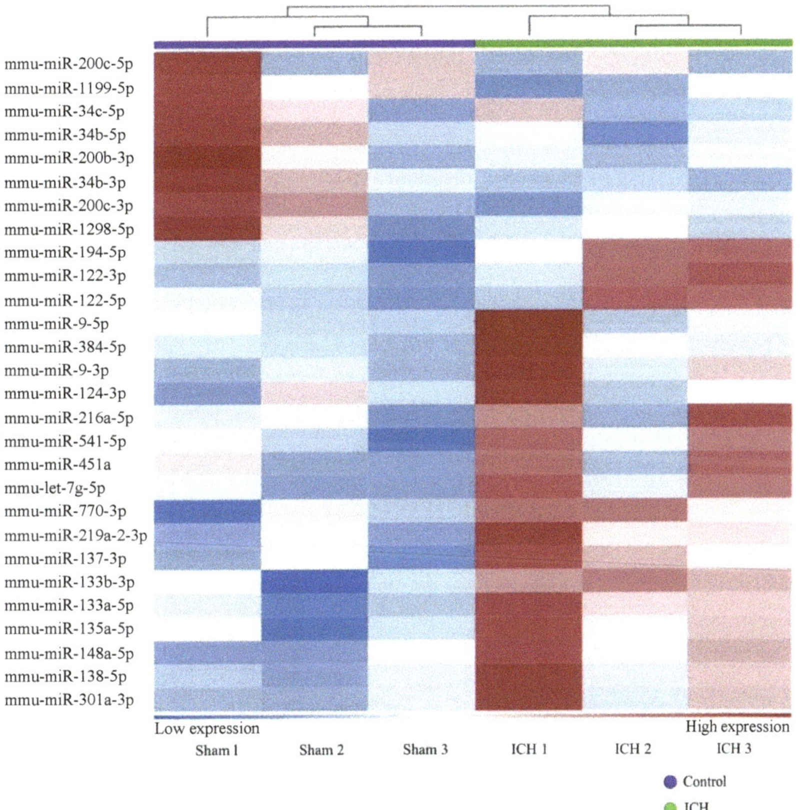

Figure 2. Heatmap representation of the differentially expressed serum miRNAs between sham and ICH. A total of 28 miRNAs exhibited a difference ($p < 0.05$) in their expression between ICH and sham. The serum levels of 20 miRNAs were found to be significantly increased and serum levels of 8 miRNAs were significantly decreased after ICH in comparison to sham (n = 3 mice per group; $p < 0.05$).

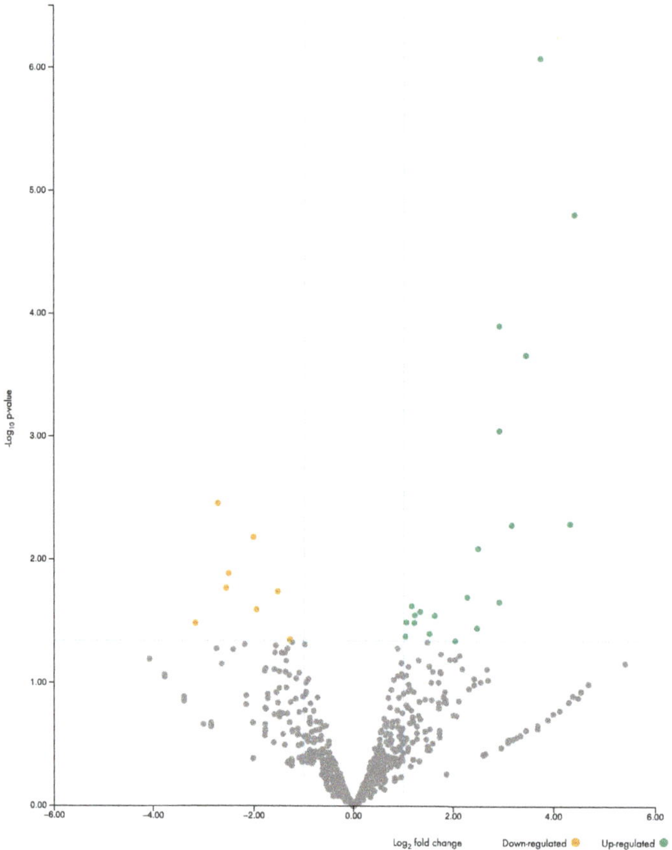

Figure 3. A volcano plot demonstrating dysregulated miRNAs after ICH, compared to sham. A total of 28 miRNAs exhibited a difference ($p < 0.05$) in their expression between ICH and sham. The serum levels of 20 miRNAs were found to be significantly increased (green dots) and serum levels of 8 miRNAs were significantly decreased (yellow dots) after ICH, in comparison to sham ($p < 0.05$; fold-change ≥ 2).

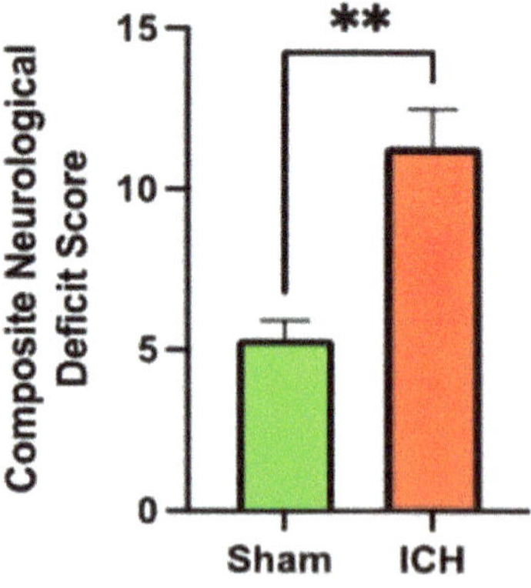

Figure 4. Induction of ICH in aged, male C57BL/6 (18–24 months) mice resulted in significant neurological deficits in comparison to sham. This was estimated using a 24-point scale, as described in methods, on day 1 post-surgery ($n = 3$ animals per group ** $p < 0.01$ vs. sham).

3.2. Functional Annotation of Differentially Expressed microRNAs

Many of the dysregulated miRNAs identified in this study play roles in various pathological conditions, as shown in Table 2. Notably, miR-122-5p, miR-9-3p, miR-9-5p, miR-137-3p, miR-1298-5p, miR-219a-2-3p, miR-384-5p, miR-124-3p, miR-34b-5p, miR-200b-3p, miR-135a-5p, miR-133b-3p, miR-1199-5p, miR-451a, miR-138-5p, miR-146a-5p, miR-200b-5p, and miR-483-5p have roles in neuroinflammation, oxidative stress, and apoptosis, which are critical processes associated with secondary brain damage after ICH.

Table 2. Pathological processes and disease states associated with dysregulated miRNAs after ICH.

MicroRNA	Cellular/Pathological Process	Disease State
miR-122-5p	Cell growth and proliferation, inflammation, oxidative stress, apoptosis	Gastric cancer [38], renal cell carcinoma [39], liver cancer [40], pancreatic ductal adenocarcinoma [41], transient ischemic attack [42]
miR-122-3p	Cell proliferation, apoptosis, cellular stress response	Chronic atrophic gastritis [43], hepatotoxicity [44]
miR-9-3p	Apoptosis, cell proliferation, oxidative stress	Traumatic brain injury [45], ischemic stroke [46], hypoxia [47]
miR-9-5p	Apoptosis, inflammation, cell growth and proliferation, autophagy	Traumatic brain injury [45], ischemic stroke [46,48], Alzheimer's disease [49], glioblastoma [50]
miR-137-3p	Oxidative stress, neuron necrosis, apoptosis	Lung cancer [51], ICH [52], traumatic brain injury [45], brachial plexus root avulsion [53]
miR-1298-5p	Proliferation, cell migration and adhesion, apoptosis, neuroinflammation	Breast cancer [54], non-small cell lung cancer [55], ischemic stroke[56], glioma [57]
miR-219a-2-3p	Oxidative stress, cell growth, apoptosis, neuroinflammation	Traumatic brain injury [58], lung cancer [59], spinal cord injury [60]
miR-384-5p	Autophagy, inflammation	Neurotoxicity [61], lung injury [62], spinal cord injury [63], diabetic encephalopathy [64]
miR-34b-3p	Metastasis, oxidative stress, cell growth	Concussion [65], leptomeningeal metastasis [66], colorectal cancer [67], renal cell carcinoma [68], breast invasive ductal carcinoma [69]
miR-124-3p	Inflammation, neuronal autophagy, apoptosis	Traumatic brain injury [70,71], ischemic stroke [72], gastric cancer [73], ICH [74–77]
miR-34c-5p	Neuroinflammation, growth, metabolism	Drug-resistant epilepsy [78], COPD [79], papillary thyroid carcinoma [80]
miR-34b-5p	Oxidative stress, apoptosis, inflammation, migration, proliferation, invasion	Parkinson's disease [81], white matter ischemic injury [82], kidney injury [83], retinoblastoma [84], B- cell acute lymphoblastic leukemia [85]
miR-200b-3p	Cell growth, proliferation, metastasis	Prion disease [86], transient ischemic attack [87], colorectal cancer [88], traumatic brain injury [89]
miR-135a-5p	Apoptosis, inflammation, metastasis, proliferation, migration	Atherosclerosis [90], temporal lobe epilepsy [91], ischemic brain injury [92], colorectal cancer [93,94], diabetic nephropathy [95]
miR-148a-5p	Inflammation, cell growth, proliferation, metabolism	Irritable bowel syndrome [96], Crohn's disease [97]
miR-133b-3p	Inflammation, neurodegeneration	Parkinson's disease [98], Central post stroke pain[99], myotonic dystrophy [100]
miR-1199-5p	Migration, metastasis	Tumor metastasis [101]
miR-133a-5p	Apoptosis	Hepatic ischemia [102]
miR-451a	Blood brain barrier permeability, cell differentiation, metastasis, inflammation, apoptosis	Cerebral ischemia [103], ICH [104], blood brain barrier dysfunction [105], glioblastoma [106], prostate cancer [107], pancreatic cancer [108], Alzheimer's disease [109], multiple sclerosis [110]
miR-138-5p	Neuroinflammation, metastasis	Breast cancer [111], cognitive impairment [112], glioblastoma [113], Parkinson's disease [114]
let-7g-5p	Cell growth and differentiation, proliferation, metastasis	Alzheimer's disease [115], gastric cancer [116], liver disease [117], glioblastoma [118]
miR-301a-3p	Inflammation, apoptosis, cell differentiation	Multiple sclerosis [119], endometriosis [120], macular degeneration [121], esophageal squamous cell carcinoma [122], heart failure [123]
miR-200c-5p	Proliferation, migration	Human hepatocellular carcinoma [124], renal cell carcinoma [125]
miR-770-3p	Metabolism	Aging [126]
miR-541-5p	Apoptosis	Carcinogenesis [127]
miR-194-5p	Proliferation, apoptosis	Epilepsy [128,129], pancreatic cancer [130], breast cancer [131], esophageal adenocarcinoma [132], mitochondrial neuro-gastrointestinal encephalomyopathy [133],
miR-200c-3p	Metastasis, inflammation, cell growth	Bladder cancer [134,135], colorectal cancer [136–138], ovarian cancer [139], knee osteoarthritis [140]
miR-216a-5p	Inflammation, cell growth, metastasis, autophagy	Alzheimer's disease [141], acute pancreatitis [142], pancreas injury [143], pituitary tumors [144]

3.3. The Comparative Analysis of miRNAs

To determine the clinical significance of our observations, we compared our results to the study reported by Wang et al. [75], which enabled the comparison of dysregulated serum miRNAs after ICH in aged mice with human-brain-enriched miRNAs and plasma miRNAs after ICH in young rodents, as demonstrated in Figure 5. We found that miR-124-3p, miR-138-5p, and miR-137-5p, which are differentially expressed in the aged mouse serum after ICH, are also enriched in the human brain tissue, implicating their potential role in ICH-associated brain injury or recovery. Also, the miR-135a-5p level was increased in the blood of both aged mice and young rats after ICH, while miR-219a-2-3p, miR-770-3p, and miR-541-5p were differentially dysregulated in aged mice and young rats after ICH. This discrepancy could be due to differences in the age of the animal models or species that were used in preclinical studies. Therefore, further research is needed to validate these findings as these microRNAs could serve as invaluable molecular targets for the diagnosis and management of ICH that occurs in the elderly.

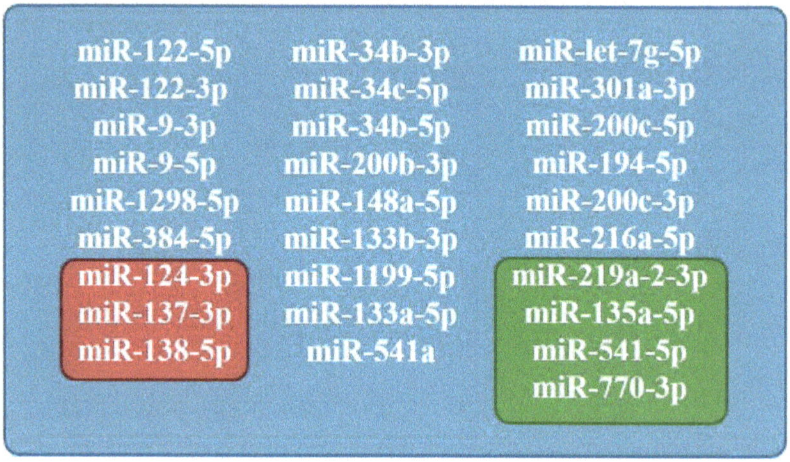

Figure 5. Comparative analysis of dysregulated serum miRNAs after ICH in aged mice (blue square) with human-brain-enriched miRNAs (red square) and dysregulated plasma miRNAs after ICH in young rats (green square). The serum microRNAs, which are human-brain-specific and differentially regulated in the plasma of young mice after ICH, are indicated.

4. Discussion

We identified 28 significantly dysregulated miRNAs in the serum of aged mice after ICH. In comparison to a previous study by Wang et al., [75], miR-124-3p, miR-137-5p, miR-138-5p, miR-219a-2-3p, miR-135a-5p, miR-541-5p, and miR-770-3p could be the most potential candidates to be tested for their roles post-ICH in the elderly. Wang et al. [75] demonstrated the dysregulation of plasma miRNAs in a young rat model of ICH. They focused primarily on miR-124 as a biomarker for ICH, but we also found increased serum levels of miR-138-5p and miR-137-5p in aged mice after ICH. A recent study documented an increased serum level of miR-137 after traumatic brain injury in patients [45]. Moreover, miR-124-3p, miR-137-5p, and miR-138-5p are human-brain-specific miRNAs [75], further demonstrating their potential to serve as serum biomarkers or therapeutic targets for intracerebral hemorrhage, warranting further investigation. Overall, the functional roles of miR-124-3p, miR-137-5p, miR-138-5p, miR-219a-2-3p, miR-135a-5p, miR-541-5p, and miR-770-3p in various pathological processes/states and their possible functions in the pathophysiology of ICH are discussed.

4.1. miR-124-3p

We found a significant increase in the level of miR-124-3p in the serum of aged mice after ICH compared to sham. This observation is consistent with a previous study, where the miR-124 level was found to be significantly increased in the plasma of ICH patients in the acute phases of injury, suggesting that miR-124 may serve as a biomarker for the diagnosis of ICH [75]. Functionally, miR-124 plays a key role in iron metabolism and neuronal cell death after ICH, and its inhibition reduced brain injury after ICH in aged mice [76], implicating the detrimental role of miR-124 after ICH. Consistently, high serum miR-124 levels were correlated with poor neurological scores in aged ICH patients [76]. Since miR-124-3p is one of the most abundant brain-specific microRNAs, studies are required to elucidate whether brain injury leads to its release into the blood plasma or serum after ICH.

In young rats, miR-124 was significantly elevated in the plasma and brain tissue, in a collagenase-injection mouse model of ICH, during the acute phase of the injury [75]. By contrast, in a blood-injection mouse model of ICH in young mice, miR-124 expression was found to be decreased in the perihematomal region of the brain [77]. Moreover, miR-124 attenuated ICH-induced inflammatory brain damage in young mice by modulating microglia polarization, implicating the neuroprotective role of miR-124 after ICH [77]. The underlying cause of this discrepancy in its expression after ICH and its function could be the difference in the ICH model and the age of animal subjects. Hence, further investigation is highly needed.

Altered serum expression of miR-124 is associated with various brain injuries. To this end, miR-124-3p was not detectable in healthy volunteers, but its increased level was observed in the serum of patients with severe traumatic brain injury [70]. Moreover, serum miR-124 is significantly enhanced in patients with acute ischemic stroke, where its expression positively correlated with infarct volume and degree of brain damage, as assessed by the National Institutes of Health Stroke Scale [145]. Overall, apart from considering miR-124-3p as a potential biomarker candidate for ICH, its precise functional role in the pathophysiology of ICH requires further validation.

4.2. miR-137-3p

Our findings show the increased level of miR-137-3p in the serum of aged mice after ICH in comparison to sham. Upregulation of miR-137-3p inhibited neuronal death, parthanatos, a type of programmed cell necrosis associated with ICH [52,146–149]. In addition, upregulation of miR-137-3p resulted in neuroprotective effects by decreasing neuronal nitric oxide synthase-positive cells and the death of motor neurons after avulsion injury to the spinal cord in rats [53]. Furthermore, an increased level of miR-137 is observed in the serum after traumatic brain injury [45]. Given the role of miR-137-3p in neuronal death and oxidative damage, further studies are warranted to explore its potential as a therapeutic target for ICH.

4.3. miR-138-5p

As per the current study, miR-138-5p levels were found to be significantly increased in the serum of aged mice after ICH compared to sham. Notably, breast cancer cell-derived miR-138-5p has been shown to inhibit M1 polarization and promote M2 polarization of macrophages [111]. It has also been proposed as a potential blood biomarker of Parkinson's disease [114]. Furthermore, miR-138-5p downregulated NLRP3 inflammasome and its downstream gene targets in lipopolysaccharide-treated rat microglia [112]. Overall, given the role of miR-138-5p in macrophage polarization and inflammation, further studies are required to elucidate its functional role after ICH.

In line with dysregulated plasma miRNAs after ICH in rats [75], levels of miR-135a-5p were increased in the serum of aged mice after ICH, but miR-219a-2-3p, miR-541-5p, and miR-770-3p were differentially dysregulated in aged mice and young rats after ICH. Based on their potential roles in ICH pathology in association with their altered expression, their functions in various pathological conditions are discussed.

4.4. miR-135a-5p

M2 microglia-derived extracellular vesicles contained elevated levels of miR-135a-5p, which reduced neuronal autophagy and ischemic brain injury in mice by inhibiting inflammasome signaling [92], suggesting its role in neuroprotection. In contrast, exercise decreased miR-135 levels in adult neural precursor cells and miR-135a-5p inhibition stimulated neurogenesis in the dentate gyrus of aged mice [150]. As well, miR-135a-5p expression in the hippocampus is increased in temporal lobe epilepsy in children [91]. miR-135a-5p mediated proapoptotic effect by inducing cellular apoptosis and reduced cell survival in temporal-lobe epilepsy [91]. Furthermore, its inhibition protected glial cells against epilepsy-induced apoptosis [151]. The miR-135a-5p expression level was significantly decreased in the serum samples of atherosclerosis patients and a mouse model of atherosclerosis [90]. Moreover, overexpression of miR-135a-5p induced a cell cycle arrest and apoptosis, and inhibited the proliferation and migration of vascular smooth muscle cells [90]. Additionally, miR-135 is a tumor suppressor and has been shown as a diagnostic biomarker of colorectal cancer [93]. Overall, given its conflicting roles, further studies are highly needed to establish its function after ICH.

4.5. miR-219a-2-3p

Upregulation of miR-219a-2-3p in tissue samples has been linked to anti-inflammatory responses, possibly by modulating NK-kB singling and promoting neuroprotection after spinal cord injury [60]. Serum-derived miR-219a-2-3p has also been shown to be a potential biomarker for traumatic brain injury in mice [58], as well as a peripheral blood biomarker for lung cancer in patients [59]. Given its potential as a biomarker in traumatic brain injury and its roles in neuroprotection and anti-inflammatory responses after a neural injury, miR-219a-2-3p needs to be explored further for its possible roles after ICH.

4.6. miR-541-5p

Upregulation of miR-541-5p has been linked to hepatocellular carcinoma [127]. miR-541 also contributes to the modulation of telomerase activity [152] and tumor suppression in non-small cell lung cancer [153]. Apart from that, its functional role after a brain pathology remains enigmatic, requiring investigation.

4.7. miR-770-3p

miR-770-3p is a biomarker for aging, as its expression was found to be increased in the serum of aged mice in comparison to young mice [126]. Apart from that, the role of miR-770-3p remains largely understudied. Therefore, further research is vital to determine its functions after ICH.

Though the study reveals several novel candidate miRNAs, some of the identified candidates could be related to ICH irrespective of age, and some could be related to ICH in the context of aging. Therefore, further studies are warranted to identify the age-dependency of those candidates. Moreover, owing to the complexity of aging, the identified candidates, whether related to ICH in an age-dependent or -independent manner, require characterization in aged animal subjects to elucidate their possible roles in apoptosis, neuroinflammation, oxidative stress and secondary brain damage after ICH.

5. Conclusions

Herein, we identified seven candidate serum miRNAs, miR-124-3p, miR-138-5p, miR-137-3p, miR-219a-2-3p, miR-135a-5p, miR541-5p, and miR-770-3p, which may have roles in the pathophysiology of ICH in the elderly, warranting further investigation. Among those, miR-124-3p, miR-138-5p, and miR-137-3p may have the greatest potential, as they are human-brain-specific miRNAs and are also implicated in neuronal apoptosis and neuroinflammation. Given the increasing prevalence of the aging population and age-related diseases, such as stroke, the miRNAs identified in this study may serve as invaluable molecular targets for future investigation post-ICH, a neurological disorder without an effective treatment option.

Author Contributions: Conceptualization, S.S.-R.; Methodology, D.-H.G.; Formal Analysis, D.R. and S.S.-R.; Investigation, D.R. and S.S.-R.; Data Curation, D.R. and S.S.-R.; Writing—Original Draft Preparation, D.R. and S.S.-R.; Writing—Review & Editing, D.R., D.A., N.W. and S.S.-R.; Supervision, S.S.-R.; Funding Acquisition, S.S.-R. All authors have read and agreed to the published version of the manuscript.

Funding: This work was supported by grants from the National Institutes of Health, R01NS107853, and R03AG077460 to S.S.-R.

Institutional Review Board Statement: Not Applicable.

Informed Consent Statement: Not Applicable.

Data Availability Statement: The data presented in this study are available on request from the corresponding author.

Acknowledgments: The authors thank Eiko Kitamura and Sam Chang at the Augusta University Integrated Genomics core facility for miRNA sequencing.

Conflicts of Interest: The authors declare that the research was conducted in the absence of any commercial or financial relationships that could be construed as a potential conflict of interest.

References

1. Caceres, J.A.; Goldstein, J.N. Intracranial hemorrhage. *Emerg. Med. Clin. N. Am.* **2012**, *30*, 771–794. [CrossRef]
2. Bonsack, F.T.; Alleyne, C.H., Jr.; Sukumari-Ramesh, S. Augmented expression of TSPO after intracerebral hemorrhage: A role in inflammation? *J. Neuroinflammation* **2016**, *13*, 151. [CrossRef] [PubMed]
3. Broderick, J.P. Efficacy of intraarterial thrombolysis of basilar artery stroke. *J. Stroke Cerebrovasc. Dis.* **1999**, *8*, III. [CrossRef]
4. Mendelow, A.D.; Gregson, B.A.; Fernandes, H.M.; Murray, G.D.; Teasdale, G.M.; Hope, D.T.; Karimi, A.; Shaw, M.D.; Barer, D.H.; the STICH Investigators. Early surgery versus initial conservative treatment in patients with spontaneous supratentorial intracerebral haematomas in the International Surgical Trial in Intracerebral Haemorrhage (STICH): A randomised trial. *Lancet* **2005**, *365*, 387–397. [CrossRef]
5. Urday, S.; Beslow, L.A.; Goldstein, D.W.; Vashkevich, A.; Ayres, A.M.; Battey, T.W.; Selim, M.H.; Kimberly, W.T.; Rosand, J.; Sheth, K.N. Measurement of perihematomal edema in intracerebral hemorrhage. *Stroke* **2015**, *46*, 1116–1119. [CrossRef] [PubMed]
6. Qureshi, A.I.; Feldmann, E.; Gomez, C.R.; Johnston, S.C.; Kasner, S.E.; Quick, D.C.; Rasmussen, P.A.; Suri, M.F.; Taylor, R.A.; Zaidat, O.O. Intracranial atherosclerotic disease: An update. *Ann. Neurol.* **2009**, *66*, 730–738. [CrossRef]
7. Duan, X.; Wang, Y.; Zhang, F.; Lu, L.; Cao, M.; Lin, B.; Zhang, X.; Mao, J.; Shuai, X.; Shen, J. Superparamagnetic Iron Oxide-Loaded Cationic Polymersomes for Cellular MR Imaging of Therapeutic Stem Cells in Stroke. *J. Biomed. Nanotechnol.* **2016**, *12*, 2112–2124. [CrossRef] [PubMed]
8. Zhang, Z.W.; Zhang, D.P.; Li, H.Y.; Wang, Z.; Chen, G. The role of nitrous oxide in stroke. *Med. Gas Res.* **2017**, *7*, 273–276.
9. Madangarli, N.; Bonsack, F.; Dasari, R.; Sukumari-Ramesh, S. Intracerebral Hemorrhage: Blood Components and Neurotoxicity. *Brain Sci.* **2019**, *9*, 316. [CrossRef] [PubMed]
10. Garton, T.; Keep, R.F.; Hua, Y.; Xi, G. Brain iron overload following intracranial haemorrhage. *Stroke Vasc. Neurol.* **2016**, *1*, 172–184. [CrossRef] [PubMed]
11. Dasari, R.; Bonsack, F.; Sukumari-Ramesh, S. Brain injury and repair after intracerebral hemorrhage: The role of microglia and brain-infiltrating macrophages. *Neurochem. Int.* **2021**, *142*, 104923. [CrossRef]
12. Zhu, H.; Wang, Z.; Yu, J.; Yang, X.; He, F.; Liu, Z.; Che, F.; Chen, X.; Ren, H.; Hong, M.; et al. Role and mechanisms of cytokines in the secondary brain injury after intracerebral hemorrhage. *Prog. Neurobiol.* **2019**, *178*, 101610. [CrossRef]
13. Vogel, C.; Marcotte, E.M. Insights into the regulation of protein abundance from proteomic and transcriptomic analyses. *Nat. Rev. Genet.* **2012**, *13*, 227–232. [CrossRef]
14. Filipowicz, W.; Bhattacharyya, S.N.; Sonenberg, N. Mechanisms of post-transcriptional regulation by microRNAs: Are the answers in sight? *Nat. Rev. Genet.* **2008**, *9*, 102–114. [CrossRef] [PubMed]
15. Cheng, Z.; Dai, Y.; Huang, W.; Zhong, Q.; Zhu, P.; Zhang, W.; Wu, Z.; Lin, Q.; Zhu, H.; Cui, L.; et al. Prognostic Value of MicroRNA-20b in Acute Myeloid Leukemia. *Front. Oncol.* **2020**, *10*, 553344. [CrossRef]
16. Zeng, L.; Liu, J.; Wang, Y.; Wang, L.; Weng, S.; Tang, Y.; Zheng, C.; Cheng, Q.; Chen, S.; Yang, G.Y. MicroRNA-210 as a novel blood biomarker in acute cerebral ischemia. *Front. Biosci.* **2011**, *3*, 1265–1272. [CrossRef]

17. Liu, D.Z.; Tian, Y.; Ander, B.P.; Xu, H.; Stamova, B.S.; Zhan, X.; Turner, R.J.; Jickling, G.; Sharp, F.R. Brain and blood microRNA expression profiling of ischemic stroke, intracerebral hemorrhage, and kainate seizures. *J. Cereb. Blood Flow Metab.* **2010**, *30*, 92–101. [CrossRef]

18. Xu, J.; Zhao, J.; Evan, G.; Xiao, C.; Cheng, Y.; Xiao, J. Circulating microRNAs: Novel biomarkers for cardiovascular diseases. *J. Mol. Med.* **2012**, *90*, 865–875. [CrossRef] [PubMed]

19. Aunin, E.; Broadley, D.; Ahmed, M.I.; Mardaryev, A.N.; Botchkareva, N.V. Exploring a Role for Regulatory miRNAs in Wound Healing during Ageing:Involvement of miR-200c in wound repair. *Sci. Rep.* **2017**, *7*, 3257. [CrossRef] [PubMed]

20. Smith-Vikos, T.; Slack, F.J. MicroRNAs and their roles in aging. *J. Cell Sci.* **2012**, *125 Pt 1*, 7–17. [CrossRef]

21. Eacker, S.M.; Dawson, T.M.; Dawson, V.L. Understanding microRNAs in neurodegeneration. *Nat. Rev. Neurosci.* **2009**, *10*, 837–841. [CrossRef] [PubMed]

22. Wagner, K.H.; Cameron-Smith, D.; Wessner, B.; Franzke, B. Biomarkers of Aging: From Function to Molecular Biology. *Nutrients* **2016**, *8*, 338. [CrossRef] [PubMed]

23. Lovelock, C.E.; Molyneux, A.J.; Rothwell, P.M. Oxford Vascular S: Change in incidence and aetiology of intracerebral haemorrhage in Oxfordshire, UK, between 1981 and 2006: A population-based study. *Lancet Neurol.* **2007**, *6*, 487–493. [CrossRef]

24. Jaul, E. Non-healing wounds: The geriatric approach. *Arch. Gerontol. Geriatr.* **2009**, *49*, 224–226. [CrossRef]

25. Huang, Y.Q.; Liu, L.; Huang, J.Y.; Chen, C.L.; Yu, Y.L.; Lo, K.; Feng, Y.Q. Prediabetes and risk for all-cause and cardiovascular mortality based on hypertension status. *Ann. Transl. Med.* **2020**, *8*, 1580. [CrossRef]

26. Persengiev, S.; Kondova, I.; Otting, N.; Koeppen, A.H.; Bontrop, R.E. Genome-wide analysis of miRNA expression reveals a potential role for miR-144 in brain aging and spinocerebellar ataxia pathogenesis. *Neurobiol. Aging* **2011**, *32*, 2316.e17–2316.e27. [CrossRef]

27. Liang, R.; Khanna, A.; Muthusamy, S.; Li, N.; Sarojini, H.; Kopchick, J.J.; Masternak, M.M.; Bartke, A.; Wang, E. Post-transcriptional regulation of IGF1R by key microRNAs in long-lived mutant mice. *Aging Cell* **2011**, *10*, 1080–1088. [CrossRef] [PubMed]

28. Ameling, S.; Kacprowski, T.; Chilukoti, R.K.; Malsch, C.; Liebscher, V.; Suhre, K.; Pietzner, M.; Friedrich, N.; Homuth, G.; Hammer, E.; et al. Associations of circulating plasma microRNAs with age, body mass index and sex in a population-based study. *BMC Med. Genom.* **2015**, *8*, 61. [CrossRef]

29. Goodall, E.F.; Leach, V.; Wang, C.; Cooper-Knock, J.; Heath, P.R.; Baker, D.; Drew, D.R.; Saffrey, M.J.; Simpson, J.E.; Romero, I.A.; et al. Age-Associated mRNA and miRNA Expression Changes in the Blood-Brain Barrier. *Int. J. Mol. Sci.* **2019**, *20*, 3097. [CrossRef]

30. Sukumari-Ramesh, S.; Alleyne, C.H.; Jr Dhandapani, K.M. Astrogliosis: A target for intervention in intracerebral hemorrhage? *Transl. Stroke Res.* **2012**, *3* (Suppl 1), 80–87. [CrossRef]

31. Sukumari-Ramesh, S.; Alleyne, C.H., Jr.; Dhandapani, K.M. Astrocyte-specific expression of survivin after intracerebral hemorrhage in mice: A possible role in reactive gliosis? *J. Neurotrauma* **2012**, *29*, 2798–2804. [CrossRef]

32. Sukumari-Ramesh, S.; Alleyne, C.H., Jr.; Dhandapani, K.M. The Histone Deacetylase Inhibitor Suberoylanilide Hydroxamic Acid (SAHA) Confers Acute Neuroprotection After Intracerebral Hemorrhage in Mice. *Transl. Stroke Res.* **2016**, *7*, 141–148. [CrossRef] [PubMed]

33. Sukumari-Ramesh, S.; Alleyne, C.H., Jr. Post-Injury Administration of Tert-butylhydroquinone Attenuates Acute Neurological Injury After Intracerebral Hemorrhage in Mice. *J. Mol. Neurosci.* **2016**, *58*, 525–531. [CrossRef] [PubMed]

34. Clark, W.; Gunion-Rinker, L.; Lessov, N.; Hazel, K. Citicoline treatment for experimental intracerebral hemorrhage in mice. *Stroke* **1998**, *29*, 2136–2140. [CrossRef]

35. Rosenberg, G.A.; Mun-Bryce, S.; Wesley, M.; Kornfeld, M. Collagenase-induced intracerebral hemorrhage in rats. *Stroke* **1990**, *21*, 801–807. [CrossRef] [PubMed]

36. Li, Q.; Han, X.; Lan, X.; Gao, Y.; Wan, J.; Durham, F.; Cheng, T.; Yang, J.; Wang, Z.; Jiang, C.; et al. Inhibition of neuronal ferroptosis protects hemorrhagic brain. *JCI Insight* **2017**, *2*, e90777. [CrossRef]

37. Aysenne, A.M.; Albright, K.C.; Mathias, T.; Chang, T.R.; Boehme, A.K.; Beasley, T.M.; Martin-Schild, S. 24-Hour ICH Score is a Better Predictor of Outcome than Admission ICH Score. *ISRN Stroke* **2013**, *2013*. [CrossRef] [PubMed]

38. Jiao, Y.; Zhang, L.; Li, J.; He, Y.; Zhang, X.; Li, J. Exosomal miR-122-5p inhibits tumorigenicity of gastric cancer by downregulating GIT1. *Int. J. Biol. Markers* **2021**, *36*, 36–46. [CrossRef]

39. Heinemann, F.G.; Tolkach, Y.; Deng, M.; Schmidt, D.; Perner, S.; Kristiansen, G.; Muller, S.C.; Ellinger, J. Serum miR-122-5p and miR-206 expression: Non-invasive prognostic biomarkers for renal cell carcinoma. *Clin. Epigenetics* **2018**, *10*, 11. [CrossRef] [PubMed]

40. Lee, H.M.; Wong, W.K.K.; Fan, B.; Lau, E.S.; Hou, Y.O.C.K.; Luk, A.O.Y.; Chow, E.Y.K.; Ma, R.C.W.; Chan, J.C.N.; Kong, A.P.S. Detection of increased serum miR-122-5p and miR-455-3p levels before the clinical diagnosis of liver cancer in people with type 2 diabetes. *Sci. Rep.* **2021**, *11*, 23756. [CrossRef]

41. Khan, I.A.; Rashid, S.; Singh, N.; Rashid, S.; Singh, V.; Gunjan, D.; Das, P.; Dash, N.R.; Pandey, R.M.; Chauhan, S.S.; et al. Panel of serum miRNAs as potential non-invasive biomarkers for pancreatic ductal adenocarcinoma. *Sci. Rep.* **2021**, *11*, 2824. [CrossRef] [PubMed]

42. Li, D.B.; Liu, J.L.; Wang, W.; Luo, X.M.; Zhou, X.; Li, J.P.; Cao, X.L.; Long, X.H.; Chen, J.G.; Qin, C. Plasma Exosomal miRNA-122-5p and miR-300-3p as Potential Markers for Transient Ischaemic Attack in Rats. *Front. Aging Neurosci.* **2018**, *10*, 24. [CrossRef] [PubMed]

43. Liu, H.; Li, P.W.; Yang, W.Q.; Mi, H.; Pan, J.L.; Huang, Y.C.; Hou, Z.K.; Hou, Q.K.; Luo, Q.; Liu, F.B. Identification of non-invasive biomarkers for chronic atrophic gastritis from serum exosomal microRNAs. *BMC Cancer* **2019**, *19*, 129. [CrossRef] [PubMed]

44. Yang, R.; Yang, F.; Huang, Z.; Jin, Y.; Sheng, Y.; Ji, L. Serum microRNA-122-3p, microRNA-194-5p and microRNA-5099 are potential toxicological biomarkers for the hepatotoxicity induced by Airpotato yam. *Toxicol. Lett.* **2017**, *280*, 125–132. [CrossRef]

45. O'Connell, G.C.; Smothers, C.G.; Winkelman, C. Bioinformatic analysis of brain-specific miRNAs for identification of candidate traumatic brain injury blood biomarkers. *Brain Inj.* **2020**, *34*, 965–974. [CrossRef]

46. Sorensen, S.S.; Nygaard, A.B.; Carlsen, A.L.; Heegaard, N.H.H.; Bak, M.; Christensen, T. Elevation of brain-enriched miRNAs in cerebrospinal fluid of patients with acute ischemic stroke. *Biomark. Res.* **2017**, *5*, 24. [CrossRef]

47. Lau, K.; Lai, K.P.; Bao, J.Y.; Zhang, N.; Tse, A.; Tong, A.; Li, J.W.; Lok, S.; Kong, R.Y.; Lui, W.Y.; et al. Identification and expression profiling of microRNAs in the brain, liver and gonads of marine medaka (Oryzias melastigma) and in response to hypoxia. *PLoS ONE* **2014**, *9*, e110698. [CrossRef] [PubMed]

48. Wang, Q.; Wang, F.; Fu, F.; Liu, J.; Sun, W.; Chen, Y. Diagnostic and prognostic value of serum miR-9-5p and miR-128-3p levels in early-stage acute ischemic stroke. *Clinics* **2021**, *76*, e2958. [CrossRef]

49. Chen, M.L.; Hong, C.G.; Yue, T.; Li, H.M.; Duan, R.; Hu, W.B.; Cao, J.; Wang, Z.X.; Chen, C.Y.; Hu, X.K.; et al. Inhibition of miR-331-3p and miR-9-5p ameliorates Alzheimer's disease by enhancing autophagy. *Theranostics* **2021**, *11*, 2395–2409. [CrossRef]

50. Luo, X.; Tu, T.; Zhong, Y.; Xu, S.; Chen, X.; Chen, L.; Yang, F. ceRNA Network Analysis Shows That lncRNA CRNDE Promotes Progression of Glioblastoma Through Sponge mir-9-5p. *Front. Genet.* **2021**, *12*, 617350. [CrossRef]

51. Li, Y.B.; Lin, L.Z.; Guan, J.S.; Chen, C.M.; Zuo, Q.; Lin, B.Q. TCM Combined Western Medicine Treatment of Advanced NSCLC: A Preliminary Study of mIRNA Expression Profiles. *Zhongguo Zhong Xi Yi Jie He Za Zhi* **2016**, *36*, 1076–1081. [PubMed]

52. Wang, J.; Kuang, X.; Peng, Z.; Li, C.; Guo, C.; Fu, X.; Wu, J.; Luo, Y.; Rao, X.; Zhou, X.; et al. EGCG treats ICH via up-regulating miR-137-3p and inhibiting Parthanatos. *Transl. Neurosci.* **2020**, *11*, 371–379. [CrossRef]

53. Tang, Y.; Fu, R.; Ling, Z.M.; Liu, L.L.; Yu, G.Y.; Li, W.; Fang, X.Y.; Zhu, Z.; Wu, W.T.; Zhou, L.H. MiR-137-3p rescue motoneuron death by targeting calpain-2. *Nitric Oxide* **2018**, *74*, 74–85. [CrossRef]

54. Zhang, J.; Hu, D. miR-1298-5p Influences the Malignancy Phenotypes of Breast Cancer Cells by Inhibiting CXCL11. *Cancer Manag. Res.* **2021**, *13*, 133–145. [CrossRef] [PubMed]

55. Guan, H.; Sun, C.; Gu, Y.; Li, J.; Ji, J.; Zhu, Y. Circular RNA circ_0003028 contributes to tumorigenesis by regulating GOT2 via miR-1298-5p in non-small cell lung cancer. *Bioengineered* **2021**, *12*, 2326–2340. [CrossRef]

56. Miao, W.; Yan, Y.; Bao, T.H.; Jia, W.J.; Yang, F.; Wang, Y.; Zhu, Y.H.; Yin, M.; Han, J.H. Ischemic postconditioning exerts neuroprotective effect through negatively regulating PI3K/Akt2 signaling pathway by microRNA-124. *Biomed. Pharmacother.* **2020**, *126*, 109786. [CrossRef]

57. Li, H.; Liu, Q.; Chen, Z.; Wu, M.; Zhang, C.; Su, J.; Li, Y.; Zhang, C. Hsa_circ_0110757 upregulates ITGA1 to facilitate temozolomide resistance in glioma by suppressing hsa-miR-1298-5p. *Cell Death Dis.* **2021**, *12*, 252. [CrossRef]

58. Ko, J.; Hemphill, M.; Yang, Z.; Beard, K.; Sewell, E.; Shallcross, J.; Schweizer, M.; Sandsmark, D.K.; Diaz-Arrastia, R.; Kim, J.; et al. Multi-Dimensional Mapping of Brain-Derived Extracellular Vesicle MicroRNA Biomarker for Traumatic Brain Injury Diagnostics. *J. Neurotrauma* **2020**, *37*, 2424–2434. [CrossRef]

59. He, Q.; Fang, Y.; Lu, F.; Pan, J.; Wang, L.; Gong, W.; Fei, F.; Cui, J.; Zhong, J.; Hu, R.; et al. Analysis of differential expression profile of miRNA in peripheral blood of patients with lung cancer. *J. Clin. Lab. Anal.* **2019**, *33*, e23003. [CrossRef] [PubMed]

60. Ma, K.; Xu, H.; Zhang, J.; Zhao, F.; Liang, H.; Sun, H.; Li, P.; Zhang, S.; Wang, R.; Chen, X. Insulin-like growth factor-1 enhances neuroprotective effects of neural stem cell exosomes after spinal cord injury via an miR-219a-2-3p/YY1 mechanism. *Aging* **2019**, *11*, 12278–12294. [CrossRef] [PubMed]

61. Ogata, K.; Sumida, K.; Miyata, K.; Kushida, M.; Kuwamura, M.; Yamate, J. Circulating miR-9* and miR-384-5p as potential indicators for trimethyltin-induced neurotoxicity. *Toxicol. Pathol.* **2015**, *43*, 198–208. [CrossRef] [PubMed]

62. Liu, X.; Gao, C.; Wang, Y.; Niu, L.; Jiang, S.; Pan, S. BMSC-Derived Exosomes Ameliorate LPS-Induced Acute Lung Injury by miR-384-5p-Controlled Alveolar Macrophage Autophagy. *Oxid. Med. Cell. Longev.* **2021**, *2021*, 9973457. [CrossRef]

63. Hachisuka, S.; Kamei, N.; Ujigo, S.; Miyaki, S.; Yasunaga, Y.; Ochi, M. Circulating microRNAs as biomarkers for evaluating the severity of acute spinal cord injury. *Spinal Cord* **2014**, *52*, 596–600. [CrossRef] [PubMed]

64. Wang, B.; Huang, J.; Li, J.; Zhong, Y. Control of macrophage autophagy by miR-384-5p in the development of diabetic encephalopathy. *Am. J. Transl. Res.* **2018**, *10*, 511–518. [PubMed]

65. Di Pietro, V.; O'Halloran, P.; Watson, C.N.; Begum, G.; Acharjee, A.; Yakoub, K.M.; Bentley, C.; Davies, D.J.; Iliceto, P.; Candilera, G.; et al. Unique diagnostic signatures of concussion in the saliva of male athletes: The Study of Concussion in Rugby Union through MicroRNAs (SCRUM). *Br. J. Sports Med.* **2021**, *55*, 1395–1404. [CrossRef]

66. Im, J.H.; Kim, T.H.; Lee, K.Y.; Gwak, H.S.; Lin, W.; Park, J.B.; Kim, J.H.; Yoo, B.C.; Park, S.M.; Kwon, J.W.; et al. Exploratory Profiling of Extracellular MicroRNAs in Cerebrospinal Fluid Comparing Leptomeningeal Metastasis with Other Central Nervous System Tumor Statuses. *J. Clin. Med.* **2021**, *10*, 4860. [CrossRef] [PubMed]

67. Peng, X.; Wang, J.; Zhang, C.; Liu, K.; Zhao, L.; Chen, X.; Huang, G.; Lai, Y. A three-miRNA panel in serum as a noninvasive biomarker for colorectal cancer detection. *Int. J. Biol. Markers* **2020**, *35*, 74–82. [CrossRef]

68. Huang, G.; Li, X.; Chen, Z.; Wang, J.; Zhang, C.; Chen, X.; Peng, X.; Liu, K.; Zhao, L.; Lai, Y.; et al. A Three-microRNA Panel in Serum: Serving as a Potential Diagnostic Biomarker for Renal Cell Carcinoma. *Pathol. Oncol. Res.* **2020**, *26*, 2425–2434. [CrossRef]

69. Chen, X.; Li, X.; Wang, J.; Zhao, L.; Peng, X.; Zhang, C.; Liu, K.; Huang, G.; Lai, Y. Breast invasive ductal carcinoma diagnosis with a three-miRNA panel in serum. *Biomark. Med.* **2021**, *15*, 951–963. [CrossRef]
70. Schindler, C.R.; Woschek, M.; Vollrath, J.T.; Kontradowitz, K.; Lustenberger, T.; Stormann, P.; Marzi, I.; Henrich, D. miR-142-3p Expression Is Predictive for Severe Traumatic Brain Injury (TBI) in Trauma Patients. *Int. J. Mol. Sci.* **2020**, *21*, 5381. [CrossRef]
71. Li, D.; Huang, S.; Yin, Z.; Zhu, J.; Ge, X.; Han, Z.; Tan, J.; Zhang, S.; Zhao, J.; Chen, F.; et al. Increases in miR-124-3p in Microglial Exosomes Confer Neuroprotective Effects by Targeting FIP200-Mediated Neuronal Autophagy Following Traumatic Brain Injury. *Neurochem. Res.* **2019**, *44*, 1903–1923. [CrossRef]
72. Qi, Z.; Zhao, Y.; Su, Y.; Cao, B.; Yang, J.J.; Xing, Q. Serum Extracellular Vesicle-Derived miR-124-3p as a Diagnostic and Predictive Marker for Early-Stage Acute Ischemic Stroke. *Front. Mol. Biosci.* **2021**, *8*, 685088. [CrossRef]
73. Jin, L.; Zhang, Z. Serum miR-3180-3p and miR-124-3p may Function as Noninvasive Biomarkers of Cisplatin Resistance in Gastric Cancer. *Clin. Lab.* **2020**, *66*. [CrossRef]
74. Fang, Y.; Hong, X. miR-124-3p Inhibits Microglial Secondary Inflammation After Basal Ganglia Hemorrhage by Targeting TRAF6 and Repressing the Activation of NLRP3 Inflammasome. *Front. Neurol.* **2021**, *12*, 653321. [CrossRef]
75. Wang, Z.; Lu, G.; Sze, J.; Liu, Y.; Lin, S.; Yao, H.; Zhang, J.; Xie, D.; Liu, Q.; Kung, H.F.; et al. Plasma miR-124 is a Promising Candidate Biomarker for Human Intracerebral Hemorrhage Stroke. *Mol. Neurobiol.* **2018**, *55*, 5879–5888. [CrossRef] [PubMed]
76. Bao, W.D.; Zhou, X.T.; Zhou, L.T.; Wang, F.; Yin, X.; Lu, Y.; Zhu, L.Q.; Liu, D. Targeting miR-124/Ferroportin signaling ameliorated neuronal cell death through inhibiting apoptosis and ferroptosis in aged intracerebral hemorrhage murine model. *Aging Cell* **2020**, *19*, e13235. [CrossRef]
77. Yu, A.; Zhang, T.; Duan, H.; Pan, Y.; Zhang, X.; Yang, G.; Wang, J.; Deng, Y.; Yang, Z. MiR-124 contributes to M2 polarization of microglia and confers brain inflammatory protection via the C/EBP-alpha pathway in intracerebral hemorrhage. *Immunol. Lett.* **2017**, *182*, 1–11. [CrossRef] [PubMed]
78. Fu, M.; Tao, J.; Wang, D.; Zhang, Z.; Wang, X.; Ji, Y.; Li, Z. Downregulation of MicroRNA-34c-5p facilitated neuroinflammation in drug-resistant epilepsy. *Brain Res.* **2020**, *1749*, 147130. [CrossRef] [PubMed]
79. Akbas, F.; Coskunpinar, E.; Aynaci, E.; Oltulu, Y.M.; Yildiz, P. Analysis of serum micro-RNAs as potential biomarker in chronic obstructive pulmonary disease. *Exp. Lung Res.* **2012**, *38*, 286–294. [CrossRef]
80. Qiu, Z.L.; Shen, C.T.; Song, H.J.; Wei, W.J.; Luo, Q.Y. Differential expression profiling of circulation microRNAs in PTC patients with non-131I and 131I-avid lungs metastases: A pilot study. *Nucl. Med. Biol.* **2015**, *42*, 499–504. [CrossRef]
81. Li, Y.; Fang, J.; Zhou, Z.; Zhou, Q.; Sun, S.; Jin, Z.; Xi, Z.; Wei, J. Downregulation of lncRNA BACE1-AS improves dopamine-dependent oxidative stress in rats with Parkinson's disease by upregulating microRNA-34b-5p and downregulating BACE1. *Cell Cycle* **2020**, *19*, 1158–1171. [CrossRef] [PubMed]
82. Baltan, S.; Sandau, U.S.; Brunet, S.; Bastian, C.; Tripathi, A.; Nguyen, H.; Liu, H.; Saugstad, J.A.; Zarnegarnia, Y.; Dutta, R. Identification of miRNAs That Mediate Protective Functions of Anti-Cancer Drugs During White Matter Ischemic Injury. *ASN Neuro* **2021**, *13*, 17590914211042220. [CrossRef]
83. Zheng, C.; Wu, D.; Shi, S.; Wang, L. miR-34b-5p promotes renal cell inflammation and apoptosis by inhibiting aquaporin-2 in sepsis-induced acute kidney injury. *Ren. Fail.* **2021**, *43*, 291–301. [CrossRef]
84. Zhang, S.; Cui, Z. MicroRNA-34b-5p inhibits proliferation, stemness, migration and invasion of retinoblastoma cells via Notch signaling. *Exp. Ther. Med.* **2021**, *21*, 255. [CrossRef]
85. Wang, Y.; Wang, Y.; Hui, H.; Fan, X.; Wang, T.; Xia, W.; Liu, L. MicroRNA expression is deregulated by aberrant methylation in B-cell acute lymphoblastic leukemia mouse model. *Mol. Biol. Rep.* **2022**, *49*, 1731–1739. [CrossRef] [PubMed]
86. Boese, A.S.; Saba, R.; Campbell, K.; Majer, A.; Medina, S.; Burton, L.; Booth, T.F.; Chong, P.; Westmacott, G.; Dutta, S.M.; et al. MicroRNA abundance is altered in synaptoneurosomes during prion disease. *Mol. Cell. Neurosci.* **2016**, *71*, 13–24. [CrossRef]
87. Zheng, J.; Wang, Z.; Li, N.; Zhang, X.; Huo, X. Synthetic role of miR-200b-3p, ABCD(2) score, and carotid ultrasound in the prediction of cerebral infarction in patients with transient ischemic attack. *Brain Behav.* **2022**, *12*, e32518. [CrossRef] [PubMed]
88. Tang, Y.; Zhao, Y.; Song, X.; Song, X.; Niu, L.; Xie, L. Tumor-derived exosomal miRNA-320d as a biomarker for metastatic colorectal cancer. *J. Clin. Lab. Anal.* **2019**, *33*, e23004. [CrossRef]
89. Osei, J.; Kelly, W.; Toffolo, K.; Donahue, K.; Levy, B.; Bard, J.; Wang, J.; Levy, E.; Nowak, N.; Poulsen, D. Thymosin beta 4 induces significant changes in the plasma miRNA profile following severe traumatic brain injury in the rat lateral fluid percussion injury model. *Expert Opin. Biol. Ther.* **2018**, *18* (Suppl 1), 159–164. [CrossRef] [PubMed]
90. Li, D.; An, Y. MiR-135a-5p inhibits vascular smooth muscle cells proliferation and migration by inactivating FOXO1 and JAK2 signaling pathway. *Pathol. Res. Pract.* **2021**, *224*, 153091. [CrossRef] [PubMed]
91. Wu, X.; Wang, Y.; Sun, Z.; Ren, S.; Yang, W.; Deng, Y.; Tian, C.; Yu, Y.; Gao, B. Molecular expression and functional analysis of genes in children with temporal lobe epilepsy. *J. Integr. Neurosci.* **2019**, *18*, 71–77. [PubMed]
92. Liu, Y.; Li, Y.P.; Xiao, L.M.; Chen, L.K.; Zheng, S.Y.; Zeng, E.M.; Xu, C.H. Extracellular vesicles derived from M2 microglia reduce ischemic brain injury through microRNA-135a-5p/TXNIP/NLRP3 axis. *Lab. Investig.* **2021**, *101*, 837–850. [CrossRef]
93. Wang, Q.; Zhang, H.; Shen, X.; Ju, S. Serum microRNA-135a-5p as an auxiliary diagnostic biomarker for colorectal cancer. *Ann. Clin. Biochem.* **2017**, *54*, 76–85. [CrossRef] [PubMed]
94. Sun, H.; Meng, Q.; Shi, C.; Yang, H.; Li, X.; Wu, S.; Familiari, G.; Relucenti, M.; Aschner, M.; Wang, X.; et al. Hypoxia-Inducible Exosomes Facilitate Liver-Tropic Premetastatic Niche in Colorectal Cancer. *Hepatology* **2021**, *74*, 2633–2651. [CrossRef]

95. Zhang, J.; Zhang, L.; Zha, D.; Wu, X. Inhibition of miRNA135a5p ameliorates TGFbeta1induced human renal fibrosis by targeting SIRT1 in diabetic nephropathy. *Int. J. Mol. Med.* **2020**, *46*, 1063–1073. [CrossRef] [PubMed]

96. Xing, Y.; Xue, S.; Wu, J.; Zhou, J.; Xing, F.; Li, T.; Nie, X. Serum Exosomes Derived from Irritable Bowel Syndrome Patient Increase Cell Permeability via Regulating miR-148b-5p/RGS2 Signaling in Human Colonic Epithelium Cells. *Gastroenterol. Res. Pract.* **2021**, *2021*, 6655900. [CrossRef]

97. Jacenik, D.; Zielinska, M.; Mokrowiecka, A.; Michlewska, S.; Malecka-Panas, E.; Kordek, R.; Fichna, J.; Krajewska, W.M. G protein-coupled estrogen receptor mediates anti-inflammatory action in Crohn's disease. *Sci. Rep.* **2019**, *9*, 6749. [CrossRef] [PubMed]

98. He, X.; Yang, L.; Huang, R.; Lin, L.; Shen, Y.; Cheng, L.; Jin, L.; Wang, S.; Zhu, R. Activation of CB2R with AM1241 ameliorates neurodegeneration via the Xist/miR-133b-3p/Pitx3 axis. *J. Cell. Physiol.* **2020**, *235*, 6032–6042. [CrossRef]

99. Guo, X.; Lu, J.; Yan, M.; Wang, Y.; Yang, Y.; Li, H.; Shen, H.; Diao, S.; Ni, J.; Lu, H.; et al. MicroRNA-133b-3p Targets Purinergic P2X4 Receptor to Regulate Central Poststroke Pain in Rats. *Neuroscience* **2022**, *481*, 60–72. [CrossRef]

100. Koutalianos, D.; Koutsoulidou, A.; Mytidou, C.; Kakouri, A.C.; Oulas, A.; Tomazou, M.; Kyriakides, T.C.; Prokopi, M.; Kapnisis, K.; Nikolenko, N.; et al. miR-223-3p and miR-24-3p as novel serum-based biomarkers for myotonic dystrophy type 1. *Mol. Ther. Methods Clin. Dev.* **2021**, *23*, 169–183. [CrossRef]

101. Diepenbruck, M.; Tiede, S.; Saxena, M.; Ivanek, R.; Kalathur, R.K.R.; Luond, F.; Meyer-Schaller, N.; Christofori, G. miR-1199-5p and Zeb1 function in a double-negative feedback loop potentially coordinating EMT and tumour metastasis. *Nat. Commun.* **2017**, *8*, 1168. [CrossRef]

102. Hao, W.; Zhao, Z.H.; Meng, Q.T.; Tie, M.E.; Lei, S.Q.; Xia, Z.Y. Propofol protects against hepatic ischemia/reperfusion injury via miR-133a-5p regulating the expression of MAPK6. *Cell Biol. Int.* **2017**, *41*, 495–504. [CrossRef]

103. Li, H.; Luo, Y.; Liu, P.; Liu, P.; Hua, W.; Zhang, Y.; Zhang, L.; Li, Z.; Xing, P.; Zhang, Y.; et al. Exosomes containing miR-451a is involved in the protective effect of cerebral ischemic preconditioning against cerebral ischemia and reperfusion injury. *CNS Neurosci. Ther.* **2021**, *27*, 564–576. [CrossRef] [PubMed]

104. Wang, X.; Hong, Y.; Wu, L.; Duan, X.; Hu, Y.; Sun, Y.; Wei, Y.; Dong, Z.; Wu, C.; Yu, D.; et al. Deletion of MicroRNA-144/451 Cluster Aggravated Brain Injury in Intracerebral Hemorrhage Mice by Targeting 14-3-3zeta. *Front. Neurol.* **2020**, *11*, 551411. [CrossRef] [PubMed]

105. Koopaei, N.N.; Chowdhury, E.A.; Jiang, J.; Noorani, B.; da Silva, L.; Bulut, G.; Hakimjavadi, H.; Chamala, S.; Bickel, U.; Schmittgen, T.D. Enrichment of the erythrocyte miR-451a in brain extracellular vesicles following impairment of the blood-brain barrier. *Neurosci. Lett.* **2021**, *751*, 135829. [CrossRef] [PubMed]

106. Wang, Z.Q.; Zhang, M.Y.; Deng, M.L.; Weng, N.Q.; Wang, H.Y.; Wu, S.X. Low serum level of miR-485-3p predicts poor survival in patients with glioblastoma. *PLoS ONE* **2017**, *12*, e0184969. [CrossRef] [PubMed]

107. Fan, B.; Jin, X.; Ding, Q.; Cao, C.; Shi, Y.; Zhu, H.; Zhou, W. Expression of miR-451a in Prostate Cancer and Its Effect on Prognosis. *Iran. J. Public Health* **2021**, *50*, 772–779. [CrossRef]

108. Chen, J.; Yao, D.; Chen, W.; Li, Z.; Guo, Y.; Zhu, F.; Hu, X. Serum exosomal miR-451a acts as a candidate marker for pancreatic cancer. *Int. J. Biol. Markers* **2022**, *37*, 74–80. [CrossRef]

109. McKeever, P.M.; Schneider, R.; Taghdiri, F.; Weichert, A.; Multani, N.; Brown, R.A.; Boxer, A.L.; Karydas, A.; Miller, B.; Robertson, J.; et al. MicroRNA Expression Levels Are Altered in the Cerebrospinal Fluid of Patients with Young-Onset Alzheimer's Disease. *Mol. Neurobiol.* **2018**, *55*, 8826–8841. [CrossRef]

110. Ebrahimkhani, S.; Vafaee, F.; Young, P.E.; Hur, S.S.J.; Hawke, S.; Devenney, E.; Beadnall, H.; Barnett, M.H.; Suter, C.M.; Buckland, M.E. Exosomal microRNA signatures in multiple sclerosis reflect disease status. *Sci. Rep.* **2017**, *7*, 14293. [CrossRef]

111. Xun, J.; Du, L.; Gao, R.; Shen, L.; Wang, D.; Kang, L.; Chen, C.; Zhang, Z.; Zhang, Y.; Yue, S.; et al. Cancer-derived exosomal miR-138–5p modulates polarization of tumor-associated macrophages through inhibition of KDM6B. *Theranostics* **2021**, *11*, 6847–6859. [CrossRef]

112. Feng, X.; Zhan, F.; Luo, D.; Hu, J.; Wei, G.; Hua, F.; Xu, G. LncRNA 4344 promotes NLRP3-related neuroinflammation and cognitive impairment by targeting miR-138-5p. *Brain Behav. Immun.* **2021**, *98*, 283–298. [CrossRef]

113. Li, Z.; Qian, R.; Zhang, J.; Shi, X. MiR-218-5p targets LHFPL3 to regulate proliferation, migration, and epithelial-mesenchymal transitions of human glioma cells. *Biosci. Rep.* **2019**, *39*, BSR20180879. [CrossRef] [PubMed]

114. Xie, S.; Niu, W.; Xu, F.; Wang, Y.; Hu, S.; Niu, C. Differential expression and significance of miRNAs in plasma extracellular vesicles of patients with Parkinson's disease. *Int. J. Neurosci.* **2020**, *132*, 673–688. [CrossRef]

115. Kumar, S.; Reddy, P.H. Are circulating microRNAs peripheral biomarkers for Alzheimer's disease? *Biochim. Biophys. Acta* **2016**, *1862*, 1617–1627. [CrossRef] [PubMed]

116. Tang, S.; Cheng, J.; Yao, Y.; Lou, C.; Wang, L.; Huang, X.; Zhang, Y. Combination of Four Serum Exosomal MiRNAs as Novel Diagnostic Biomarkers for Early-Stage Gastric Cancer. *Front. Genet.* **2020**, *11*, 237. [CrossRef]

117. Calvopina, D.A.; Chatfield, M.D.; Weis, A.; Coleman, M.A.; Fernandez-Rojo, M.A.; Noble, C.; Ramm, L.E.; Leung, D.H.; Lewindon, P.J.; Ramm, G.A. MicroRNA Sequencing Identifies a Serum MicroRNA Panel, Which Combined With Aspartate Aminotransferase to Platelet Ratio Index Can Detect and Monitor Liver Disease in Pediatric Cystic Fibrosis. *Hepatology* **2018**, *68*, 2301–2316. [CrossRef] [PubMed]

118. Dong, L.; Li, Y.; Han, C.; Wang, X.; She, L.; Zhang, H. miRNA microarray reveals specific expression in the peripheral blood of glioblastoma patients. *Int. J. Oncol.* **2014**, *45*, 746–756. [CrossRef]

119. Selmaj, I.; Cichalewska, M.; Namiecinska, M.; Galazka, G.; Horzelski, W.; Selmaj, K.W.; Mycko, M.P. Global exosome transcriptome profiling reveals biomarkers for multiple sclerosis. *Ann. Neurol.* **2017**, *81*, 703–717. [CrossRef]

120. Huang, Y.; Zhu, L.; Li, H.; Ye, J.; Lin, N.; Chen, M.; Pan, D.; Chen, Z. Endometriosis derived exosomal miR-301a-3p mediates macrophage polarization via regulating PTEN-PI3K axis. *Biomed. Pharmacother.* **2022**, *147*, 112680. [CrossRef]

121. ElShelmani, H.; Brennan, I.; Kelly, D.J.; Keegan, D. Differential Circulating MicroRNA Expression in Age-Related Macular Degeneration. *Int. J. Mol. Sci.* **2021**, *22*, 12321. [CrossRef]

122. Lu, Q.; Wang, X.; Zhu, J.; Fei, X.; Chen, H.; Li, C. Hypoxic Tumor-Derived Exosomal Circ0048117 Facilitates M2 Macrophage Polarization Acting as miR-140 Sponge in Esophageal Squamous Cell Carcinoma. *OncoTargets Ther.* **2020**, *13*, 11883–11897. [CrossRef] [PubMed]

123. Yang, J.; Yang, X.S.; Fan, S.W.; Zhao, X.Y.; Li, C.; Zhao, Z.Y.; Pei, H.J.; Qiu, L.; Zhuang, X.; Yang, C.H. Prognostic value of microRNAs in heart failure: A meta-analysis. *Medicine* **2021**, *100*, e27744. [CrossRef]

124. Li, Y.; Bai, W.; Zhang, J. MiR-200c-5p suppresses proliferation and metastasis of human hepatocellular carcinoma (HCC) via suppressing MAD2L1. *Biomed. Pharmacother.* **2017**, *92*, 1038–1044. [CrossRef]

125. Xu, C.; Liang, H.; Zhou, J.; Wang, Y.; Liu, S.; Wang, X.; Su, L.; Kang, X. lncRNA small nucleolar RNA host gene 12 promotes renal cell carcinoma progression by modulating the miR200c5p/collagen type XI alpha1 chain pathway. *Mol. Med. Rep.* **2020**, *22*, 3677–3686. [PubMed]

126. Lee, E.K.; Jeong, H.O.; Bang, E.J.; Kim, C.H.; Mun, J.Y.; Noh, S.; Gim, J.A.; Kim, D.H.; Chung, K.W.; Yu, B.P.; et al. The involvement of serum exosomal miR-500-3p and miR-770-3p in aging: Modulation by calorie restriction. *Oncotarget* **2018**, *9*, 5578–5587. [CrossRef]

127. Livingstone, M.C.; Johnson, N.M.; Roebuck, B.D.; Kensler, T.W.; Groopman, J.D. Profound changes in miRNA expression during cancer initiation by aflatoxin B1 and their abrogation by the chemopreventive triterpenoid CDDO-Im. *Mol. Carcinog.* **2017**, *56*, 2382–2390. [CrossRef] [PubMed]

128. Niu, X.; Zhu, H.L.; Liu, Q.; Yan, J.F.; Li, M.L. MiR-194-5p serves as a potential biomarker and regulates the proliferation and apoptosis of hippocampus neuron in children with temporal lobe epilepsy. *J. Chin. Med. Assoc.* **2021**, *84*, 510–516. [CrossRef]

129. An, N.; Zhao, W.; Liu, Y.; Yang, X.; Chen, P. Elevated serum miR-106b and miR-146a in patients with focal and generalized epilepsy. *Epilepsy Res.* **2016**, *127*, 311–316. [CrossRef]

130. van der Sijde, F.; Homs, M.Y.V.; van Bekkum, M.L.; van den Bosch, T.P.P.; Bosscha, K.; Besselink, M.G.; Bonsing, B.A.; de Groot, J.W.B.; Karsten, T.M.; Groot Koerkamp, B.; et al. Serum miR-373-3p and miR-194-5p Are Associated with Early Tumor Progression during FOLFIRINOX Treatment in Pancreatic Cancer Patients: A Prospective Multicenter Study. *Int. J. Mol. Sci.* **2021**, *22*, 10902. [CrossRef] [PubMed]

131. Huo, D.; Clayton, W.M.; Yoshimatsu, T.F.; Chen, J.; Olopade, O.I. Identification of a circulating microRNA signature to distinguish recurrence in breast cancer patients. *Oncotarget* **2016**, *7*, 55231–55248. [CrossRef]

132. Chiam, K.; Wang, T.; Watson, D.I.; Mayne, G.C.; Irvine, T.S.; Bright, T.; Smith, L.; White, I.A.; Bowen, J.M.; Keefe, D.; et al. Circulating Serum Exosomal miRNAs As Potential Biomarkers for Esophageal Adenocarcinoma. *J. Gastrointest. Surg.* **2015**, *19*, 1208–1215. [CrossRef]

133. Mencias, M.; Levene, M.; Blighe, K.; Bax, B.E.; Project Group. Circulating miRNAs as Biomarkers for Mitochondrial Neuro-Gastrointestinal Encephalomyopathy. *Int. J. Mol. Sci.* **2021**, *22*, 3681. [CrossRef] [PubMed]

134. Wen, Z.; Huang, G.; Lai, Y.; Xiao, L.; Peng, X.; Liu, K.; Zhang, C.; Chen, X.; Li, R.; Li, X.; et al. Diagnostic panel of serum miR-125b-5p, miR-182-5p, and miR-200c-3p as non-invasive biomarkers for urothelial bladder cancer. *Clin. Transl. Oncol.* **2022**, *24*, 909–918. [CrossRef] [PubMed]

135. Lin, G.B.; Zhang, C.M.; Chen, X.Y.; Wang, J.W.; Chen, S.; Tang, S.Y.; Yu, T.Q. Identification of circulating miRNAs as novel prognostic biomarkers for bladder cancer. *Math. Biosci. Eng.* **2019**, *17*, 834–844. [CrossRef]

136. Ardila, H.J.; Sanabria-Salas, M.C.; Meneses, X.; Rios, R.; Huertas-Salgado, A.; Serrano, M.L. Circulating miR-141-3p, miR-143-3p and miR-200c-3p are differentially expressed in colorectal cancer and advanced adenomas. *Mol. Clin. Oncol.* **2019**, *11*, 201–207. [CrossRef] [PubMed]

137. Jiang, Y.; Ji, X.; Liu, K.; Shi, Y.; Wang, C.; Li, Y.; Zhang, T.; He, Y.; Xiang, M.; Zhao, R. Exosomal miR-200c-3p negatively regulates the migraion and invasion of lipopolysaccharide (LPS)-stimulated colorectal cancer (CRC). *BMC Mol. Cell Biol.* **2020**, *21*, 48. [CrossRef] [PubMed]

138. Huang, G.; Wei, B.; Chen, Z.; Wang, J.; Zhao, L.; Peng, X.; Liu, K.; Lai, Y.; Ni, L. Identification of a four-microRNA panel in serum as promising biomarker for colorectal carcinoma detection. *Biomark. Med.* **2020**, *14*, 749–760. [CrossRef] [PubMed]

139. Wang, W.; Wu, L.R.; Li, C.; Zhou, X.; Liu, P.; Jia, X.; Chen, Y.; Zhu, W. Five serum microRNAs for detection and predicting of ovarian cancer. *Eur. J. Obstet. Gynecol. Reprod. Biol. X* **2019**, *3*, 100017. [CrossRef]

140. Lai, Z.; Cao, Y. Plasma miR-200c-3p, miR-100-5p, and miR-1826 serve as potential diagnostic biomarkers for knee osteoarthritis: Randomized controlled trials. *Medicine* **2019**, *98*, e18110. [CrossRef]

141. Shao, P. MiR-216a-5p ameliorates learning-memory deficits and neuroinflammatory response of Alzheimer's disease mice via regulation of HMGB1/NF-kappaB signaling. *Brain Res.* **2021**, *1766*, 147511. [CrossRef]

142. Kusnierz-Cabala, B.; Nowak, E.; Sporek, M.; Kowalik, A.; Kuzniewski, M.; Enguita, F.J.; Stepien, E. Serum levels of unique miR-551-5p and endothelial-specific miR-126a-5p allow discrimination of patients in the early phase of acute pancreatitis. *Pancreatology* **2015**, *15*, 344–351. [CrossRef]

143. Erdos, Z.; Barnum, J.E.; Wang, E.; DeMaula, C.; Dey, P.M.; Forest, T.; Bailey, W.J.; Glaab, W.E. Evaluation of the Relative Performance of Pancreas-Specific MicroRNAs in Rat Plasma as Biomarkers of Pancreas Injury. *Toxicol. Sci.* **2020**, *173*, 5–18. [CrossRef] [PubMed]
144. Lee, Y.J.; Kang, C.W.; Oh, J.H.; Kim, J.; Park, J.P.; Moon, J.H.; Kim, E.H.; Lee, S.; Kim, S.H.; Ku, C.R.; et al. Downregulation of miR-216a-5p and miR-652-3p is associated with growth and invasion by targeting JAK2 and PRRX1 in GH-producing pituitary tumours. *J. Mol. Endocrinol.* **2021**, *68*, 51–62. [CrossRef]
145. Ji, Q.; Ji, Y.; Peng, J.; Zhou, X.; Chen, X.; Zhao, H.; Xu, T.; Chen, L.; Xu, Y. Increased Brain-Specific MiR-9 and MiR-124 in the Serum Exosomes of Acute Ischemic Stroke Patients. *PLoS ONE* **2016**, *11*, e0163645. [CrossRef]
146. Porte Alcon, S.; Gorojod, R.M.; Kotler, M.L. Regulated Necrosis Orchestrates Microglial Cell Death in Manganese-Induced Toxicity. *Neuroscience* **2018**, *393*, 206–225. [CrossRef]
147. Satoh, M.; Date, I.; Nakajima, M.; Takahashi, K.; Iseda, K.; Tamiya, T.; Ohmoto, T.; Ninomiya, Y.; Asari, S. Inhibition of poly(ADP-ribose) polymerase attenuates cerebral vasospasm after subarachnoid hemorrhage in rabbits. *Stroke* **2001**, *32*, 225–231. [CrossRef] [PubMed]
148. Fan, Y.; Yan, G.; Liu, F.; Rong, J.; Ma, W.; Yang, D.; Yu, Y. Potential role of poly (ADP-ribose) polymerase in delayed cerebral vasospasm following subarachnoid hemorrhage in rats. *Exp. Ther. Med.* **2019**, *17*, 1290–1299. [CrossRef]
149. Bao, X.; Wu, G.; Hu, S.; Huang, F. Poly(ADP-ribose) polymerase activation and brain edema formation by hemoglobin after intracerebral hemorrhage in rats. *Acta Neurochir. Suppl.* **2008**, *105*, 23–27.
150. Pons-Espinal, M.; Gasperini, C.; Marzi, M.J.; Braccia, C.; Armirotti, A.; Potzsch, A.; Walker, T.L.; Fabel, K.; Nicassio, F.; Kempermann, G.; et al. MiR-135a-5p Is Critical for Exercise-Induced Adult Neurogenesis. *Stem Cell Rep.* **2019**, *12*, 1298–1312. [CrossRef] [PubMed]
151. Wang, Y.; Yang, Z.; Zhang, K.; Wan, Y.; Zhou, Y.; Yang, Z. miR-135a-5p inhibitor protects glial cells against apoptosis via targeting SIRT1 in epilepsy. *Exp. Ther. Med.* **2021**, *21*, 431. [CrossRef] [PubMed]
152. Hrdlickova, R.; Nehyba, J.; Bargmann, W.; Bose, H.R., Jr. Multiple tumor suppressor microRNAs regulate telomerase and TCF7, an important transcriptional regulator of the Wnt pathway. *PLoS ONE* **2014**, *9*, e86990. [CrossRef] [PubMed]
153. Lu, Y.J.; Liu, R.Y.; Hu, K.; Wang, Y. MiR-541-3p reverses cancer progression by directly targeting TGIF2 in non-small cell lung cancer. *Tumour Biol.* **2016**, *37*, 12685–12695. [CrossRef] [PubMed]

 *biomedicines*

Review

Circulating Biomarkers for Cancer Detection: Could Salivary microRNAs Be an Opportunity for Ovarian Cancer Diagnostics?

Marzia Robotti [1,†], **Francesca Scebba** [2,†] **and Debora Angeloni** [2,3,*]

[1]   Graduate School in Translational Medicine, Scuola Superiore Sant'Anna, Via G. Moruzzi, 56124 Pisa, Italy
[2]   Health Science Interdisciplinary Center, Scuola Superiore Sant'Anna, Via G. Moruzzi, 56124 Pisa, Italy
[3]   The Institute of Biorobotics, Scuola Superiore Sant'Anna, Via G. Moruzzi, 56124 Pisa, Italy
*   Correspondence: d.angeloni@santannapisa.it; Tel.: +39-050-3153092
†   These authors contributed equally to this work.

Abstract: MicroRNAs (miRNAs) are small non-coding RNAs with the crucial regulatory functions of gene expression at post-transcriptional level, detectable in cell and tissue extracts, and body fluids. For their stability in body fluids and accessibility to sampling, circulating miRNAs and changes of their concentration may represent suitable disease biomarkers, with diagnostic and prognostic relevance. A solid literature now describes the profiling of circulating miRNA signatures for several tumor types. Among body fluids, saliva accurately reflects systemic pathophysiological conditions, representing a promising diagnostic resource for the future of low-cost screening procedures for systemic diseases, including cancer. Here, we provide a review of literature about miRNAs as potential disease biomarkers with regard to ovarian cancer (OC), with an excursus about liquid biopsies, and saliva in particular. We also report on salivary miRNAs as biomarkers in oncological conditions other than OC, as well as on OC biomarkers other than miRNAs. While the clinical need for an effective tool for OC screening remains unmet, it would be advisable to combine within a single diagnostic platform, the tools for detecting patterns of both protein and miRNA biomarkers to provide the screening robustness that single molecular species separately were not able to provide so far.

Keywords: ovarian cancer; biomarker; saliva; urine; ascites; blood; microRNA; biofluids; early diagnosis; screening

Citation: Robotti, M.; Scebba, F.; Angeloni, D. Circulating Biomarkers for Cancer Detection: Could Salivary microRNAs Be an Opportunity for Ovarian Cancer Diagnostics? . *Biomedicines* **2023**, *11*, 652. https://doi.org/10.3390/biomedicines11030652

Academic Editors: Milena Rizzo and Elena Levantini

Received: 26 September 2022
Revised: 2 February 2023
Accepted: 16 February 2023
Published: 21 February 2023

1. Introduction

We reviewed the current literature to understand what opportunities exist in terms of circulating microRNA (miRNA), specifically salivary miRNAs, as biomarkers (BMs) for diagnostic/prognostic approaches to oncological diseases in general, and to ovarian cancer (OC) in particular. We realized that currently there are not miRNA BMs in clinical use for OC screening.

First, we provide a general introduction on the subject of OC and the state of the art in terms of screening opportunities. Section 2 introduces the subject of miRNAs, showing their recognized potential as disease biomarkers (BM). In Section 3, we analyze relevant literature on the potential of liquid biopsies for the detection of miRNAs in OC, with a specific focus on saliva (Section 4) to discuss the pros and cons of this biofluid, whose molecular profile resembles blood, with possible advantageous simplifications. For the sake of completeness, and to support the hypothesis that the search for salivary miRNAs could be useful in OC, we also reviewed the state of the art of salivary miRNAs in oncological conditions other than OC (Section 5), and, to support the validity of saliva as a useful biofluid for BMs search, we provide an excursus (Section 6) on OC candidate salivary BMs other than miRNAs. Finally, to compare saliva with other possible biofluids as a source of miRNA BMs, we reviewed the literature on this type of candidate BMs of OC in other biofluids (Section 7).

Overall, this review aims to provide an informed understanding of the need for more research on the subject of reliable BMs for OC screening, because presently, the clinical need for an early, reliable, relatively inexpensive diagnostic tool for OC screening in the population remains unmet. Specifically, we propose saliva as a convenient bioanalyte, possibly for the combined search of candidate BMs of different nature, such as miRNA and proteins.

OC, although rare, is the first cause of death for gynecological cancers and the fifth most common cause of cancer-related death for women of industrialized countries [1]. The poor outcome is mostly associated with the lack of specific symptoms and of effective screening strategies. Despite increasingly radical surgical approaches and huge efforts put into new, targeted, therapeutic agents, the prognosis for patients with OC has hardly improved in the past three decades and two thirds of women still die within ten years from diagnosis [2]. Five-year survival is less than 20% in women diagnosed with advanced-stage (stage III or IV) invasive epithelial OC, but exceeds 90% in those detected at stage I [3], suggesting that success in curing OC relies on early diagnosis. Efforts have therefore focused on diagnosing early-stage or low-volume disease through risk prediction, prevention, and screening. Worldwide, there is inconsistency in availability of and access to treatment for OC, and the outcomes are complicated because of the complexities of the disease in terms of epidemiology, genetic features, and histopathology, all contributing to the still poor understanding of OC, especially in comparison with other oncological diseases [4].

In fact, OC is heterogeneous, comprising several disease types and subtypes [5,6]. The extra-ovarian originations of epithelial OCs contribute to the intricacies of the disease. Over the past decade, a dualistic pathway of epithelial ovarian carcinogenesis has emerged: Type I and Type II, that differ in epidemiology, etiology, and treatment. Therapeutic strategies may not be equally effective for all. Therefore, success in treating OC also requires to identify new BMs, not only to detect OC early, at a time when outcomes could be improved, but also to allow a better stratification of patients with full blown disease for more efficacious, personalized therapy [7–9].

Years ago, the first biomarker (BM) identified and proposed as a blood test for women with OC was the carcinoembryonic antigen (CEA) [10]. However, the current literature agrees that multi-BM panels might perform better than a single BM for a personalized treatment in the context of precision medicine in general [9,11] and for early detection of OC in particular [12].

At present time, despite decades of international efforts to identify panels of blood-borne BMs suitable for screening, no one superior to serum cancer antigen 125 (CA125) has reached clinical practice, except for human epididymis protein 4 (HE4), which is still second best to CA125 [13]. Their combined analysis within the risk of ovarian malignancy (ROMA) algorithm has improved the diagnostic accuracy for OC. However, ROMA is not applicable in screening procedures for early detection [9,13,14]. In addition to ROMA, the OVA1 test, a multivariate in vitro assay for five proteomic BMs, is used as an aid to the pre-chirurgical definition of a pelvic mass, but, again, it does not possess the characteristics to be applied on large-scale, prevention screening [15].

Currently, the combination of CA125 serum analysis and transvaginal ultrasonography (TVS) is the most utilized tool for the initial evaluation of suspect cases, even though CA125 lacks specificity (as it rises due to several physiological and pathological conditions in addition to cancer [16]), and TVS can identify adnexal masses, but is less reliable in differentiating benign from malignant tumors [17].

Over time, other serum BMs have been evaluated in OC together with CA125 [18], and several strategies have been developed by combining the analysis of multiple circulating proteins [19,20].

The literature shows that proteins expressed in OC cells can be retrieved in serum and urine [11,21]. However, only a few of several reported candidate BMs were validated for clinical practice [22], this discrepancy being ascribed to different reasons [11,23].

Therefore, regardless of the enormous progress in analytical techniques, the insights on the biochemical and molecular mechanisms of OC and the description in the literature of several candidate BMs, up to now, the clinical need for an effective test able to perform early, reproducible, noninvasive, and possibly inexpensive diagnosis of OC remains completely unanswered [24].

Moreover, despite recent encouraging data on sensitivity and cost-effectiveness of multimodal analysis, screening for OC in the general population is not recommended due to the lack of a definitive mortality benefit. This has been reinforced in the latest recommendation from the U.S. Preventative Task Force (USPSTF) [25] and the U.K. National Screening Committee (UK NSC) [4].

Several recent studies aim at exploring other possible sources of BMs in addition to blood as analyte, and to proteins as BMs. For example, there are studies dealing with tumor DNA detection, also for methylation profiling, in liquid cytology samples from vagina and endocervix, including routinely collected cervical screening specimens, vaginal self-swab and tampons, and uterine lavage samples. Moreover, improvements in imaging methods are pursued to include refinements of TVS, Doppler flow, microbubble contrast-enhanced TVS, and photo-acoustic imaging, all of which allow high-resolution detection of angiogenesis with the potential to detect neovascularization in early cancers [26–28]. In addition, multi-omics technologies (genomics, transcriptomics, proteomics, and metabolomics) hold the possibility of discovering new informative BMs [29,30] in signature panels. Currently, important questions in this field are concerned with the type of sample source selected, the analytical technique that leads to the most valuable results, whether it is advisable to look for one specific BM type (e.g., proteins or nucleic acids) or whether to combine the results of different analytical procedures (e.g., proteins and nucleic acids in parallel).

In summary, while on one hand the state of the art in OC screening recommends against screening for fear of adverse complication in false-positive women undergoing surgery, on the other hand, personalized medicine requires an individualized approach and reliable tools for screening procedures to detect the disease at an early, curable stage, to improve patient stratification and therapeutic regimes, and to spare unnecessary, even possibly hazardous treatments and save lives and costs, as invoked by authoritative editorials [31,32].

2. MicroRNAs as Potential Disease Biomarkers

Among molecules with potential BM characteristics, miRNAs have recently emerged as powerful keys to a better understanding of molecular mechanisms controlling cancer cells and their niche. MiRNAs are small molecules (18–25 nucleotides) of single-strand, non-coding RNA (ncRNA) that are found in the transcriptome of all living beings and regulate gene expression at the transcriptional and post-transcriptional level [33] (Figure 1, lower part).

The first evidence that miRNAs may have diagnostic and therapeutic potential was presented soon after the identification of the first miRNAs in humans [34]. The study showed correlation between the loss of miR-15 and miR-16 and the occurrence of B-cell leukemia [35]. Three years later, the homozygous deletion of the gene coding for Dicer, an enzyme essential for miRNA biogenesis, showed that miRNA loss of function disrupted prenatal development of the murine embryo [36], providing the first genetic evidence of miRNAs' vital role in development. The first indication that miRNAs may become easily accessible BMs for cancer diagnosis and prognosis came three years later when miRNAs were isolated from patient serum [37,38] and their profiling revealed specific patterns across different groups of diseases [39]. Following studies confirmed specific miRNA signatures in many types of human diseases, including different cancers [40–42]. MiRNA signature in blood is comparable to that of the tumor of origin [43] and specific miRNAs were associated with prostate cancer (PrC) [38], lung cancer (LC), colorectal (CRC) [44–46], breast cancer (BC), gastric cancer (GC) and OC [47–49]. Indeed, miRNA profiling revealed different patterns in different histological tumor subtypes of OC, a highly heterogeneous disease, as already mentioned [23,49–54].

Several clinical trials were initiated, based on the suggested applicability of miRNAs for cancer diagnostics and prognostics [55]. In fact, miRNAs meet most of the criteria for being ideal BMs, i.e., accessibility, specificity, and sensitivity.

Despite the lack of standardized protocols and use in current clinical practice, miRNAs hold the potential to become a routine approach in the development of personalized medicine in the future.

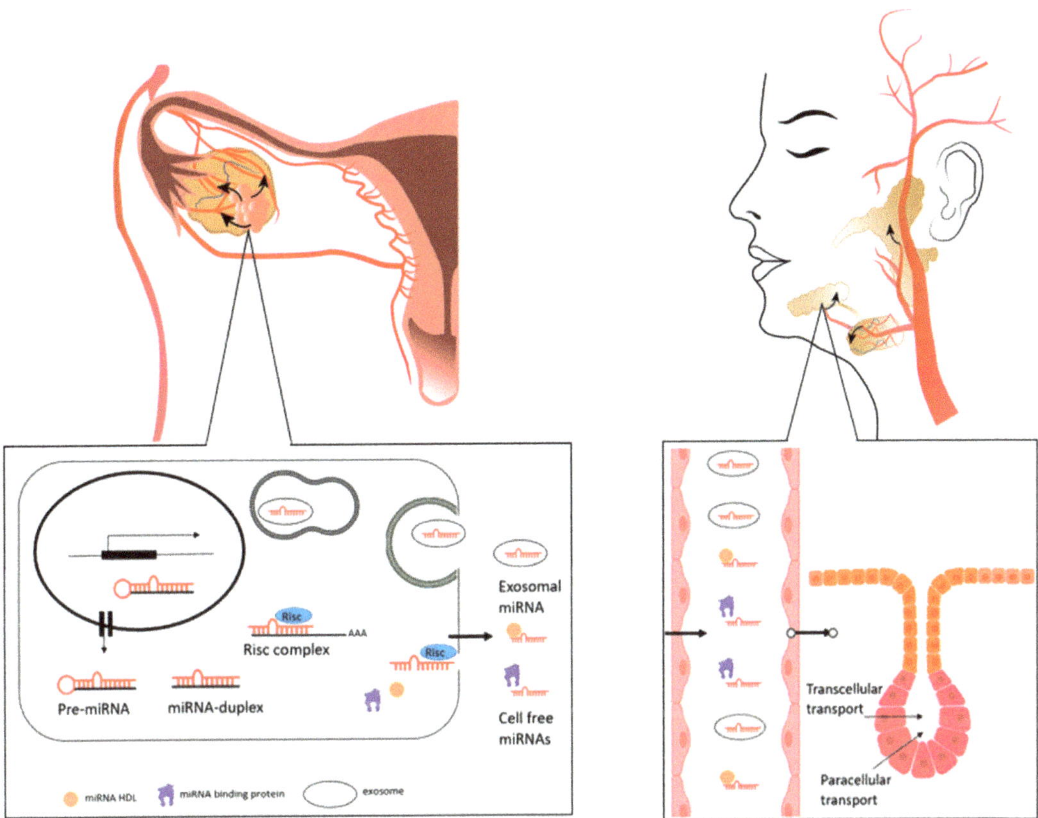

Figure 1. Anatomical and molecular visualization of miRNAs trafficking between OC tissue to bloodstream and saliva. Direction of molecular trafficking between ovarian tissue and bloodstream (**upper left**), and between blood and salivary glands (**upper right**) is indicated by arrows. There is an intense exchange of molecules from the bloodstream to the ovary and vice versa: the ovary is highly vascularized so that it is supplied with hormones and nutrients needed for its high metabolism. At the same time, small molecules, such as microRNAs, are released from the tissue, also under tumor conditions. The composition of blood therefore influences salivary composition. The salivary fluid is produced by the local vascular bed in the acinar region and through the duct system released into the oral cavity. At cell level (**lower left**), miRNAs generated in the cell are released within exosomes or associated with miRNA binding proteins (such AGO2) or high-density lipoprotein (HDL). They are stable and protected by RNase degradation. Intra- and extracellular mechanisms allow these molecules to be transported from blood to saliva (**lower right**), so they can be found in the oral cavity reflecting a distant physiological or pathological condition.

3. Liquid Biopsies for Detection of miRNAs in Ovarian Cancer

Tissue biopsy is the gold standard to evaluate molecular features of tumors. However, in the case of OC, tissue biopsies should be avoided because puncture can cause cancer cells dissemination into the peritoneal cavity, promoting peritoneal metastasis. On the contrary, liquid biopsy represents a safe choice, also enabling serial sampling over time [29]. Body fluids contain several types of molecules: circulating tumor cells, circulating nucleic acids, and extracellular vesicles. As miRNAs pass from tissues to blood and are stable in body fluids [56], circulating miRNAs are suitable BMs for OC detection and staging [57–60], (Figure 1, lower part).

Circulating miRNAs can be contained within extracellular vesicles (EVs), which are released by normal and tumor cells and, depending on their size, divided into three main groups: exosomes, microvesicles, and apoptotic bodies [61]. Cancer-derived exosomal miRNAs have captured the interest of many researchers for their role in promoting cancer through angiogenesis and metastasis. Cancer cells may produce more exosomes than normal cells; for this reason, exosomes could be informative regarding patients' health [62].

However, miRNAs can also be found in the EV-free fraction, where they are complexed to RNA-binding proteins, lipids, or lipoproteins, in extracellular fluids [63].

Both types of miRNAs are resistant to RNase degradation, temperature, and pH changes [38].

4. Saliva: An Informative Analyte for miRNA Detection, with Practical Advantages

Among body fluid analytes, saliva offers some yet unexplored advantages. Saliva collection can be carried out without medical intervention and does not present risks associated with even minimally invasive procedures; the sample, properly stabilized, can be stored and shipped from even remote collection sites to high-tech testing sites with minimal costs. Saliva released by the major salivary glands consists of 99% water containing inorganic and organic species including secretion and putrefaction products, lipids, over 2400 proteins, metabolites, components of the microbiome and abundant, stable extracellular coding- and non-coding RNA species, among which are a wide variety of miRNAs [64]. Saliva represents blood very faithfully, even with possible advantages. For example, storage is simpler; saliva does not coagulate, it is stable for 24 h at room temperature and for a week at 4 °C. In addition, it may be collected at no risk several times a day to monitor therapy with repeated sampling [65]. All the above makes saliva a relatively simple, accurate, easy, safe, and economical material to be tested for clinically significant molecules.

Some of the molecules characterized in saliva are candidate BMs for cancer diagnosis, prognosis, drug monitoring, and pharmacogenetic studies, although just a few of such candidates were validated in multicenter studies, with large sample size and standardized protocols [66].

The salivary transcriptome has been entirely characterized in 2012 [67]. It includes long and small RNA species. Salivary miRNAs are more stable and discriminatory than mRNAs. They are abundant, and fit the profile of other body fluids [68,69], which makes them good BM candidates for systemic diseases. In fact, MiRNAs associated with CRC, esophageal, and pancreatic cancer (PC) were isolated from saliva [70,71], showing that profiling miRNA expression could allow for the detection of a distant neoplasia while discriminating between different cancer types [23,72]. For all these reasons, several current research efforts are focused on the detection of salivary miRNAs, although we need to better understand their biogenesis and how to recognize their tissue of origin [73].

Kits for Salivary Diagnostics

Indeed, several diagnostic kits that use saliva as the biofluid test have already been developed and are commercially available. The kits encompass different uses; SARS-CoV-2 has received emergency use authorization from the U.S. Food and Drug Administration

(FDA) in the last two years, and kits for the search in oral fluids of other viruses such as HIV, HPV, HSV, or other infectious agents (e.g., Candida albicans) are also available (The ADA Science & Research Institute, LLC (ADASRI) for Oral health). Lin-Zhi International, Inc. (LZI) (https://www.lin-zhi.com/oral-fluid-eia, accessed on 20 February 2023) is a manufacturer of in vitro diagnostic reagents for oral fluid screening (in addition to urine) for the detection of drugs of abuse. Salivary miRNA diagnostic tests for ASD (autism spectrum disease) have also been validated (NIH ClinicalTrials.gov, https://www.clinicaltrials.gov/ct2/show/NCT05418023, accessed on 20 February 2023). Salimetrics, LLC (https://salimetrics.com/assay-kits/, accessed on 20 February 2023) produces salivary ELISA kits (hormones, COVID, cytochines) and Salignostics (https://www.salignostics.com/saliva-diagnostics/, accessed on 20 February 2023) produces approved salivary tests for COVID and pregnancy and is developing tests for cardiac risk, malaria, and Helicobacter pylori.

5. Salivary miRNAs in Oncological Conditions

To show the overall interest and potential of salivary miRNA in disease detection well beyond pathologies of the oral cavity [74], in this section, we provide the state of the art regarding microRNA dysregulation detected in saliva in association with systemic oncological conditions. With this objective, we reviewed articles published in the last ten years (2012 to January 2023), including case control or cohort studies for diagnostic biomarkers, review article, and meta-analysis. We did not set, as a criterion, a minimum number of patients recruited in the study. We considered only miRNAs associated to cancer types with the primary site distant from the oral cavity; therefore, studies evaluating deregulated miRNAs in patients with pharyngeal, esophageal, salivary glands, and oral cavity diseases were excluded. Similarly, studies concerning neurodegenerative diseases, microbial infections, and drug/hormone response therapy were excluded. Studies on salivary miRNAs for forensic purposes were excluded as well.

Multiple papers highlight qualitative or quantitative differences in miRNA composition between different body fluids [75,76]. At the same time, specific miRNAs are found in all body fluids analyzed. There are some inconsistencies among papers dealing with salivary miRNAs in the same oncological condition, and this can be ascribed to different reasons: the studies were performed on relatively small sample sizes, methods were often different, and subject groups were not always overlapping. Nevertheless, some miRNAs were pointed out by different authors as putative BMs for the same or for different oncological conditions [75,77–79]. Overall, literature agrees on the promising potential of salivary miRNAs as BMs for cancer detection [80].

PC, the seventh leading cause of cancer-related deaths worldwide, is the most studied in terms of salivary miRNAs. Similar to OC, early-stage PC rarely causes any symptoms, and for this reason it remains one of the most undiagnosed and lethal malignant neoplasms [23]. Carbohydrate antigen 19-9 (CA19-9) is routinely used for prognosis and to monitor the disease [81]; however, it often fails to detect precancerous or early-stage lesions because of inadequate sensitivity and specificity. Therefore, for better prevention or treatment of PC, just as for OC, we have the urgent need to find new technologies for earlier detection. The search for miRNA biomarkers is advancing, although, to date, none has yet reached clinical applications [81]. Papers dealing with salivary miRNAs BMs for PC [75–77,82,83], CRC [78,84], LiC [79,85], LC [86], BC [87], GC [88,89], and PrC [90,91] are summarized in Table 1. MiRNAs were detected both as cell free or as exosomal miRNAs, both in saliva alone as well as in other parallel biofluids. Different techniques have been utilized for the discovery phase (microarrays or sequencing) and real-time quantitative PCR (QRT-PCR) was used for validation. Panels of up- or down-regulated miRNAs have been evaluated for their diagnostic ability; in some cases, sensitivity and sensibility have reached high scores (Table 1). Xie et al., 2014, profiled with a miRNome microarray 2006 mature miRNA sequences from saliva samples of 8 patients with resectable PC, and of 8 healthy subjects; they validated significant candidates with QRT-PCR on the same discovery sample set, and finally analyzed the expression level of 10 selected miRNAs on an independent cohort of 40 patients with PC, 20 with benign pancreatic tumors (BPT), and 40 healthy controls. They showed that miR-3679-5p is significantly downregulated and miR-940 significantly upregulated, in PC versus BPT, and in PC versus BPT and controls. No significant differences were observed between BPT and healthy controls [82]. Humeau et al., 2015, reported a salivary miRNA signature for PC, including miR-21, miR-23a, miR-23b, and miR-29c, that was significantly overexpressed in PC patients compared to healthy subjects. MiR-210 and let-7c were up-regulated in pancreatitis patients compared to controls; miR-216 was upregulated in PC compared to pancreatitis. The authors xenografted human-derived PC cells in athymic mice and found that miR-21 was higher in saliva from tumor-bearing mice, while the other three remained low [77]. Machida et al., 2016, analyzed the expression of miR-1246, miR-3976, miR-4306, and miR-4644 (previously reported as PC BMs in serum exosomes [83]) in salivary exosomes of pancreatobiliary tract cancer patients. MiR-1246 and miR-4644 were significantly over-expressed in salivary exosomes of pancreatobiliary tract cancer patients compared to controls [92].

The last two publications regarding salivary miRNAs in PC highlight an important discrepancy to take note of. The level of miR-1246, analyzed in serum, urine, and saliva, was significantly higher in the serum and urine of cancer patients as opposed to healthy subjects, but salivary levels did not differ significantly between the two groups [75]. Similarly, the analysis of six miRNAs (miR-21, miR-155, miR-196a, miR-200b, miR-376a, and miR-34a) in serum and saliva samples showed that, although miR-21 and miR-34a were differentially expressed in serum samples of pancreatic ductal adenocarcinoma patients, and the same miRNAs were detectable in saliva, their levels did not change significantly among cancer patients and controls [76]. These studies were performed on relatively small sample sizes and only one ethnic group was included. As already mentioned for OC studies, methods are still not adequately standardized for miRNA research in PC.

Excluding the pathologies of the oral cavity and PC, other systemic conditions were investigated for miRNA representation in saliva [74]. Sazanov et al., 2017, focused on the expression miR-21 in blood and saliva samples of patients with distal colorectal cancer (CRC) at different stages. It was found that miR-21 was higher in CRC patients than controls, with diagnostic sensitivity and specificity in saliva higher than in blood [84]. Rapado-González et al., 2019, analyzed only saliva samples of CRC and adenoma patients and of healthy subjects. Through validation phases, the authors selected five miRNAs upregulated in CRC compared to healthy controls (Table 1). Building a COX regression analysis with the clinico-pathological parameters, they found that the high level of the five miRNAs, and the concentration of CEA protein, correlated with a higher risk of progression [78]. In liver cancer (LiC), Petkevich et al., 2021 compared the exosomal with the exosomal-free fraction of blood or saliva. They included HCV-related liver cirrhosis, primary liver cancer patients, and healthy volunteers and confirmed in saliva, as already found in plasma, the differential expression of three miRNAs in liver cancer (LiC) patients compared to controls. (Table 1). The saliva exosomal fraction was found to be the best performing [79]. Mariam et al., 2022, with an RNA-seq study in saliva based on 20 hepatocellular cancer (HCC) and 19 cirrhosis patients, found that 283 salivary miRNAs were significantly downregulated in HCC. Machine learning identified a combination of 10 miRNAs as good indicators of LiC. Of interest here, mir-92b, mir-548i-2, and mir-548l were found differentially expressed both in saliva and in HCC tissue samples compared to cirrhotic liver tissue samples [85].

LC, BC, PrC, and GC are the four other oncological conditions for which salivary microRNAs have been investigated. Yang et al., 2020, profiled salivary miRNAs in four groups of lung cancer patients: 57 patients with malignant pleural effusion (MPE); 33 patients with benign pleural effusion (BPE); 50 patients with a diagnosis of malignancy without pleural effusion (MT), and 49 healthy controls (HCs). The discovery phase with microarray was performed on six salivary samples from three MPE patients and three HCs and showed 29 upregulated and 48 downregulated miRNAs in MPE compared to controls. The validation phase performed with QRT-PCR highlighted two miRNAs to be up- and down-regulated in MPE patients respectively, (Table 1) [86]. Koopaie et al., 2021, analyzed with QRT-PCR, 41 BC patients and 39 healthy subjects. MiR-21 was significantly upregulated in BC samples relative to controls, but not among disease stages [87]. Li F. et al., 2018, analyzed salivary samples of 10 early-stage GC patients and 10 non-GC controls, with TaqMan Human miRNA array. Four miRNAs (Table 1) made the best BM panel [88].

Nanographene Oxide (nGO), a novel method to detect circulating oncomiRs, was used by Hizir et al. in PrC. The authors analyzed exogenous miR-21 and miR-141 in saliva and other body fluids using the two-dimensional surface of nGO. The two miRNAs reflected the disease stage: miR-141 was higher in advanced PrC, while miR-21 was higher in early-stage cancer. The copies of circulating miRNAs in patient specimens are lower than the concentration used in this study, thus, the sensitivity of the assay needs to be improved. However, the method is fast and easy, can be performed with a spectrofluorometer, and potentially allows the reveal of PrC patients at different stages with a non-invasive approach [90].

Table 1. Salivary miRNAs in oncological conditions. PC: pancreatic cancer, BPT: benign pancreatic tumor, PBTC: pancreatobiliary tract cancer, HS: Healthy subjects, CRC: colorectal cancer, LiC: liver cancer, HCC: epatocellular carcinoma, MPE: malignant pleural effusion, BC: breast cancer, GC: gastric cancer. Each row provides information regarding the citation in parenthesis at the far left end of the row.

Reference	Source (Analytes)	miRNAs	Cancer Type	N° of Subjects (Cancer vs Control)	Status	Sensitivity	Specificity	AUC
[82]	cell free	miR-3679-5p miR-940	PC	40 vs 40	↓ ↑	72.5%	70.0%	0.750
[77]	cell free	miR-21 miR-23a miR-23b miR-29c	PC	7 vs 4	↑ ↑ ↑ ↑	71.4% 85.7% 85.7% 57.0%	100% 100% 100% 100%	- - - -
[92]	exosomes	miR-1246 miR-4644	PBTC	14 vs 13	↑ ↑	83.3%	92.3%	0.833
[75]	cell free	miR-1246	PC	41 vs 30	-	91.0%	26.7%	0.480
[84]	cell free	miR-21	CRC	34 vs 34	↑	97.0%	91.0%	-
[78]	cell free	miR-186-5p miR-29a-3p miR-29c-3p miR-766-3p miR-491-5p	CRC	51 vs 37	↑ ↑ ↑ ↑ ↑	72.0%	66.7%	0.754
[79]	exosomes cell free	miR-21-5p miR-122-5p miR-221-3p	LiC	24 vs 21	↑ ↓ ↑	66.0%	78.0%	0.770
[85]	cell free	mir-1262 miR-1262 mir-216a mir-484 mir-30d miR-216a-5p miR-30d-5p miR-484 mir-10401 miR-454-3p	HCC	20 vs 19 Cirrhosis	-	83.0%	68.0%	0.780

Table 1. Cont.

Reference	Source (Analytes)	miRNAs	Cancer Type	N° of Subjects (Cancer vs Control)	Status	Sensitivity	Specificity	AUC
[86]	cell free	miR-4484 miR-3663-3p	MPE	57 vs 49	↑ ↓	82.2%	74.1%	0.802
[87]	cell free	miR-21	BC	41 vs 39	↑	100%	100%	1.0
[88]	cell free	miR-140-5p miR-374a miR-454 miR-15b	GC	100 vs 100	↑ ↑ ↑ ↑	- - - -	- - - -	0.700 0.650 0.630 0.650

6. Candidate Salivary Biomarkers of Ovarian Cancer Other Than miRNAs

The urgency of finding reliable BMs for early OC detection has prompted many researchers to focus on saliva as an advantageous biofluid in terms of collection, procedure, and cost-effectiveness [93]. Several studies compared saliva with other solid or liquid biopsies for the sake of methodological cross validation, to evaluate the use of saliva as the analyte of choice.

Lee et al., 2012, explored mRNAs expression in saliva of OC patients. The study design was based on: (i) an initial profiling with microarray assay; (ii) validation with QRT-PCR of the preliminary findings obtained on a discovery cohort and on an additional, independent cohort; and (iii) logistic regression analysis of QRT-PCR results. AGPAT1, B2M, IER3, IL1B, and BASP1 mRNAs were identified as those providing the highest discriminatory power for OC diagnosis. The findings were hindered by the small sample size and the lack of benign tumor control group [93]. Yang et al., 2021, evaluated the same salivary mRNAs expression profile in a Chinese population of 140 OC patients and 140 control individuals. Concomitantly, the CEA protein concentration was measured in blood. With machine-learning methods, the authors identified a novel panel of disease predictors. The validation phase on an independent cohort of 60 OC patients and 60 healthy subjects resulted in 85.0% of sensitivity and 88.3% of specificity, while measuring CEA concentration in blood, and BASP1 and IER3 levels in saliva. The mRNAs of interest were preselected from Lee et al. [93] who collected samples from Korean rather than Chinese women [94]. Li et al., 2018, assessed the potential use of immediate early response gene X-1 (IEX-1) transcript as OC BM. In OC, IEX-1 is down-regulated, thereby acting as a tumor suppressor. Its expression was quantified in saliva and blood samples with RT-qPCR. Three groups of patients were involved in the study: 26 patients of epithelial ovarian cancer (EOC), 37 women with benign ovarian tumors (BOT), and 55 controls. IEX-1 expression in blood and saliva was lower in EOC compared to BOT and controls. No significant differences were found in IEX-1 expression between BOT and controls. Diagnostic efficacy of IEX-1 was high in blood and medium in saliva but, of interest here, the results showed that both biofluids are suitable for BM detection. Saliva might have a higher specificity but lower sensitivity in discriminating EOC from BOT [95].

Mass spectrometry (MS) proved to be a powerful method to analyze fluids without extensive chemical preparation of samples. Tajmul et al., 2018, applied it in search for an OC signature in salivary proteins. On a proteome-wide scale, they analyzed quantitative differences in saliva using two-dimensional difference gel electrophoresis (2D-DIGE) combined with matrix-assisted laser desorption/ionization time of flight (MALDI-TOF) MS. They found 44 differentially expressed proteins, among which Lipocalin-2, indoleamine-2, 3-dioxygenase1 (IDO1), and S100A8 had the highest statistical scores. The three proteins, known to be involved in pathways of cancer progression and metastasis, were validated with Western blot and Elisa. Finally, salivary proteomics was combined with tissue-based transcriptomics. Immunohistochemistry, microarray, and QRT-PCR confirmed the upregulation of these three candidate proteins in ovarian tissue. This signature holds a good diagnostic potential even though it would require validation in a larger population set [96].

Zermeño-Nava et al., 2018, applied surface-enhanced Raman spectroscopy (SERS) to measure levels of sialic acid (SA) in saliva samples collected from 37 women with benign ovarian adnexal masses (observed by TVS), and from 15 OC patients. Salivary SA levels were statistically different among the two groups, with a threshold corresponding to SA concentration > 15.5 mg/dL. However, SA cannot be considered a specific BM for OC: it is useful in the clinical scenario for diagnosing patients with adnexal mass, but not for screening the general female population [97].

Bel'skaya et al., 2019, used infrared (IR) Fourier spectroscopy to assess changes of the lipid profile in saliva in endometrial and OC patients. Three groups of women were involved: 51 ovarian and endometrial cancer patients, 26 positive controls of non-ovarian or endometrial cancer patients, and 30 healthy women. The authors identified a significantly reduced intensity of the absorption bands 2923/2957 cm^{-1} in OC and endometrial cancer patients [98].

Despite the methodological limitations due to the small size of patient groups, these works suggest promising directions for developing new diagnostic methods. The idea behind them all was to reduce unnecessary biopsies using non-invasive, saliva-based protocols for the early detection of OC [93].

Currently, studies exploring saliva for detection, diagnosis, or stratification of OC are quite heterogeneous. Transcriptomics and proteomics have been applied for an unbiased approach; however, some degree of partiality was still present in the selection of study participants. In addition, often the study design missed the positive control group (benign tumor or other cancer types). Further, discrepancies exist in the methods for processing samples (e.g., supernatant of saliva or whole saliva), all of which introduces some degree of fragmentation that hampers our comprehension of results. Finally, combining the analysis of saliva with other liquid or solid biopsies is certainly advantageous to better understand the disease; however, they are not so relevant in the search for effective BMs, regardless of their biological role.

7. Candidate miRNA Biomarkers of Ovarian Cancer in Other Biofluids

A multi-marker approach could improve the diagnosis of a heterogeneous disease such as OC, and microRNA are particularly suitable to build a multi-marker model. By improving methods and standardization procedures, miRNAs could become a routine tool to profile patients and select therapeutic interventions [99].

MiRNAs were characterized in biofluids of OC patients including blood (serum or plasma), urine, ascites, and utero-tubal lavage [100]. Currently, blood-derived biofluids still remain the gold standard for liquid biopsies; they have been extensively investigated in OC and in other oncological conditions as well. Here, we provide a comparison of the most representative results from a blood-based analysis and refer to other recent review articles for a more detailed knowledge of the literature [101–103].

7.1. Blood-Derived Biofluids

Several articles and systematic reviews have been written about OC and microRNAs in blood-derived biofluids since the diagnostic relevance of microRNAs detected in serum, plasma, and whole blood samples was reported by Resknick and colleagues for the first time in 2008 [104]. They profiled CF circulating miRNAs from serum samples of OC patients and healthy subjects. In the same year, Taylor and Taylor [43] focused on miRNAs content in OC-derived exosomes compared to benign controls in order to validate their potential as diagnostic BMs. Since then, several research groups have tried to find the most representative miRNA signature for pathogenesis of OC [100].

Hulstaert et al. in 2021, in a meta-analysis of the literature to identify candidate RNA BMs in body fluids for early diagnosis of OC, highlighted seventy-five RNAs candidates.

Only ten of them were then considered good candidates because of their differential expression in at least two studies. Five miRNAs were up-regulated in OC patients compared to healthy subjects (miR-21, the miR-200 family, miR-205, miR-10a, and miR-346), and the other five (miR-122, miR-193a, miR-223, miR-126, and miR-106b) were down-regulated. Heterogeneity among all the reviewed publications was the main limitation for the authors' work, especially concerning the methodologies [102].

Occasionally, some inconsistencies are found in the literature; they could result from relatively small size of samples, different ethnic groups recruited, or from the type of control groups (healthy subjects or benign disease patients) [103]. Additionally, pre-analytical parameters for sample preparation (blood volume, centrifugation speed, duration) can deeply affect the recovery of miRNAs from blood samples [102]. For example, plasma and serum undergo two distinct collection processes and they are differently affected by hemolysis. In plasma samples, coagulation is prevented by adding EDTA and, through centrifugation at high speeds for a long time, platelet-depleted plasma can be easily obtained. In turn, in serum samples, clot-activation causes the release of different biological molecules (also miRNAs) from platelets [105]. Even if, for the above reason, plasma should be considered the sample of choice to detect circulating CF-miRNAs, hemolysis can in both cases affect the accuracy of serum or plasma-based tests. Shah et al., 2016, evidenced the need of quantifying hemolysis as an essential step before measuring circulating miRNAs as potential diagnostic BMs [106].

Additionally, experimental approaches differ considerably among different studies: exosomal miRNAs are usually processed with ultracentrifugation methods or with specific isolation kits. Different methods could be also used for RNA isolation (extraction solutions with or without passage in columns) with varying performance. Then, among high-throughput methods, miRNAs can be profiled with microarrays or next-generation sequencing (NGS), followed by validation with QRT-PCR [100].

The initial screening analyses and criteria of miRNAs selection for subsequent validation differ among the reviewed articles. Three key examples follow. Kim et al. studied the expression level of seven exosomal miRNAs in the serum of OC patients. The seven candidates had been previously identified with high-throughput profiling studies as the most differentially expressed in OC tissues [107].

Elias et al. applied a neural network analysis to RNA-sequencing data from 179 serum samples. In an innovative way, they analyzed miRNA-seq data with a specific algorithm for discriminating OC patients from the other groups involved in the study [108].

Kumar et al. performed a miRNA screening analyzing the methylation status of genomic DNA. The differentially methylated regions of miRNA gene promoters led to identifying three hypomethylated regions through QRT-PCR on tissues and matched serum samples [109].

Another considerable constraint is represented by the use of endogenous controls for miRNA normalization, because currently the selection of reference genes has not yet been standardized. The most used spike-in control is cel-mir-39 which is added into the lysate, and its quantification used to evaluate the success of the isolation procedure. However, endogenous controls can also be used for circulating miRNAs, for example, small nuclear and nucleolar RNA such as U6, RNU44, RNU43, and RNU48. Several authors proposed other internal controls that were more stable in the biofluids of their interest, making comparisons across different articles even more difficult [105].

Another level of complexity is due to the dual possibility of isolating miRNAs as CF or as extracellular vesicle (EV)-contained molecules. In fact, EVs are released from many different cell types, not only cancer cells. Therefore, different extraction and detection methods might influence the efficacy of discriminating among normal- or cancer-derived EVs [110], with issues related to standard BM concentrations and small sample sizes [111]. In addition, most of the published studies on exosomal miRNAs do not use high-throughput discovery approaches as a first step. Instead, they mostly select candidate exosomal miRNAs from the literature and then proceed with validation or functional assays [105].

7.2. Other Biofluids

This review's authors selected candidates miRNAs characterized in biofluids other than blood for OC liquid biopsy.

Urine. Urinary miRNAs have been mostly explored in urological and gynecological diseases. As an advantage, urine is generated near to the OC site of origin and its collection is non-invasive [112]. Detectable miRNA level in urine is lower compared to blood. Probably most circulating miRNAs are reabsorbed by kidneys through a not yet clearly understood mechanism, or they are destroyed in the urinary tract by high levels of RNases [112]. In addition, some authors stressed that the supernatant fraction, but not the exosomal fraction, exhibit a diagnostic potential. Currently, there is not a consensus yet around this topic; it will be critical to define whether the exosomal or the supernatant fraction perform better to yield a representative signature of urinary miRNA [113,114].

Záveský et al., 2015, worked on miRNA in urine samples of pre/post-surgery endometrial and EOC patients. Analysis of the urine supernatant fraction show that miRNAs were differently expressed between pathological and healthy samples. On the contrary, while working with the exosomal RNAs, the authors did not find any difference for any miRNA under analysis [113,114].

Zhou et al., 2015, profiled miRNAs in the urine of three groups of subjects: ovarian serous adenocarcinoma patients (OSA), 26 patients with benign gynecological disease, and 30 healthy controls. They revealed a higher miR-30a-5p expression in ovarian serous adenocarcinoma patients (OSA) urines versus both controls and benign ovarian specimens. Further, comparing urine from OSA, GC, CRC, and healthy controls, they found miR-30a-5p upregulation in OSA and lower expression in the other cancer types, suggesting a diagnostic BM value for miR-30a-5p in OC. Moreover, the authors found miR-30a-5p upregulation in OSA tissue samples and cancer cell lines. Surgical removal of OSA affects urinary level of miR-30a-5p, further reinforcing the hypothesis of its ovarian origin [115].

Berner et al., 2022, detected differentially expressed miRNAs in urine, in OC cell culture, and cell culture supernatant. The authors confirmed the small total amount of miRNA that could be found in urine as one crucial issue for miRNA end-point usability [116]. Záveský et al., 2019, profiled the expression level of eight selected miRNAs, comparing tissue, ascites, and urine samples. Their work represented well the difficulties provided by urine. Some miRNAs were down-regulated in both ascitic fluid and tumor tissues in OC patients, while extracellular urine-derived miRNAs were not differentially expressed [117].

Ascites. Despite the fact that most of the studies are still based on RNA isolation from tissue, cell lines, or blood-derived biofluids, ascites is mentioned frequently in the scientific literature concerning OC, especially if compared with urine. In fact, more than other cancers, OC mostly disseminates in the peritoneal cavity, also because of the primary tumor site. The ascites, intraperitoneal liquid accumulation, is a symptom found mostly in OC patients at advanced stages, but sometimes it also occurs early on. It contains malignant tumor cells together with lymphocytes, mesothelial cells, macrophages, fibroblasts, and other cell types, that create a tumor microenvironment of soluble growth factors and cytokines. Tissue and ascites comparison for miRNAs expression level showed coherent results, confirming the close relation between the solid tumor and its dissemination in ascites [117]. Yet, the lack of systematic validationdoes not allow miRNA profiling from ascites as a novel diagnostic, prognostic, and predictive method. There are several problems related to the use of peritoneal effusions: (1) ascites (or effusion in general) is not always found in OC patients, especially at early stage; (2) effusion collection is an invasive procedure; (3) the components found in ascites are not evenly distributed, so variability between samples increases, making reproducibility even more difficult [118]; (4) there is still a gap in standardizing methods for sampling, separating extracellular fraction and isolating RNA; and (5) different studies employ different control groups (benign tumor lavages, plasma, or tissue of healthy controls) [119]. Ascites is considered an indicator of poor prognosis because it associates with tumor progression and chemoresistance [119].

Many studies have investigated miRNA expression level in ascites for the diagnosis of OC. Záveský et al., 2019, performed the first large-scale expression profiling of 754 miRNAs, working on ascitic fluid extracellular fraction (or ascites derived lavages) of high-grade serous OC patients and control plasma samples. After a screening phase with TaqMan array, seven miRNAs were validated with QRT-PCR. MiR-1290 and members of miR-200 family were found overexpressed in ascites compared to control plasma [119]. In a follow-up study, the same authors collected tissue samples and ascites, focusing on eight selected candidates (miR-203a-3p, miR-204-5p, miR-451a, miR-185-5p, miR-135b-5p, miR-182-5p, miR-200b-3p, and miR-1290) and validated six of them [117]. Yammamoto et al., 2018, examined miRNAs expression in EV from OC ascites, focusing on selected candidates. Six miRNAs were found significantly decreased in OC ascites compared to benign peritoneal fluids [120].

Vaksman et al., 2014, worked on exosomal miRNAs from effusion supernatants of OC patients. They started with a TaqMan array-based screening of miRNA from pooled exosomes derived from the effusion supernatants of OC patients, of tumor cells (positive control group), and of reactive mesothelium (negative control). 99 miRNAs showed higher expression in exosomes from OC effusion supernatants. MiR-21, miR-23a, and miR-29a were associated with poor prognosis. In vitro and in vivo assays were performed with selected miRNAs [121]. Similarly, Mitra and colleagues, 2021, characterized EVs from the ascites of OC patients to understand how EVs miRNA can influence growth, migration, and invasion in vitro and ex vivo assays. EVs from ascitic supernatants of patients with high-grade serous OC (HGSOC) or benign disease were profiled and then validated with QRT-PCR [122].

Wang and colleagues, 2022, identified an EV miRNA signature based on eight miRNAs (miR-1246, miR-1290, miR-483, miR-429, miR-34b-3p, miR-34c-5p, miR-145-5p, and miR-449a), analyzing plasma and ascites from malignant, high-grade serous OC patients and peritoneal fluids from benign gynecologic diseases patients. The authors showed how EV from malignant ascites increase the aggressive phenotypes of OC cells in 2D and 3D models. Gain and loss of function assays for the upregulated miR-1246 and miR-1290 showed their capacity of improving cell migration and invasiveness. Authors underlined the difficulties in obtaining ascites, and concomitantly to collect both plasma and ascites from the same patient [123].

Uterine cavity biofluids. The origin of high-grade serous carcinoma is still unclear. Since it could arise from the epithelium of the fallopian tube fimbriae, some authors explored the diagnostic value of utero-tubal lavage. This collection method allows to sample cells, or their secreted biological products, on the fimbriae. It is a safe and minimally invasive procedure: a saline solution is flushed in the uterine cavity and fallopian tube, then it is aspirated back from the gynecologic tract. Hulstaert et al., 2022, for the first time, sequenced the transcriptome of utero-tubal lavage and found upregulation of 300 mRNAs, mainly involved in cycle regulation and proliferation. Furthermore, they found 41 miRNAs more expressed in OC patients compared to healthy subjects. Five microRNAs were found to be associated to OC pathogenesis [124]. Skryabin et al., 2022, conducted the first exosomal miRNA profiling by small RNA-seq on uterine aspirates (UA) from EOC patients and healthy donors. The results showed significant differences for more than 57 miRNAs and three of them have been validated and confirmed for their up-/down-regulation in OC including miRNAs previously found associated with OC [125].

Proximal liquid biopsy and miRNAs detection remain insufficiently explored, although promising as a diagnostic method because they could be performed during routine gynecological visits. Moreover, the close relation with the site of origin of OC suggests the possibility to detect the disease when it is still at the initial stage. However, this is not applicable as a large-scale and rapid screening method because it is relatively time-consuming and requires specialized medical support.

To summarize the most relevant information, Table 2 reports an excursus on candidate miRNA BMs of OC in body fluids other than blood.

Table 2. Biofluids other than blood or saliva for OC liquid biopsy and miRNAs BMs search (as selected by this review's authors). EOC: FTC: fallopian tube cancer; HS: healthy subject; BGD: benign gynecological disease; AUC: area under the ROC curve. Each row provides information regarding the citation in parenthesis at the far left end of the row.

Reference	Biofluid	Source (Analytes)	miRNAs	Status	Case	Controls	AUC (95% CI)
[113]	urine	cell free	miR-92 miR-106b miR-100 miR-200b	↑ ↓ ↓ ↑	6 EOC FTC	13 HS	1.00 (0.815–1.000) 0.97 (0.764–1.000) 0.85 (0.601–0.970) 1.00 (0.782–1.000)
[115]	urine	exosomes	miR-30a-5p	↑	39 OSA	26 BGD 30 HS	0.86 (0.709–1.016)
[116]	urine	cell free	miR-15a let-7a	↑ ↓	13 OC	17 HS	
[117]	Ascite vs Plasma	exosomes	miR-203-3p miR-204-5p miR-135b-5p miR-182-5p miR-451a	↑ ↑ ↑ ↑ ↓	12 OC	12 HS	1.000 (0.782–1.000) 1.000 (0.782–1.000) 1.000 (0.782–1.000) 0.964 (0.725–1.000) 0.964 (0.725–1.000)
[120]	Ascite vs Benign peritoneal fluids	exosomes	let-7b. miR-23b miR-29a miR-30d miR-205 miR-720	↓ ↓ ↓ ↓ ↓ ↓	8 OC	10 HS	
[122]	Ascite and plasma	exosomes	miR-200c-3p miR-18a-5p miR-1246 miR-1290 miR-100- 5p miR125b-3p	↑ ↑ ↑ ↑ ↓ ↓	5 OC	2 BGD	
[123]	Ascite vs Benign peritoneal fluids	exosomes	miR-1246 miR-1290	↑ ↑	78 OC	72 BGD	
[124]	Uterine cavity fluids	Cell free	let-7d-5p miR-203a miR-200b miR-200c miR-191	↑ ↑ ↑ ↑ ↑	26 OC	48 BGD	
[125]	Uterine cavity fluids	exosomes	miR-451a miR-199a-3p miR-375-3p	↓ ↓ ↑	5 EOC	5 HS	

8. Conclusion and Future Perspectives

Current literature shows that saliva released by the major salivary glands contains various systemic BMs, thereby accurately reflecting pathophysiological conditions in humans. For this reason, salivary diagnostics were indicated by recent, authoritative editorials as a major resource for future diagnostics and for the low-cost screening procedures of systemic diseases including cancer [126].

MiRNAs present several advantages as possible BMs, over other molecules, for their accessibility, strong stability, and resistance to degradation in body fluids. More and more articles describe specific miRNA signatures in several tumor types [74]. Therefore, measuring quickly and accurately small variations of miRNAs concentration in body fluids may offer diagnostic and prognostic opportunities. International efforts aimed at a deeper understanding of circulating miRNA function and standardization of the miRNA analysis field (e.g., Extracellular RNA Communication Consortium, ERCC; https://exrna.org/ (accessed on 15 September 2022), or CANCER-ID), (www.cancerid.eu, accessed on 15 September 2022) [reviewed in Valihracha et al., 2019 [127], will help in providing more solid ground for biomedical exploitation of miRNAs.

Up to now, circulating miRNAs for OC detection have been mainly investigated in blood, a biofluid more complex than saliva in terms of collection, processing, and composition, where the high number of proteins represent a critical methodological step [128], as well as in other biofluids (Table 2). Salivary miRNAs, for their hitherto mentioned characteristics, might offer better performances compared to other fluids' BMs [80].

In conclusion, the current lack of salivary tests for OC, whether based on miRNAs or other BMs, is a challenge to win so as to spare lives from OC and limit social costs. Current research efforts are ongoing to identify a pattern of salivary miRNAs suitable as BMs for OC diagnostics (D. Angeloni, personal communication). Auspicial results would make a possible alternative to the current options for excluding OC that still rely on gynecologist check-up, TVS, and blood test. Overall, this procedure could be expensive and difficult to manage, especially in remote areas or in disadvantaged socio-economic conditions.

For such a low-prevalence cancer type, a screening test would require a sensitivity for asimptomatic women > 75% and a specificity > 99.6% [129]. The currently used biomarkers for the blood test of OC are CA125 and HE4. They have been extensively studied and we refer to Dochez et al., 2019, for a summary of the sensitivity and specificity of the two protein biomarkers [130]. However, a pattern of good candidate miRNA BMs should reach at least the above suggested values.

With regard to the technological approach, the current on-chip technologies combined with photonic biosensors make it possible to envisage a strategy based on bringing together the molecular probes for detecting patterns of BMs within diagnostic devices that could speed up timing and reduce the costs of screening pertinent sections of the population to decrease mortality from OC [131]. Although such devices do not exist for clinical use yet, several proofs of concept exist, both for proteins [132,133] and RNA [134]. With further effort, one could imagine that combining the patterns of different BM types (e.g protein and RNA) could provide the screening robustness that single molecular types were not able to provide so far.

Author Contributions: Conceptualization, D.A.; methodology, D.A., M.R. and F.S.; resources, D.A.; writing—original draft preparation, M.R. and F.S.; writing—review and editing, D.A., M.R. and F.S.; supervision, D.A.; project administration, D.A.; funding acquisition, D.A. All authors have read and agreed to the published version of the manuscript.

Funding: This research was funded by the Italian Ministry of University and Research under the grant "Fondo per la promozione e lo sviluppo delle politiche del Programma nazionale per la ricerca (PNR) 2021", project title: Samarcanda, to D.A. The APC was funded under the same grant. M.R. was supported by a graduate fellowship from Scuola Sant'Anna.

Institutional Review Board Statement: Not applicable, this article reviews existing literature.

Informed Consent Statement: Not applicable, this review article does not include new human studies.

Data Availability Statement: This review article does not include original data. All statements are properly referred to citations.

Acknowledgments: The Authors thank Giulia Robotti for preparing the figures, and Ivana Barravecchia for critical reading of the manuscript. We acknowledge Biorender (https://www.biorender.com/) as a source of three icons used for Figure 1, which for the rest is an original composition. The citations of biotechnological companies and products do not constitute any type of endorsement.

Conflicts of Interest: The authors declare no conflict of interest.

References

1. Sung, H.; Ferlay, J.; Siegel, R.L.; Laversanne, M.; Soerjomataram, I.; Jemal, A.; Bray, F. Global Cancer Statistics 2020: GLOBOCAN Estimates of Incidence and Mortality Worldwide for 36 Cancers in 185 Countries. *CA Cancer J. Clin.* **2021**, *71*, 209–249. [CrossRef]
2. Reid, B.M.; Permuth, J.B.; Sellers, T.A. Epidemiology of Ovarian Cancer: A Review. *Cancer Biol. Med.* **2017**, *14*, 9–32. [CrossRef] [PubMed]
3. Siegel, R.L.; Miller, K.D.; Jemal, A. Cancer Statistics, 2018. *CA Cancer J. Clin.* **2018**, *68*, 7–30. [CrossRef]
4. Jacobs, I.J.; Menon, U.; Ryan, A.; Gentry-Maharaj, A.; Burnell, M.; Kalsi, J.K.; Amso, N.N.; Apostolidou, S.; Benjamin, E.; Cruickshank, D.; et al. Ovarian Cancer Screening and Mortality in the UK Collaborative Trial of Ovarian Cancer Screening (UKCTOCS): A Randomised Controlled Trial. *Lancet* **2016**, *387*, 945–956. [CrossRef] [PubMed]
5. Kossaï, M.; Leary, A.; Scoazec, J.Y.; Genestie, C. Ovarian Cancer: A Heterogeneous Disease. *Pathobiology* **2018**, *85*, 41–49. [CrossRef]
6. Berek, J.S.; Crum, C.; Friedlander, M. Cancer of the Ovary, Fallopian Tube, and Peritoneum. *Int. J. Gynecol. Obstet.* **2012**, *119*, S118–S129. [CrossRef]
7. Buys, S.S.; Partridge, E.; Black, A.; Johnson, C.C.; Lamerato, L.; Isaacs, C.; Reding, D.J.; Greenlee, R.T.; Yokochi, L.A.; Kessel, B.; et al. Effect of Screening on Ovarian Cancer Mortality: The Prostate, Lung, Colorectal and Ovarian (PLCO) Cancer Screening Randomized Controlled Trial. *JAMA J. Am. Med. Assoc.* **2011**, *305*, 2295–2302. [CrossRef]
8. Zhang, Z.; Chan, D.W. The Road from Discovery to Clinical Diagnostics: Lessons Learned from the First FDA-Cleared In Vitro Diagnostic Multivariate Index Assay of Proteomic Biomarkers. *Cancer Epidemiol. Biomark. Prev.* **2010**, *19*, 2995–2999. [CrossRef]
9. Lokich, E.; Palisoul, M.; Romano, N.; Craig Miller, M.; Robison, K.; Stuckey, A.; Disilvestro, P.; Mathews, C.; Granai, C.O.; Lambert-Messerlian, G.; et al. Assessing the Risk of Ovarian Malignancy Algorithm for the Conservative Management of Women with a Pelvic Mass. *Gynecol. Oncol.* **2015**, *139*, 248–252. [CrossRef]
10. Gold, P.; Freedman, S.O. Demonstration of Tumor-Specific Antigens in Human Colonic Carcinomata By Immunological Tolerance and Absorption Techniques. *J. Exp. Med.* **1965**, *121*, 439–462. [CrossRef]
11. Lee, J.M.; Kohn, E.C. Proteomics as a Guiding Tool for More Effective Personalized Therapy. In Proceedings of the Annals of Oncology, Milan, Italy, 8–12 October 2010; Volume 21.
12. Mor, G.; Visintin, I.; Lai, Y.; Zhao, H.; Schwartz, P.; Rutherford, T.; Yue, L.; Bray-Ward, P.; Ward, D.C. Serum Protein Markers for Early Detection of Ovarian Cancer. *Proc. Natl. Acad. Sci. USA* **2005**, *102*, 7677–7682. [CrossRef] [PubMed]
13. van Gorp, T.; Cadron, I.; Despierre, E.; Daemen, A.; Leunen, K.; Amant, F.; Timmerman, D.; de Moor, B.; Vergote, I. HE4 and CA125 as a Diagnostic Test in Ovarian Cancer: Prospective Validation of the Risk of Ovarian Malignancy Algorithm. *Br. J. Cancer* **2011**, *104*, 863–870. [CrossRef]
14. Li, F.; Tie, R.; Chang, K.; Wang, F.; Deng, S.; Lu, W.; Yu, L.; Chen, M. Does Risk for Ovarian Malignancy Algorithm Excel Human Epididymis Protein 4 and CA125 in Predicting Epithelial Ovarian Cancer: A Meta-Analysis. *BMC Cancer* **2012**, *12*, 258. [CrossRef]
15. Fung, E.T. A Recipe for Proteomics Diagnostic Test Development: The OVA1 Test, from Biomarker Discovery to FDA Clearance. *Clin. Chem.* **2010**, *56*, 327–329. [CrossRef]
16. Zhang, M.; Cheng, S.; Jin, Y.; Zhao, Y.; Wang, Y. Roles of CA125 in Diagnosis, Prediction, and Oncogenesis of Ovarian Cancer. *Biochim. Biophys. Acta Rev. Cancer* **2021**, *1875*, 188503. [CrossRef]
17. Kamal, R.; Hamed, S.; Mansour, S.; Mounir, Y.; Sallam, S.A. Ovarian Cancer Screening-Ultrasound; Impact on Ovarian Cancer Mortality. *Br. J. Radiol.* **2018**, *91*, 20170571. [CrossRef]
18. Elorriaga, M.Á.; Neyro, J.L.; Mieza, J.; Cristóbal, I.; Llueca, A. Biomarkers in Ovarian Pathology: From Screening to Diagnosis. Review of the Literature. *J. Pers. Med.* **2021**, *11*, 1115. [CrossRef] [PubMed]
19. Moore, R.G.; Blackman, A.; Miller, M.C.; Robison, K.; DiSilvestro, P.A.; Eklund, E.E.; Strongin, R.; Messerlian, G. Multiple Biomarker Algorithms to Predict Epithelial Ovarian Cancer in Women with a Pelvic Mass: Can Additional Makers Improve Performance? *Gynecol. Oncol.* **2019**, *154*, 150–155. [CrossRef] [PubMed]
20. Russell, M.R.; Graham, C.; D'Amato, A.; Gentry-Maharaj, A.; Ryan, A.; Kalsi, J.K.; Whetton, A.D.; Menon, U.; Jacobs, I.; Graham, R.L.J. Diagnosis of Epithelial Ovarian Cancer Using a Combined Protein Biomarker Panel. *Br. J. Cancer* **2019**, *121*, 483–489. [CrossRef] [PubMed]
21. Hays, J.L.; Kim, G.; Giuroiu, I.; Kohn, E.C. Proteomics and Ovarian Cancer: Integrating Proteomics Information into Clinical Care. *J. Proteom.* **2010**, *73*, 1864–1872. [CrossRef] [PubMed]
22. Aktas, B.; Kasimir-Bauer, S.; Wimberger, P. Utility of Mesothelin, L1CAM and Afamin as Biomarkers in Primary Ovarian Cancer. *Anticancer Res.* **2013**, *33*, 329–336. [PubMed]
23. Setti, G.; Pezzi, M.E.; Viani, M.V.; Pertinhez, T.A.; Cassi, D.; Magnoni, C.; Bellini, P.; Musolino, A.; Vescovi, P.; Meleti, M. Salivary MicroRNA for Diagnosis of Cancer and Systemic Diseases: A Systematic Review. *Int. J. Mol. Sci.* **2020**, *21*, 907. [CrossRef]
24. Ueland, F. A Perspective on Ovarian Cancer Biomarkers: Past, Present and Yet-To-Come. *Diagnostics* **2017**, *7*, 14. [CrossRef] [PubMed]
25. Grossman, D.C.; Curry, S.J.; Owens, D.K.; Barry, M.J.; Davidson, K.W.; Doubeni, C.A.; Epling, J.W.; Kemper, A.R.; Krist, A.H.; Kurth, A.E.; et al. Screening for Ovarian Cancer US Preventive Services Task Force Recommendation Statement. *JAMA J. Am. Med. Assoc.* **2018**, *319*, 588–594.

26. González-Guerrero, A.B.; Maldonado, J.; Dante, S.; Grajales, D.; Lechuga, L.M. Direct and Label-Free Detection of the Human Growth Hormone in Urine by an Ultrasensitive Bimodal Waveguide Biosensor. *J. Biophotonics* **2017**, *10*, 61–67. [CrossRef] [PubMed]

27. Jiu, L.; Hogervorst, M.A.; Vreman, R.A.; Mantel-Teeuwisse, A.K.; Goettsch, W.G. Understanding Innovation of Health Technology Assessment Methods: The IHTAM Framework. *Int. J. Technol. Assess. Health Care* **2022**, *38*, e16. [CrossRef]

28. Barrett, J.E.; Jones, A.; Evans, I.; Reisel, D.; Herzog, C.; Chindera, K.; Kristiansen, M.; Leavy, O.C.; Manchanda, R.; Bjørge, L.; et al. The DNA Methylome of Cervical Cells Can Predict the Presence of Ovarian Cancer. *Nat. Commun.* **2022**, *13*, 448. [CrossRef]

29. Xiao, Y.; Bi, M.; Guo, H.; Li, M. Multi-Omics Approaches for Biomarker Discovery in Early Ovarian Cancer Diagnosis-NC-ND License. *EBioMedicine* **2022**, *79*, 104001. [CrossRef]

30. Gahlawat, A.W.; Witte, T.; Haarhuis, L.; Schott, S. A Novel Circulating MiRNA Panel for Non-Invasive Ovarian Cancer Diagnosis and Prognosis. *Br. J. Cancer* **2022**, *127*, 1550–1556. [CrossRef]

31. Drescher, C.W.; Anderson, G.L. The yet Unrealized Promise of Ovarian Cancer Screening. *JAMA Oncol.* **2018**, *4*, 456–457. [CrossRef]

32. Longo, D.L. Personalized Medicine for Primary Treatment of Serous Ovarian Cancer. *N. Engl. J. Med.* **2019**, *381*, 2471–2474. [CrossRef]

33. Ying, S.Y.; Chang, D.C.; Lin, S.L. The MicroRNA (MiRNA): Overview of the RNA Genes That Modulate Gene Function. *Mol. Biotechnol.* **2008**, *38*, 257–268. [CrossRef] [PubMed]

34. Pasquinelli, A.E.; Reinhart, B.J.; Slack, F.; Martindale, M.Q.; Kuroda, M.I.; Maller, B.; Hayward, D.C.; Ball, E.E.; Degnan, B.; Müller, P.; et al. Degnan Bernard Conservation of the Sequence and Temporal Expression of Let-7 Heterochronic Regulatory RNA. *Nature* **2000**, *408*, 86–89. [CrossRef] [PubMed]

35. Calin, G.A.; Dumitru, C.D.; Shimizu, M.; Bichi, R.; Zupo, S.; Noch, E.; Aldler, H.; Rattan, S.; Keating, M.; Rai, K.; et al. Frequent Deletions and Down-Regulation of Micro-RNA Genes MiR15 and MiR16 at 13q14 in Chronic Lymphocytic Leukemia. *Proc. Natl. Acad. Sci. USA* **2002**, *99*, 15524–15529. [CrossRef] [PubMed]

36. Yang, W.J.; Yang, D.D.; Na, S.; Sandusky, G.E.; Zhang, Q.; Zhao, G. Dicer Is Required for Embryonic Angiogenesis during Mouse Development. *J. Biol. Chem.* **2005**, *280*, 9330–9335. [CrossRef] [PubMed]

37. Lawrie, C.H.; Gal, S.; Dunlop, H.M.; Pushkaran, B.; Liggins, A.P.; Pulford, K.; Banham, A.H.; Pezzella, F.; Boultwood, J.; Wainscoat, J.S.; et al. Detection of Elevated Levels of Tumour-Associated MicroRNAs in Serum of Patients with Diffuse Large B-Cell Lymphoma. *Br. J. Haematol.* **2008**, *141*, 672–675. [CrossRef]

38. Mitchell, P.S.; Parkin, R.K.; Kroh, E.M.; Fritz, B.R.; Wyman, S.K.; Pogosova-Agadjanyan, E.L.; Peterson, A.; Noteboom, J.; O'briant, K.C.; Allen, A.; et al. Circulating MicroRNAs as Stable Blood-Based Markers for Cancer Detection. *Proc. Natl. Acad. Sci. USA* **2008**, *105*, 10513–10518. [CrossRef]

39. Chen, X.; Ba, Y.; Ma, L.; Cai, X.; Yin, Y.; Wang, K.; Guo, J.; Zhang, Y.; Chen, J.; Guo, X.; et al. Characterization of MicroRNAs in Serum: A Novel Class of Biomarkers for Diagnosis of Cancer and Other Diseases. *Cell Res.* **2008**, *18*, 997–1006. [CrossRef]

40. Larrea, E.; Sole, C.; Manterola, L.; Goicoechea, I.; Armesto, M.; Arestin, M.; Caffarel, M.M.; Araujo, A.M.; Araiz, M.; Fernandez-Mercado, M.; et al. New Concepts in Cancer Biomarkers: Circulating MiRNAs in Liquid Biopsies. *Int. J. Mol. Sci.* **2016**, *17*, 627. [CrossRef]

41. Matsuzaki, J.; Ochiya, T. Circulating MicroRNAs and Extracellular Vesicles as Potential Cancer Biomarkers: A Systematic Review. *Int. J. Clin. Oncol.* **2017**, *22*, 413–420. [CrossRef] [PubMed]

42. Izzotti, A.; Carozzo, S.; Pulliero, A.; Zhabayeva, D.; Ravetti, J.L.; Bersimbaev, R. Extracellular MicroRNA in Liquid Biopsy: Applicability in Cancer Diagnosis and Prevention. *Am. J. Cancer Res.* **2016**, *7*, 1461–1493.

43. Taylor, D.D.; Gercel-Taylor, C. MicroRNA Signatures of Tumor-Derived Exosomes as Diagnostic Biomarkers of Ovarian Cancer. *Gynecol. Oncol.* **2008**, *110*, 13–21. [CrossRef] [PubMed]

44. Lin, P.-Y.; Yu, S.-L.; Yang, P.-C. MicroRNA in Lung Cancer. *Br. J. Cancer* **2010**, *103*, 1144–1148. [CrossRef]

45. Wang, Q.Z.; Xu, W.; Habib, N.; Xu, R. Potential Uses of MicroRNA in Lung Cancer Diagnosis, Prognosis, and Therapy. *Curr. Cancer Drug Targets* **2009**, *9*, 572–594. [CrossRef]

46. Slaby, O.; Svoboda, M.; Michalek, J.; Vyzula, R. MicroRNAs in Colorectal Cancer: Translation of Molecular Biology into Clinical Application. *Mol. Cancer* **2009**, *8*, 102. [CrossRef] [PubMed]

47. Wang, W.; Luo, Y. ping MicroRNAs in Breast Cancer: Oncogene and Tumor Suppressors with Clinical Potential. *J. Zhejiang Univ. Sci. B* **2015**, *16*, 18–31. [CrossRef]

48. Liu, H.-S.; Xiao, H.-S. MicroRNAs as Potential Biomarkers for Gastric Cancer. No. 2012AA020103. *World J. Gastroenterol.* **2014**, *20*, 12007–12017. [CrossRef] [PubMed]

49. Kinose, Y.; Sawada, K.; Nakamura, K.; Kimura, T. The Role of MicroRNAs in Ovarian Cancer. *Biomed Res. Int.* **2014**, *2014*, 249393. [CrossRef]

50. Nam, E.J.; Yoon, H.; Kim, S.W.; Kim, H.; Kim, Y.T.; Kim, J.H.; Kim, J.W.; Kim, S. MicroRNA Expression Profiles in Serous Ovarian Carcinoma. *Clin. Cancer Res.* **2008**, *14*, 2690–2695. [CrossRef]

51. Dahiya, N.; Morin, P.J. MicroRNAs in Ovarian Carcinomas. *Endocr. Relat. Cancer* **2010**, *17*, F77. [CrossRef]

52. Kö Bel, M.; Kalloger, S.E.; Boyd, N.; Mckinney, S.; Mehl, E.; Palmer, C.; Leung, S.; Bowen, N.J.; Ionescu, D.N.; Rajput, A.; et al. Ovarian Carcinoma Subtypes Are Different Diseases: Implications for Biomarker Studies. *PLoS Med.* **2008**, *5*, e232. [CrossRef]

53. Zuberi, M.; Mir, R.; Das, J.; Ahmad, I.; Javid, J.; Yadav, P.; Masroor, M.; Ahmad, S.; Ray, P.C.; Saxena, A. Expression of Serum MiR-200a, MiR-200b, and MiR-200c as Candidate Biomarkers in Epithelial Ovarian Cancer and Their Association with Clinicopathological Features. *Clin. Transl. Oncol.* **2015**, *17*, 779–787. [CrossRef] [PubMed]

54. Zhang, Z.; Bast, R.C.; Yu, Y.; Li, J.; Sokoll, L.J.; Rai, A.J.; Rosenzweig, J.M.; Cameron, B.; Wang, Y.Y.; Meng, X.-Y.; et al. Three Biomarkers Identified from Serum Proteomic Analysis for the Detection of Early Stage Ovarian Cancer. *Cancer Res.* **2004**, *64*, 5882–5890. [CrossRef]

55. Anfossi, S.; Babayan, A.; Pantel, K.; Calin, G.A. Clinical Utility of Circulating Non-Coding RNAs—An Update. *Nat. Rev. Clin. Oncol.* **2018**, *15*, 541–563. [CrossRef] [PubMed]

56. Ravegnini, G.; de Iaco, P.; Gorini, F.; Dondi, G.; Klooster, I.; de Crescenzo, E.; Bovicelli, A.; Hrelia, P.; Perrone, A.M.; Angelini, S. Role of Circulating Mirnas in Therapeutic Response in Epithelial Ovarian Cancer: A Systematic Revision. *Biomedicines* **2021**, *9*, 1316. [CrossRef]

57. Zheng, H.; Liu, J.-Y.; Song, F.-J.; Chen, K.-X. Advances in Circulating MicroRNAs as Diagnostic and Prognostic Markers for Ovarian Cancer. *Cancer Biol. Med.* **2013**, *10*, 123–130. [CrossRef] [PubMed]

58. Montagnana, M.; Benati, M.; Danese, E. Circulating Biomarkers in Epithelial Ovarian Cancer Diagnosis: From Present to Future Perspective. *Ann. Transl. Med.* **2017**, *5*, 276. [CrossRef]

59. Weber, J.A.; Baxter, D.H.; Zhang, S.; Huang, D.Y.; Huang, K.H.; Lee, M.J.; Galas, D.J.; Wang, K. The MicroRNA Spectrum in 12 Body Fluids. *Clin. Chem.* **2010**, *56*, 1733–1741. [CrossRef]

60. Godoy, P.M.; Bhakta, N.R.; Barczak, A.J.; Cakmak, H.; Fisher, S.; MacKenzie, T.C.; Patel, T.; Price, R.W.; Smith, J.F.; Woodruff, P.G.; et al. Large Differences in Small RNA Composition Between Human Biofluids. *Cell Rep.* **2018**, *25*, 1346–1358. [CrossRef]

61. Nik Mohamed Kamal, N.N.S.B.; Shahidan, W.N.S. Non-Exosomal and Exosomal Circulatory MicroRNAs: Which Are More Valid as Biomarkers? *Front. Pharmacol.* **2020**, *10*, 1500. [CrossRef]

62. Preethi, K.A.; Selvakumar, S.C.; Ross, K.; Jayaraman, S.; Tusubira, D.; Sekar, D. Liquid Biopsy: Exosomal MicroRNAs as Novel Diagnostic and Prognostic Biomarkers in Cancer. *Mol. Cancer* **2022**, *21*, 54. [CrossRef] [PubMed]

63. Sohel, M.H. Extracellular/Circulating MicroRNAs: Release Mechanisms, Functions and Challenges. *Achiev. Life Sci.* **2016**, *10*, 175–186. [CrossRef]

64. Roblegg, E.; Coughran, A.; Sirjani, D. Saliva: An all-rounder of our body. *Eur J Pharm Biopharm.* **2019**, *142*, 133–141. [CrossRef] [PubMed]

65. Chojnowska, S.; Baran, T.; Wilińska, I.; Sienicka, P.; Cabaj-Wiater, I.; Knaś, M. Human Saliva as a Diagnostic Material. *Adv. Med. Sci.* **2018**, *63*, 185–191. [CrossRef] [PubMed]

66. Rapado-González, Ó.; Majem, B.; Muinelo-Romay, L.; López-López, R.; Suarez-Cunqueiro, M.M. Cancer Salivary Biomarkers for Tumours Distant to the Oral Cavity. *Int. J. Mol. Sci.* **2016**, *17*, 1531. [CrossRef] [PubMed]

67. Spielmann, N.; Ilsley, D.; Gu, J.; Lea, K.; Brockman, J.; Heater, S.; Setterquist, R.; Wong, D.T.W. The Human Salivary RNA Transcriptome Revealed by Massively Parallel Sequencing. *Clin. Chem.* **2012**, *58*, 1314–1321. [CrossRef] [PubMed]

68. Bahn, J.H.; Zhang, Q.; Li, F.; Chan, T.M.; Lin, X.; Kim, Y.; Wong, D.T.W.; Xiao, X. The Landscape of MicroRNA, Piwi-Interacting RNA, and Circular RNA in Human Saliva. *Clin. Chem.* **2015**, *61*, 221–230. [CrossRef]

69. Pfaffe, T.; Cooper-White, J.; Beyerlein, P.; Kostner, K.; Punyadeera, C. Diagnostic Potential of Saliva: Current State and Future Applications. *Clin. Chem.* **2011**, *57*, 675–687. [CrossRef]

70. Rapado-González, Ó.; Martínez-Reglero, C.; Salgado-Barreira, Á.; Takkouche, B.; López-López, R.; Suárez-Cunqueiro, M.M.; Muinelo-Romay, L. Salivary Biomarkers for Cancer Diagnosis: A Meta-Analysis. *Ann. Med.* **2020**, *52*, 131–144. [CrossRef]

71. Streckfus, C.F.; Guajardo-Edwards, C. The Use of Salivary as a Biometric Tool to Determine the Presence of Carcinoma of the Breast Among Women. *Biometrics* 2011. [CrossRef]

72. Nonaka, T.; Wong, D.T.W. Saliva-Exosomics in Cancer: Molecular Characterization of Cancer-Derived Exosomes in Saliva. In *Enzymes*; Academic Press: Cambridge, MA, USA, 2017; Volume 42, pp. 125–151.

73. Lin, X.; Lo, H.C.; Wong, D.T.W.; Xiao, X. Noncoding RNAs in Human Saliva as Potential Disease Biomarkers. *Front. Genet.* **2015**, *6*, 175. [CrossRef] [PubMed]

74. Rapado-González, Ó.; Majem, B.; Muinelo-Romay, L.; álvarez-Castro, A.; Santamaría, A.; Gil-Moreno, A.; López-López, R.; Suárez-Cunqueiro, M.M. Human Salivary MicroRNAs in Cancer. *J. Cancer* **2018**, *9*, 638–649. [CrossRef] [PubMed]

75. Ishige, F.; Hoshino, I.; Iwatate, Y.; Chiba, S.; Arimitsu, H.; Yanagibashi, H.; Nagase, H.; Takayama, W. MIR1246 in Body Fluids as a Biomarker for Pancreatic Cancer. *Sci. Rep.* **2020**, *10*, 8723. [CrossRef]

76. Alemar, B.; Izetti, P.; Gregório, C.; Macedo, G.S.; Antonio, M.; Castro, A.; Osvaldt, A.B.; Matte, U.; Ashton-Prolla, P. MiRNA-21 and MiRNA-34a Are Potential Minimally Invasive Biomarkers for the Diagnosis of Pancreatic Ductal Adenocarcinoma. *Pancreas* **2016**, *45*, 84–92. [CrossRef]

77. Humeau, M.; Vignolle-Vidoni, A.; Sicard, F.; Martins, F.; Bournet, B.; Buscail, L.; Torrisani, J.; Cordelier, P. Salivary MicroRNA in Pancreatic Cancer Patients. *PLoS ONE* **2015**, *10*, e0130996. [CrossRef]

78. Rapado-González, Ó.; Majem, B.; Álvarez-Castro, A.; Díaz-Peña, R.; Abalo, A.; Suárez-Cabrera, L.; Gil-Moreno, A.; Santamaría, A.; López-López, R.; Muinelo-Romay, L.; et al. A Novel Saliva-Based Mirna Signature for Colorectal Cancer Diagnosis. *J. Clin. Med.* **2019**, *8*, 2029. [CrossRef] [PubMed]

79. Petkevich, A.A.; Abramov, A.A.; Pospelov, V.I.; Malinina, N.A.; Kuhareva, E.I.; Mazurchik, N.V.; Tarasova, O.I. Exosomal and Non-Exosomal MiRNA Expression Levels in Patients with HCV-Related Cirrhosis and Liver Cancer. *Oncotarget* **2021**, *12*, 1697. [CrossRef]
80. Ding, Y.; Ma, Q.; Liu, F.; Zhao, L.; Wei, W. The Potential Use of Salivary Mirnas as Promising Biomarkers for Detection of Cancer: A Meta-Analysis. *PLoS ONE* **2016**, *11*, e0166303. [CrossRef] [PubMed]
81. Gablo, N.A.; Prochazka, V.; Kala, Z.; Slaby, O.; Kiss, I. Cell-Free MicroRNAs as Non-Invasive Diagnostic and Prognostic Biomarkers in Pancreatic Cancer. *Curr. Genom.* **2019**, *20*, 569–580. [CrossRef]
82. Xie, Z.; Yin, X.; Gong, B.; Nie, W.; Wu, B.; Zhang, X.; Huang, J.; Zhang, P.; Zhou, Z.; Li, Z. Salivary MicroRNAs Show Potential as a Noninvasive Biomarker for Detecting Resectable Pancreatic Cancer. *Cancer Prev. Res.* **2015**, *8*, 165–173. [CrossRef]
83. Madhavan, B.; Yue, S.; Galli, U.; Rana, S.; Gross, W.; Müller, M.; Giese, N.A.; Kalthoff, H.; Becker, T.; Büchler, M.W.; et al. Combined Evaluation of a Panel of Protein and MiRNA Serum-Exosome Biomarkers for Pancreatic Cancer Diagnosis Increases Sensitivity and Specificity. *Int. J. Cancer* **2015**, *136*, 2616–2627. [CrossRef] [PubMed]
84. Sazanov, A.A.; Kiselyova, E.V.; Zakharenko, A.A.; Romanov, M.N.; Zaraysky, M.I. Plasma and Saliva MiR-21 Expression in Colorectal Cancer Patients. *J. Appl. Genet.* **2017**, *58*, 231–237. [CrossRef] [PubMed]
85. Mariam, A.; Miller-Atkins, G.; Moro, A.; Rodarte, A.I.; Siddiqi, S.; Acevedo-Moreno, L.A.; Brown, J.M.; Allende, D.S.; Aucejo, F.; Rotroff, D.M. Salivary MiRNAs as Non-Invasive Biomarkers of Hepatocellular Carcinoma: A Pilot Study. *PeerJ* **2022**, *9*, e12715. [CrossRef] [PubMed]
86. Yang, Y.; Ma, L.; Qiao, X.; Zhang, X.; Dong, S.F.; Wu, M.T.; Zhai, K.; Shi, H.Z. Salivary MicroRNAs Show Potential as Biomarkers for Early Diagnosis of Malignant Pleural Effusion. *Transl. Lung Cancer Res.* **2020**, *9*, 1247–1257. [CrossRef] [PubMed]
87. Koopaie, M.; Kolahdooz, S.; Fatahzadeh, M.; Manifar, S. Salivary Biomarkers in Breast Cancer Diagnosis: A Systematic Review and Diagnostic Meta-Analysis. *Cancer Med.* **2022**, *11*, 2644–2661. [CrossRef] [PubMed]
88. Li, F.; Yoshizawa, J.M.; Kim, K.M.; Kanjanapangka, J.; Grogan, T.R.; Wang, X.; Elashoff, D.E.; Ishikawa, S.; Chia, D.; Liao, W.; et al. Discovery and Validation of Salivary Extracellular RNA Biomarkers for Noninvasive Detection of Gastric Cancer. *Clin. Chem.* **2018**, *64*, 1513–1521. [CrossRef]
89. Lopes, C.; Chaves, J.; Ortigão, R.; Dinis-Ribeiro, M.; Pereira, C. Gastric Cancer Detection by Non-Blood-Based Liquid Biopsies: A Systematic Review Looking into the Last Decade of Research. *United Eur. Gastroenterol. J.* **2022**, *11*, 114–130. [CrossRef]
90. Hizir, M.S.; Balcioglu, M.; Rana, M.; Robertson, N.M.; Yigit, M.V. Simultaneous Detection of Circulating OncomiRs from Body Fluids for Prostate Cancer Staging Using Nanographene Oxide. *ACS Appl. Mater. Interfaces* **2014**, *6*, 14772–14778. [CrossRef]
91. Luedemann, C.; Reinersmann, J.L.; Klinger, C.; Degener, S.; Dreger, N.M.; Roth, S.; Kaufmann, M.; Savelsbergh, A. Prostate Cancer-Associated MiRNAs in Saliva: First Steps to an Easily Accessible and Reliable Screening Tool. *Biomolecules* **2022**, *12*, 1366. [CrossRef]
92. Machida, T.; Tomofuji, T.; Maruyama, T.; Yoneda, T.; Ekuni, D.; Azuma, T.; Miyai, H.; Mizuno, H.; Kato, H.; Tsutsumi, K.; et al. MIR 1246 and MIR-4644 in Salivary Exosome as Potential Biomarkers for Pancreatobiliary Tract Cancer. *Oncol. Rep.* **2016**, *36*, 2375–2381. [CrossRef]
93. Lee, Y.H.; Kim, J.H.; Zhou, H.; Kim, B.W.; Wong, D.T. Salivary Transcriptomic Biomarkers for Detection of Ovarian Cancer: For Serous Papillary Adenocarcinoma. *J. Mol. Med.* **2012**, *90*, 427–434. [CrossRef]
94. Yang, J.; Xiang, C.; Liu, J. Clinical Significance of Combining Salivary MRNAs and Carcinoembryonic Antigen for Ovarian Cancer Detection. *Scand. J. Clin. Lab. Investig.* **2021**, *81*, 39–45. [CrossRef]
95. Li, Y.; Tan, C.; Liu, L.; Han, L. Significance of Blood and Salivary IEX-1 Expression in Diagnosis of Epithelial Ovarian Carcinoma. *J. Obstet. Gynaecol. Res.* **2018**, *44*, 764–771. [CrossRef]
96. Tajmul, M.; Parween, F.; Singh, L.; Mathur, S.R.; Sharma, J.B.; Kumar, S.; Sharma, D.N.; Yadav, S. Identification and Validation of Salivary Proteomic Signatures for Non-Invasive Detection of Ovarian Cancer. *Int. J. Biol. Macromol.* **2018**, *108*, 503–514. [CrossRef] [PubMed]
97. Zermeño-Nava, J.D.J.; Martínez-Martínez, M.U.; Rámirez-De-Ávila, A.L.; Hernández-Arteaga, A.C.; García-Valdivieso, M.G.; Hernández-Cedillo, A.; José-Yacamán, M.; Navarro-Contreras, H.R. Determination of Sialic Acid in Saliva by Means of Surface-Enhanced Raman Spectroscopy as a Marker in Adnexal Mass Patients: Ovarian Cancer vs Benign Cases. *J. Ovarian Res.* **2018**, *11*, 61. [CrossRef] [PubMed]
98. Bel'skaya, L.V.; Sarf, E.A.; Solomatin, D.V.; Kosenok, V.K. Analysis of the Lipid Profile of Saliva in Ovarian and Endometrial Cancer by IR Fourier Spectroscopy. *Vib. Spectrosc.* **2019**, *104*, 102944. [CrossRef]
99. Condrat, C.E.; Thompson, D.C.; Barbu, M.G.; Bugnar, O.L.; Boboc, A.; Cretoiu, D.; Suciu, N.; Cretoiu, S.M.; Voinea, S.C. MiRNAs as Biomarkers in Disease: Latest Findings Regarding Their Role in Diagnosis and Prognosis. *Cells* **2020**, *9*, 276. [CrossRef]
100. Nakamura, K.; Sawada, K.; Yoshimura, A.; Kinose, Y.; Nakatsuka, E.; Kimura, T. Clinical Relevance of Circulating Cell-Free MicroRNAs in Ovarian Cancer. *Mol. Cancer* **2016**, *15*, 48. [CrossRef] [PubMed]
101. Yoshida, K.; Yokoi, A.; Kato, T.; Ochiya, T.; Yamamoto, Y. The Clinical Impact of Intra- and Extracellular MiRNAs in Ovarian Cancer. *Cancer Sci.* **2020**, *111*, 3435–3444. [CrossRef] [PubMed]
102. Hulstaert, E.; Morlion, A.; Levanon, K.; Vandesompele, J.; Mestdagh, P. Candidate RNA Biomarkers in Biofluids for Early Diagnosis of Ovarian Cancer: A Systematic Review. *Gynecol. Oncol.* **2021**, *160*, 633–642. [CrossRef]
103. Montazerian, M.; Yasari, F.; Aghaalikhani, N. Ovarian Extracellular MicroRNAs as the Potential Non-Invasive Biomarkers: An Update. *Biomed. Pharmacother.* **2018**, *106*, 1633–1640. [CrossRef]

104. Resnick, K.E.; Alder, H.; Hagan, J.P.; Richardson, D.L.; Croce, C.M.; Cohn, D.E. The Detection of Differentially Expressed MicroRNAs from the Serum of Ovarian Cancer Patients Using a Novel Real-Time PCR Platform. *Gynecol. Oncol.* **2009**, *112*, 55–59. [CrossRef] [PubMed]

105. Shiao, M.S.; Chang, J.M.; Lertkhachonsuk, A.A.; Rermluk, N.; Jinawath, N. Circulating Exosomal MiRNAs as Biomarkers in Epithelial Ovarian Cancer. *Biomedicines* **2021**, *9*, 1433. [CrossRef] [PubMed]

106. Shah, J.S.; Soon, P.S.; Marsh, D.J. Comparison of Methodologies to Detect Low Levels of Hemolysis in Serum for Accurate Assessment of Serum MicroRNAs. *PLoS ONE* **2016**, *11*, e0153200. [CrossRef] [PubMed]

107. Kim, S.; Choi, M.C.; Jeong, J.Y.; Hwang, S.; Jung, S.G.; Joo, W.D.; Park, H.; Song, S.H.; Lee, C.; Kim, T.H.; et al. Serum Exosomal MiRNA-145 and MiRNA-200c as Promising Biomarkers for Preoperative Diagnosis of Ovarian Carcinomas. *J. Cancer* **2019**, *10*, 1958–1967. [CrossRef]

108. Elias, K.M.; Fendler, W.; Stawiski, K.; Fiascone, S.J.; Vitonis, A.F.; Berkowitz, R.S.; Frendl, G.; Konstantinopoulos, P.; Crum, C.P.; Kedzierska, M.; et al. Diagnostic Potential for a Serum MiRNA Neural Network for Detection of Ovarian Cancer. *Ellife* **2017**, *6*, e28932. [CrossRef]

109. Kumar, V.; Gupta, S.; Chaurasia, A.; Sachan, M. Evaluation of Diagnostic Potential of Epigenetically Deregulated MiRNAs in Epithelial Ovarian Cancer. *Front. Oncol.* **2021**, *11*, 681872. [CrossRef]

110. Carollo, E.; Paris, B.; Samuel, P.; Pantazi, P.; Bartelli, T.F.; Dias-Neto, E.; Brooks, S.A.; Pink, R.C.; Francisco Carter, D.R. Detecting Ovarian Cancer Using Extracellular Vesicles: Progress and Possibilities. *Biochem. Soc. Trans.* **2019**, *47*, 295–304. [CrossRef]

111. Zheng, X.; Li, X.; Wang, X. Extracellular Vesicle-Based Liquid Biopsy Holds Great Promise for the Management of Ovarian Cancer. *Biochim. Biophys. Acta Rev. Cancer* **2020**, *1874*, 188395. [CrossRef]

112. Gasparri, M.L.; Casorelli, A.; Bardhi, E.; Besharat, A.R.; Savone, D.; Ruscito, I.; Farooqi, A.A.; Papadia, A.; Mueller, M.D.; Ferretti, E.; et al. Beyond Circulating MicroRNA Biomarkers: Urinary MicroRNAs in Ovarian and Breast Cancer. *Tumor Biol.* **2017**, *39*, 1010428317695525. [CrossRef] [PubMed]

113. Zavesky, L.; Jandakova, E.; Turyna, R.; Langmeierova, L.; Weinberger, V.; Minar, L. Supernatant versus Exosomal Urinary MicroRNAs. Two Fractions with Different Outcomes in Gynaecological Cancers. *Neoplasma* **2016**, *63*, 121–132. [CrossRef]

114. Záveský, L.; Jandáková, E.; Turyna, R.; Langmeierová, L.; Weinberger, V.; Záveská Drábková, L.; Hůlková, M.; Hořínek, A.; Dušková, D.; Feyereisl, J.; et al. Evaluation of Cell-Free Urine MicroRNAs Expression for the Use in Diagnosis of Ovarian and Endometrial Cancers. A Pilot Study. *Pathol. Oncol. Res.* **2015**, *21*, 1027–1035. [CrossRef]

115. Zhou, J.; Gong, G.; Tan, H.; Dai, F.; Zhu, X.; Chen, Y.; Wang, J.; Liu, Y.; Chen, P.; Wu, X.; et al. Urinary MicroRNA-30a-5p Is a Potential Biomarker for Ovarian Serous Adenocarcinoma. *Oncol. Rep.* **2015**, *33*, 2915–2923. [CrossRef] [PubMed]

116. Berner, K.; Hirschfeld, M.; Weiß, D.; Rücker, G.; Asberger, J.; Ritter, A.; Nöthling, C.; Jäger, M.; Juhasz-Böss, I.; Erbes, T. Evaluation of Circulating MicroRNAs as Non-Invasive Biomarkers in the Diagnosis of Ovarian Cancer: A Case–Control Study. *Arch. Gynecol. Obstet.* **2022**, *306*, 151–163. [CrossRef] [PubMed]

117. Záveský, L.; Jandáková, E.; Weinberger, V.; Minář, L.; Hanzíková, V.; Dušková, D.; Drábková, L.Z.; Hořínek, A. Ovarian Cancer: Differentially Expressed MicroRNAs in Tumor Tissue and Cell-Free Ascitic Fluid as Potential Novel Biomarkers. *Cancer Investig.* **2019**, *37*, 440–452. [CrossRef] [PubMed]

118. Nicolè, L.; Cappello, F.; Cappellesso, R.; VandenBussche, C.J.; Fassina, A. MicroRNA Profiling in Serous Cavity Specimens: Diagnostic Challenges and New Opportunities. *Cancer Cytopathol.* **2019**, *127*, 493–500. [CrossRef] [PubMed]

119. Záveský, L.; Jandáková, E.; Weinberger, V.; Minář, L.; Hanzíková, V.; Dušková, D.; Záveská Drábková, L.; Svobodová, I.; Hořínek, A. Ascites-Derived Extracellular MicroRNAs as Potential Biomarkers for Ovarian Cancer. *Reprod. Sci.* **2019**, *26*, 510–522. [CrossRef]

120. Yamamoto, C.M.; Oakes, M.L.; Murakami, T.; Muto, M.G.; Berkowitz, R.S.; Ng, S.W. Comparison of Benign Peritoneal Fluid- and Ovarian Cancer Ascites-Derived Extracellular Vesicle RNA Biomarkers. *J. Ovarian Res.* **2018**, *11*, 20. [CrossRef]

121. Vaksman, O.; Tropé, C.; Davidson, B.; Reich, R. Exosome-Derived MiRNAs and Ovarian Carcinoma Progression. *Carcinogenesis* **2014**, *35*, 2113–2120. [CrossRef]

122. Mitra, A.; Yoshida-Court, K.; Solley, T.N.; Mikkelson, M.; Yeung, C.L.A.; Nick, A.; Lu, K.; Klopp, A.H. Extracellular Vesicles Derived from Ascitic Fluid Enhance Growth and Migration of Ovarian Cancer Cells. *Sci. Rep.* **2021**, *11*, 9149. [CrossRef]

123. Wang, W.; Jo, H.; Park, S.; Kim, H.; Kim, S.I.; Han, Y.; Lee, J.; Seol, A.; Kim, J.; Lee, M.; et al. Integrated Analysis of Ascites and Plasma Extracellular Vesicles Identifies a MiRNA-Based Diagnostic Signature in Ovarian Cancer. *Cancer Lett.* **2022**, *542*, 215735. [CrossRef]

124. Hulstaert, E.; Levanon, K.; Morlion, A.; Van Aelst, S.; Christidis, A.A.; Zamar, R.; Anckaert, J.; Verniers, K.; Bahar-Shany, K.; Sapoznik, S.; et al. RNA Biomarkers from Proximal Liquid Biopsy for Diagnosis of Ovarian Cancer. *Neoplasia* **2022**, *24*, 155–164. [CrossRef]

125. Skryabin, G.O.; Komelkov, A.V.; Zhordania, K.I.; Bagrov, D.V.; Vinokurova, S.V.; Galetsky, S.A.; Elkina, N.V.; Denisova, D.A.; Enikeev, A.D.; Tchevkina, E.M. Extracellular Vesicles from Uterine Aspirates Represent a Promising Source for Screening Markers of Gynecologic Cancers. *Cells* **2022**, *11*, 1064. [CrossRef]

126. Kaczor-Urbanowicz, K.E.; Wei, F.; Rao, S.L.; Kim, J.; Shin, H.; Cheng, J.; Tu, M.; Wong, D.T.W.; Kim, Y. Clinical Validity of Saliva and Novel Technology for Cancer Detection. *Biochim. Biophys. Acta Rev. Cancer* **2019**, *1872*, 49–59. [CrossRef]

127. Valihrach, L.; Androvic, P.; Kubista, M. Circulating MiRNA Analysis for Cancer Diagnostics and Therapy. *Mol. Asp. Med.* **2020**, *72*, 100825. [CrossRef]

128. Zhao, Y.; Song, Y.; Yao, L.; Song, G.; Teng, C. Circulating MicroRNAs: Promising Biomarkers Involved in Several Cancers and Other Diseases. *DNA Cell Biol.* **2017**, *36*, 77–94. [CrossRef] [PubMed]
129. Nebgen, D.R.; Lu, K.H.; Bast, R.C. Novel Approaches to Ovarian Cancer Screening. *Curr. Oncol. Rep.* **2019**, *21*, 75. [CrossRef] [PubMed]
130. Dochez, V.; Caillon, H.; Vaucel, E.; Dimet, J.; Winer, N.; Ducarme, G. Biomarkers and Algorithms for Diagnosis of Ovarian Cancer: CA125, HE4, RMI and ROMA, a Review. *J. Ovarian Res.* **2019**, *12*, 28. [CrossRef] [PubMed]
131. Soler, M.; Calvo-Lozano, O.; Estevez, M.-C.; Lechuga, L.M. Nanophotonic Biosensors Driving Personalized Medicine. *Opt. Photonics News* **2020**, *31*, 24–31. [CrossRef]
132. Marin, Y.; Velha, P.; Faralli, S.; Di Pasquale, F.; Oton, C.J. Oton Micro-Interferometers on Chip for Sensing Applications. In *Optical Sensors*; Optical Sensors and Sensing Congress (ES, FTS, HISE, Sensors); Optica Publishing Group: Washington, DC, USA, 2018.
133. Iqbal, M.; Gleeson, M.A.; Spaugh, B.; Tybor, F.; Gunn, W.G.; Hochberg, M.; Baehr-Jones, T.; Bailey, R.C.; Gunn, L.C. Label-Free Biosensor Arrays Based on Silicon Ring Resonators and High-Speed Optical Scanning Instrumentation. *IEEE J. Sel. Top. Quantum Electron.* **2010**, *16*, 654–661. [CrossRef]
134. Scheler, O.; Kindt, J.T.; Qavi, A.J.; Kaplinski, L.; Glynn, B.; Barry, T.; Kurg, A.; Bailey, R.C. Label-Free, Multiplexed Detection of Bacterial TmRNA Using Silicon Photonic Microring Resonators. *Biosens. Bioelectron.* **2012**, *36*, 56–61. [CrossRef] [PubMed]

Review

Occurrence, Role, and Challenges of MicroRNA in Human Breast Milk: A Scoping Review

Adrianna Kondracka [1], Paulina Gil-Kulik [2], Bartosz Kondracki [3,*], Karolina Frąszczak [4], Anna Oniszczuk [5], Magda Rybak-Krzyszkowska [6], Jakub Staniczek [7], Anna Kwaśniewska [1] and Janusz Kocki [2]

[1] Department of Obstetrics and Pathology of Pregnancy, Medical University of Lublin, 20-059 Lublin, Poland
[2] Department of Clinical Genetics, Medical University of Lublin, 11 Radziwillowska Str., 20-080 Lublin, Poland
[3] Department of Cardiology, Medical University of Lublin, 20-059 Lublin, Poland
[4] Department of Oncological Gynecology and Gynecology, Medical University of Lublin, 20-059 Lublin, Poland
[5] Department of Inorganic Chemistry, Medical University of Lublin, 20-059 Lublin, Poland
[6] Department of Obstetrics and Perinatology, University Hospital, 30-688 Krakow, Poland
[7] Department of Gynecology, Obstetrics and Gynecologic Oncology, Medical University of Silesia, 40-055 Katowice, Poland
*   Correspondence: kondracki.bartosz@gmail.com

Abstract: MicroRNAs are non-coding segments of RNA involved in the epigenetic modulation of various biological processes. Their occurrence in biological fluids, such as blood, saliva, tears, and breast milk, has drawn attention to their potential influence on health and disease development. Hundreds of microRNAs have been isolated from breast milk, yet the evidence on their function remains inconsistent and inconclusive. The rationale for the current scoping review is to map the evidence on the occurrence, characterization techniques, and functional roles of microRNAs in breast milk. The review of the sources of this evidence highlights the need to address methodological challenges to achieve future advances in understanding microRNAs in breast milk, particularly their role in conditions such as neoplasms. Nonetheless, remarkable progress has been made in characterizing the microRNA profiles of human breast milk.

Keywords: microRNA; human breast milk; polymerase chain reaction (PCR); microRNA characterization

Citation: Kondracka, A.; Gil-Kulik, P.; Kondracki, B.; Frąszczak, K.; Oniszczuk, A.; Rybak-Krzyszkowska, M.; Staniczek, J.; Kwaśniewska, A.; Kocki, J. Occurrence, Role, and Challenges of MicroRNA in Human Breast Milk: A Scoping Review. *Biomedicines* **2023**, *11*, 248. https://doi.org/10.3390/biomedicines11020248

Academic Editors: Milena Rizzo and Elena Levantini

Received: 25 November 2022
Revised: 12 January 2023
Accepted: 13 January 2023
Published: 18 January 2023

1. Introduction

It has been shown that breastfeeding brings multiple nutritional, developmental, immunological, and biological benefits to infants, making breast milk the most appropriate food for newborns. Numerous studies have shown a positive association between the duration of breastfeeding and variables such as intelligence quotient, income, and educational performance. In addition, the physical health benefits of breastfeeding include a lower incidence of childhood illnesses, such as gastroenteritis, pneumonia, and asthma. The World Health Organization recommends that infants should be exclusively breastfed for about the first six months of their life [1]. Apart from demonstrating the advantages of breast milk, research on the subject has also focused on identifying the mechanisms underlying these characteristics. As a result, some features of breastfeeding have been explained by the identification of molecules and other factors present in breast milk that confer certain characteristics. For instance, the presence of secretory IgA and lactoferrin has been linked to the immunogenic potential of human breastmilk, although the specific mechanism of their transfer to the infant remains poorly understood [2]. Recent studies have reported that human milk contains factors such as non-coding RNA species that are capable of regulating the genetic development of infants' biological systems [2].

MicroRNAs (miRNAs) are a group of regulatory RNA species capable of modulating the activation of certain mRNA targets to influence a wide range of biological processes. Segments of mRNA consist of short nucleotide sequences, between 19 and 24, and are derived from hairpin precursors [3]. The biogenesis of miRNAs is controlled by RNA polymerase II to generate primary microRNAs, which are modified by various proteins from the RNAse III family to produce shorter microRNAs. MicroRNAs are known to be ubiquitous, occurring in multiple tissue types, cells, and such biological fluids as blood, saliva, and tears, as well as breast milk. The molecules regulate many processes, including organogenesis, immune activity, and cellular metabolism, as well as cell differentiation [2]. Recently the presence of other small RNA species in body fluids has been reported, including circular RNAs (circRNAS), long non-coding RNAs, transfer RNAs (tRNAs), and small nuclear RNAs (snRNAs) [3,4]. The role of these non-coding RNA species in breast milk and other fluids remains poorly understood and is the subject of ongoing research on human breast milk. The current research on the subject has highlighted the role of breast milk microRNAs in infant biological system development. While more roles of microRNAs are still being discovered, the most established role is the regulation of specific genes involved in infant immune system development [4]. As a result of their expression in multiple body fluids and involvement in various biological processes, micro-RNAs have the potential to be used as non-invasive biomarkers for monitoring diseases in which their levels change.

Multiple studies have been published on breast milk characteristics and composition, including the occurrence of non-coding RNAs in animal and human breast milk [1,2,4]. These studies cover a wide range of subjects, from the concentration of specific RNA species to the function of these molecules in breast milk. The most commonly reported role of microRNAs in human breast milk is immune modulation. The potential for these molecules to play a role in the immune system development pathways in infants has been considered. Despite these recent advances, the extent and types of sources of evidence for the occurrence and role of microRNAs in human breast milk remain limited [4]. Therefore, a scoping review is necessary to map the scholarly studies on the subject to identify emerging evidence sources and novel methodological approaches to the study of the subject. The scoping review undertaken here aims to map the literature on this subject and highlight potential research gaps and future trends in its scholarly investigation. The following research questions were formulated to guide the scoping review:

1.   What microRNA subtypes occur in human breast milk and what are their functions?
2.   What conditions affect the quantities and types of microRNA in breast milk?

2. Methods

2.1. Protocol

While a formal review protocol was not published a priori in this study, the methodological approach was informed by the Preferred Reporting Items for Systematic Reviews and Meta-analysis Protocols (PRISMA-P), as well as scoping review guidelines from the Joanna Briggs Institute. The reviewer revised and adapted the protocol components to align with the review aims and based the report items on the Preferred Reporting Items for Systematic reviews and Meta-Analyses extension for Scoping Reviews (PRISMA-ScR) Checklist, as well as its elaboration by Tricco et al. [5].

2.2. Data Sources

A comprehensive electronic search of the following databases to identify sources of evidence on the subject was conducted by the reviewer: PubMed (Medline), The Cochrane Library, Web of Science, and Scopus. No additional sites were searched for unpublished sources or grey literature due to the focus on peer-reviewed, methodologically rigorous sources of evidence. The search strategy involved various combinations of MeSH terms and Boolean operators combining multiple formats and synonyms of two principal concepts: microRNA and breast milk. The specific keywords used to retrieve sources from the databases were microRNA, miRNA, human breast milk, breastfeeding, and lactation. each concept was expressed using multiple synonyms and abbreviations, which were operationalized using the Boolean operators OR and AND, to ensure that all relevant sources were retrieved from the target databases. This search strategy was slightly adapted to comply with the features offered by each database or optimized to give the widest range of results limited to sources relevant to the topic. The search string used in the PubMed database, for instance, was as follows:

(microRNA OR miRNA OR qRNA or lncRNA) AND (breastmilk OR breast milk OR human breast milk OR human breastmilk OR colostrum OR breastfeeding OR lactation).

2.3. Eligibility Criteria

As a scoping review, the study adopted inclusive eligibility criteria to allow the identification of the largest number of relevant sources of evidence on the role of microRNAs in human breast milk. The analysis included the detection, characterization, quantification, and functional analysis of microRNA components in human milk, which enabled retrieval of sources that aligned with the review's aim of identifying microRNA types, quantities, and functions. The exclusion criteria included papers published in languages other than English, a specification that was necessary to ensure the accuracy of the review by avoiding potential changes due to translation. In addition, narrative review articles, book chapters, and conference papers were excluded from the review to ensure that only high-level primary evidence sources were included.

2.4. Source Selection

The retrieved sources were screened by the reviewers in two phases. The first phase involved the removal of duplicates, which was performed using free online software, and then double-checked manually. After duplicate removal, the titles, abstracts and keywords of sources were screened to identify those that mentioned the two main concepts and to eliminate non-English sources. In the second phase, the abstracts of sources retrieved by title were screened to determine the relevance to the review topic and identify the methodology used, particularly whether microRNA analysis was performed. The screening of abstracts also helped the reviewers identify articles whose methodologies aligned with the aims and inclusion criteria of the review. Articles whose methodology was not clearly evident were screened by the reviewer using the Section 2. Finally, the reference sections of the final list of articles were assessed to identify potential sources that could be included in the review (See Figure 1).

2.5. Data Extraction and Charting

A standardized data abstraction chart was developed and used for extracting relevant information from the sources. The table included the most relevant information on each study, including the publication year, study location, aims, participants or sample sizes, and findings. These characteristics were also relevant to the subsequent synthesis of results from all included studies (see Table 1).

2.6. Synthesis of Results

The results from the included studies were presented in a narrative synthesis, outlining the key types of studies, analysis methods, types of non-coding RNAs in breast milk, functions of microRNAs in milk, as well as future trends. The results synthesis also included the categorization of the goals and a review of the methodological techniques used by the researchers.

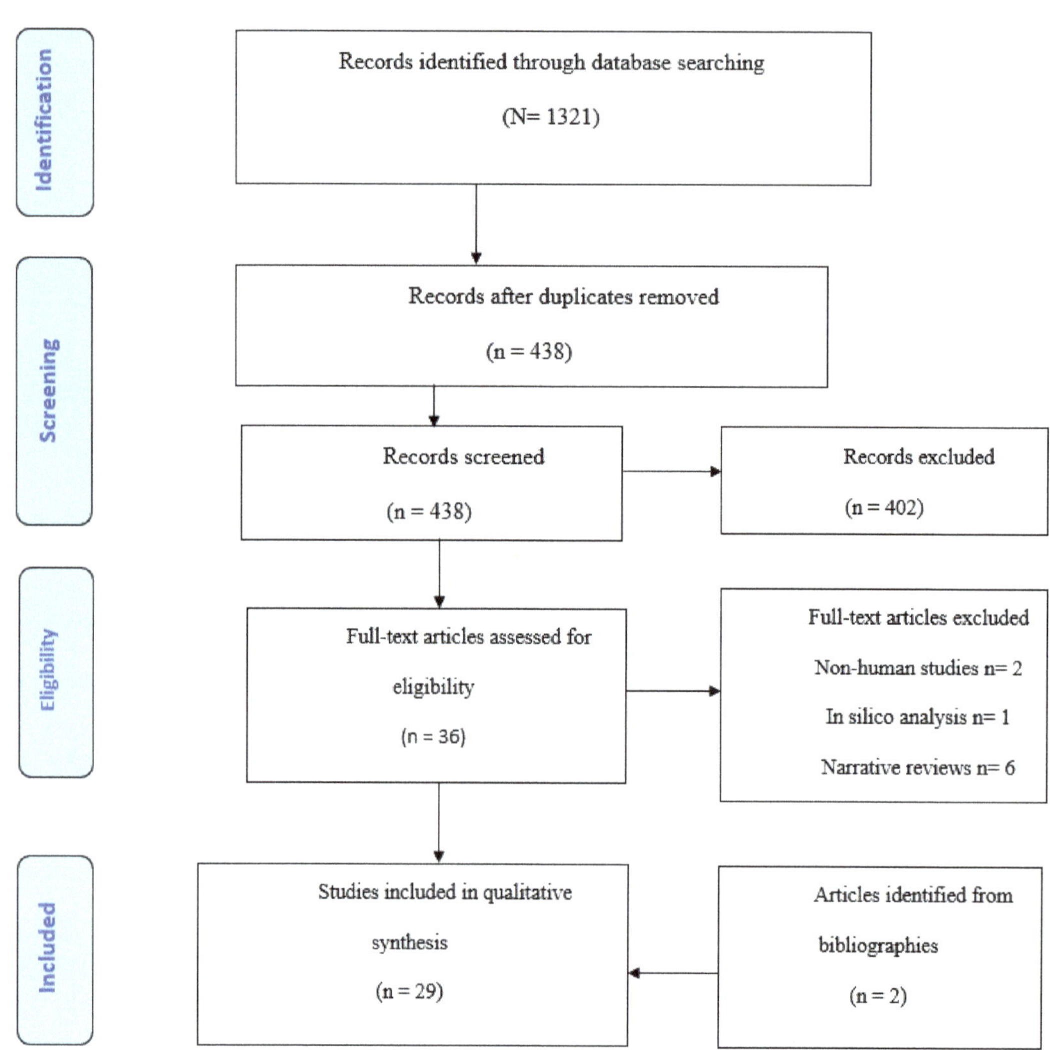

Figure 1. Flowchart: selection of sources.

Table 1. Summary of studies.

First Author, Year (Country)	Study Aims	Methods	Sample Size	Findings
1. microRNA EXTRACTION TECHNIQUES				
Ahlberg, 2021 (Sweden) [6]	To determine the efficiency of five RNA extraction kits in retrieving microRNA from human breast milk.	Five internationally available RNA extraction kits were used to retrieve the microRNA hsa-miR-39-3p and an exogenous control microRNA from four samples of skim milk. Total and specific RNA extracts were quantified using a Qubit microRNA assay kit, and performance was compared using statistical analysis in GraphPad Prism.	Five international kits were compared to determine the extraction efficiency: Promega, Zymo, Norgen RNA, Norgen Cell, and Sigma-Aldrich.	Two of the extraction kits, Promega and Zymo, had the highest yields of mRNA extracts. Variations in kit performance highlight the need for a standardized protocol for research on mRNA in milk.
Alsaweed, 2015 (Australia) [7]	To standardize microRNA isolation from three fractions of centrifuged human milk using different methods.	Human milk samples were fractionated, and microRNA content was extracted from each layer (lipid, skim milk and cells) using 8 commercially available kits.	49 milk samples from 29 lactating women.	All three layers of fractionated milk had microRNA content, with the cellular and lipid layers having the highest concentrations.
2. TYPES AND QUANTITIES OF ISOLATED microRNAS				
Alsaweed, 2016 (Australia) [8]	To determine the types of microRNAs present in human milk after different lactation periods.	Milk samples were collected at 2, 4, and 6 months of lactation. Bioinformatics analysis of microRNAs extracted from the lipid and cell layers was conducted to identify known and predicted microRNAs.	Breast milk was obtained from 10 lactating women	MicroRNA content in cell fractions was higher than in lipid fractions. The most common microRNA species from both columns was let-7f-5p. Across the three lactation stages, the profile (amount and species) of microRNA in breast milk changed significantly, suggesting adaptation to infant's needs.

Table 1. *Cont.*

First Author, Year (Country)	Study Aims	Methods	Sample Size	Findings
Bozack, 2020 (USA) [9]	To investigate the association between maternal psychological stress and the microRNA amount and quantity expressed in breast milk.	Psychological stress was assessed using Life Stressor Checklist-Revised, while microRNA profiling was performed using the TaqMan Open Array Human miRNA panel. Statistical analysis based on regression was used to determine association.	Breast milk was obtained from 80 lactating women	744 microRNAs were detected in total. The quantity of microRNA in breast milk was negatively correlated with psychological stress scores. The types of microRNA extracted also differed depending on psychological stress, with six species demonstrating the highest association with high stress scores.
Carney, 2017 (USA) [10]	To determine the correlation between gestational age at delivery and maternal milk microRNA profiles.	MicroRNA content from breast milk from mothers of infants born prematurely was compared with breast milk collected from mothers of term babies.	Colostrum and hind milk samples were obtained from the total of 44 mothers.	The quantities of nine microRNAs differed across the two groups. The affected microRNAs had target genes related to metabolism and lipid formation.
Floris, 2015 (France) [11]	To investigate changes in breast milk microRNA content in a 24-h period.	MicroRNA content from samples collected at different times were analyzed to determine expression of 4 reference species.	84 milk samples were collected at different times from 22 healthy mothers.	Two microRNA species, hsa-miR-16-5p, hsa-miR-21-5p, were stably expressed throughout the day, supporting their potential for use as endogenous reference species in future studies.
Munch, 2013 (USA) [12]	To determine the amount and types of microRNAs in human breast milk.	Deep sequence analysis of human breast milk to identify known and novel microRNAs.	5 lactating women	Generated sequences matched 308 known mature microRNAs targeting 9074 genes; and 21 novel microRNAs. A number of the novel microRNAs had origins from enriched foods.
Rubio, 2018 (Spain) [13]	To characterize the profiles of small RNA species in milk and plasma.	Isolation and quantitation of clusters of small RNAs from plasma and breast milk samples.	15 healthy postpartum mothers	Both milk and plasma expressed microRNAs, piwi-interfering RNAs (piRNAs), tRNAs, and small nucleolar RNAs. The concentrations of different small RNAs varied between the two biofluids.

Table 1. *Cont.*

First Author, Year (Country)	Study Aims	Methods	Sample Size	Findings
3. EFFECTS OF PROCESSING, TIMING, ENVIRONMENTAL, AND MATERNAL CONDITIONS				
Hicks, 2022 (USA) [14]	To identify the most abundant microRNAs in milk and determine their variation with time.	Deep sequence analysis to identify microRNAs in samples collected at different times.	503 samples obtained from 192 mothers	The quantity of most microRNA species increases with breast milk maturity, with nearly half being influenced by maternal diet.
Leiferman, 2019 (USA) [15]	To assess exosome and microRNA expression in human breast milk and commercial infant formulas.	RNA sequencing of fresh and stored breast milk and infant formulas	Breast milk was obtained from 5 lactating mothers.	Fresh human breast milk had 21 microRNAs stored in exosomes, while infant formulas had none. Storage at 4 °C resulted in a 49% decrease in the microRNA content in breast milk.
Raymond, 2021 (France) [16]	To characterize the microRNA profile of breast milk from week 2 to the second month post-partum	Next generation sequencing of breast milk samples collected at different times.	44 mothers	685 microRNAs were isolated, 35 of which were present in stable quantities throughout the lactation period.
Shiff, 2019 (Israel) [17]	To investigate the differences in the microRNA profiles of breast milk collected from mothers of premature and term infants.	Quantitative real-time PCR detection and comparison of four common microRNA species	38 healthy mothers	MicroRNA 320 occurred more commonly in colostrum of term infant mothers, while microRNA 148a occurred more commonly in premature human breast milk.
Shah, 2021 (USA) [18]	To examine the association between maternal obesity/overweight and expression of specific microRNAs in breast milk.	Real-time quantitative PCR for 6 specific microRNAs, followed by regression analysis to determine association with different maternal characteristics.	60 lactating mothers	Two of the six microRNAs, miR-148a and miR-30b, were lower in overweight mothers than in normal weight mothers.
Smyczynska, 2020 (Poland) [19]	To determine the effects of pasteurization on the microRNA content in human breast milk.	Milk samples were pasteurized using either Holder pasteurization or high-pressure processing, or left unpasteurized. Subsequently, they were analyzed for microRNA content using quantitative sequencing techniques.	Milk samples were collected from 3 volunteers	Pasteurization resulted in significant reduction in microRNA content in the breast milk samples.

Table 1. *Cont.*

First Author, Year (Country)	Study Aims	Methods	Sample Size	Findings
Wang, 2022 (USA) [20]	To assess the feasibility of microRNA analysis in frozen human breast milk.	Breast milk samples were frozen at -80° C, then analyzed for microRNAs and exosomes using PCR and immunoblot techniques.	Milk samples were obtained from three mothers of preterm infants.	While freezing resulted in significant reductions in microRNA and exosome levels, analysis was still feasible.
Wu, 2020 (China) [1]	To investigate the expression of microRNA in human colostrum.	Microarray analysis of colostrum samples and prediction of microRNA targets.	18 lactating volunteers	The microRNA composition of human milk changes through the lactation period. 49 microRNAs were consistently found in colostrum in higher concentrations than in milk.
4. ROLE OF microRNAS IN DISEASES				
Yang, 2020 (China) [21]	To determine whether microRNAs in breast milk contribute to breast milk jaundice.	Flow cytometry, Western blotting, and nanoparticle tracking were used to monitor exosomes in breast milk, and association analysis was used to find association with breast milk jaundice.	32 mother-infant dyads	Based on the expression of a few microRNA species, the evidence supports a role for microRNAs in breast milk jaundice.
Pisano, 2020 (USA) [22]	To determine whether extracellular vesicles (EVs) from human breast milk can protect against experimentally induced necrotizing (NEC) enterocolitis in mice.	Lactating mice were randomized to either breastfeeding without NEC, NEC with no breastfeeding, NEC with intraperitoneal EVs, and NEC with oral EVs. Histological analysis was done to determine the extent of NEC.	Not indicated	Breastfed rats had 0% NEC, while 62% of those with experimentally induced disease experienced it. Intraperitoneal EV reduced NEC to 29%, while oral administration reduced it to 11.9%.
Gu, 2014 (China) [23]	To determine the association between milk stasis and breast carcinogenesis and the potential role of microRNAs as biomarkers.	Milk samples from patients with milk stasis and patients with both stasis and breast cancer were analyzed for microRNA profiles.	A total of 20 patients	MicroRNA profiles differed between samples from patients with and without a neoplasm. In addition, microRNAs with roles in tumor suppression and oncogenesis had pro-neoplastic expression profiles.
Simpson, 2015 (Norway) [24]	To investigate the potential mediating role of breast milk microRNAs in the anti-dermatitis effects of probiotic supplements.	Small RNA sequencing for microRNAs in breast milk samples, followed by statistical analysis to determine correlations between expression profiles and probiotic effects.	54 lactating mothers	The biological significance of microRNAs in atopic dermatitis was not apparent.

Table 1. *Cont.*

First Author, Year (Country)	Study Aims	Methods	Sample Size	Findings
5. PHYSIOLOGICAL FUNCTIONS OF microRNAS				
Yun, 2021 (South Korea) [25]	To explore microRNA profiles of human, bovine, and caprine milk.	RNA sequencing of fractionated milk samples.	Milk samples were obtained from 2 human volunteers, 6 cows, and 3 goats.	Several dietary micro-RNAs are expressed in human, cow, and caprine milk. These species are mostly involved in immune regulation.
Zhou, 2012 (China) [26]	To investigate microRNAs in human breast milk.	Deep sequence analysis of samples for microRNA identification.	Breast milk was collected from 4 healthy mothers.	602 unique microRNAs were isolated from the milk samples. Resistance to harsh conditions suggests transfer to infants via the digestive tract.
Kahn, 2018 (USA) [27]	To determine the effect of in vitro digestion on microRNA from breast milk of mothers of preterm infants.	Exosomes from milk samples were isolated and lysed using commercial RNAses, exposed to in vitro digestive enzymes, and subsequently sequenced to determine surviving contents.	Milk samples were collected from 10 mothers.	MicroRNAs that are specific to preterm infant milk were not affected by in vitro digestion, suggesting increased susceptibility of preterm infants to maternal microRNA-driven epigenetic effects.
Na, 2015 (China) [28]	To screen for presence of 5 immune-related microRNAs in human, goat, and cow milk.	Real-time PCR of breast milk samples to identify MiR-146, miR-155, miR-181a, miR-223, and miR-150 species.	Milk samples were collected from 3 human volunteers, 2 lactating goats, and 3 dairy cows.	All five microRNA species were found in human and goat milk, and four were present in cow milk.
Karlsson, 2016 (USA) [29]	Screening for developmentally important lncRNAs in breast milk.	Quantitative analysis was used to determine the presence of specific non-coding RNAs in breast milk samples.	Breast milk was collected from 30 participants.	55 of the 87 non-coding RNAs screened for were found, demonstrating a possible role in childhood development.
Kosaka, 2010 (Japan) [2]	To determine the importance of breast milk microRNAs in infant immune regulation.	MicroRNA content of breast milk was screened for species known to play a role in immune function.	6 breastfeeding women provided milk samples	Breast milk has high levels of immune-regulating microRNAs in the first 6 months of lactation. These microRNAs are stable in acidic conditions, allowing for dietary intake.

Table 1. *Cont.*

First Author, Year (Country)	Study Aims	Methods	Sample Size	Findings
Kupsco, 2021 (USA) [30]	To characterize breast milk microRNA in population cohort and evaluate association with maternal characteristics.	Small RNA PCR sequencing of human milk samples to identify the most common microRNA species, followed by association analysis with several maternal characteristics.	Breast milk was obtained from 364 mothers	A total of 1523 microRNAs were detected in at least two samples. Three of the most common clusters demonstrated common ontogeny in pathways enriched by such systems as endocrine development and neurodevelopment. Association of microRNA expression with maternal BMI, time of milk collection, and smoking was demonstrated.
Lukasik, 2017 (Poland) [31]	To screen human breast milk for presence of specific plant-derived microRNAs.	Real-time quantitative reverse transcription PCR of human breast milk samples to detect 5 specific plant-derived microRNAs.	6 lactating mothers	Two of the 5 screened microRNAs were isolated in breast milk samples. These microRNAs are believed to be acquired from ingested plant-based foods, and to have a role in infant development.
Velez-Ixta, 2022 (Mexico) [3]	To assess the expression of immunoregulatory microRNAs in human breast milk.	Expression of 5 specific microRNAs was assessed using quantitative PCR in milk samples and infant formula.	60 lactating mothers	All five screened microRNAs (miR-146b-5p, miR148a-3p, miR155-5p, mir181a-5p, and mir200a-3p) were detected in human breast milk, but only very small concentrations of the species were found in infant formulas.

3. Results

3.1. Selection of Sources

The initial database search returned a total of 1321 results, which were filtered in the first phase of the selection strategy involving duplicate removal and exclusion of non-English entries, as well as those dealing with unrelated topics. In the second phase, a total of 36 sources underwent full-text analysis for relevance and methodological rigor, resulting in the exclusion of nine sources, six of which were narrative reviews, six studies on only non-human breast milk, and one in silico analysis. Screening of the reference sections led to the identification of two more sources, the inclusion of which resulted in a total of 29 articles included in the scoping review (Figure 1).

3.2. Study Characteristics

The studies included in this review were published between 2010 and 2022, the majority of which were conducted in 2021 (five studies) and 2022 (five studies). The most commonly investigated topic was the microRNA profile of human breast milk, using an explorative approach to determine what types, concentrations, and functional categories of microRNAs and other non-coding RNAs are found in human milk [8,11,12,14,30] (See Table 1). As shown in Table 1, the studies on the subject can be categorized into five groups based on the research focus. Studies that focused on microRNA extraction techniques highlighted the methodological approaches used in collecting and processing breast milk for microRNA analysis, isolating the exosomes, and quantifying the microRNA content. The methodologies used for microRNA quantification are an important aspect of the subject as they appear to account for a significant number of inconsistencies in findings. Importantly, as noted by Ahlberg (2021), standardization of quantification techniques is a priority for research on the subject, the success of which has implications for the potential future use of microRNAs as disease markers. The second category of studies focused on the types and quantities of microRNAs in milk. A review of findings from these studies confirmed that a large variety of microRNA subtypes occur in breast milk, with their quantities varying unpredictably and their role remaining poorly understood. The most commonly reported microRNAs, however, occur quite consistently, regardless of the processing techniques or time during the lactation period. These include miR-148a-3p, miR-30b-5p, and miR-200a. The third category of studies focused on how breast milk processing techniques, such as pasteurization, time during lactation, environmental conditions, including temperature, maternal characteristics, such as obesity, and infant maturity, affect the type and amounts of microRNAs isolated from milk. Here, the most significant results were that microRNA types and quantities vary during the lactation period, depend on the source of milk (breast milk versus infant formulas), the mother's weight, and on pasteurization, which reduces the microRNA content in breast milk. The fourth research category explored the role of microRNAs in disease, which was perhaps the least researched aspect of the subject. Nonetheless, the studies in this group showed that breast milk microRNAs play a role in the development of neonatal jaundice and necrotizing enterocolitis and serve as a potential biomarker in breast cancer. The fifth category of publications focused on the physiological functions of microRNAs in breast milk. It has long been apparent that these RNA species play an important role in the regulation of genetic development in infants. The most prominent role of microRNAs is the regulation of gene expression in the development of the immune system in infants, although the specific genes involved are poorly characterized. In addition, microRNAs in breast milk perform epigenetic modulation of certain genes involved in infant brain development, physical cellular development, and endocrine system development. Importantly, studies have demonstrated that microRNAs may be transferred to the infant through intestinal absorption due to the species' hardiness and ability to resist digestion (Table 1).

3.3. MicroRNA Isolation and Quantification

Various methodological techniques were applied in different studies depending on the aims. Techniques, such as centrifugation and fractionation, were described or referred to in most sources, as most studies focused on isolating microRNAs from milk samples. In addition, microRNA detection was conducted using two closely related approaches, real-time PCR [17,18,28,30,31] and deep sequence analysis [12,14,26]. A number of expression kits were described, but the TaqMan miRNA kit was the most popular reverse transcription kit used in the studies [6].

3.4. Types and Functions of microRNAs

While most researchers focused exclusively on human breast milk, a few performed comparative analyses of microRNA profiles in other species, including cows and goats [25,28]. The concentration and types of microRNAs isolated from breast milk varied widely, yet several species were identified as the most abundant. These included miR-148a-3p, miR-30b-5p, and miR-200a [15,23,26,31]. Similarly, researchers attributed various functions to the identified microRNAs, ranging from regulation of genes that control immune function [2,3,20,25,28] to epigenetic modulation of genes that determine physical health and infant neurodevelopment [29–31] (Table 1).

4. Discussion

MicroRNAs are small gene segments involved in the epigenetic regulation of multiple pathways. Their presence in several biofluids has been demonstrated, including in plasma and, as the abovementioned studies showed, in breast milk. Quantifying and characterizing microRNAs and other non-coding exons have attracted researchers' attention because of their potential application in the detection, prevention, and treatment of health disorders, including nutritional conditions and neoplasms. MicroRNAs in breast milk appear to have specific functions, most notably to modulate the development of the infant immune system. Breastmilk provides a unique and convenient pathway for microRNAs to be transferred to the infant, where they appear to be absorbed in the intestinal mucosa and where their epigenetic modifying effects are employed. Importantly, the available literature does not provide evidence on the specific cellular targets and the mechanisms involved in epigenetic modulation in relation to the immune system. Similarly, the role of microRNAs in neurodevelopment, which is relevant to the functional maturation of this system in newborns, has been suggested in the literature, while its exact mechanisms remain relatively unexplored.

The literature provides insights not only into the functional implication of microRNAs, but also highlights the variety of methodological approaches used to conduct the isolation and characterization of the molecules. The typical process of isolating breast milk microRNAs includes collecting and storing milk samples, centrifugation, fractionation, and selection of a target phase, as well as molecule isolation, using methods such as reverse transcription and quantification. Different types of equipment and methodological specifications are used to conduct each of these processes, but there is no standardized approach or choice of tools to ensure uniformity. Methodological approaches to the isolation and quantification of microRNAs are, however, somewhat limited. In the current studies considered, the segments are typically identified in a three-step process involving extraction through sample centrifugation and fraction, reverse transcription, and quantification [3,18,31]. Several methods are used for microRNA quantification, the most common being quantitative PCR [3,18,28,31]. Microarray techniques [1,7] and small RNA sequencing [12,14,16] enable the processing of large amounts of genetic material at the same time. The development of deep sequencing techniques has enhanced the sensitivity of gene-detection systems, enabling the identification of minute amounts of material in samples. The performance of each of these methods, and how it varies depending on the sample source and processing method, are important considerations in order to standardize methodological approaches to microRNA characterization. Currently, practical microRNA characterization relies on a combination of different methods and tools. Standardized identification and quantification are crucial to assign therapeutic target and pathologic marker roles to unique microRNA molecules.

An important aspect of the empirical research on this subject is to identify the specific microRNAs that consistently occur in breast milk and to quantify them in order to build a profile of normal breast milk microRNA content. Unfortunately, this goal is not easy to achieve since a large number of microRNA subtypes occur in breast milk and some of them are not as predictable as others. Various concentrations of microRNAs reported in different studies appear to be even more challenging, probably due to the use of differing methodologies. At this point, a wide range of microRNAs have been found in human breast milk. While different researchers typically isolate various microRNA species, a few segments are consistently present in breast milk, including miR-148a-3p, miR-30b-5p, and miR-200a [1,13,23,26].

Logically, the next step in the investigation is to determine the role of these commonly occurring microRNAs, which can be challenging. Determining the functional role of microRNAs in breast milk relies on the identification of known microRNAs in the biofluids to provide insight into the systems targeted by the microRNAs. As indicated by multiple researchers, the functions performed by microRNAs in breast milk are similarly heterogenous, with the best characterized role being the regulation of genes involved in infant immune system development [2,3,20,25,28]. In addition, researchers have suggested that microRNAs in breast milk are involved in regulating genes responsible for neurodevelopment [30], endocrine function, and protection from such diseases as necrotizing enterocolitis [22] and atopic dermatitis [24]. Interestingly, microRNAs have been found to cause some diseases too, including breast milk jaundice [21] and breast carcinoma [23]. While further research is needed to confirm the role of microRNAs in the development of these conditions, it is already known that these segments have the potential to be used as biomarkers and treatment targets in patients.

One area of research that needs additional investigation is the association between microRNA expression and various maternal and environmental characteristics. A significant number of studies have already been conducted to evaluate the impact of maternal diet, weight, BMI, post-partum period, age, and other factors on the amount and types of microRNA expressed in milk [9,10,14,18,25,30]. However, the findings from such investigations showed inconsistencies in factors found to impact microRNA expression. Similarly, the mechanism of transfer of microRNAs to infants needs to be further investigated; a growing number of studies suggest intestinal observation of unaltered segments due to high resistance to hostile conditions [2,19,26,27]. Overall, many studies have been conducted on breast milk microRNA, providing a considerable amount of evidence on previously unclear mechanisms. However, much research still needs to be undertaken before practically useful evidence is found. A notable limitation of the published studies was small sample sizes and the lack of a standardized approach to microRNA extraction and quantification.

5. Conclusions

The potential applications of microRNAs in preventive medicine and treatment have driven extensive research on their occurrence and role in various biofluids. The current scoping review has highlighted the identification of an increasing number of microRNAs in breast milk. Their isolation and comparison with known functional microRNAs have enabled researchers to speculate on what roles they may play in the infant and in the lactating mother. The best understood functional role of breastmilk microRNAs is the regulation of the development of the infant immune system, which is supported by the occurrence of specific types of microRNAs known to be involved in immune system development and regulation. By searching for specific microRNAs with known functions, researchers have also identified the role of the molecules in neurodevelopment. Interestingly, the expression of microRNAs in breast milk has been shown to be associated with certain diseases occurring in infants and mothers, including breast neoplasms and neonatal jaundice. While the evidence for this association is uncertain, it represents a potential first step towards the use of the molecules in disease monitoring. The methodological techniques applied in the isolation, characterization, and functional analysis of microRNAs in biofluids have important implications for the results obtained. Among the studies reviewed, various techniques were applied in the processing, fractionation, and quantification of microR-NAs, pointing to possible challenges in standardizing the detection and analysis of these molecules. Future research should be targeted at investigating the pathophysiological bases of the different functions of microRNAs in breast milk, as well as developing and adopting standard isolation and characterization methodologies.

Author Contributions: Conceptualization, A.K. (Adrianna Kondracka), B.K., P.G.-K., M.R.-K., J.S. and A.K. (Anna Kwaśniewska); methodology, A.K. (Adrianna Kondracka), B.K., K.F., A.O., J.K. and A.K. (Anna Kwaśniewska); software, B.K., J.K., K.F. and J.S.; validation, A.K. (Adrianna Kondracka), A.O., P.G.-K., M.R.-K. and A.K. (Anna Kwaśniewska); writing—original draft preparation, A.K. (Adrianna Kondracka), B.K. and J.K.; writing—review and editing, P.G.-K., K.F., A.O., M.R.-K., J.S. and A.K. (Anna Kwaśniewska); visualization, A.K. (Adrianna Kondracka), B.K.; supervision, A.K. (Anna Kwaśniewska) and B.K. All authors have read and agreed to the published version of the manuscript.

Funding: This research received no external funding.

Institutional Review Board Statement: Not applicable.

Informed Consent Statement: Not applicable.

Data Availability Statement: Not appliable.

Conflicts of Interest: The authors declare no conflict of interest.

References

1. Wu, F.; Zhi, X.; Xu, R.; Liang, Z.; Wang, F.; Li, X.; Li, Y.; Sun, B. Exploration of microrna profiles in human colostrum. *Ann. Transl. Med.* **2020**, *8*, 1170. [CrossRef] [PubMed]
2. Kosaka, N.; Izumi, H.; Sekine, K.; Ochiya, T. MicroRNA as a new immune-regulatory agent in Breast Milk. *Silence* **2010**, *1*, 7. [CrossRef]
3. Vélez-Ixta, J.M.; Benítez-Guerrero, T.; Aguilera-Hernández, A.; Martínez-Corona, H.; Corona-Cervantes, K.; Juárez-Castelán, C.J.; Rangel-Calvillo, M.N.; García-Mena, J. Detection and quantification of immunoregulatory mirnas in human milk and infant milk formula. *BioTech* **2022**, *11*, 11. [CrossRef]
4. Carrillo-Lozano, E.; Sebastián-Valles, F.; Knott-Torcal, C. Circulating micrornas in breast milk and their potential impact on the infant. *Nutrients* **2020**, *12*, 3066. [CrossRef]
5. Tricco, A.C.; Lillie, E.; Zarin, W.; O'Brien, K.K.; Colquhoun, H.; Levac, D.; Moher, D.; Peters, M.D.; Horsley, T.; Weeks, L.; et al. Prisma extension for scoping reviews (PRISMA-SCR): Checklist and explanation. *Ann. Intern. Med.* **2018**, *169*, 467–473. [CrossRef]
6. Ahlberg, E.; Jenmalm, M.C.; Tingö, L. Evaluation of five column-based isolation kits and their ability to extract MIRNA from human milk. *J. Cell. Mol. Med.* **2021**, *25*, 7973–7979. [CrossRef]
7. Alsaweed, M.; Hepworth, A.R.; Lefèvre, C.; Hartmann, P.E.; Geddes, D.T.; Hassiotou, F. Human milk microrna and total RNA differ depending on milk fractionation. *J. Cell. Biochem.* **2015**, *116*, 2397–2407. [CrossRef]
8. Alsaweed, M.; Lai, C.T.; Hartmann, P.E.; Geddes, D.T.; Kakulas, F. Human milk cells and lipids conserve numerous known and novel mirnas, some of which are differentially expressed during lactation. *PLoS ONE* **2016**, *11*, e0152610. [CrossRef]
9. Bozack, A.K.; Colicino, E.; Rodosthenous, R.; Bloomquist, T.R.; Baccarelli, A.A.; Wright, R.O.; Wright, R.J.; Lee, A.G. Associations between maternal lifetime stressors and negative events in pregnancy and breast milk-derived extracellular vesicle micrornas in the programming of Intergenerational Stress Mechanisms (prism) pregnancy cohort. *Epigenetics* **2020**, *16*, 389–404. [CrossRef]
10. Carney, M.C.; Tarasiuk, A.; DiAngelo, S.L.; Silveyra, P.; Podany, A.; Birch, L.L.; Paul, I.M.; Kelleher, S.; Hicks, S.D. Metabolism-related micrornas in maternal breast milk are influenced by premature delivery. *Pediatr. Res.* **2017**, *82*, 226–236. [CrossRef]
11. Floris, I.; Billard, H.; Boquien, C.Y.; Joram-Gauvard, E.; Simon, L.; Legrand, A.; Boscher, C.; Rozé, J.C.; Bolaños-Jiménez, F.; Kaeffer, B. MIRNA analysis by quantitative PCR in preterm human breast milk reveals daily fluctuations of HSA-Mir-16-5P. *PLoS ONE* **2015**, *10*, e0140488. [CrossRef] [PubMed]
12. Munch, E.M.; Harris, R.A.; Mohammad, M.; Benham, A.L.; Pejerrey, S.M.; Showalter, L.; Hu, M.; Shope, C.D.; Maningat, P.D.; Gunaratne, P.H. Transcriptome profiling of microrna by Next-Gen Deep Sequencing reveals known and novel Mirna species in the lipid fraction of human breast milk. *PLoS ONE* **2013**, *8*, e50564. [CrossRef] [PubMed]

13. Rubio, M.; Bustamante, M.; Hernandez-Ferrer, C.; Fernandez-Orth, D.; Pantano, L.; Sarria, Y.; Piqué-Borras, M.; Vellve, K.; Agramunt, S.; Carreras, R.; et al. Circulating mirnas, isomirs and small RNA clusters in human plasma and breast milk. *PLoS ONE* **2018**, *13*, e0193527. [CrossRef] [PubMed]
14. Hicks, S.D.; Confair, A.; Warren, K.; Chandran, D. Levels of breast milk micrornas and other non-coding RNAS are impacted by milk maturity and maternal diet. *Front. Immunol.* **2022**, *12*, 5754. [CrossRef]
15. Leiferman, A.; Shu, J.; Upadhyaya, B.; Cui, J.; Zempleni, J. Storage of extracellular vesicles in human milk, and MicroRNA profiles in human milk exosomes and infant formulas. *J. Pediatr. Gastroenterol. Nutr.* **2019**, *69*, 235–238. [CrossRef]
16. Raymond, F.; Lefebvre, G.; Texari, L.; Pruvost, S.; Metairon, S.; Cottenet, G.; Zollinger, A.; Mateescu, B.; Billeaud, C.; Picaud, J.C.; et al. Longitudinal human milk mirna composition over the first 3 mo of lactation in a cohort of healthy mothers delivering term infants. *J. Nutr.* **2021**, *152*, 94–106. [CrossRef]
17. Shiff, Y.E.; Reif, S.; Marom, R.; Shiff, K.; Reifen, R.; Golan-Gerstl, R. Mirna-320A is less expressed and MIRNA-148a more expressed in preterm human milk compared to term human milk. *J. Funct. Foods* **2019**, *57*, 68–74. [CrossRef]
18. Shah, K.B.; Chernausek, S.D.; Garman, L.D.; Pezant, N.P.; Plows, J.F.; Kharoud, H.K.; Demerath, E.W.; Fields, D.A. Human milk exosomal microrna: Associations with maternal overweight/obesity and infant body composition at 1 month of life. *Nutrients* **2021**, *13*, 1091. [CrossRef]
19. Smyczynska, U.; Bartlomiejczyk, M.A.; Stanczak, M.M.; Sztromwasser, P.; Wesolowska, A.; Barbarska, O.; Pawlikowska, E.; Fendler, W. Impact of processing method on donated human breast milk microrna content. *PLoS ONE* **2020**, *15*, e0236126. [CrossRef]
20. Wang, H.; Wu, D.; Sukreet, S.; Delaney, A.; Belfort, M.B.; Zempleni, J. Quantitation of exosomes and their MicroRNA cargos in Frozen Human Milk. *JPGN Rep.* **2022**, *3*, e172. [CrossRef]
21. Yang, L.; Hu, R.; Li, J.; Mo, X.; Xu, L.; Shen, N.; Sheng, W.; Li, Y. Exosomal micrornas in human breast milk: Potential effect on neonatal breast milk jaundice. *Res. Sq.* **2020**, PPR223039.
22. Pisano, C.; Galley, J.; Elbahrawy, M.; Wang, Y.; Farrell, A.; Brigstock, D.; Besner, G.E. Human breast milk-derived extracellular vesicles in the protection against experimental necrotizing enterocolitis. *J. Pediatr. Surg.* **2020**, *55*, 54–58. [CrossRef] [PubMed]
23. Gu, Y.Q.; Gong, G.; Xu, Z.L.; Wang, L.Y.; Fang, M.L.; Zhou, H.; Xing, H.; Wang, K.R.; Sun, L. MIRNA profiling reveals a potential role of milk stasis in breast carcinogenesis. *Int. J. Mol. Med.* **2014**, *33*, 1243–1249. [CrossRef] [PubMed]
24. Simpson, M.R.; Brede, G.; Johansen, J.; Johnsen, R.; Storrø, O.; Sætrom, P.; Øien, T. Human breast milk mirna, maternal probiotic supplementation and atopic dermatitis in offspring. *PLoS ONE* **2015**, *10*, e0143496. [CrossRef] [PubMed]
25. Yun, B.; Kim, Y.; Park, D.J.; Oh, S. Comparative analysis of dietary exosome-derived micrornas from human, bovine and Caprine Colostrum and mature milk. *J. Anim. Sci. Technol.* **2021**, *63*, 593–602. [CrossRef]
26. Zhou, Q.; Li, M.; Wang, X.; Li, Q.; Wang, T.; Zhu, Q.; Zhou, X.; Wang, X.; Gao, X.; Li, X. Immune-related micrornas are abundant in breast milk exosomes. *Int. J. Biol. Sci.* **2012**, *8*, 118–123. [CrossRef]
27. Kahn, S.; Liao, Y.; Du, X.; Xu, W.; Li, J.; Lönnerdal, B. Exosomal MicroRNAs in milk from mothers delivering preterm infants survive in vitro digestion and are taken up by human intestinal cells. *Mol. Nutr. Food Res.* **2018**, *62*, 1701050. [CrossRef]
28. Na, R.S.; GX, E.; Sun, W.; Sun, X.W.; Qiu, X.Y.; Chen, L.P.; Huang, Y.F. Expressional analysis of immune-related mirnas in breast milk. *Genet. Mol. Res.* **2015**, *14*, 11371–11376. [CrossRef]
29. Karlsson, O.; Rodosthenous, R.S.; Jara, C.; Brennan, K.J.; Wright, R.O.; Baccarelli, A.A.; Wright, R.J. Detection of long non-coding RNAS in human breastmilk extracellular vesicles: Implications for early child development. *Epigenetics* **2016**, *11*, 721–729. [CrossRef]
30. Kupsco, A.; Prada, D.; Valvi, D.; Hu, L.; Petersen, M.S.; Coull, B.; Grandjean, P.; Weihe, P.; Baccarelli, A.A. Human milk extracellular vesicle MIRNA expression and associations with maternal characteristics in a population-based cohort from the Faroe Islands. *Sci. Rep.* **2021**, *11*, 5840. [CrossRef]
31. Lukasik, A.; Brzozowska, I.; Zielenkiewicz, U.; Zielenkiewicz, P. Detection of plant mirnas abundance in human breast milk. *Int. J. Mol. Sci.* **2017**, *19*, 37. [CrossRef] [PubMed]

biomedicines

Article

Dietary Improvement during Lactation Normalizes miR-26a, miR-222 and miR-484 Levels in the Mammary Gland, but Not in Milk, of Diet-Induced Obese Rats

Catalina A. Pomar [1,2,3], Pedro Castillo [1,2,3], Andreu Palou [1,2,3], Mariona Palou [1,2,3,*] and Catalina Picó [1,2,3]

[1] Laboratory of Molecular Biology, Nutrition and Biotechnology, Group of Nutrigenomics, Biomarkers and Risk Evaluation, University of the Balearic Islands, 07122 Palma, Spain; c.pomar@uib.es (C.A.P.); pedro.castillo@uib.es (P.C.); andreu.palou@uib.es (A.P.); cati.pico@uib.es (C.P.)
[2] Health Research Institute of the Balearic Islands (IdISBa), 07010 Palma, Spain
[3] CIBER de Fisiopatología de la Obesidad y Nutrición, Instituto de Salud Carlos III, 28029 Madrid, Spain
* Correspondence: mariona.palou@uib.es; Tel.: +34-971172373

Citation: Pomar, C.A.; Castillo, P.; Palou, A.; Palou, M.; Picó, C. Dietary Improvement during Lactation Normalizes miR-26a, miR-222 and miR-484 Levels in the Mammary Gland, but Not in Milk, of Diet-Induced Obese Rats. *Biomedicines* **2022**, *10*, 1292. https://doi.org/10.3390/biomedicines10061292

Academic Editors: Milena Rizzo and Elena Levantini

Received: 8 May 2022
Accepted: 24 May 2022
Published: 31 May 2022

Publisher's Note: MDPI stays neutral with regard to jurisdictional claims in published maps and institutional affiliations.

Abstract: We aimed to evaluate in rats whether the levels of specific miRNA are altered in the mammary gland (MG) and milk of diet-induced obese dams, and whether improving maternal nutrition during lactation attenuates such alterations. Dams fed with a standard diet (SD) (control group), with a Western diet (WD) prior to and during gestation and lactation (WD group), or with WD prior to and during gestation but moved to SD during lactation (Rev group) were followed. The WD group showed higher miR-26a, miR-222 and miR-484 levels than the controls in the MG, but the miRNA profile in Rev animals was not different from those of the controls. The WD group also displayed higher miR-125a levels than the Rev group. Dams of the WD group, but not the Rev group, displayed lower mRNA expression levels of *Rb1* (miR-26a's target) and *Elovl6* (miR-125a's target) than the controls in the MG. The WD group also presented lower expression of *Insig1* (miR-26a's target) and *Cxcr4* (miR-222's target) than the Rev group. However, both WD and Rev animals displayed lower expression of *Vegfa* (miR-484's target) than the controls. WD animals also showed greater miR-26a, miR-125a and miR-222 levels in the milk than the controls, but no differences were found between the WD and Rev groups. Thus, implementation of a healthy diet during lactation normalizes the expression levels of specific miRNAs and some target genes in the MG of diet-induced obese dams but not in milk.

Keywords: miRNA; mammary gland; lactation; obesogenic diet

1. Introduction

Breastfeeding in mammals has evolved over millions of years to provide more than what is traditionally considered nutrition. In fact, breast milk contains not only macronutrients, minerals and vitamins, which can provide the energy and factors necessary for growth and development, but it also provides a large variety of bioactive compounds [1]. Such compounds mainly include immune factors, hormones and non-coding small RNAs—such as microRNAs (miRNAs)—that can regulate cellular signaling pathways, affect organ development and protect against infection and inflammation, among other functions, with subsequent health outcomes [2–4]. Regarding miRNAs, evidence suggests that breast milk is a rich source of these compounds [5], which are produced and secreted into milk primarily through endogenous synthesis in the mammary epithelium, with minor contributions from the maternal blood circulation [6]. Interestingly, specific miRNAs in milk are modulated by maternal diet and maternal weight [7–9] and, once consumed, they may exert an effect on the target mRNAs in specific tissues of the descendants [3,10,11]. For example, there is indirect evidence that specific miRNAs in human breast milk affected by maternal overweight/obesity may influence infants' body mass index [8]. Moreover, certain miRNAs that are differentially expressed in breast milk from mothers with Type

1 diabetes, compared with healthy mothers, are associated with immunomodulatory effects [12], so that such alterations may affect the development of the immune system in the infants. Furthermore, alterations in the miRNA profile associated with maternal diet and/or status might also be expected to have autocrine or paracrine effects on maternal mammary function itself.

Previous studies have pointed out the importance of miRNA expression dynamics in mammary gland (MG) development and lactation [13,14]. In this sense, it has been described that the development of the MG can be affected by maternal diet. Lactating animals fed with an obesogenic diet showed an impaired lactogenesis and abnormal mammary morphology, leading to an inflammatory process [15,16]. However, the specific mechanisms involved are unclear. We have previously described how rats fed on an obesogenic diet during lactation displayed lower miR-26a levels in milk compared with dams fed a standard diet, which was in accordance with the lower expression of this specific miRNA in the MG at the end of lactation [17,18]. Therefore, alterations in the miRNA profile in the MG may be detected in the milk and reflect maternal diet and/or maternal physiological state. Thus, considering the potential impact of miRNAs on infant development, the study of how maternal nutrition and/or obesity states could affect miRNA expression in the MG and, consequently, the miRNA profile in milk, and whether such alterations may be prevented, is of great interest in the field of infant nutrition.

We have previously shown that maternal intake of an obesogenic diet during lactation in diet-induced obese rats, rather than maternal excess of adiposity (in the absence of dietary alterations during lactation), appears to be the main contributor to alterations in the macronutrient content and the levels of metabolic hormones in milk [19]. Interestingly, we have also shown that dietary normalization during lactation in diet-induced obese dams attenuates the early detrimental programming effects in the offspring [19]. Therefore, in the present study, we have analyzed a set of miRNAs (miR-26a, miR-27a, miR-30d, miR-99b, miR-125a, miR-181a, miR-200a, miR-200b, miR-222, miR-320a, miR-331 and miR-484), which have been described as being dysregulated in certain metabolic disorders [20–29], in the MG and milk of the aforementioned animal models. These consisted of diet-induced obese dams that continued on an obesogenic diet during lactation or had their diet improved during this period. We hypothesized that normalizing the diet during lactation in obese rats could potentially prevent changes in MG function and restore the miRNA profile in milk caused by the consumption of an obesogenic diet during these critical periods: gestation and lactation.

2. Materials and Methods

2.1. Study Design and Sample Collection

The study was performed on samples from a previously published study [19]. The animal protocol was approved by the Bioethical Committee of the University of the Balearic Islands (Exp. 2018/13/AEXP, 23 January 2019). Virgin female Wistar rats, housed at a controlled temperature (22 °C), with a 12-h light–dark period and with free access to food and water, were fed with a standard chow diet (SD; 3.3 kcal·g^{-1}, with 8.4% calories from fat, 72.4% from carbohydrates and 19.3% from protein; Safe, Augy, France) (control group), or a high-fat and high-sucrose diet (Western diet, WD; 4.7 kcal·g^{-1}, with 40.0% calories from fat, 43.0% from carbohydrates and 17.0% from proteins; Research Diets, New Brunswick, NJ, USA) for one month prior to being bred with male rats (this period was referred as pre-gestation). Pregnant dams were housed individually and continued with their assigned diets during the gestation period. On postnatal Day 1, the litters were equated to 10 pups per dam, 5 males and 5 females when possible. Throughout lactation, the dams of the control group continued with the SD ($n = 8$), but those fed with the WD either continued with the WD diet (Western diet group: WD group, $n = 9$) or were exposed to the SD during this period (reversion group: Rev group; $n = 10$). At three time points of lactation (Days 5, 10 and 15), milk samples from the dams were collected. For milk extraction, dams were separated from their pups for 2–3 h, then, 4 IU kg^{-1} of body weight of oxytocin (Facilpart, Laboratory Syva S.A.U, León, España) was administered intraperitoneally, and the dams

Table 1. Body weight and body fat content of the control, Western diet and reversion groups prior to gestation, on Day 17 of gestation and at the end of the lactation period. Data are expressed as the mean ± SEM of 8–10 animals per group. Statistics: MD, effect of maternal diet (one-way ANOVA). Least significant difference (LSD) post-hoc test, a ≠ b. These results were previously published in [19].

Phenotypic Traits		Control	Western Diet	Reversion	ANOVA
	Body weight (g)	223 ± 9	253 ± 7	252 ± 11	-
Prior to gestation	Body fat (g)	29.1 ± 2.1 a	51.5 ± 4.7 b	52.9 ± 6.6 b	MD
	Body fat (%)	12.4 ± 0.8 a	20.0 ± 1.3 b	20.4 ± 1.7 b	MD
Day 17 of gestation	Body weight (g)	309 ± 9	322 ± 8	324 ± 14	-
	Body weight (g)	304 ± 9 a	283 ± 5 b	311 ± 7 a	MD
Day 21 of lactation	Body fat (g)	29.5 ± 2.4 a	32.0 ± 2.2 a	41.3 ± 3.9 b	MD
	Body fat (%)	9.7 ± 0.8 a	11.3 ± 0.7 ab	13.2 ± 1 b	MD

3.2. miRNA Profile in the Mammary Gland

The levels of selected miRNAs (miR-26a, miR-27a, miR-30d, miR-99b, miR-125a, miR-181a, miR-200a, miR-200b, miR-320a, miR-331 and miR-484) in the MG of dams at weaning were analyzed (Figure 1A). Dams of the WD group displayed higher expression levels of miR-26a and miR-222 in MG compared with the control (29% and 47%, respectively) and Rev (23% and 28%, respectively) groups (one-way ANOVA, post-hoc analysis). miR-484 expression levels in the MG of WD animals were increased compared with the controls (87%), whereas the expression levels in Rev rats were not different from those in the control and WD groups (one-way ANOVA, post-hoc analysis). In addition, the WD group displayed greater miR-125a expression in MG compared with the Rev group (69%) but not compared with the controls (one-way ANOVA, post-hoc analysis). Correlation analyses between differentially expressed miRNAs in the MG that were significant (Pearson's correlation) are shown in Figure 1B. miR-26a, miR-222 and miR-484 levels were positively correlated with each other (miR-222 and miR-26a (r = 0.552, *p* = 0.003), miR-222 and miR-484 (r = 0.616, *p* = 0.001), miR-484 and miR-26a (r = 0.530, *p* = 0.004), and miR-125a levels correlated positively with miR-222 levels (r = 0.587, *p* = 0.002).

3.3. Expression of Validated Target Genes in the Mammary Gland

Next, we examined whether the expression of a subset of target genes for the altered miRNAs was affected in the MG. Specifically, the gene expression of insulin induced gene 1 (*Insig1*), phosphatase and tensin homolog (*Pten*) and RB transcriptional corepressor 1 (*Rb1*) (targets of miR-26a [17,34]); ELOVL fatty acid elongase 6 (*Elovl6*) and signal transducer and activator of transcription 3 (*Stat3*) (targets of miR-125a [35,36]); cyclin-dependent kinase inhibitor 1B (*Cdkn1b*) and C-X-C motif chemokine receptor 4 (*Cxcr4*) (targets of miR-222 [37,38]) and vascular endothelial growth factor A (*Vegfa*) (a target of miR-484 [39]) were selected, as they have been previously validated as target genes (Figure 2). The WD group but not the Rev group displayed lower mRNA expression levels of *Rb1* (27.1%) and *Elovl6* (71.9%) than the controls (one-way ANOVA, post-hoc analysis). In addition, WD animals displayed lower mRNA expression levels of *Insig1* (57%) and *Cxcr4* (47%) than the Rev group (Mann–Whitney U test). Interestingly, both WD and Rev animals displayed lower mRNA expression levels of *Vegfa* than the controls (39% and 42%, respectively) (one-way ANOVA, post-hoc analysis).

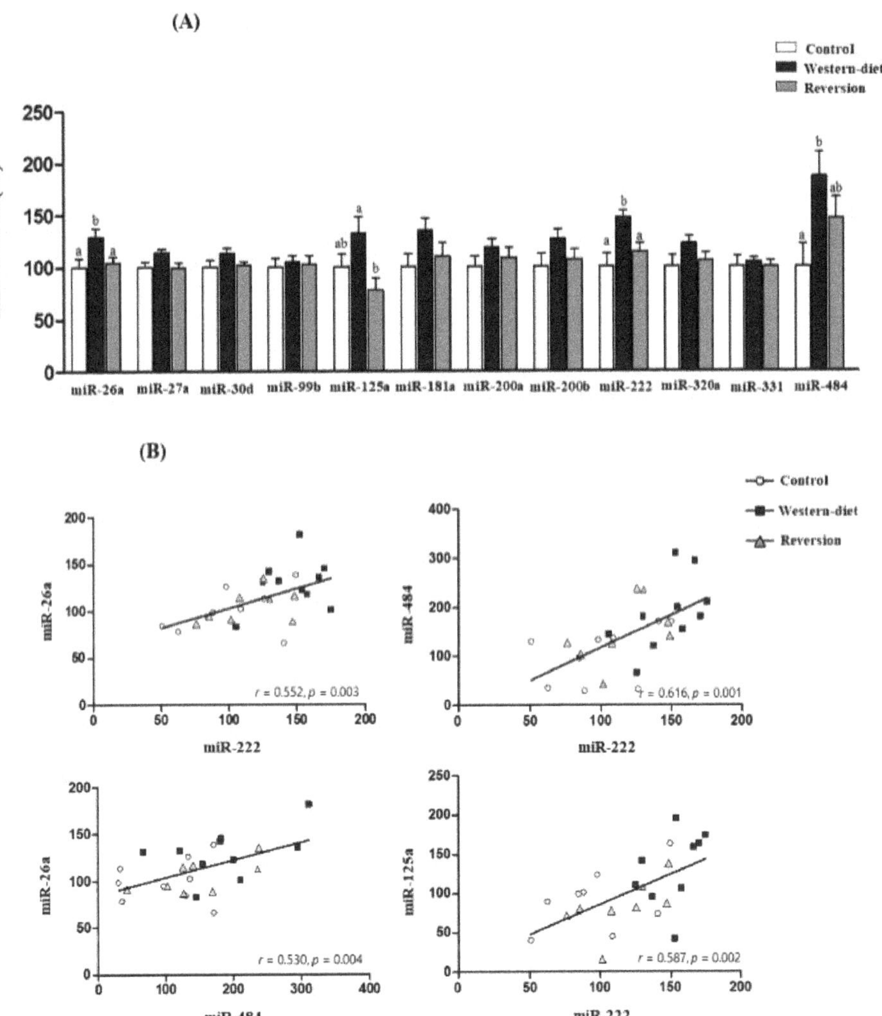

Figure 1. (A) miRNA levels in the mammary gland on Day 21 of lactation in control, Western diet and reversion animals. **(B)** Correlations between altered miRNAs in the mammary gland. The selected miRNA levels (miR-26a, miR-27a, miR-30d, miR-99b, miR-125a, miR-181a, miR-200a, miR-200b, miR-222, miR-320a, miR-331 and miR-484) were measured by RT-qPCR. Values are presented as percentages versus the levels of the controls. Data are expressed as the mean $\pm$ SEM of 8–10 animals per group. Statistics: Least significant difference (LSD) post-hoc test, a $\neq$ b (effect of maternal diet, one-way ANOVA). The correlations were assessed by Pearson's correlation (2-tailed). Pearson's correlation coefficients (r) and *p*-values are indicated.

3.4. *In Silico Prediction of Altered Pathways in the Mammary Gland*

To gain insight into the potential role of those miRNAs (miR-26a, miR-222 and miR-484), whose levels were altered in the MG of dams of the WD group but not of the Rev group, and to ascertain in which pathways could be involved, putative target genes were searched with the online algorithm for miRNA target prediction, TargetScan. The resulting list of 1699 predicted target genes recognized by ClueGO were used to assess the altered WikiPathways (Supplementary Table S2). The pathway analysis identified eight significantly altered pathways with *p*-values of <0.0005 (Table 2), namely the VEGFA-VEGFR2 signaling pathway (WP:3888), endoderm differentiation (WP:2853), the brain-derived neurotrophic factor (BDNF) signaling pathway (WP:2380), insulin signaling (WP:481), the mesodermal commitment pathway (WP:2857), ErbB signaling (WP:673), melanoma (WP:4685) and hematopoietic stem cell gene regulation by the GABP alpha/beta complex (WP:3657). It is remarkable that 74 target genes were related to the VEGFA-VEGFR2 signaling pathway (16.9% of the associated genes), 38 were related to endoderm differentiation (26.0% of the associated genes), 36 were related to the brain-derived neurotrophic factor (BDNF) signaling pathway (25.0 % of the associated genes) and 36 were related to the insulin signaling pathway (22.4% of the associated genes). These results suggest the potential dysregulation of the abovementioned pathways due to miRNA dysregulation in WD animals, but not in Rev, compared with the controls.

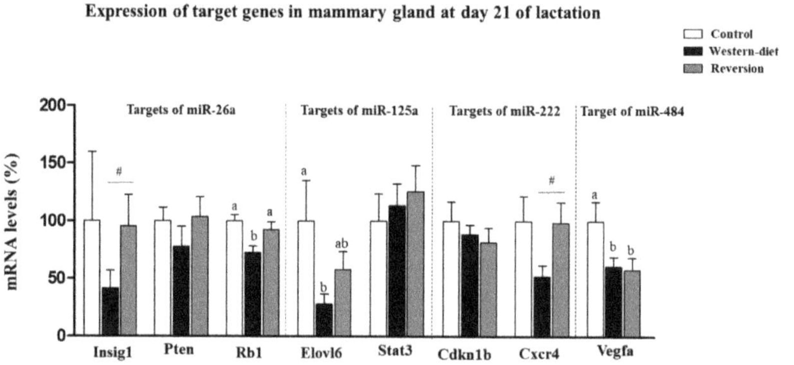

Expression of target genes in mammary gland at day 21 of lactation

Figure 2. Expression of target genes of altered miRNAs (miR-26a, miR-125a, miR-222 and miR-484) in the mammary gland on Day 21 of lactation in control, Western diet and reversion dams. mRNA levels were measured by RT-qPCR. The genes determined were insulin induced gene 1 (*Insig1*), phosphatase and tensin homolog (*Pten*), RB transcriptional corepressor 1 (*Rb1*), ELOVL fatty acid elongase 6 (*Elovl6*), signal transducer and activator of transcription 3 (*Stat3*), cyclin-dependent kinase inhibitor 1B (*Cdkn1b*), C-X-C motif chemokine receptor 4 (*Cxcr4*) and vascular endothelial growth factor A (*Vegfa*). Values are presented as percentages versus the levels of the controls. Data are expressed as the mean ± SEM of 8–10 animals per group. Statistics: Least significant difference (LSD) post-hoc test, a ≠ b (effect of maternal diet, one-way ANOVA). #, Western diet versus reversion dams (*p* < 0.05, Mann–Whitney U-test).

Table 2. Altered WikiPathways based on the predicted target genes of miR-26a, miR-222 and miR-484. Detailed information on the GO terms, *p*-values after Bonferroni's step-down correction and their associated genes (% and number) are indicated. The analysis was conducted using the ClueGo+Clupedia app in Cytoscape [32,33].

GO ID	GO Term	Corrected *p*-Value	% Associated Genes	No. Genes
WP:3888	VEGFA-VEGFR2 signaling pathway	0.0002	16.9	74
WP:2853	Endoderm differentiation	<0.0001	26.0	38
WP:2380	Brain-derived neurotrophic factor (BDNF) signaling pathway	<0.0001	25.0	36
WP:481	Insulin signaling	0.0003	22.4	36
WP:2857	Mesodermal commitment pathway	0.0004	22.3	35
WP:673	ErbB signaling pathway	0.0001	28.6	26
WP:4685	Melanoma	0.0002	30.9	21
WP:3657	Hematopoietic stem cell gene regulation by the GABP alpha/beta Complex	0.0004	50.0	11

3.5. miRNA in Milk at Different Time Points of Lactation

In view of the differences among the groups found in the levels of certain miRNAs in the MG and considering that miRNA from the MG can be released into the milk, the levels of the altered miRNAs were analyzed in milk at three time points of lactation (Days 5, 10 and 15) (Figure 3). Regarding miR-26a, its levels were higher in milk on Days 10 and 15 compared with the levels on Day 5 of lactation (repeated-measures ANOVA, post-hoc analysis). On Day 10, both the WD and Rev groups showed higher levels of miR-26a in the milk than controls (one-way ANOVA, post-hoc analysis). miR-125a levels increased in WD and Rev animals during lactation, but not in the controls (repeated-measures ANOVA, post-hoc analysis). On Day 15, both WD and Rev animals displayed greater levels of miR-125a in milk than the controls (one-way ANOVA, post-hoc analysis). miR-222 levels in milk increased progressively in all dams (repeated-measures ANOVA, post-hoc analysis). On Day 15 of lactation, WD animals displayed greater levels in milk than the controls (Mann–Whitney U-test). However, on Day 5, dams of the Rev group showed greater miR-222 levels than the controls, while levels in the WD group were not different from those of the control and Rev groups (one-way ANOVA, post-hoc analysis). Regarding miR-484, a different pattern was observed in the milk of the WD and Rev groups compared with the controls. miR-484 levels in the milk increased on Day 10 in all groups, compared with the levels on Day 5, and decreased on Day 15 in the controls but not in the WD and Rev groups (repeated-measures ANOVA, post-hoc analysis). On Day 15, Rev rats displayed greater miR-484 levels in milk, while the levels in WD animals were not different from those of the control and Rev groups (one-way ANOVA, post-hoc analysis).

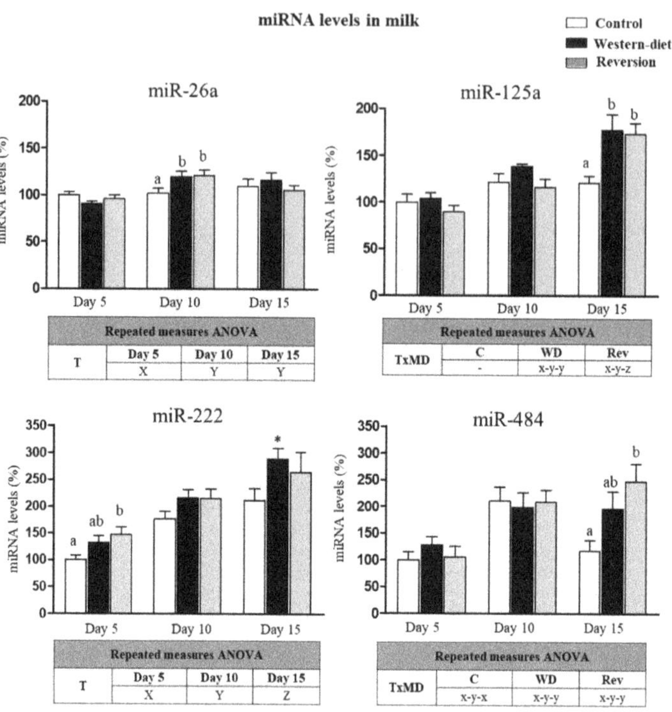

Figure 3. miRNA levels in milk on Days 5, 10 and 15 of lactation in the control, Western diet and reversion groups. Milk samples were collected at different time points of lactation (Days 5, 10 and 15). The selected miRNA levels (miR-26a, miR-125a, miR-222 and miR-484) were measured by RT-qPCR. Values are presented as percentages versus the levels of the controls on Day 5. Data are expressed as the mean $\pm$ SEM of 8–10 animals per group. Statistics: T, effect of time; TxMD, interactive effect between time and maternal diet (repeated-measures ANOVA). Least significant difference (LSD)post-hoc test, $X \neq Y \neq Z$ (effect of time, repeated-measures ANOVA), $x \neq y \neq z$ (effect of time, one-way ANOVA) and $a \neq b$ (effect of maternal diet, one-way ANOVA). *, Western diet versus control dams ($p < 0.05$, Mann–Whitney U-test).

The correlations among the miRNAs analyzed in milk that were significant (Pearson's correlation) are shown in Figure 4. miR-222 correlated positively with miR-26a ($r = 0.303$, $p = 0.006$), miR-125a ($r = 0.460$, $p < 0.001$) and miR-484 ($r = 0.472$, $p < 0.001$). miR-125a also correlated positively with miR-26a ($r = 0.329$, $p = 0.003$) and miR-484 ($r = 0.472$, $p < 0.001$).

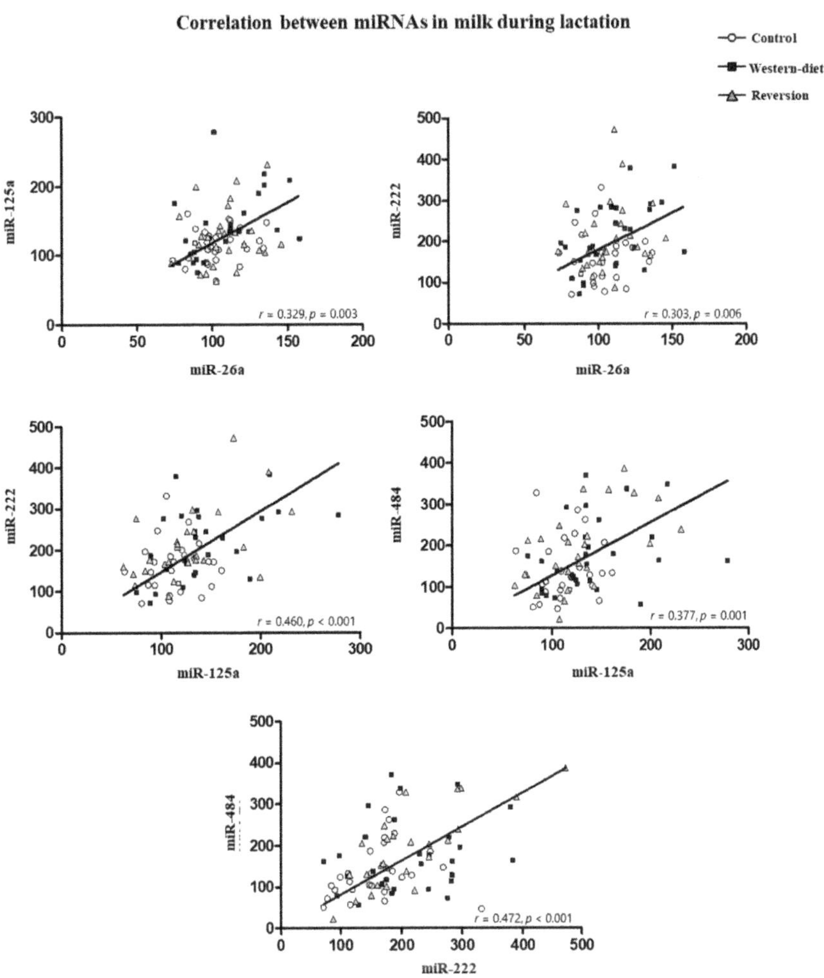

Figure 4. Correlation between miRNAs in milk during lactation. The correlations were assessed by Pearson's correlation (2-tailed). Pearson's correlation coefficients (*r*) and *p*-values are indicated.

4. Discussion

MG development takes place from the fetal stage and continues during critical periods of life, such as pregnancy and lactation [40]. The intake of an obesogenic diet during lactation and/or maternal obesity have been shown to affect the development and function of the MG [15,41], in addition to having an impact on the metabolic health of the offspring [42]. However, the underlying mechanisms and whether such alterations can be prevented by improving maternal diet during lactation have not been fully elucidated. Considering that miRNA production by the MG may be modulated by maternal conditions [7–9], and that they may exert an effect on target mRNAs in the offspring [17,18], the present study aimed to investigate potential alterations in the levels of selected miRNAs in the MG and in the milk of diet-induced obese rats, and whether such changes could be prevented by dietary improvements during lactation.

We show here that obese rats on an obesogenic diet during lactation exhibited, at the end of lactation, greater expression levels of miR-26a, miR-222 and miR-484 in MG, but dietary improvement during lactation normalized their levels to those of the controls, despite these animals maintaining excess adiposity. It is also remarkable that these miRNAs were mutually correlated. Regarding miR-26a, its role in the MG has hardly been explored, while increased or decreased levels in different tissues have been linked to states of obesity and/or diabetes. Specifically, overweight humans and obese mouse models have been shown to exhibit decreased expression levels of miR-26a in the liver compared with lean individuals [43]. In fact, the silencing of endogenous miR-26a impairs insulin sensitivity, and enhances glucose production and fatty acid synthesis [43], whereas overexpression of miR-26a in mice fed a high-fat diet prevents obesity-induced metabolic complications, improves insulin sensitivity and decreases hepatic glucose production and fatty acid synthesis [43]. In contrast, mothers with Type 1 diabetes during lactation [44] and children with newly diagnosed Type 1 diabetes [28] display greater plasma miR-26a levels in comparison with their controls. Moreover, plasma miR-26a levels were positively correlated with glycated hemoglobin in children with Type 1 diabetes at diagnosis [45]. Here, we show that obese dams on an obesogenic diet (prior to mating and during gestation and lactation) (the WD group) displayed, at the end of lactation, greater miR-26a levels in the MG than the controls, and also in milk on Day 10. Of note, dietary improvements during lactation in obese rats (Rev group) normalized miR-26a levels in the MG but not their levels in milk. Unlike the present results, we previously described that nursing rats fed a cafeteria diet only during lactation displayed lower expression levels of miR-26a in the MG at the end of lactation compared with control dams, together with lower miR-26a levels in milk on Day 15 of lactation and no changes on Day 10 of lactation [17,18]. The differences in the results obtained regarding miR-26a levels in the milk between both studies could be attributed to the differences in the dietary intervention, particularly in terms of the period and type of intervention. In the present study, we evaluated changes in diet-induced obese rats, with or without Western diet exposure during lactation, whereas in the previous study [18], the cafeteria diet was offered only during lactation. Further research is needed to explain such differences among animal models, but these results suggest that the expression of miR-26a in the MG and its levels in milk are highly sensitive to maternal conditions, including diet and/or metabolic alterations.

To determine the functional significance of changes in the miR-26a levels in the MG, we studied the expression levels of validated target genes (*Insig1*, *Pten* and *Rb1* [17,34]) in the MG. Of interest, the greater levels of miR-26a in the MG of WD rats were accompanied by a decrease in the mRNA levels of *Rb1*. In addition, dietary improvement during lactation in Rev animals led to normalization of their expression levels, as well as an increase in the mRNA levels of *Insig1*, in comparison with the levels present in the WD group, which agreed with the normalization of miR-26a levels in Rev animals.

INSIG1 is a regulator of intracellular lipid metabolism, and abnormal expression of *Insig1* is widely involved in various lipid disorders [46]. Specifically, INSIG1 reduces the plasma levels of free cholesterol (by controlling the activation of sterol regulatory element-binding proteins and the degradation of 3-hydroxy-3-methylglutaryl-coenzyme A reductase) and protects β cells against lipotoxicity (by suppressing lipid droplet accumulation and fatty acid synthesis) [46]. Although the role of INSIG1 is well studied in hepatocytes and adipocytes, their role in MG is less known. In this regard, Wang et al. have described the functional relationship between the miR-26 family and its target gene, Insig1, in the regulation of milk fat synthesis in mammary epithelial cells in goats [34]. Moreover, overexpression of INSIG1 in mammary epithelial cells of buffalo has been associated with a reduction in the triglyceride content in these cells, suggesting a key role of INSIG1 in the regulation of milk fat synthesis in the MG [47]. Therefore, in the present study, the lower expression levels of *Insig1* in the MG of WD animals compared with the Rev group could be associated with a greater milk fat synthesis in the MG of obese dams that continued to be exposed to the obesogenic diet during lactation. In addition, the changes in the abundance of miR-26a were in accordance with the expression profile of another of its target genes, *Rb1*. The encoded protein, retinoblastoma Protein 1 (RB1), is a nuclear phosphoprotein that is critical in the regulation of cell cycle progression [48]. Therefore, the decreased *Rb1* mRNA levels found in the MG of obese dams exposed to an obesogenic diet during lactation, but not in obese dams after the dietary improvement, could tentatively affect cell proliferation capacity. This has not been directly assessed in these animals, but a trend toward a higher weight of the MG was observed in the WD rats compared with the controls (a 20% increase), although the results did not reach statistical significance ($p = 0.110$, Mann–Whitney U-test).

miR-222 levels in the MG were also increased in WD animals but were normalized in animals of the Rev group. Elevated miR-222 levels have been associated with obesity and/or diabetes, and with insulin resistance in both human and animal studies. Specifically, circulating miR-222 levels have been reported to be increased in obese and in Type 2 diabetic patients, while metformin treatment was shown to restore its levels [49–52]. Pregnant women with gestational diabetes mellitus (GDM) also displayed greater expression of miR-222 in omental adipose tissue [53], and greater plasma levels [54] versus control pregnant women. Using animal models, Ono et al. found that miR-222 levels were increased in the livers of mice fed a high-fat/high-sucrose diet, and this increase was associated with an impairment in insulin signaling [55]. In addition, overexpression of the miR-221/222 family was shown to impair insulin production and secretion by β-cells and resulted in glucose intolerance in mice [56]. Therefore, alterations in the levels of miR-222 in the MG may be related to alterations in insulin signaling. In fact, computational predictions of the target genes of the three miRNAs (miR-26a, miR-222 and miR-484) whose levels were altered in the WD group but not in the Rev group indicated a relevant role of these miRNAs modulating the expression of genes involved in insulin signaling. One of the validated target genes for miR-222 is the C-X-C chemokine receptor type 4 (Cxcr4) [37], a transmembrane receptor for C-X-C motif chemokine ligand 12 (CXCL12) [57]. The interaction between CXCL12 and its receptor, CXCR4, induces downstream signaling involved in chemotaxis, cell survival and proliferation [57]. Although the role of CXCL12 in diabetes is complex, the CXCL12/CXCR4 axis in adipose tissue has been associated with the production of proinflammatory cytokines and, finally, systemic insulin resistance [58]. It has also been reported that miR-222 levels were increased while *Cxcr4* mRNA levels were decreased in the placentas of women and mice with GDM [37]. Interestingly, the silencing of miR-222 was shown to suppress the inflammatory response and stimulate insulin sensitivity in mice with GDM by promoting the expression of *Cxcr4* [37]. Therefore, in the present study, it could be speculated that the presence of increased miR-222 levels in the MG of WD animals compared with both control and Rev animals, accompanied by the presence of lower mRNA *Cxcr4* levels in comparison with the Rev group, could contribute to a proinflammatory and insulin resistance state in these dams, and would be in accordance with the presence of elevated plasma insulin levels, as previously described [19]. Interestingly, the fact that miR-222 levels and the mRNA levels of its target gene *Cxcr4* were restored in the MG of obese dams fed an SD during lactation supports the relevance of this miRNA and interest in a nutritional intervention during lactation as a strategy to prevent such alterations in the MG associated with dietary obesity and, in turn, the proinflammatory and insulin-resistant states in these dams.

In line with what shown in the MG, the WD group also presented higher miR-222 levels in milk on Day 15 of lactation in comparison with their controls. This is in agreement with what was previously described in rats fed a cafeteria diet during lactation [18]. However, on Day 5 of lactation, the profile was somewhat different, with Rev animals showing higher miR-222 levels than the controls, and WD animals showing intermediate levels. Therefore, miR-222 levels in the milk appear to reflect both maternal diet and obesity status. In addition, it should also be highlighted that the secretion of this miRNA in the milk seems to be a time-dependent regulated process, since its levels increased progressively during lactation in all groups of dams, regardless of maternal conditions, which is consistent with our previous results [18].

miR-484 levels were also significantly higher in the MG of diet-induced obese dams that were maintained on a WD during lactation. Elevated serum levels of miR-484 have been described in individuals with coronary artery disease [59]. They were also upregulated in children with obesity compared with their normal-weight counterparts [60]. Computational functional analysis performed with miRNAs that were altered in WD dams has suggested a relevant role of the miRNAs modulating the expression of genes involved in the VEGFA-VEGFR2 signaling pathway, which is associated with obesity-related metabolic diseases [61]. Specifically, miR-484 has been shown to inhibit the expression of VEGF-A and VEGF-B [39]. VEGF-A is a growth factor that binds to members of the VEGF tyrosine kinase receptor (VEGFR) family on the surface of vascular endothelial cells to induce the proliferation and migration of vascular endothelial cells, and is essential for angiogenesis [61]. VEGF and its receptors have been shown to be upregulated in the MG during pregnancy and lactation [62,63]. This factor seems to be essential for MG differentiation and milk production [64]. Inactivation of VEGF in the MG epithelium in transgenic mice resulted in an impaired secretory activity of the epithelial cells, ultimately leading to reduced milk secretion and malnutrition in offspring [64]. Thus, VEGF must be considered as a factor that controls the correct function of the MG in the lactating state. However, as we show here, the production of this factor by the MG is decreased in obese dams, and its mRNA levels were not recovered by dietary normalization, despite the partial normalization of miR-484 levels in Rev animals. This could be tentatively related to the higher body fat content that Rev rats showed at the end of lactation compared with the controls, despite being on a SD. Notably, miR-484 levels in the milk also followed different patterns during lactation, since the levels experienced a peak on Day 10 of lactation in control rats, but the levels remained high in the WD and Rev groups. Therefore, the regulation of milk miR-484 levels and the expression of its target gene *Vegfa* in the MG appear to be associated primarily with excessive fat accumulation, rather than dietary conditions. However, more research is needed to further assess specific factors affecting the regulation of miR-484 and its target genes in the MG, and the potential consequences in the offspring.

Finally, WD animals displayed a greater miR-125a expression in MG compared with Rev animals. Circulating miR-125a levels have been found to be increased in hyperlipidemic and hyperglycemic patients [65]. In addition, miR-125a and miR-125b were overexpressed in the adipose tissue of obese patients with Type 2 diabetes compared with subjects with the same body mass index but without Type 2 diabetes [66]. Functional analysis indicated that miR-125a could negatively regulate elongase of very long chain fatty acids 6 (ELOVL6), which catalyzes the rate-limiting step in the elongation cycle, exerting a key function in milk fat synthesis [35,67]. Expression levels of the *Elovl6* gene in the MG have been shown to be increased during lactation as an adaptation to ensure the adequate concentration of long-chain polyunsaturated fatty acids required by the newborn [68]. Therefore, decreased expression of *Elovl6* in the MG of WD animals but not in Rev animals could be tentatively related with alterations in fatty acid metabolism in the MG, with potential consequences on the lipid composition of milk. However, despite the changes found in the expression of some of the target genes, a limitation of this study is that protein expression levels have not been measured. This analysis could provide relevant information, and deserves to be addressed in future studies.

In conclusion, here, we show a subset of miRNAs and target genes in the MG that could mediate, at least in part, alterations in lactating MG function due to maternal intake of an obesogenic diet. Interestingly, the implementation of a healthy diet during lactation in diet-induced obese rats attenuates most of these alterations, which highlights the importance of maternal diet during lactation. However, it is noteworthy that the dietary intervention did not fully normalize the altered miRNA levels in the milk. Further research is needed to better understand the determinants of the milk miRNA profile, in addition to the contribution of the MG, as well as to study the dynamics of the MG during lactation in relation to miRNA production.

Supplementary Materials: The following supporting information can be downloaded at: https://www.mdpi.com/article/10.3390/biomedicines10061292/s1. Table S1: Nucleotide sequences of the primers. Table S2: List of putative target genes of miR-26a, miR-222 and miR-484 searched with TargetScan.

Author Contributions: Conceptualization, M.P., A.P. and C.P.; methodology, C.A.P. and P.C.; investigation, C.A.P., P.C. and M.P.; writing—original draft preparation, C.A.P. and C.P.; writing—review and editing C.A.P., P.C., M.P., A.P. and C.P.; funding acquisition, A.P. and C.P. All authors have read and agreed to the published version of the manuscript.

Funding: This research was supported by Proyecto PGC2018-097436-B-I00 financed by MCIN/AEI/1-0.13039/501100011033/ and by "FEDER Una manera de hacer Europa". The Research Group Nutrigenomics, Biomarkers and Risk Evaluation (NuBE) receives financial support from Instituto de Salud Carlos III, Centro de Investigación Biomédica en Red Fisiopatología de la Obesidad y Nutrición, CIBERobn, and is a member of the European Research Network of Excellence NuGO (The European Nutrigenomics Organization, EU Contrac No. FP6-506360.

Institutional Review Board Statement: The animal protocol was approved by the Bioethical Committee of the University of the Balearic Islands (Exp. 2018/13/AEXP, 23 January 2019).

Conflicts of Interest: The authors declare no conflict of interest.

References

1. Victora, C.G.; Bahl, R.; Barros, A.J.; França, G.V.; Horton, S.; Krasevec, J.; Murch, S.; Sankar, M.J.; Walker, N.; Rollins, N.C.; et al. Breastfeeding in the 21st century: Epidemiology, mechanisms, and lifelong effect. *Lancet* **2016**, *387*, 475–490. [CrossRef]

2. Lönnerdal, B. Bioactive Proteins in Human Milk: Health, Nutrition, and Implications for Infant Formulas. *J. Pediatr.* **2016**, *173*, S4–S9. [CrossRef] [PubMed]

3. Melnik, B.C.; Schmitz, G. MicroRNAs: Milk's epigenetic regulators. *Best Pract. Res. Clin. Endocrinol. Metab.* **2017**, *31*, 427–442. [CrossRef] [PubMed]

4. Andreas, N.J.; Kampmann, B.; Mehring Le-Doare, K. Human breast milk: A review on its composition and bioactivity. *Early Hum. Dev.* **2015**, *91*, 629–635. [CrossRef]

5. Weber, J.A.; Baxter, D.H.; Zhang, S.; Huang, D.Y.; Huang, K.H.; Lee, M.J.; Galas, D.J.; Wang, K. The microRNA spectrum in 12 body fluids. *Clin. Chem.* **2010**, *56*, 1733–1741. [CrossRef]

6. Alsaweed, M.; Lai, C.T.; Hartmann, P.E.; Geddes, D.T.; Kakulas, F. Human milk miRNAs primarily originate from the mammary gland resulting in unique miRNA profiles of fractionated milk. *Sci. Rep.* **2016**, *6*, 20680. [CrossRef]

7. Munch, E.M.; Harris, R.A.; Mohammad, M.; Benham, A.L.; Pejerrey, S.M.; Showalter, L.; Hu, M.; Shope, C.D.; Maningat, P.D.; Gunaratne, P.H.; et al. Transcriptome profiling of microRNA by Next-Gen deep sequencing reveals known and novel miRNA species in the lipid fraction of human breast milk. *PLoS ONE* **2013**, *8*, e50564. [CrossRef]

8. Zamanillo, R.; Sánchez, J.; Serra, F.; Palou, A. Breast Milk Supply of MicroRNA Associated with Leptin and Adiponectin Is Affected by Maternal Overweight/Obesity and Influences Infancy BMI. *Nutrients* **2019**, *11*, 2589. [CrossRef]

9. Xi, Y.; Jiang, X.; Li, R.; Chen, M.; Song, W.; Li, X. The levels of human milk microRNAs and their association with maternal weight characteristics. *Eur. J. Clin. Nutr.* **2016**, *70*, 445–449. [CrossRef]

10. Alsaweed, M.; Hartmann, P.E.; Geddes, D.T.; Kakulas, F. MicroRNAs in Breastmilk and the Lactating Breast: Potential Immunoprotectors and Developmental Regulators for the Infant and the Mother. *Int. J. Environ. Res. Public Health* **2015**, *12*, 13981–14020. [CrossRef]

11. McNeill, E.M.; Hirschi, K.D. Roles of Regulatory RNAs in Nutritional Control. *Annu. Rev. Nutr.* **2020**, *40*, 77–104. [CrossRef] [PubMed]

12. Mirza, A.H.; Kaur, S.; Nielsen, L.B.; Størling, J.; Yarani, R.; Roursgaard, M.; Mathiesen, E.R.; Damm, P.; Svare, J.; Mortensen, H.B.; et al. Breast Milk-Derived Extracellular Vesicles Enriched in Exosomes From Mothers with Type 1 Diabetes Contain Aberrant Levels of microRNAs. *Front. Immunol.* **2019**, *10*, 2543. [CrossRef]

13. Avril-Sassen, S.; Goldstein, L.D.; Stingl, J.; Blenkiron, C.; Le Quesne, J.; Spiteri, I.; Karagavriilidou, K.; Watson, C.J.; Tavaré, S.; Miska, E.A.; et al. Characterisation of microRNA expression in post-natal mouse mammary gland development. *BMC Genom.* **2009**, *10*, 548. [CrossRef] [PubMed]

14. Alsaweed, M.; Lai, C.T.; Hartmann, P.E.; Geddes, D.T.; Kakulas, F. Human Milk Cells and Lipids Conserve Numerous Known and Novel miRNAs, Some of Which Are Differentially Expressed during Lactation. *PLoS ONE* **2016**, *11*, e0152610. [CrossRef] [PubMed]

15. Flint, D.J.; Travers, M.T.; Barber, M.C.; Binart, N.; Kelly, P.A. Diet-induced obesity impairs mammary development and lactogenesis in murine mammary gland. *Am. J. Physiol. Endocrinol. Metab.* **2005**, *288*, E1179–E1187. [CrossRef] [PubMed]

16. Hernandez, L.L.; Grayson, B.E.; Yadav, E.; Seeley, R.J.; Horseman, N.D. High fat diet alters lactation outcomes: Possible involvement of inflammatory and serotonergic pathways. *PLoS ONE* **2012**, *7*, e32598. [CrossRef] [PubMed]

17. Pomar, C.A.; Serra, F.; Palou, A.; Sánchez, J. Lower miR-26a levels in breastmilk affect gene expression in adipose tissue of offspring. *FASEB J.* **2021**, *35*, e21924. [CrossRef]

18. Pomar, C.A.; Castro, H.; Picó, C.; Serra, F.; Palou, A.; Sánchez, J. Cafeteria Diet Consumption during Lactation in Rats, Rather than Obesity Per Se, alters miR-222, miR-200a, and miR-26a Levels in Milk. *Mol. Nutr. Food Res.* **2019**, *63*, e1800928. [CrossRef]

19. Pomar, C.; Castillo, P.; Palou, M.; Palou, A.; Picó, C. Implementation of a healthy diet to lactating rats attenuates the early detrimental programming effects in the offspring born to obese dams. Putative relationship with milk hormone levels. *J. Nutr. Biochem.* 2022; *proof in printing.*

20. Zarrinpar, A.; Gupta, S.; Maurya, M.R.; Subramaniam, S.; Loomba, R. Serum microRNAs explain discordance of non-alcoholic fatty liver disease in monozygotic and dizygotic twins: A prospective study. *Gut* **2016**, *65*, 1546–1554. [CrossRef]

21. Nunez Lopez, Y.O.; Garufi, G.; Seyhan, A.A. Altered levels of circulating cytokines and microRNAs in lean and obese individuals with prediabetes and type 2 diabetes. *Mol. Biosyst.* **2016**, *13*, 106–121. [CrossRef]

22. Lin, H.; Tas, E.; Børsheim, E.; Mercer, K.E. Circulating miRNA Signatures Associated with Insulin Resistance in Adolescents with Obesity. *Diabetes Metab. Syndr. Obes.* **2020**, *13*, 4929–4939. [CrossRef] [PubMed]

23. Carreras-Badosa, G.; Bonmatí, A.; Ortega, F.J.; Mercader, J.M.; Guindo-Martínez, M.; Torrents, D.; Prats-Puig, A.; Martinez-Calcerrada, J.M.; de Zegher, F.; Ibáñez, L.; et al. Dysregulation of Placental miRNA in Maternal Obesity Is Associated With Pre- and Postnatal Growth. *J. Clin. Endocrinol. Metab.* **2017**, *102*, 2584–2594. [CrossRef] [PubMed]

24. Khalyfa, A.; Kheirandish-Gozal, L.; Bhattacharjee, R.; Khalyfa, A.A.; Gozal, D. Circulating microRNAs as Potential Biomarkers of Endothelial Dysfunction in Obese Children. *Chest* **2016**, *149*, 786–800. [CrossRef] [PubMed]

25. Hulsmans, M.; Sinnaeve, P.; Van der Schueren, B.; Mathieu, C.; Janssens, S.; Holvoet, P. Decreased miR-181a expression in monocytes of obese patients is associated with the occurrence of metabolic syndrome and coronary artery disease. *J. Clin. Endocrinol. Metab.* **2012**, *97*, E1213–E1218. [CrossRef] [PubMed]

26. Wei, G.; Yi, S.; Yong, D.; Shaozhuang, L.; Guangyong, Z.; Sanyuan, H. miR-320 mediates diabetes amelioration after duodenal-jejunal bypass via targeting adipoR1. *Surg. Obes. Relat. Dis.* **2018**, *14*, 960–971. [CrossRef]

27. Oses, M.; Margareto Sanchez, J.; Portillo, M.P.; Aguilera, C.M.; Labayen, I. Circulating miRNAs as Biomarkers of Obesity and Obesity-Associated Comorbidities in Children and Adolescents: A Systematic Review. *Nutrients* **2019**, *11*, 2890. [CrossRef]

28. Nielsen, L.B.; Wang, C.; Sørensen, K.; Bang-Berthelsen, C.H.; Hansen, L.; Andersen, M.L.; Hougaard, P.; Juul, A.; Zhang, C.Y.; Pociot, F.; et al. Circulating levels of microRNA from children with newly diagnosed type 1 diabetes and healthy controls: Evidence that miR-25 associates to residual beta-cell function and glycaemic control during disease progression. *Exp. Diabetes Res.* **2012**, *2012*, 896362. [CrossRef]

29. Dantas da Costa E Silva, M.E.; Polina, E.R.; Crispim, D.; Sbruzzi, R.C.; Lavinsky, D.; Mallmann, F.; Martinelli, N.C.; Canani, L.H.; Dos Santos, K.G. Plasma levels of miR-29b and miR-200b in type 2 diabetic retinopathy. *J. Cell. Mol. Med.* **2019**, *23*, 1280–1287. [CrossRef]

30. Costé, É.; Rouleux-Bonnin, F. The crucial choice of reference genes: Identification of miR-191-5p for normalization of miRNAs expression in bone marrow mesenchymal stromal cell and HS27a/HS5 cell lines. *Sci. Rep.* **2020**, *10*, 17728. [CrossRef]

31. Agarwal, V.; Bell, G.W.; Nam, J.W.; Bartel, D.P. Predicting effective microRNA target sites in mammalian mRNAs. *Elife* **2015**, *4*, e05005. [CrossRef]

32. Bindea, G.; Mlecnik, B.; Hackl, H.; Charoentong, P.; Tosolini, M.; Kirilovsky, A.; Fridman, W.H.; Pagès, F.; Trajanoski, Z.; Galon, J. ClueGO: A Cytoscape plug-in to decipher functionally grouped gene ontology and pathway annotation networks. *Bioinformatics* **2009**, *25*, 1091–1093. [CrossRef] [PubMed]

33. Bindea, G.; Galon, J.; Mlecnik, B. CluePedia Cytoscape plugin: Pathway insights using integrated experimental and in silico data. *Bioinformatics* **2013**, *29*, 661–663. [CrossRef]

34. Wang, H.; Luo, J.; Zhang, T.; Tian, H.; Ma, Y.; Xu, H.; Yao, D.; Loor, J.J. MicroRNA-26a/b and their host genes synergistically regulate triacylglycerol synthesis by targeting the INSIG1 gene. *RNA Biol.* **2016**, *13*, 500–510. [CrossRef] [PubMed]

35. Du, J.; Xu, Y.; Zhang, P.; Zhao, X.; Gan, M.; Li, Q.; Ma, J.; Tang, G.; Jiang, Y.; Wang, J.; et al. MicroRNA-125a-5p Affects Adipocytes Proliferation, Differentiation and Fatty Acid Composition of Porcine Intramuscular Fat. *Int. J. Mol. Sci.* **2018**, *19*, 501. [CrossRef] [PubMed]

36. Xu, Y.; Du, J.; Zhang, P.; Zhao, X.; Li, Q.; Jiang, A.; Jiang, D.; Tang, G.; Jiang, Y.; Wang, J.; et al. MicroRNA-125a-5p Mediates 3T3-L1 Preadipocyte Proliferation and Differentiation. *Molecules* **2018**, *23*, 317. [CrossRef]

37. Zhang, H.; Luan, S.; Xiao, X.; Lin, L.; Zhao, X.; Liu, X. Silenced microRNA-222 suppresses inflammatory response in gestational diabetes mellitus mice by promoting CXCR4. *Life Sci.* **2021**, *266*, 118850. [CrossRef]

38. Fornari, F.; Gramantieri, L.; Ferracin, M.; Veronese, A.; Sabbioni, S.; Calin, G.A.; Grazi, G.L.; Giovannini, C.; Croce, C.M.; Bolondi, L.; et al. MiR-221 controls CDKN1C/p57 and CDKN1B/p27 expression in human hepatocellular carcinoma. *Oncogene* **2008**, *27*, 5651–5661. [CrossRef]

39. Zhao, Z.; Shuang, T.; Gao, Y.; Lu, F.; Zhang, J.; He, W.; Qu, L.; Chen, B.; Hao, Q. Targeted delivery of exosomal miR-484 reprograms tumor vasculature for chemotherapy sensitization. *Cancer Lett.* **2022**, *530*, 45–58. [CrossRef]

40. Ivanova, E.; Le Guillou, S.; Hue-Beauvais, C.; Le Provost, F. Epigenetics: New Insights into Mammary Gland Biology. *Genes* **2021**, *12*, 231. [CrossRef]

41. Hue-Beauvais, C.; Faulconnier, Y.; Charlier, M.; Leroux, C. Nutritional Regulation of Mammary Gland Development and Milk Synthesis in Animal Models and Dairy Species. *Genes* **2021**, *12*, 523. [CrossRef]

42. Pomar, C.A.; van Nes, R.; Sánchez, J.; Picó, C.; Keijer, J.; Palou, A. Maternal consumption of a cafeteria diet during lactation in rats leads the offspring to a thin-outside-fat-inside phenotype. *Int. J. Obes.* **2017**, *41*, 1279–1287. [CrossRef] [PubMed]

43. Fu, X.; Dong, B.; Tian, Y.; Lefebvre, P.; Meng, Z.; Wang, X.; Pattou, F.; Han, W.; Lou, F.; Jove, R.; et al. MicroRNA-26a regulates insulin sensitivity and metabolism of glucose and lipids. *J. Clin. Investig.* **2015**, *125*, 2497–2509. [CrossRef] [PubMed]

44. Frørup, C.; Mirza, A.H.; Yarani, R.; Nielsen, L.B.; Mathiesen, E.R.; Damm, P.; Svare, J.; Engelbrekt, C.; Størling, J.; Johannesen, J.; et al. Plasma Exosome-Enriched Extracellular Vesicles From Lactating Mothers with Type 1 Diabetes Contain Aberrant Levels of miRNAs During the Postpartum Period. *Front. Immunol.* **2021**, *12*, 744509. [CrossRef] [PubMed]

45. Garavelli, S.; Bruzzaniti, S.; Tagliabue, E.; Prattichizzo, F.; Di Silvestre, D.; Perna, F.; La Sala, L.; Ceriello, A.; Mozzillo, E.; Fattorusso, V.; et al. Blood Co-Circulating Extracellular microRNAs and Immune Cell Subsets Associate with Type 1 Diabetes Severity. *Int. J. Mol. Sci.* **2020**, *21*, 477. [CrossRef] [PubMed]

46. Ouyang, S.; Mo, Z.; Sun, S.; Yin, K.; Lv, Y. Emerging role of Insig-1 in lipid metabolism and lipid disorders. *Clin. Chim. Acta* **2020**, *508*, 206–212. [CrossRef] [PubMed]

47. Fan, X.; Qiu, L.; Teng, X.; Zhang, Y.; Miao, Y. Effect of INSIG1 on the milk fat synthesis of buffalo mammary epithelial cells. *J. Dairy Res.* **2020**, *87*, 349–355. [CrossRef]

48. Goodrich, D.W.; Wang, N.P.; Qian, Y.W.; Lee, E.Y.; Lee, W.H. The retinoblastoma gene product regulates progression through the G1 phase of the cell cycle. *Cell* **1991**, *67*, 293–302. [CrossRef]

49. Ahmed, F.W.; Bakhashab, S.; Bastaman, I.T.; Crossland, R.E.; Glanville, M.; Weaver, J.U. Anti-Angiogenic miR-222, miR-195, and miR-21a Plasma Levels in T1DM Are Improved by Metformin Therapy, Thus Elucidating Its Cardioprotective Effect: The MERIT Study. *Int. J. Mol. Sci.* **2018**, *19*, 3242. [CrossRef]

50. Ortega, F.J.; Mercader, J.M.; Moreno-Navarrete, J.M.; Rovira, O.; Guerra, E.; Esteve, E.; Xifra, G.; Martínez, C.; Ricart, W.; Rieusset, J.; et al. Profiling of circulating microRNAs reveals common microRNAs linked to type 2 diabetes that change with insulin sensitization. *Diabetes Care* **2014**, *37*, 1375–1383. [CrossRef]

51. Coleman, C.B.; Lightell, D.J.; Moss, S.C.; Bates, M.; Parrino, P.E.; Woods, T.C. Elevation of miR-221 and -222 in the internal mammary arteries of diabetic subjects and normalization with metformin. *Mol. Cell Endocrinol.* **2013**, *374*, 125–129. [CrossRef]

52. Villard, A.; Marchand, L.; Thivolet, C.; Rome, S. Diagnostic Value of Cell-free Circulating MicroRNAs for Obesity and Type 2 Diabetes: A Meta-analysis. *J. Mol. Biomark. Diagn.* **2015**, *6*, 251. [CrossRef] [PubMed]

53. Shi, Z.; Zhao, C.; Guo, X.; Ding, H.; Cui, Y.; Shen, R.; Liu, J. Differential expression of microRNAs in omental adipose tissue from gestational diabetes mellitus subjects reveals miR-222 as a regulator of ERα expression in estrogen-induced insulin resistance. *Endocrinology* **2014**, *155*, 1982–1990. [CrossRef] [PubMed]

54. Filardi, T.; Catanzaro, G.; Grieco, G.E.; Splendiani, E.; Trocchianesi, S.; Santangelo, C.; Brunelli, R.; Guarino, E.; Sebastiani, G.; Dotta, F.; et al. Identification and Validation of miR-222-3p and miR-409-3p as Plasma Biomarkers in Gestational Diabetes Mellitus Sharing Validated Target Genes Involved in Metabolic Homeostasis. *Int. J. Mol. Sci.* **2022**, *23*, 4276. [CrossRef] [PubMed]

55. Ono, K.; Igata, M.; Kondo, T.; Kitano, S.; Takaki, Y.; Hanatani, S.; Sakaguchi, M.; Goto, R.; Senokuchi, T.; Kawashima, J.; et al. Identification of microRNA that represses IRS-1 expression in liver. *PLoS ONE* **2018**, *13*, e0191553. [CrossRef]

56. Fan, L.; Shan, A.; Su, Y.; Cheng, Y.; Ji, H.; Yang, Q.; Lei, Y.; Liu, B.; Wang, W.; Ning, G.; et al. MiR-221/222 Inhibit Insulin Production of Pancreatic β-Cells in Mice. *Endocrinology* **2020**, *161*, bqz027. [CrossRef]

57. Vidaković, M.; Grdović, N.; Dinić, S.; Mihailović, M.; Uskoković, A.; Arambašić Jovanović, J. The Importance of the CXCL12/CXCR4 Axis in Therapeutic Approaches to Diabetes Mellitus Attenuation. *Front. Immunol.* **2015**, *6*, 403. [CrossRef]

58. Kim, D.; Kim, J.; Yoon, J.H.; Ghim, J.; Yea, K.; Song, P.; Park, S.; Lee, A.; Hong, C.P.; Jang, M.S.; et al. CXCL12 secreted from adipose tissue recruits macrophages and induces insulin resistance in mice. *Diabetologia* **2014**, *57*, 1456–1465. [CrossRef]

59. Gongol, B.; Marin, T.; Zhang, J.; Wang, S.C.; Sun, W.; He, M.; Chen, S.; Chen, L.; Li, J.; Liu, J.H.; et al. Shear stress regulation of miR-93 and miR-484 maturation through nucleolin. *Proc. Natl. Acad. Sci. USA* **2019**, *116*, 12974–12979. [CrossRef]

60. Marzano, F.; Faienza, M.F.; Caratozzolo, M.F.; Brunetti, G.; Chiara, M.; Horner, D.S.; Annese, A.; D'Erchia, A.M.; Consiglio, A.; Pesole, G.; et al. Pilot study on circulating miRNA signature in children with obesity born small for gestational age and appropriate for gestational age. *Pediatr. Obes.* **2018**, *13*, 803–811. [CrossRef]

61. Wu, L.E.; Meoli, C.C.; Mangiafico, S.P.; Fazakerley, D.J.; Cogger, V.C.; Mohamad, M.; Pant, H.; Kang, M.J.; Powter, E.; Burchfield, J.G.; et al. Systemic VEGF-A neutralization ameliorates diet-induced metabolic dysfunction. *Diabetes* **2014**, *63*, 2656–2667. [CrossRef]

62. Nishimura, S.; Maeno, N.; Matsuo, K.; Nakajima, T.; Kitajima, I.; Saito, H.; Maruyama, I. Human lactiferous mammary gland cells produce vascular endothelial growth factor (VEGF) and express the VEGF receptors, Flt-1 AND KDR/Flk-1. *Cytokine* **2002**, *18*, 191–198. [CrossRef] [PubMed]

63. Pepper, M.S.; Baetens, D.; Mandriota, S.J.; Di Sanza, C.; Oikemus, S.; Lane, T.F.; Soriano, J.V.; Montesano, R.; Iruela-Arispe, M.L. Regulation of VEGF and VEGF receptor expression in the rodent mammary gland during pregnancy, lactation, and involution. *Dev. Dyn.* **2000**, *218*, 507–524. [CrossRef]

64. Rossiter, H.; Barresi, C.; Ghannadan, M.; Gruber, F.; Mildner, M.; Födinger, D.; Tschachler, E. Inactivation of VEGF in mammary gland epithelium severely compromises mammary gland development and function. *FASEB J.* **2007**, *21*, 3994–4004. [CrossRef]

65. Simionescu, N.; Niculescu, L.S.; Sanda, G.M.; Margina, D.; Sima, A.V. Analysis of circulating microRNAs that are specifically increased in hyperlipidemic and/or hyperglycemic sera. *Mol. Biol. Rep.* **2014**, *41*, 5765–5773. [CrossRef]

66. Brovkina, O.; Nikitin, A.; Khodyrev, D.; Shestakova, E.; Sklyanik, I.; Panevina, A.; Stafeev, I.; Menshikov, M.; Kobelyatskaya, A.; Yurasov, A.; et al. Role of MicroRNAs in the Regulation of Subcutaneous White Adipose Tissue in Individuals With Obesity and Without Type 2 Diabetes. *Front. Endocrinol.* **2019**, *10*, 840. [CrossRef] [PubMed]

67. Shi, H.B.; Wu, M.; Zhu, J.J.; Zhang, C.H.; Yao, D.W.; Luo, J.; Loor, J.J. Fatty acid elongase 6 plays a role in the synthesis of long-chain fatty acids in goat mammary epithelial cells. *J. Dairy Sci.* **2017**, *100*, 4987–4995. [CrossRef] [PubMed]

68.   Rodriguez-Cruz, M.; Sánchez, R.; Sánchez, A.M.; Kelleher, S.L.; Sánchez-Muñoz, F.; Maldonado, J.; López-Alarcón, M. Participation of mammary gland in long-chain polyunsaturated fatty acid synthesis during pregnancy and lactation in rats. *Biochim. Biophys. Acta* **2011**, *1811*, 284–293. [CrossRef]

Article

Extraction-Free Absolute Quantification of Circulating miRNAs by Chip-Based Digital PCR

Yuri D'Alessandra [1], Vincenza Valerio [1], Donato Moschetta [1,2], Ilaria Massaiu [1], Michele Bozzi [1], Maddalena Conte [3,4], Valentina Parisi [3], Michele Ciccarelli [5], Dario Leosco [3], Veronika A. Myasoedova [1] and Paolo Poggio [1,*]

[1] Centro Cardiologico Monzino IRCCS, 20138 Milan, Italy; yuri.dalessandra@bio-techne.com (Y.D.); vincenza.valerio@cardiologicomonzino.it (V.V.); donato.moschetta@cardiologicomonzino.it (D.M.); ilaria.massaiu@cardiologicomonzino.it (I.M.); michele.bozzi@cardiologicomonzino.it (M.B.); veronika.myasoedova@cardiologicomonzino.it (V.A.M.)
[2] Dipartimento di Scienze Farmacologiche e Biomolecolari, Università degli Studi di Milano, 20122 Milan, Italy
[3] Department of Translational Medical Sciences, University of Naples Federico II, 80138 Naples, Italy; maddalena-conte@libero.it (M.C.); valentina.parisi@unina.it (V.P.); dleosco@unina.it (D.L.)
[4] Casa di Cura San Michele, 81024 Maddaloni, Italy
[5] Department of Medicine, Surgery and Dentistry, University of Salerno, 84084 Fisciano, Italy; mciccarelli@unisa.it
*   Correspondence: paolo.poggio@cardiologicomonzino.it; Tel.: +39-02-5800-2853

Abstract: Circulating microRNAs (miRNA) have been proposed as specific biomarkers for several diseases. Quantitative Real-Time PCR (RT-qPCR) is the gold standard technique currently used to evaluate miRNAs expression from different sources. In the last few years, digital PCR (dPCR) emerged as a complementary and accurate detection method. When dealing with gene expression, the first and most delicate step is nucleic-acid isolation. However, all currently available protocols for RNA extraction suffer from the variable loss of RNA species due to the chemicals and number of steps involved, from sample lysis to nucleic acid elution. Here, we evaluated a new process for the detection of circulating miRNAs, consisting of sample lysis followed by direct evaluation by dPCR in plasma from healthy donors and in the cardiovascular setting. Our results showed that dPCR is able to detect, with high accuracy, low-copy-number as well as highly expressed miRNAs in human plasma samples without the need for RNA extraction. Moreover, we assessed a known myocardial infarction-related miR-133a in acute myocardial infarct patients vs. healthy subjects. In conclusion, our results show the suitability of the extraction-free quantification of circulating miRNAs as disease markers by direct dPCR.

Keywords: microRNA; digital PCR; biomarkers; extraction-free; plasma

Citation: D'Alessandra, Y.; Valerio, V.; Moschetta, D.; Massaiu, I.; Bozzi, M.; Conte, M.; Parisi, V.; Ciccarelli, M.; Leosco, D.; Myasoedova, V.A.; et al. Extraction-Free Absolute Quantification of Circulating miRNAs by Chip-Based Digital PCR. *Biomedicines* **2022**, *10*, 1354. https://doi.org/10.3390/ biomedicines10061354

Academic Editors: Milena Rizzo and Elena Levantini

Received: 17 May 2022
Accepted: 3 June 2022
Published: 8 June 2022

1. Introduction

MiRNAs are small endogenous non-coding RNAs involved in the regulation of gene expression by the modulation of messenger RNAs (mRNAs) translation and stability. They influence vital cellular processes such as responses to external stimuli, proliferation, differentiation, and apoptosis [1], and their dysregulation is associated with various pathologies [2]. In recent years, miRNAs emerged as possible specific biomarkers of several conditions because of their stable expression in almost all body fluids and due to the development of accurate and quantitative techniques for their detection [3]. In particular, several groups investigated the potential use of circulating miRNAs in the diagnostic and/or prognostic setting of cardiovascular diseases in order to formulate tailored therapeutic approaches [4]. However, the use of circulating miRNAs as biomarkers can be hypothesized only if reliable and accurate methods for quantification are devised. Quantitative reverse transcription PCR (RT-qPCR) represents, to date, the method of choice for nucleic acid quantification.

Nevertheless, one of the major drawbacks of this technique is the normalization issue. Indeed, there is a lack of consensus about both normalization methods and the choice of reliable endogenous reference miRNAs [5,6]. Digital PCR (dPCR) is a recent technology that allows nucleic acid measurement based on the segregation of the sample into thousands of separate micro-reactions of defined volume [7]. After PCR, each partition is analyzed for amplicon presence/absence, and the number of molecules in each reaction is estimated by applying the Poisson distribution [8]. When compared to RT-qPCR, the dPCR technique was described as being able to perform absolute quantification, thus bypassing normalization issues [9–11]. Moreover, when coping with low amounts of nucleic acids, it presents higher precision [12] and sensitivity [13], possibly because of its increased resistance to PCR inhibitors [14].

Interestingly, several works comparing dPCR performance with RT-qPCR in evaluating circulating miRNAs expression indicated an almost equal sensitivity for the two methods. At the same time, though, there was an almost undisputed consensus about the superior accuracy and precision of dPCR [15–17]. In addition to detection efficiency, one of the most important steps in mRNA (and miRNA) expression experiments is nucleic acid extraction. This step is usually conducted by means of specific kits and/or chemicals, and the selection of one specific method can greatly affect the following results [18–21].

In the present study, we assessed the reliability of chip-based dPCR in detecting circulating miRNAs with high accuracy and precision by evaluating six differently expressed plasma miRNAs, which we previously found to have low, medium, and high expression levels [22], in samples from healthy subjects. In addition, we tested if chip-based dPCR could be applied to a clinical setting evaluating miR-133a in non-ST-segment-elevation-myocardial-infarction (NSTEMI) patients, previously found to be upregulated upon myocardial infarction in human plasma samples [23,24].

2. Materials and Methods

2.1. Human Specimens

The Institutional Review Board and the Ethical Committee of Centro Cardiologico Monzino IRCCS (university hospital) approved this study (CCM 1068). The investigation conformed to the principles outlined in the Declaration of Helsinki (1964).

Peripheral blood samples were collected into EDTA-coated tubes (Vacutainer Systems, Becton Dickinson, Franklin Lakes, NJ, USA), kept on ice, and centrifuged at $3000 \times g$ for 10 min at 4 °C within 30 min after being drawn. Plasma was separated, centrifuged again to precipitate remaining cells, aliquoted, and stored at -80 °C until analyses were performed.

2.2. RNA Extraction and miRNA Reverse Transcription

Total RNA extraction was performed from 200 μL of plasma/sample using the Total RNA Purification Plus Kit (Norgen Biotek Corp., Thorold, ON, USA). Final elution was performed using 50 μL of Elution Solution A, as indicated by the manufacturer's protocol. In the case of the "no-extraction" protocol, 200 μL of plasma from the same samples were incubated at room temperature with proteinase K (200 μg/mL, final concentration) for 15 min with occasional flicking, followed by 5 min at 75 °C in agitation.

Since the quantification of circulating miRNAs is not possible due to the technological limitations of actual tools, 2 μL of RNA from each sample were used for two-step PCR amplification with the TaqMan Advanced miRNA cDNA Synthesis Kit (Thermo Fisher Scientific, Waltham, MA, USA; A28007) following the manufacturer's instructions, regardless of "extraction" or "no-extraction" protocols. In brief, we followed the described steps without modifications. (1) Poly(a) tailing addition: polyadenylation at 37 °C for 45 min followed by incubation at 65 °C to stop the reaction. (2) 5′-end Adaptor ligation: 16 °C for 1 h. (3) Reverse transcription (RT): 42 °C for 15 min, using a universal RT primer included in the kit, followed by incubation at 85 °C for 5 min to stop the reaction. (4) miRNAs universal pre-amplification (using proprietary primers included in the kit): enzyme activation at 95 °C for 5 min, denaturation at 95 °C for 3 s and annealing/extension at 60 °C for 30 s

(14 cycles), and finally, incubation at 99 °C for 10 min to stop the reaction. cDNA samples were stored at −80 °C until PCR further analyses. A synthetic miRNA from *C. elegans*, cel-miR-54-3p, which is not present in human samples, was used as a spike-in control to assess dPCR efficiency.

2.3. Chip-Based Digital PCR

Chip-based dPCR was performed on a QuantStudio 3D Digital PCR System platform composed of the QuantStudio 3D Instrument (Thermo Fisher Scientific; 4489084), the Dual Flat Block GeneAmp PCR System 9700 (Thermo Fisher Scientific; 4484078), and the QuantStudio 3D Digital PCR Chip Loader (Thermo Fisher Scientific; 4482592). dPCR was performed according to the manufacturer's instructions. In particular, enzyme activation was performed at 96 °C for 10 min, denaturation at 98 °C for 30 s, followed by annealing/extension at 56 °C for 2 min (40 cycles), and then final extension at 60 °C for 2 min. Analysis was executed with the online version of the QuantStudio 3D AnalysisSuite (Thermo Fisher Cloud, Waltham, MA, USA). The evaluated copy numbers for each miRNA are expressed as the mean of Log10 of the copy number/µL within a 95% confidence interval (CI). Upper and lower limits of the 95% CI are indicated within square brackets.

Primers and probes purchased from Thermo Fisher Scientific were labelled with FAM dyes and used to evaluate the expression of candidate miRNAs (Supplementary Table S1).

2.4. Statistical Analysis

We compared each gene expression technique using a correlation analysis: Pearson's correlation coefficient (R), the R^2 coefficient, and the *p*-value were computed. We performed this kind of analysis on the averaged expression values across all samples, and each value was previously "mean-centered". The analysis in the clinical setting, comparing healthy subjects and cardiovascular disease patients, was performed using the Student's *t*-test.

3. Results

3.1. Assessment of dPCR Efficiency

As a first step, we assessed dPCR efficiency in detecting a synthetic miRNA designed from nematode *C. elegans* cel-miR-54-3p at different dilutions (10^5 to 10^1 copies/µL). Linear regression analysis between expected and assessed copy numbers for miR-54-3p showed an R^2 of 0.999 ($p < 0.0001$; Figure 1A), thus indicating the suitability of dPCR in assessing with precision the expression of selected miRNAs, even in a "very-low abundance" setting. Next, we assessed whether extraction-free RNA from human plasma could be used for direct miRNA detection by dPCR. To this aim, we analyzed miRNA expression in undiluted, 1:10, and 1:100 dilutions of "no extraction" plasma samples. Three miRNAs were selected (miR-223-3p, miR-15b-5p, and miR-128-3p) because of their high, intermediate, and low expression levels in human plasma based on our previous experiences (not shown). Our experiments showed that all miRNAs were detectable regardless of dilution conditions by dPCR (Figure 1B).

3.2. Optimization of dPCR Conditions for Plasma miRNAs Detection in the Absence of RNA Extraction

The process of miRNAs expression assessment involves an amplification step (preAmp) after the conversion of RNA to cDNA, followed by dilution. Since we did not perform RNA extraction, we decided to evaluate whether different dilution settings could affect miRNA detection. Thus, after assessing the suitability of undiluted "no extraction" plasma miRNAs for dPCR, we investigated the detection of plasma miRNAs usually presenting expression levels ranging from low to high in standard conditions. Thus, we evaluated the copy number/µL of six different circulating miRNAs (miR-1180-3p, miR-128-3p, miR-186-5p, miR-451a, miR-15b-5p, and miR-223-3p). Different serial preAmp dilutions for each miRNA (1:100, 1:1000, and 1:10,000) were prepared in order to identify those most suitable for each specific target. As depicted in Figure 2A, our results showed very high overlap in terms of copies/μL, regardless of dilution, for all evaluated miRNAs. Given its low expression, miR-1180-3p was not tested beyond the 1:1000 dilution. The selected working conditions, based on chip quality (i.e., threshold = 0.6; Supplementary Figure S1) and precision (the lowest value; Supplementary Table S2), for each miRNA were identified as: 1:100 for miR-186-5p, miR-128-3p, miR-451a, and miR-1180-3p; 1:1000 for miR-15b-5p and miR-223-5p.

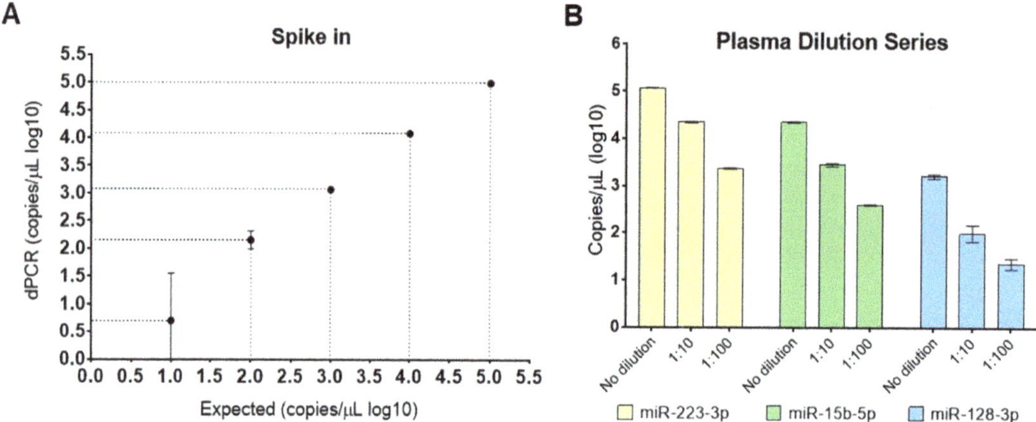

Figure 1. Digital PCR capability to detect miRNA in a wide range of concentrations. (**A**) Regression analysis between expected (x-axis) and dPCR-assessed (y-axis) expression of serial dilutions (10^5 to 10^1 copies/μL) of synthetic cel-miR-54-3p in triplicate. The data are expressed as Log10 (miRNA copies/μL of the sample), and the bars represent the 95% confidence interval. (**B**) Expression levels of undiluted, 1:10, and 1:100 dilutions of miR-223-3p, miR-15b-5p, and miR-128-3p in one healthy subject in triplicate. No RNA extraction was performed on the plasma sample. The data are expressed as Log10(miRNA copies/μL of the sample), and error bars represent the 95% confidence interval.

Following the evaluation of dPCR-mediated detection of extraction-free plasma miR-NAs, we investigated whether we could obtain consistent results in terms of reproducibility. Hence, we assessed the expression of miR-1180-3p, miR-128-3p, miR-186-5p, miR-451a, miR-15b-5p, and miR-223-5p in plasma samples from 10 healthy donors. As shown in Figure 2B, we obtained good results in terms of reproducibility and consistency for all miRNAs, none of which presented a widely sparse distribution. Consistently with previous results, miR-1180-3p showed the lowest expression, while miR-223-5p was the one with the highest copy number/μL. Further analyses conducted on all evaluated miRNAs indicated a very good correlation between the expression levels of all miRNAs in all samples, with a mean Pearson coefficient of 0.97 (Figure 2C).

To further corroborate our analyses, we evaluated the consistency of our detection method upon three different technical replicates executed at different times. We focused on the two low-expression miRNAs, namely miR-1180-3p and -186-5p, in five no-RNA-extraction samples upon three distinct dPCR runs conducted on different days. The results were highly reproducible for both miRNAs, with miR-186-5p presenting a mean coefficient of variation (CV) of 3.1%, while miR-1180-3p demonstrated an even higher accuracy, with a mean CV of 2.9% for the three runs (Supplementary Table S3).

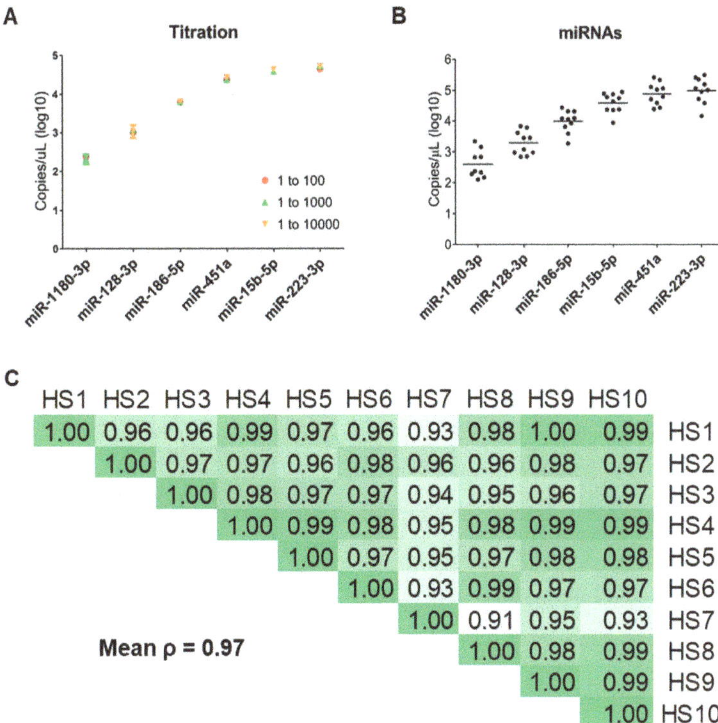

Figure 2. Digital PCR settings for miRNA detection in human plasma without RNA extraction. (**A**) Expression levels of miR-1180-3p, miR-128-3p, miR-186-5p, miR-451a, miR-15b-5p, and miR-223-5p at different dilutions in a plasma sample without RNA extraction. The data are expressed as Log10 (miRNA copies/µL of the sample), and error bars represent the 95% confidence interval. Each color and shape indicate the different dilution conditions. (**B**) Dot plot depicting the expression levels of miR-1180-3p, miR-128-3p, miR-186-5p, miR-451a, miR-15b-5p, and miR-223-5p in ten plasma samples without RNA extraction. The data are expressed as Log10 (miRNA copies/µL of the sample). Horizontal bars represent mean values. (**C**) Correlation matrix conducted on the expression of levels of miR-1180-3p, miR-128-3p, miR-186-5p, miR-451a, miR-15b-5p, and miR-223-5p in no-extraction samples from ten healthy subjects (HS). Green-shade intensity is directly proportional to the mean Pearson correlation coefficient.

3.3. Evaluation of RNA Extraction Influence on Plasma miRNA Expression

RNA-based studies usually involve nucleic acid extraction from a specific substrate as a starting point. Total RNA extraction can be achieved by means of different methods, which can consist of lysis followed by either nucleic acid precipitation or binding to affinity columns or magnetic beads. Each of the listed methods presents advantages and disadvantages, but all have different impacts on the final composition of the RNA "populations" obtained. Since column-based extraction is one of the most adopted approaches in the case of liquid samples, we decided to compare its outcome in terms of plasma miRNA expression with the "no-extraction" approach (proteinase K). In particular, we assessed the plasma expression of miR-1180-3p, miR-128-3p, miR-186-5p, miR-451a, miR-15b-5p, and miR-223-5p by dPCR in a plasma sample obtained from a healthy subject undergoing either column-based extraction or "no-extraction" protocol. Figure 3A depict the results in terms of Log10 of copy number/µL for all evaluated miRNAs. Both "conditions" showed the same hierarchic order in terms of miRNAs expression (ordered from lowest to highest expression), although some discrepancies seem to present. However, the difference in terms of yield when looking at the detected levels of miRNAs upon column-based RNA extraction vs. the no-extraction protocol is negligible. Indeed, the mean miRNAs Log10 expression ratio between the two protocols is 1.08, indicating a good relationship between the two approaches. However, two miRNAs, namely miR-128-3p and miR-223-3p, showed slightly different results when comparing the two protocols with a ratio of 1.21 and 1.12, respectively. Nevertheless, we observed a very good correlation ($p = 0.003$) between the results obtained in terms of expression from the two protocols (Figure 3B).

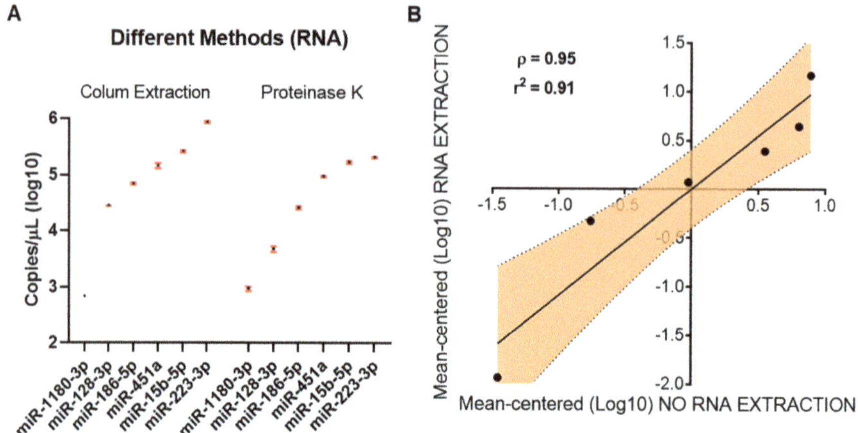

Figure 3. Comparison of column-based RNA extraction (Colum Extraction) or no-extraction (Proteinase K) protocols. (**A**) Normalized copies of column-based RNA extraction and no-extraction protocols on starting volume (50 µL and 200 µL, respectively). The level of each miRNA was calculated as Log10 of copy number/µL. Black dots represent the mean value, while error bars (red) represent the 95% confidence interval. (**B**) Correlation analysis between plasma miRNA expression after either column-based RNA extraction and no-extraction protocols. The level of each miRNA was calculated as Log10 of copy number/µL and reported as mean-centered.

3.4. No-Extraction dPCR-Based miRNA Detection in the Clinical Setting

After demonstrating the suitability of dPCR to detect even low-abundance circulating miRNAs in healthy subjects, we decided to challenge the "clinical arena". Thus, we investigated the expression levels of cardiac disease-related miR-133a-3p in no-RNA-extraction plasma of NSTEMI patients ($n = 6$) in comparison to healthy subjects ($n = 6$). As shown in Figure 4, in keeping with the literature, our data demonstrated a higher expression for miR-133-3p in NSTEMI patients (2.07 [2.02, 2.13] copies/µL) vs. controls (1.33 [1.21, 1.46] copies/µL, $p = 0.002$).

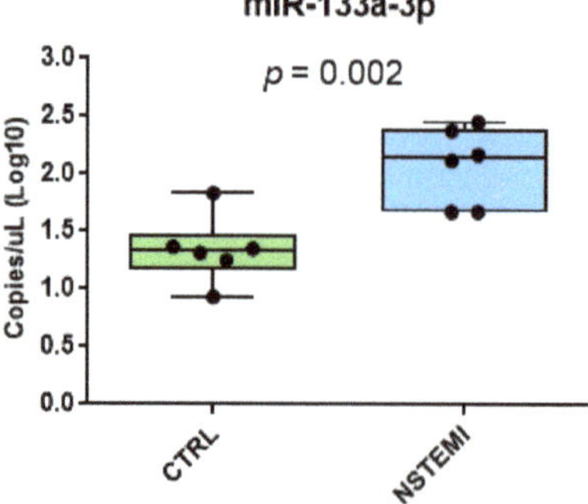

Figure 4. Plasma miRNA evaluation in a clinical setting. Plasma miR-133a-3p expression in healthy subjects (CTRL) and non-ST-segment-elevation-myocardial-infarction (NSTEMI) patients employing the proteinase K protocol. The level of the miRNA was calculated as Log10 of copy number/µL.

4. Discussion

This is, to the best of our knowledge, the first work reporting on the extraction-free detection of circulating miRNAs by dPCR. We think that our data can lead to at least two important conclusions. The first is that we were able to demonstrate that the extraction-free approach is viable and valid for dPCR even in a context where the quantity of miRNAs present in the substrate is very low, as in the case of plasma. The second relies on the fact that dPCR is a suitable tool for the direct and rapid quantification of circulating miRNAs in the clinical setting. Indeed, previous studies already demonstrated the suitability of dPCR to assess the modulation of circulating miRNAs and their possible use as biomarkers in the disease context [25,26].

A recent work by Su and colleagues [27] described the application of an extraction-free method to assess the expression of circulating miRNAs by "canonical" RT-qPCR in coronary artery disease. However, it was demonstrated that droplet-digital PCR has greater precision and improved day-to-day reproducibility over RT-qPCR with similar sensitivity [15]. In addition, the end-point approach employed by dPCR could offer better performances being less prone to suffer from differences in sample quality and more resilient to PCR inhibitors [15].

In keeping with previous literature, we showed that RNA extraction could be avoided without affecting the sensitivity of dPCR in detecting even the lowest-abundant miRNAs. Indeed, our data clearly indicate that, upon a proper assessment of the best conditions for each miRNA, dPCR is a suitable tool to analyze the expression of poorly expressed miRNAs in plasma samples even without RNA extraction. We also observed a negligible overall difference in terms of yield when looking at the detected levels of miRNAs upon column-based RNA extraction vs. the no-extraction protocol. Nevertheless, two miRNAs showed slightly different results when comparing the two protocols. This discrepancy could be explained by a different "compartmentalization" of the analyzed miRNAs in the plasma. Indeed, the use of proteinase K in the no-extraction protocol led to the release of protein-bound miRNAs but could be less effective on other plasma-miRNA vehicles, such as extracellular vesicles and lipoproteins and other known transporters [28]. However, an advantage of avoiding the extraction steps is represented by the reduction in terms of expense (i.e., not using extraction reagents or RNA affinity columns, which adds to the total cost of miRNA evaluation) and working time (i.e., jumping directly to the PCR mix step saving about 30 min per batch). It is clear that dPCR is unquestionably far from being a high throughput method. Nevertheless, the combination of absolute quantitation with high reproducibility, without the generation of a standard curve, and lack of extraction could allow the detection of miRNAs by dPCR even in the clinical setting.

We must also consider that the emerging role of miRNAs as potential biomarkers for clinical applications requires the standardization of miRNA processing [29]. To date, the most controversial issue related to miRNA level quantification is represented by the normalization step since it is well known that this step could greatly influence the results [29]. Our data indicate that the evaluation of miRNA levels in human plasma via direct and absolute dPCR quantification could be obtained without a normalizer. Indeed, we could see a significant increment in miR-133a-3p levels in plasma obtained from NSTEMI patients without normalizing it to any other miRNA, exogenous or endogenous [23,24]. In addition, it was shown that in NSTEMI patients, miR-133a-3p was only detected in serum and not in plasma by RT-qPCR, indicating a different pattern of circulating miRNA expression in these patients [30]. However, our data, showing the detection of this specific miRNA in the plasma of NSTEMI patients, corroborate the hypothesis that the use of a high precision technique (i.e., dPCR) could be implemented in clinical laboratory analyses.

A major limitation of our study is represented by the small sample size while setting up dilution conditions pre-, post-RT, pre-amp, and validation. However, we must point out that on many occasions, different conditions led to overlapping results and that inter-day reproducibility was very high. In addition, once characterization was finished, we were able to assess the expression of several miRNAs in plasma both with and without extraction with success and covering results.

5. Conclusions

Our results showed that dPCR is able to detect, with high accuracy, low-copy-number as well as highly expressed miRNAs in human plasma samples without the need for RNA extraction. Moreover, the ability to assess a known myocardial infarction-related miRNA in acute myocardial infarct patients vs. healthy subjects without the use of a normalizer encourages the use of this high precision technique in the clinical setting. In conclusion, our results show the suitability of extraction-free quantification of circulating miRNAs as disease markers by direct dPCR.

Supplementary Materials: The following supporting information can be downloaded at: https://www.mdpi.com/article/10.3390/biomedicines10061354/s1, Supplementary Figure S1: miRNA detection in human plasma by digital PCR; Supplementary Table S1: miRNA used with respective mature sequence and manufacturer codes; Supplementary Table S2: Digital PCR settings for miRNA detection in human plasma without RNA extraction; Supplementary Table S3: Detection of expression levels of low-abundance miR-186-5p, and miR-1180-3p is highly consistent at different times across multiple samples.

Author Contributions: Data curation, Y.D. and V.A.M.; Formal analysis, I.M.; Funding acquisition, P.P.; Investigation, M.C. (Michele Ciccarelli), V.A.M. and P.P.; Methodology, V.V., D.M., M.C. (Maddalena Conte), V.P., M.C. (Michele Ciccarelli) and D.L.; Project administration, V.A.M.; Supervision, P.P.; Writing—original draft, Y.D. and P.P.; Writing—review and editing, V.V., D.M., I.M., M.B., M.C. (Maddalena Conte), V.P., M.C. (Michele Ciccarelli) and D.L. All authors have read and agreed to the published version of the manuscript.

Funding: This work was supported by Gigi e Pupa Ferrari ONLUS [FPF-14] and the Italian Ministry of Health [GR-2018-12366423].

Institutional Review Board Statement: The Institutional Review Board and the Ethical Committee of Centro Cardiologico Monzino IRCCS (university hospital) approved this study (CCM 1068). The investigation conformed to the principles outlined in the Declaration of Helsinki (1964).

Informed Consent Statement: Informed consent was obtained from all subjects involved in the study.

Data Availability Statement: Data generated for this study are available upon reasonable request to the corresponding author P.P. (paolo.poggio@cardiologicomonzino.it).

Conflicts of Interest: The authors declare no conflict of interest.

References

1. Ambros, V. The functions of animal microRNAs. *Nature* **2004**, *431*, 350–355. [CrossRef] [PubMed]
2. Williams, A.E. Functional aspects of animal microRNAs. *Cell Mol. Life Sci.* **2008**, *65*, 545–562. [CrossRef] [PubMed]
3. Weber, J.A.; Baxter, D.H.; Zhang, S.; Huang, D.Y.; Huang, K.H.; Lee, M.J.; Galas, D.J.; Wang, K. The microRNA spectrum in 12 body fluids. *Clin. Chem.* **2010**, *56*, 1733–1741. [CrossRef]
4. Schulte, C.; Karakas, M.; Zeller, T. microRNAs in cardiovascular disease—Clinical application. *Clin. Chem. Lab. Med.* **2017**, *55*, 687–704. [CrossRef]
5. Peltier, H.J.; Latham, G.J. Normalization of microRNA expression levels in quantitative RT-PCR assays: Identification of suitable reference RNA targets in normal and cancerous human solid tissues. *RNA* **2008**, *14*, 844–852. [CrossRef] [PubMed]
6. Sun, Y.; Zhang, K.; Fan, G.; Li, J. Identification of circulating microRNAs as biomarkers in cancers: What have we got? *Clin. Chem. Lab. Med.* **2012**, *50*, 2121–2126. [CrossRef]
7. Day, E.; Dear, P.H.; McCaughan, F. Digital PCR strategies in the development and analysis of molecular biomarkers for personalized medicine. *Methods* **2013**, *59*, 101–107. [CrossRef]
8. Sanders, R.; Huggett, J.F.; Bushell, C.A.; Cowen, S.; Scott, D.J.; Foy, C.A. Evaluation of digital PCR for absolute DNA quantification. *Anal. Chem.* **2011**, *83*, 6474–6484. [CrossRef]
9. Llano-Diez, M.; Ortez, C.I.; Gay, J.A.; Alvarez-Cabado, L.; Jou, C.; Medina, J.; Nascimento, A.; Jimenez-Mallebrera, C. Digital PCR quantification of miR-30c and miR-181a as serum biomarkers for Duchenne muscular dystrophy. *Neuromuscul. Disord.* **2017**, *27*, 15–23. [CrossRef]

10. Beheshti, A.; Vanderburg, C.; McDonald, J.T.; Ramkumar, C.; Kadungure, T.; Zhang, H.; Gartenhaus, R.B.; Evens, A.M. A Circulating microRNA Signature Predicts Age-Based Development of Lymphoma. *PLoS ONE* **2017**, *12*, e0170521. [CrossRef]

11. Ferracin, M.; Lupini, L.; Salamon, I.; Saccenti, E.; Zanzi, M.V.; Rocchi, A.; Da Ros, L.; Zagatti, B.; Musa, G.; Bassi, C.; et al. Absolute quantification of cell-free microRNAs in cancer patients. *Oncotarget* **2015**, *6*, 14545–14555. [CrossRef] [PubMed]

12. Brunetto, G.S.; Massoud, R.; Leibovitch, E.C.; Caruso, B.; Johnson, K.; Ohayon, J.; Fenton, K.; Cortese, I.; Jacobson, S. Digital droplet PCR (ddPCR) for the precise quantification of human T-lymphotropic virus 1 proviral loads in peripheral blood and cerebrospinal fluid of HAM/TSP patients and identification of viral mutations. *J. Neurovirol.* **2014**, *20*, 341–351. [CrossRef] [PubMed]

13. Zhao, S.; Lin, H.; Chen, S.; Yang, M.; Yan, Q.; Wen, C.; Hao, Z.; Yan, Y.; Sun, Y.; Hu, J.; et al. Sensitive detection of Porcine circovirus-2 by droplet digital polymerase chain reaction. *J. Vet. Diagn. Investig.* **2015**, *27*, 784–788. [CrossRef] [PubMed]

14. Dingle, T.C.; Sedlak, R.H.; Cook, L.; Jerome, K.R. Tolerance of droplet-digital PCR vs real-time quantitative PCR to inhibitory substances. *Clin. Chem.* **2013**, *59*, 1670–1672. [CrossRef] [PubMed]

15. Hindson, C.M.; Chevillet, J.R.; Briggs, H.A.; Gallichotte, E.N.; Ruf, I.K.; Hindson, B.J.; Vessella, R.L.; Tewari, M. Absolute quantification by droplet digital PCR versus analog real-time PCR. *Nat. Methods* **2013**, *10*, 1003–1005. [CrossRef]

16. Campomenosi, P.; Gini, E.; Noonan, D.M.; Poli, A.; D'Antona, P.; Rotolo, N.; Dominioni, L.; Imperatori, A. A comparison between quantitative PCR and droplet digital PCR technologies for circulating microRNA quantification in human lung cancer. *BMC Biotechnol.* **2016**, *16*, 60. [CrossRef]

17. Maheshwari, Y.; Selvaraj, V.; Hajeri, S.; Yokomi, R. Application of droplet digital PCR for quantitative detection of Spiroplasma citri in comparison with real time PCR. *PLoS ONE* **2017**, *12*, e0184751. [CrossRef]

18. Deng, M.Y.; Wang, H.; Ward, G.B.; Beckham, T.R.; McKenna, T.S. Comparison of six RNA extraction methods for the detection of classical swine fever virus by real-time and conventional reverse transcription-PCR. *J. Vet. Diagn. Investig.* **2005**, *17*, 574–578. [CrossRef]

19. McAlexander, M.A.; Phillips, M.J.; Witwer, K.W. Comparison of Methods for miRNA Extraction from Plasma and Quantitative Recovery of RNA from Cerebrospinal Fluid. *Front. Genet.* **2013**, *4*, 83. [CrossRef]

20. Monleau, M.; Bonnel, S.; Gostan, T.; Blanchard, D.; Courgnaud, V.; Lecellier, C.H. Comparison of different extraction techniques to profile microRNAs from human sera and peripheral blood mononuclear cells. *BMC Genom.* **2014**, *15*, 395. [CrossRef]

21. Baran-Gale, J.; Kurtz, C.L.; Erdos, M.R.; Sison, C.; Young, A.; Fannin, E.E.; Chines, P.S.; Sethupathy, P. Addressing Bias in Small RNA Library Preparation for Sequencing: A New Protocol Recovers MicroRNAs that Evade Capture by Current Methods. *Front. Genet.* **2015**, *6*, 352. [CrossRef] [PubMed]

22. Songia, P.; Chiesa, M.; Valerio, V.; Moschetta, D.; Myasoedova, V.A.; D'Alessandra, Y.; Poggio, P. Direct screening of plasma circulating microRNAs. *RNA Biol.* **2018**, *15*, 1268–1272. [CrossRef] [PubMed]

23. Kaur, A.; Mackin, S.T.; Schlosser, K.; Wong, F.L.; Elharram, M.; Delles, C.; Stewart, D.J.; Dayan, N.; Landry, T.; Pilote, L. Systematic review of microRNA biomarkers in acute coronary syndrome and stable coronary artery disease. *Cardiovasc. Res.* **2020**, *116*, 1113–1124. [CrossRef] [PubMed]

24. D'Alessandra, Y.; Devanna, P.; Limana, F.; Straino, S.; Di Carlo, A.; Brambilla, P.G.; Rubino, M.; Carena, M.C.; Spazzafumo, L.; De Simone, M.; et al. Circulating microRNAs are new and sensitive biomarkers of myocardial infarction. *Eur. Heart J.* **2010**, *31*, 2765–2773. [CrossRef] [PubMed]

25. Conte, D.; Verri, C.; Borzi, C.; Suatoni, P.; Pastorino, U.; Sozzi, G.; Fortunato, O. Novel method to detect microRNAs using chip-based QuantStudio 3D digital PCR. *BMC Genom.* **2015**, *16*, 849. [CrossRef]

26. Robinson, S.; Follo, M.; Haenel, D.; Mauler, M.; Stallmann, D.; Heger, L.A.; Helbing, T.; Duerschmied, D.; Peter, K.; Bode, C.; et al. Chip-based digital PCR as a novel detection method for quantifying microRNAs in acute myocardial infarction patients. *Acta Pharmacol. Sin.* **2018**, *39*, 1217–1227. [CrossRef]

27. Su, M.; Niu, Y.; Dang, Q.; Qu, J.; Zhu, D.; Tang, Z.; Gou, D. Circulating microRNA profiles based on direct S-Poly(T)Plus assay for detection of coronary heart disease. *J. Cell Mol. Med.* **2020**, *24*, 5984–5997. [CrossRef]

28. Boon, R.A.; Vickers, K.C. Intercellular transport of microRNAs. *Arterioscler. Thromb. Vasc. Biol.* **2013**, *33*, 186–192. [CrossRef]

29. Faraldi, M.; Gomarasca, M.; Banfi, G.; Lombardi, G. Free Circulating miRNAs Measurement in Clinical Settings: The Still Unsolved Issue of the Normalization. *Adv. Clin. Chem.* **2018**, *87*, 113–139. [CrossRef]

30. Mompeon, A.; Ortega-Paz, L.; Vidal-Gomez, X.; Costa, T.J.; Perez-Cremades, D.; Garcia-Blas, S.; Brugaletta, S.; Sanchis, J.; Sabate, M.; Novella, S.; et al. Disparate miRNA expression in serum and plasma of patients with acute myocardial infarction: A systematic and paired comparative analysis. *Sci. Rep.* **2020**, *10*, 5373. [CrossRef]

Review

Exosomal MicroRNAs as Novel Cell-Free Therapeutics in Tissue Engineering and Regenerative Medicine

Eric Z. Zeng [1], Isabelle Chen [1,2], Xingchi Chen [1] and Xuegang Yuan [1,3,*]

1. Department of Chemical and Biomedical Engineering, FAMU-FSU College of Engineering, Florida State University, Tallahassee, FL 32310, USA
2. Los Altos High School, Los Altos, CA 94022, USA
3. Department of Pathology & Laboratory Medicine, David Geffen School of Medicine, University of California-Los Angeles (UCLA), Los Angeles, CA 95616, USA
* Correspondence: xy13b@fsu.edu or yuanxg1989@g.ucla.edu

Abstract: Extracellular vesicles (EVs) are membrane-bound vesicles (50–1000 nm) that can be secreted by all cell types. Microvesicles and exosomes are the major subsets of EVs that exhibit the cell–cell communications and pathological functions of human tissues, and their therapeutic potentials. To further understand and engineer EVs for cell-free therapy, current developments in EV biogenesis and secretion pathways are discussed to illustrate the remaining gaps in EV biology. Specifically, microRNAs (miRs), as a major EV cargo that exert promising therapeutic results, are discussed in the context of biological origins, sorting and packing, and preclinical applications in disease progression and treatments. Moreover, advanced detection and engineering strategies for exosomal miRs are also reviewed. This article provides sufficient information and knowledge for the future design of EVs with specific miRs or protein cargos in tissue repair and regeneration.

Keywords: extracellular vesicles; microRNA; biogenesis; cargo sorting; tissue repair and regeneration

Citation: Zeng, E.Z.; Chen, I.; Chen, X.; Yuan, X. Exosomal MicroRNAs as Novel Cell-Free Therapeutics in Tissue Engineering and Regenerative Medicine. *Biomedicines* **2022**, *10*, 2485. https://doi.org/10.3390/biomedicines10102485

Academic Editors: Milena Rizzo, Elena Levantini and Christos K. Kontos

Received: 17 July 2022
Accepted: 22 September 2022
Published: 5 October 2022

Publisher's Note: MDPI stays neutral with regard to jurisdictional claims in published maps and institutional affiliations.

1. Introduction

In general, all types of cells are able to secrete membrane-bound vesicles both in vivo and in vitro, which are broadly termed as extracellular vesicles (EVs). Initially, researchers found these lipid bilayer-enclosed secreted vesicles circulating across mammalian tissues/fluids and identified them as cellular debris or platelet dust [1]. Early studies of EV functions demonstrated their ability to remove cellular waste and lyse cellular compartments. Nowadays, intercellular communication is acknowledged as a major function of EVs. Due to the nature of "cell-secretion", heterogeneity is a critical characteristic of EVs. Based on isolation and size characterization, EVs can be broadly (and roughly) classified as apoptotic bodies (ApoBs, ~500–5000 nm), microvesicles (MVs, ~100–1000 nm), and exosomes (~40–150 nm). ApoBs are distinctive populations originating from dying cells as a hallmark of apoptosis. At the final stage of apoptosis, cells disassemble into an abundance of ApoBs containing cellular fragments, which are precisely phagocytosed by macrophages, parenchymal cells, or neoplastic cells for degradation [2]. Little is known about ApoBs as therapeutic agents despite the fact that no inflammation or cytotoxicity is established by ApoBs. The major focus of ApoBs research concerns drug delivery vessels or diagnostics, as they carry a large number of proteins, lipids, RNA, and DNA molecules [3]. In this review, we mainly focus on MVs and exosomes, as an increasing body of evidence demonstrates their potential in disease diagnosis and therapeutic design. Specifically, exosomal micro-RNAs (exo-miRs), as one of the major EV cargos, are discussed in detail through EV biogenesis, secretion, and cargo sorting and packaging. Moreover, advanced studies on diagnostics, therapeutic applications, and bioengineering strategies of exo-miRNAs are also reviewed to provide insights for the future design of cell-free therapies with EVs and exo-miRs.

2. EV Biogenesis

2.1. Microvesicle Biogenesis

Depending on the cell type, culture conditions, and isolation strategies, a mixture of exosome and MVs can always co-exist, though the percentage may vary. This heterogeneity is because MVs and exosomes share similar characteristics at a certain range of sizes (Figure 1), which makes isolation and purification difficult. Based on size characteristics, MVs vary from 100–500 nm (ectosome), but can reach up to 1000 nm (which are referred to as oncosomes and are identified in highly invasive cancer cells) [4,5].

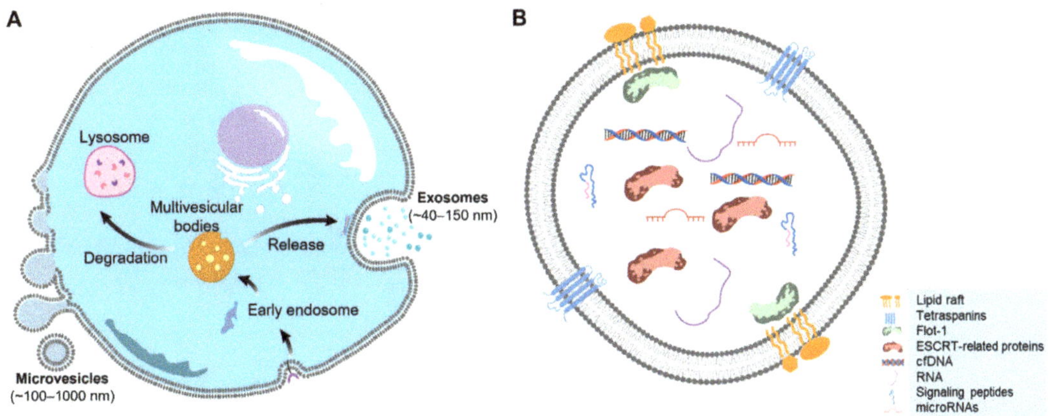

Figure 1. Extracellular vesicles (EV) subtypes based on size, biogenesis difference, and composition. (**A**) The major types of EVs are microvesicles (MVs, diameter: ~100–1000 nm) and exosomes (diameter: ~40–150 nm). The biogenesis of MVs and exosomes is generally different with certain shared pathways. (**B**) General EV compositions and cargos include peptides, lipids, proteins, genetic materials, and metabolites. MicroRNAs (miRs) are the major cargo in EVs and exhibit extensive functions in vitro and in vivo.

Both MVs and exosomes are distinctive populations compared to apoptotic bodies secreted through cell apoptosis [6]. Although sharing the similar membrane budding release, the biogenesis of MVs is quite different from that of exosomes. MVs undergo fission and directly bud outward at the plasma membrane (PM) of cells. However, the heterogeneity of MVs (with a wide range of sizes from 100–1000 nm) suggests that multiple mechanisms could be involved during membrane shedding, such as phospholipid site alteration or PM blebs. These alterations and blebs keep expanding to generate outward membrane curvature for MV budding out from the lipid sites, where cytoskeleton reorganization is commonly observed [7–11].

The cytoskeleton rearrangement results in the generation of the budding neck and eventually rupture of the membrane to release MVs following Ca^{2+} level changes (Figure 2A) [12]. Several substances on the cell membrane have been identified for MV biogenesis: (1) phosphatidylserine—the most abundant anionic phospholipid in the cell membrane—can move from the inner to the outer leaflet of the PM via enzymatic reactions of flippases (ATP-driven); (2) floppases, which generate uneven force for membrane bending; and (3) ATP-independent scramblases, which redistribute phosphatidylserine and stabilize membrane rigidity [7,11,12]. In another case, acid sphingomyelinase (a-SMase), a lipid metabolic enzyme, mediates the P2X7 (an ATP receptor, generally found to be highly expressed in immune cells)-dependent release of large vesicles in glial cells (microglia and astrocytes), as the activation of P2X7 induces the translocation of a-SMase from lysosomes to the PM outer leaflet and further alters PM physical conditions [13,14]. This translocation of a-SMase can be achieved by stimulation of other receptors to enhance MV biogenesis [15]. In addition, MVs are mostly enriched with lipid rafts (containing cholesterol), which are associated with the proteins responsible for cytoskeleton reorganization (e.g., calpain, which is regulated by extracellular Ca^{2+}, and ADP-ribosylation factor 6 (Arf-6), which is synchronized with the RhoA and ROCK signaling pathways for oncosome release) [16,17].

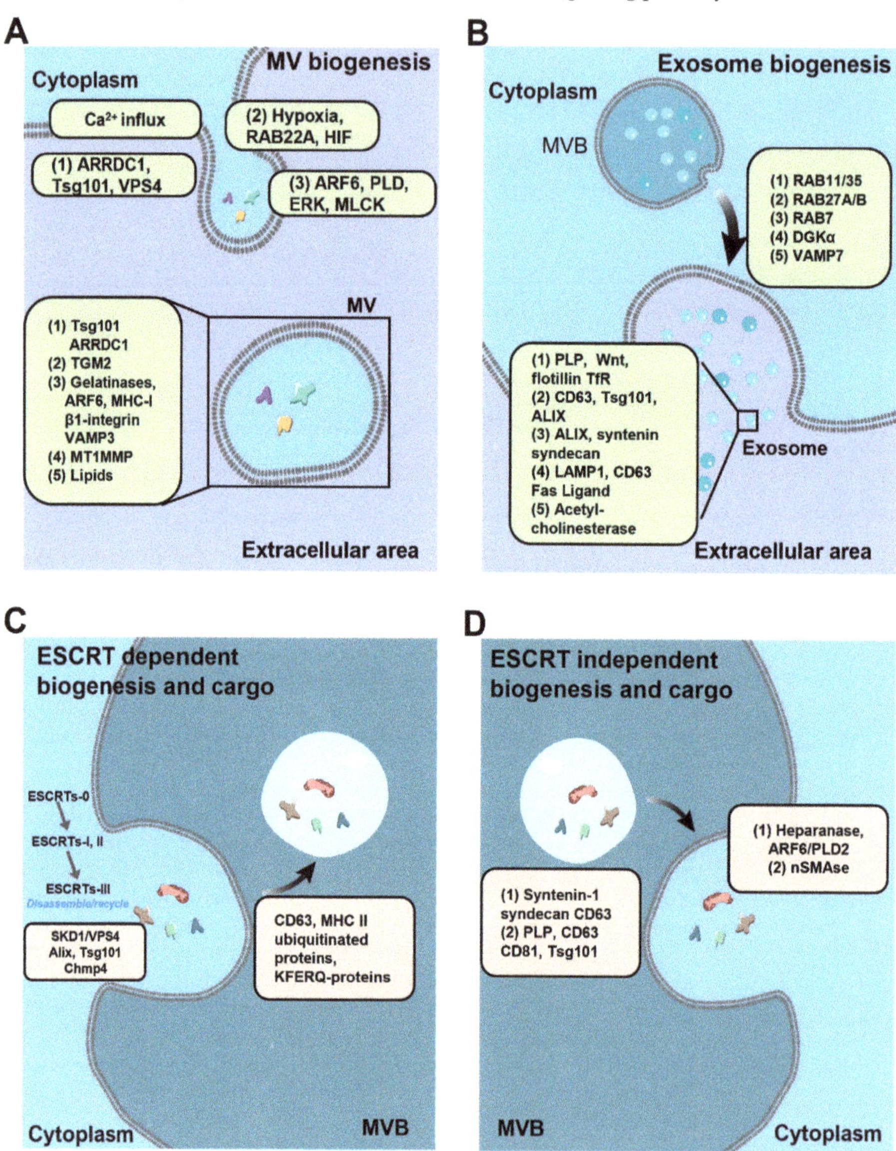

Figure 2. EV biogenesis pathways. (**A**) MV biogenesis. Plasma membrane rearranges at specific sites following Ca^{2+} influx, which recruits enzymes such as scramblase and floppase. ARRDC1 regulates ESCRT proteins TSG101 in an ATP-required manner (VPS4) to release MVs. Other modifications such as hypoxia-induced factors (HIF) and ARF6-stimulated PLD-ERK activation of myosin light chain kinase (MLCK) can also induce MV biogenesis. Associated proteins and molecules can be found in secreted MVs. (**B**) Exosome biogenesis and associated proteins promote MVB fusion to PM and regulate specific cargo packaging. (**C**) ESCRT-dependent exosome biogenesis. ESCRTs act in a stepwise manner to generate exosomes in cytosol and regulate protein cargo packaging. (**D**) ESCRT-independent pathways and associated proteins.

Another mechanism for MV biogenesis involves endosomal sorting complexes required for transport (ESCRT) proteins and cytoskeleton interactions. Arrestin domain-containing protein 1 (ARRDC1), which is an accessory protein of tetrapeptide PSAP motifs, acts as an adaptor that binds the PM to produce MVs. TSG101 relocates from the endosomes to the PM and binds to ARRDC1 through tetrapeptide PSAP motifs and promotes the release of MVs. Thus, these MVs contain TSG101, ARRDC1, and other late exosomal markers such as CD63 and LAMP1 [18]. The ESCRT-dependent biogenesis of MVs requires ATPases, such as vacuolar protein-sorting-associated protein 4 (VPS4), to enable the final pinch-off of the membrane and release of MVs [19–21]. In summary, MV subpopulations with unique cargo and functions may exist and different cell types and culture conditions can further complicate the heterogeneity of MVs. The proteins involved in MV biogenesis are summarized in Table 1.

Table 1. Proteins involved in MV biogenesis.

Proteins	Location	Functions	Ref
ARF1	Golgi apparatus and shedding microvesicles	Regulation of matrix degradation by directly acting on the structures associated with invasiveness—invadopodia maturation and the shedding of membrane-derived microvesicles	[10]
ARF6	Plasma membrane, cytosol, and endosomal membranes	Regulating the actomyosin-based membrane abscission mechanism to control the shedding of microvesicle in tumor cells	[7]
Rab22a	Nonclathrin-derived Endosomes, budding microvesicles	Increasing microvesicle shedding in human breast cancer under hypoxic conditions and knockdown of RAB22A impairs breast cancer metastasis	[22]
RhoA	Membrane and cytosol	Involved in microvesicle biogenesis through regulation of myosin light chain phosphatase. required for microvesicle shedding	[11]
ARRDC1	Plasma membrane	ARRDC1-mediated relocalization of TSG101 may alter endosomal trafficking and sorting and signal transduction by receptors subjected to endosomal sorting mechanisms	[23]
DIAPH3	Plasma membrane, Microtubules/microvilli	DIAPH3 silencing also promotes shedding of extracellular vesicles (EV) containing bioactive cargo and increases proliferation of recipient tumor cells, and suppresses proliferation of human macrophages and peripheral blood mononuclear cells	[24,25]
Myosin-1a	Plasma membrane	Enterocyte microvilli containing Myosin-1a are active vesicle-generating organelles	[26]

2.2. Exosome Biogenesis

Exosomes are considered as small EVs (Figure 1A). Cargo analysis reveals the diversity of their contents, which include membrane receptors, soluble proteins, lipids, RNAs, metabolites, and organelles, leading to functional variance in recipient cells (Figure 1B). As shown in Figure 1A, exosomes originate at the internal endosomal membranes of multivesicular bodies (MVBs). Initiated by the endo-lyososomal pathway or endocytosis, early endosomes are generated from the PM, they bud into cytosol for maturation, and then are packed with MVBs or multivesicular endosomes (MVEs) regulated in a specific protein-cargo manner (Figure 2B). The inward budding of the endosomal membrane in MVBs/MVEs results in the accumulation of intraluminal vesicles (ILVs, precursors of exosome). The accumulated ILVs are released into the extracellular environment upon the fusion of MVBs/MVEs with the PM, now termed as exosomes [27,28]. Although it is difficult to dissect exosome biogenesis into clearly separated pathways, the current knowledge classifies exosome biogenesis into two categories: ESCRT-dependent and ESCRT-independent pathways.

2.2.1. ESCRT-Dependent Pathways

ESCRT proteins represent the major machinery that regulates both MV and exosome biogenesis and can be grouped into five distinct complexes: ESCRTs -0, -I, -II, -III, and Vps4 [19,29], together with the accessory protein ALIX (Table 2) [30]. As reviewed above, TSG101 participates in the abscission of the vesicle bud from the PM to promote MV secretion, which is identified as an ESCRT-I complex process [23].

Table 2. Proteins involved in ESCRT pathways.

Complex		Location	Cargo Sorting	Functions	Ref
ESCRTs-0	HRS	VHS, FYVE, P(S/T)XP, GAT domain and coiled-coil core, clathrin-binding	Binding to/clustering with ubiquitinated cargo for delivery into MVBs, and recruits clathrin, ubiquitin ligases, and deubiquitinating enzymes, and almost certainly has other functions as well	Clustering of Ub cargo, MVB biogenesis	[31]
	STAM1/2	VHS, UIM, SH3, GAT domain and coiled-coil core			[32]
ESCRTs-I	TSG101	UEV, PRD, stalk, headpiece	Binding ubiquitinated cargo, ESCRT-0, ESCRT1, BRO1 and viral proteins	Membrane budding, MVB biogenesis, viral budding, replication and cytolinesis	[33,34]
	HVPS28	headpiece, Vps28 CTD	ESCRT-0, ESCRT1, BRO1 and viral proteins		[35]
	VPS37	basic helix, stalk, headpiece	Membrane binding		[36]
	hMVB12	stalk, headpiece ("UMA domain"), MAPB	N/A		[37]
ESCRTs-II	EAP20/VPS25	Winged-helix	Binding ubiquitinated cargo, binding to human ESCRT-I	The essential partner of ESCRT-I in MVB biogenesis and budding formation, membrane budding	[38]
	EAP30/VPS22	basic helix, Winged-helix	Forming nearly equivalent interactions with the two Vps25 molecules		[39]
	EAP45/VPS36	Winged-helix, GLUE,	Binding PI containing membranes, ubiquitinated cargo and ESCRT-1-i		[40]
ESCRTs-III	CHMP2/VPS2	MIM1	Recruits VPS4, initiates ESCRT disassembly	Membrane scission	[41]
	CHMP3/VPS24	weak MIM1	Caps Snf7 polymer, recruits VPS2		[41]
	CHMP4/SNF7	weak MIM2	Main driver of membrane scission, bind Bro1		[42]
	CHMP6/VPS20	MIM2	Binding ESCRT-II and Doa4, acts as nucleator of Snf7 polymer		[41]
VPS4	SKD1/VPS4	MIT, AAA	AAA ATPase disassembles ESCRT-III, active function in MVB membrane scission	Vps4 solubilizes ESCRT-III subunits at the cost of ATP hydrolysis. LIP5 binds to Vps4 and promotes its oligomerization, activity, and ESCRT-III binding	[43]
	LIP5	MIT	Binding vps4 to promote ESCRT-III recycling		[44]

For exosome formation, all the ESCRT complexes relocate at the endosomal membrane of late endosomes and function in a stepwise manner to drive cargo sorting/packaging, vesicle budding, and fission (Figure 2C).

The ESCRTs-0 complex initiates the process at the endosomal membrane after being recruited by phosphatidylinositol 3-phosphate. It recognizes and sequesters ubiquitylated proteins such as clathrin, ubiquitin, and other activated growth factors receptors [45]. Two major subunits of ESCRTs-0, HRS, and STAM1/2, bind ubiquitinated cargos and may be responsible for different subpopulations of exosomes [30]. Depletions of HRS and STAM1/2 have less effect on early endosomes but enlarge the MVBs, partially explaining the decrease during the production of small-size EVs [46]. Clearly, different subunits of ESCRTs-0 regulate specific exosome subpopulations.

The ubiquitinated proteins and receptors are then passed along to ESCRTs-I and -II complexes, which provide ubiquitin-interaction domains to sort ubiquitinated cargos. ESCRTs-I and -II are mainly responsible for membrane deformation to accumulate ILVs [47]. Depletion of ESCRT-I protein TSG101 leads to an altered exosome protein profile (enriched CD63 and MHC-II negative vesicles) [30]. In addition, depletion of TSG101 also alters early endosome morphology by causing vacuolar domain alteration and further inhibits MVB formation [46]. Interestingly, overexpressed TSG101, however, also inhibited EV production, especially in an exosome size range of 30–100 nm [48].

Finally, ESCRTs-I and -II recruit ESCRTs-III monomers in the cytosol and reassemble them in active complexes (charged multivesicular body proteins, i.e., CHMPs) at the endosomal membrane via interactions with Alix and ESCRTs-I complexes. The activated complexes form filaments/spirals that drive the final step of ILV biogenesis, including cargo crowding, membrane deformation/budding, neck tightening, and scission of ILVs [49,50]. Then, ESCRTs-III spirals disassemble into small filaments and monomers, which are recycled in cytosol. Knock-out of ESCRTs-III protein CHMP1 resulted in reduced formation of MVBs but with an enlarged morphology [51]. VPS4/SKD1 disassembles and recycles ESCRTs-III complexes, while deficient ATPase caused enlarged MVBs and the accumulation of nonreleasable particles [52,53].

ESCRTs also interact with accessory proteins, namely, Syntenin, Syndecan, and Alix, for exosome biogenesis. Syntenin, as a cytoplasmic adapter protein, is found to interact directly with Alix via protein motifs [54]. Syntenin also binds to Syndecans at their cytosolic tails, and Alix connects the Syndecan–Syntenin complex to ESCRTs-III to eventually form ILVs. Direct evidence of this process can be observed in the fact that exosomes derived from MCF-7 cells perturbated with heparanase (the only mammalian enzyme that cleaves heparan sulfate of oligomerized Syndecans) exhibit different Syntenin-1, Syndecan, and CD63 protein profiles [55]. Moreover, recent studies revealed that the formation of exosomes via Syntenin/Syndecan/Alix/ESCRT-III is regulated by Arf6 and its effectors, phospholipase D2 (PLD2), which is translocated from PM to MVB lumen and enriched on secreted exosomes [56,57]. Alix/ESCRTs-III interactions are considered as ESCRT-independent in some studies [36,40,41].

2.2.2. ESCRT-Independent Pathways

Interestingly, even with the complete abolishment of ESCRT functions, a certain level of ILV formation and exosome secretion still remains (although with altered subpopulation), which indicates that there is an ESCRT-independent pathway for exosome biogenesis (Figure 2D) [58,59]. For example, inhibition of neutral sphingomyelinase 2 (nSMase2) leads to the decreased production of sphingolipid ceramide-enriched exosome [60]. Other lipid interactions, such as sphingosine1-phosphate with metabolized ceramide, promote exosome release and cholesterol redistribution, thus influence cargo packaging [61–63]. The tetraspanin family, a series of proteins, can regulate the dynamic membrane domains, and influence exosome biogenesis independent of ESCRTs. For instance, CD63 is particularly enriched in exosomes and regulates endosomal cargo targeting and sorting, as well as protein trafficking and packing into exosomes [64–68]. Other tetraspanin proteins, such as CD9, CD81, CD82, Tspan6, and Tspan8, exhibit different mechanisms at different steps of exosome formation [69–73]. In summary, both the ESCRT-dependent and ESCRT-independent pathways are equally important for exosome biogenesis and operate simultaneously. Future investigations of EV biogenesis are required in order to fine-tune these biogenesis pathways for EV engineering.

3. Exo-miRNA Loading and Sorting in EVs

Typically, a complex cargo profile can be found in EVs regardless of cell type and culture conditions. On the other hand, the cargo profile varies dynamically depending on the cellular microenvironment and tissue origins. Therefore, understanding the cargo sorting mechanism is critical for engineering therapeutic EVs by manipulating culture conditions. As mentioned above, ESCRT complex is responsible for EV biogenesis and cargo recognition by providing distinct ubiquitin-binding motifs. For example, the ESCRTs-0 complex, with both HRS and STAM1/2 subunits, can bind to ubiquitin by its ubiquitin-interacting motif and recognize polyubiquitinated proteins [74,75]. Moreover, ESCRTs-0 also binds to the clathrin heavy chain via its clathrin box motif [45,76]. Similarly, ESCRTs-I and -II also provide ubiquitin-binding domains, which are not identified in ESCRTs-III [47,77]. Interestingly, EV secretion with functional cargos still occurs with the complete deletion of ESCRTs, implying other components, such as a lipid raft and ceramide, may also regulate the protein sorting process independent of ESCRTs [30,58,78–81].

Another important set of cargos in EVs are nucleic acids including DNAs, mRNAs, and miRs. It is highly possible that the cytoplasmic DNA fragments generated by the nucleus or mitochondria (due to DNA damage repair or DNA metabolism) are directly encapsulated into EVs [82–84]. However, other studies indicated lower levels of DNA cargo in nontransformed cell lines or in the circulatory system of healthy people compared to cancer cell lines and cancer patient cells [85,86]. Besides the extrusion of damaged genetic materials to maintain cellular homeostasis, exosomal DNAs also contribute to immunomodulatory functions in cancer therapy and could act as liquid biopsy markers for diagnosis [85,87–90]. However, little is known concerning their extensive functions in stem cell therapy and tissue development.

MiRs (with lengths of 19–24 nt) play important roles in inhibiting the expressions of target protein-coding genes and fit well with the function of EVs. MiRs were reported to make up the highest proportion of nucleic acids enriched in EV cargo along with other RNAs including mRNA, ribosomal RNA, long noncoding RNAs, and circular RNAs [91–95]. Interestingly, deep sequencing revealed that EVs generally had a distinguished miR profile compared to their parent cells, implying the regulated sorting rather than random packaging of miRs in EVs [75,77,80,81].

Obviously, cellular/cytosol abundance of miRs is associated with their sorting into EVs [96], although the mechanism is still not well understood. Based on the current knowledge and evidence, several pathways have been proposed as the mechanisms for miR sorting into EVs (Figure 3A):

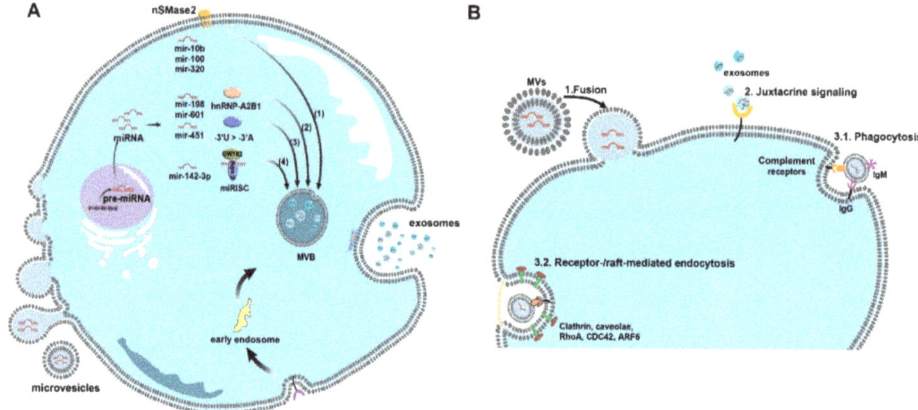

Figure 3. Exosomal cargo sorting and EV uptake by recipient cells. (**A**) MicroRNAs are sorted through different mechanisms/pathways into MVs and exosomes. Nucleus-released miRs in cytosol can be directly packed into MVs. For the exosome (MVB in cytosol), (1) nSMase2 regulates ceramide biosynthesis for selective miR sorting; (2) hnRNP proteins bind to specific miR motifs for sorting; (3) the modification of noncoding RNAs regulates 3′-end adenylated miR isoforms; (4) GW182 and Ago2 co-localized with MVB accumulate in the miR-induced silencing complex (miRISC). (**B**) Potential EV uptake mechanisms by recipient cells. MVs can directly fuse with the plasma membrane (PM) to deliver exosomal cargo. EVs can also bind to specific sites on the PM to exert juxtacrine signaling to activate intracellular pathways. Alternatively, EVs can be phagocytosed or endocytosed via specific receptor mediation on recipient cells.

(1) The modulation of lipid biosynthesis could influence both EV biogenesis and miR sorting. As reviewed above, ceramide is critical for exosome release and protein targeting in EVs. Disrupting ceramide biosynthesis by the inhibition of nSMase-2 leads to a reduction in miR-16 and miR-146a levels in EVs [80,97]. The modulation of nSMase-2 has been established for perturbing the sorting of other miRs in EVs, such as miR-10b, miR-100, and miR-320 [64,84,85]. However, alteration of exosomal miRs by modulating lipids is risky since the integrity of EVs can be significantly impacted as lipids are crucial in EV biogenesis.

(2) MiR sorting into EVs is also dependent on the specific sequence/motifs interactions with binding proteins. In Jurkat cell-secreted exosomes, over 70% of the exo-miRs have a GGAG motif (or an extra seed sequence) in the 3'-portion of miR, which is a binding site for sumoylated heterogeneous nuclear ribonucleoprotein (hnRNP), or hnRNP-A2B1 specifically [98]. Other types of hnRNP proteins can bind to specific motifs of miRs and then regulate miR sorting into EVs [98,99]. In addition, RNA-binding proteins, such as Y-box protein 1 (YBP-1), also regulate miR sorting into EVs, although the binding motifs are not identified [100,101].

(3) Similar to the specific binding motifs, the 3'-end nontemplate sequence in miRs also plays certain roles in cargo sorting into EVs. For example, 3'-end uridylated miR isoforms are mainly expressed in exosomes, while 3'-end adenylated miR isoforms are relatively enriched in B cells [102]. Although this post-transcriptional modification of noncoding RNA seems to drive cytoplasmic Y RNA sorting, more evidence is required to elucidate its general role in sorting other miRs.

(4) MiR sorting could also be mediated by miR-induced silencing complex (miRISC). The major components of miRISC found in monocytes are miRs, miR-repressible mRNAs, and GW bodies (GW182 and Ago2) co-localized with MVBs, all of which were determined by immunofluorescent staining of RISC-MVB markers [103,104]. Studies also utilized the inhibition of ESCRT to block the turnover of MVBs into lysosomes, which leads to the accumulation of miRISC. On the other hand, disrupting MVB formation causes the loss of miRISC and relieves miR-mediated gene silencing [103]. These are the first pieces of evidence showing that miRISC and MVBs are both physically and functionally associated. Further studies indicate that knockout of Ago2 could eliminate or decrease the expression of certain exosomal miRs, such as miR-451, miR-150, and miR-142-3p in HEK293T cells [91]. Moreover, Ago2 can sometimes be expressed in exosomes [105]. In addition, elevating the cellular levels of miR-repressible mRNAs also contributes to the enrichment of target miRs in MVBs and facilitates miR sorting [96]. This evidence may imply that miRISC is involved in miR sorting into EVs. However, to better modulate EV content and the miR profile via miRISC, more investigations are needed.

4. Mechanism for EV Uptake by Recipient Cells and Exosomal miRNA Functions

Cell–cell communication represents the most important role of EVs in cellular events and tissue development. The heterogeneity of cargo, surface components, and sizes influence the uptake of EVs by recipient cells. EVs can interfere with cellular pathways and behavior by binding to the target cell surface without delivering any cargo. One widely observed example is the activation of T lymphocytes [106]. During the immune response, B lymphocyte-secreted EVs with major histocompatibility complex (MHC) class II-enriched compartments can directly activate antigen-specific MHC class II-restricted T cells without delivery of cargo [107]. Similarly, dendritic cells also secrete EVs with MHC-peptide complexes for the activation of T lymphocytes, although different EV subtypes exert different capacity for activation [108]. The mechanism of this direct binding has encouraged research on manufacturing EV mimics with functional surface markers that regulate immune responses. However, the main interest in the functionality of EVs is the delivery of cargo to the recipient/target cells. Thus, understanding the uptake of EVs and the fate of the delivered exosomal cargo is critical for the future design of EV-based therapies.

In general, three major interactions exist between EVs and recipient cells and these interactions are highly dependent on the specificity of EVs (Figure 3B). For instance, directly binding and docking on the recipient cell PM is likely to be regulated by Tetraspanin proteins, adhesion molecules (e.g., integrins, ICAMs, and lectins), lipids, proteoglycan, and the extracellular matrix [70,109–112]. After binding, EVs can stay at the membrane without delivering cargo as discussed above, where EVs act as ligands or intracellular signal mediators to regulate recipient cell behaviors [113]. This juxtracrine fashion of interaction eventually ends up with the release of binding EVs instead of internalization. In other cases, EVs enter the recipient cells by phagocytosis, macropinocytosis, or receptor-mediated endocytosis, where endosomes are the destination for cargo delivery. Alternatively, EVs can also enter the recipient cells by fusion with the cytoplasm membrane to directly release the intraluminal cargo inside recipient cells [114–122]. The ultimate fate of EVs is degradation by lysosomes to recycle the compartment for re-secretion [123]. The internal trafficking of EVs also requires cellular components from recipient cells, although the specific mechanism is unknown.

5. Engineering and Therapeutic Strategies with Exosomal miRs in Regenerative Medicine

5.1. Exo-miR from Mesenchymal Stem Cells (MSCs) in Bone-Associated Regeneration

After being incorporated by recipient cells, EV cargos such as exo-miRNAs exhibit extensive regulations on various cellular behaviors in different tissue and potential therapeutic efficacy in disease models (Figure 4). For instance, mesenchymal stem cell (MSC)-derived exosomes have drawn major attention due to their broad therapeutic impacts in multiple diseases and the exo-miRs of MSC-derived EVs were proposed as the major bioactive compartments for those promising results. In bone-associated diseases, exosomal miR-150-3p promotes osteoblast proliferation and differentiation in osteoporosis and establishes potential targets in osteoporosis treatment [124]. Another study points out that EVs from MSCs during different osteogenic stages induced bone formation differently, due to the fact that their miR profile (such as miR-31, -144, and -221 as negative regulators) was sequentially changed from the expansion stage to osteogenic differentiation stage [125]. For cartilage regeneration, exosomal miR-92a-3p could regulate chondrogenesis and extend cartilage development and homeostasis by targeting WNT5A for osteoarthritis treatment [126]. MiR-148a and -29b enriched in Wharton's jelly mesenchymal stem cell-derived (WJMSC) EVs promote cartilage repair by regulating lineage commitment towards chondrogenesis instead of hypertrophic phenotype [127]. These studies suggest that regeneration may come from MSC paracrine effect rather than direct osteogenesis or chondrogenesis.

5.2. Exo-miR from MSCs in Cancer Treatment

In cancer models, MSC EVs also demonstrated potential regulations. Specifically, exosomal miR-139-5p inhibits bladder tumorigenesis by targeting the polycomb repressor complex 1 and miR-15a delays carcinoma progression by inhibiting spalt-like transcription factor 4 [128,129]. EVs enriched with miR-497 showed effective inhibition of tumor growth and angiogenesis [130]. On the other hand, exo-miRs (e.g., miR-21, miR-155, miR-146a, miR-148a, and miR-494) derived from tumor cells or activated macrophages can promote angiogenesis and immune escape to facilitate cancer metastasis in new studies [131–134]. Thus, exo-miRs can generally act as biomarkers in cancer diagnosis and prognosis. This complex functional diversity of exo-miRs suggests the needs for thorough evaluations of specific miRs in a case-by-case manner before defining the exo-miR profile for cancer treatments.

5.3. Exo-miR in Alzheimer's Disease Pathology and Treatment

Another potential application of MSC-derived EVs is for Alzheimer's disease (AD). For instance, WJMSCs produced exosomes enriched with miR-29a, which specifically target histone deacetylase 4 (HDAC4, an elevated marker in AD patients). A reduction in Aβ expression and improved cognitive recovery were observed after MSC EVs treatment, and a significant decrease in nuclear HDAC4, with a certain amount of HDAC4-related gene fluctuation [135,136]. Similarly, miR-29-enriched EVs also reduced the toxic effects of Amyloid β (Aβ) peptide and partially recovered cognitive impairment in rat AD models [137]. In another study, exosomal miR-21 was found to be enriched in EVs from hypoxia-preconditioned MSCs and ameliorated cognitive decline by regulating neuroinflammation and synaptic damage [138]. Interestingly, MSCs can be manipulated by altering in vitro culture conditions (e.g., hypoxia or 3D aggregation) and certain cellular changes will be captured in their secreted EVs [138,139]. Injection of 3D hMSC-derived EVs with a spectrum of upregulated miRs (such as miR-21, miR-22, and miR-1246) effectively prevented cognitive declines in AD mice [140]. Moreover, miR-21-5p from human urine-derived stem cell-derived EVs attenuated Rett syndrome, an early cognitive loss and neurologic deterioration disease, through the inhibition of Eph receptor A4 (Epha4) and its downstream signaling TEX [141]. Pathologically, cerebral EVs have been proved to contain amyloid precursor protein (APP) and C-terminal fragments (CTT), which all contribute to Aβ and tau protein accumulation [117]. In addition, EVs also carry neurotoxins and inflammation molecules, which further facilitate AD progression. Understanding the biological roles of EVs in disease pathology provides the possibility of designing EV-based vaccinations with exo-miRs, such as cell-free cancer immunotherapy (Figure 4) [142,143]. These studies indicate the existing connections between AD progression and exo-miRs, although the exact mechanisms need to be elucidated to achieve optimized miR cargo for AD therapy.

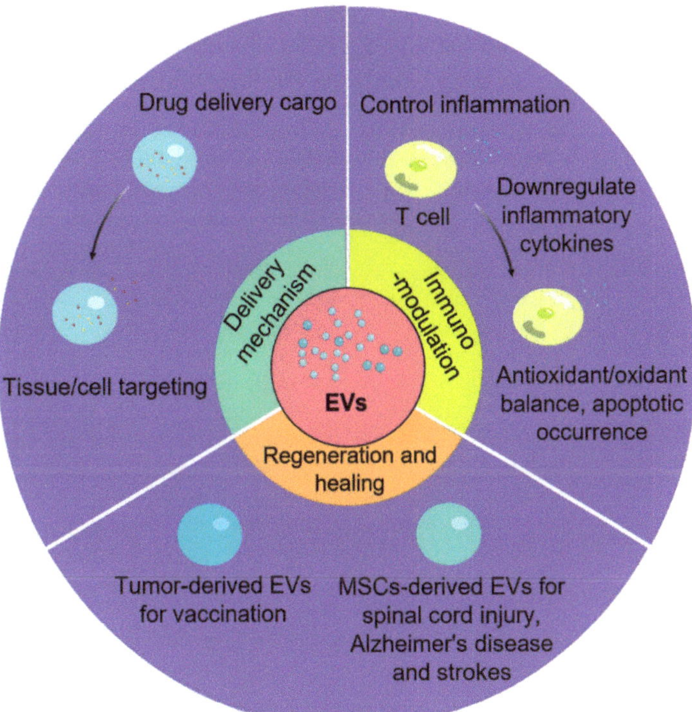

Figure 4. Potential engineering strategies in EV therapeutics. Based on their bioactivity and biostructure, EVs or EV mimics have been engineered to deliver therapeutic molecules. The immunomodulatory potentials of EVs have been applied for the relief of inflammatory sites or tissue damage. EVs from specific cell sources can be directly used as a cell-free therapy or for vaccination purposes. With the understanding of EV biogenesis and cargo sorting, the desired size subpopulation and cargo profile can be engineered via altering the cellular microenvironment of in vitro cultures.

5.4. Exo-miR in Spinal Cord Injury and Treatment

In spinal cord injury (SCI), neurological repairs require the regulation of miRs for neurogenesis, neural differentiation, and neural tube formation. A set of circulating exo-miRs have been revealed to be upregulated (miR-9, -124a, -7, -125a/b, and -375) or downregulated (miR-291-3p, -183, -92, -200b/c, and -382-5p) during this process in rodent models [144,145]. Therapeutically, MSC-derived EVs containing miR-126, -133b, -21, or -199a-3p/145-5p have shown potential therapeutic benefits to neuron regeneration and immunomodulation to promote functional recovery, possibly due to the complex regulatory mechanisms on RhoA, STAT3, and the Pi3k/Pten/Akt axis, etc. [144,146–149]. Moreover, EVs derived from neurons, microglia, and oligodendrocytes also provide neuroprotective effects via miRs after SCI, namely, miR-21 [150,151], -124-3p [152], -9, and -19a [153]. With these potential cargo candidates, artificial EVs or EV-mimic particles can be optimized for SCI treatment [154].

5.5. Exo-miR from MSCs in Ischemic Diseases

For ischemic diseases, EVs and miRNAs also exhibited promising therapeutic efficacy. For instance, ischemic stroke caused by oxygen and nutrient deprivation leads to severe lesions and neurological damage. MSC EVs loaded with miR-138-5p (specifically targeting lipocalin) prevented further astrocyte damage by oxygen/glucose deprivation (OGD) after endocytosis and thus alleviated lesion damage in ischemic mice [155]. Activating transcription factor 3 (ATF3) was upregulated in a rodent stroke model and later identified as a target of miR-221-3p. MSC-derived EVs loaded with miR-221-3p could be recognized by neurons, reduced local inflammation, and eased cellular death caused by OGD, via suppression of ATF3 expression in neurons [156]. Similarly, exosomal miR-146a-5p also modulated neuroinflammation via microglia in ischemic stroke, and the proposed target is the IRAK/TRAF6 signaling pathway. MiR-133b was found to be transferred by MSC EVs, promoted neural plasticity, and enhanced neurite outgrowth in MSC-based stroke treatment [157,158]. Those studies demonstrate the important role of EVs and miRs in neural tissue regeneration. In ischemic cardiomyocyte injury, one study showed that MSC EVs treatment could inhibit cardiomyocyte apoptosis caused by hypoxia/reoxygenation. Exosomal miR-486-5p targeted Pten in this process, thus activating the Pi3k/Akt survival pathway and extending the protective effect in mice I/R (ischemia/reperfusion)-injured myocardium [159]. Another study revealed that miR-126 improved cell survival in neonatal rat ventricular cardiomyocytes cultured under H_2O_2 and $CoCl_2$. MiR-126 binds to ERRFI1 protein to exhibit antioxidative effects and restore mitochondrial fitness, thus improving resistance to I/R in vivo [160]. MiR-182 in mouse bone marrow-derived MSC EVs was also proposed to regulate macrophage polarization by downregulating TLR4 and NF-κB, thus attenuating myocardial injury. Both miRNA-133a and miRNA-141 from MSC exosomes exhibited myocardial protection through the downregulation (via suppressing mastermind-like 1) and upregulation (via PTEN) of β-catenin, respectively [161,162]. These studies have aroused broad interests in the complicated miR profile and functions in EVs, which may lead to many applications in therapeutic engineering, although further validation is required.

5.6. Exo-miR Detection and EV Biomanufacturing

Recently, researchers have pushed forward the detection sensitivity and variety of exo-miRs with novel engineering strategies, such as microfluidic chips, to enrich EVs and the in situ detection of exo-miRs with catalyzed hairpin assembly [163]; nanochannel biosensors coated with functional peptide nucleic acids to achieve charge alteration for EV capture and high resolution detection [164]; and self-assembled tetrahedral DNA nanolabel-based electrochemical sensors to selectively detect exo-miRs [165]. A new subpopulation of EVs (e.g., exomere, <50 nm) was even identified recently with asymmetric-flow field-flow fractionation (AF4), which further elucidated the dynamic composition of proteins, small RNAs, and lipids in EVs [166,167]. Besides the popular delivery system with EVinspired liposome to load doxorubicin and kartogenin or other therapeutic drugs [168], fully synthetic/engineered EVs with the precisely controlled composition of miR cargo and the specific rations between miRs (i.e., stoichiometry) were engineered for wound healing and keratinocyte function, achieving similar effects compared to natural EVs [169]. With increased knowledge of exo-miRs and other cargo profiles in EVs, further modified/engineered EVs with therapeutic biomolecules are a promising alternative for next generation targeting therapies in tissue engineering and regenerative medicine. Engineering EVs for biomanufacturing is also critical for translational applications. While the cellular source of EVs is potentially case/disease dependent, the general implication for EV production is to scale-up and boost yield. Recent studies demonstrated that traditional suspension and microcarrier-based bioreactors are able to support the scale-up of cells and EVs [170]. When cultured in larger scale vertical-wheel bioreactors (0.5 Liter vs. 0.1 Liter), hMSC-EV production was well maintained or slightly boosted with similar proteomics and metabolomics, despite the shear stress impact on cultured cells. A similar bioprocessing approach has been tested in a hollow fiber bioreactor [171]. Moreover, a nonadherent culture of MSCs under a WAVE bioreactor also boosted EV yield threefold in a recent study [139]. The 3D aggregation process of hMSCs significantly altered EV production and the cargo profile compared to a 2D adherent culture, exerting different cargo profiles and advanced functions in immunomodulation and rejuvenation. In addition to the 3D aggregation of hMSCs for EV production [172–174], the application of a wave motion bioreactor enables the closed-system scale-up of EV bioprocessing and biomanufacturing. Interestingly, these studies also indicate the potential engineering strategy of EVs by manipulating the in vitro microenvironment.

6. Summary

The EV world is complex and fascinating. The complexity of EV heterogeneity, biogenesis, and important cargo miRs encourages us to further explore the biology in order to further understand the nature of vesicles. This review summarizes the basic concepts and current knowledge on exo-miRs and their sorting mechanisms in the parent cells for designing and engineering therapeutic cargos in EV-based therapies. Applications and preclinical studies indicate the potential therapeutic effects of exo-miRs in multiple disease models. Continuous studies are needed as our knowledge evolves in the EV field with advanced technologies and experimental strategies. As exo-miRs have been shown to participate in disease progression in ischemia and neurodegeneration, they are promising for potential cell-free therapies using fully synthetic EVs with defined miR and protein cargo neurological restoration.

Author Contributions: E.Z.Z. wrote the initial manuscript under X.Y.'s guidance. I.C. helped some section and reviewed the manuscript. X.C. helped the references and figures. X.Y. initiated the study, wrote, revised and finalized the manuscript. All authors have read and agreed to the published version of the manuscript.

Funding: Research reported in this publication was partially supported by the National Institutes of Health (USA) under Award Number R01NS125016. The content is solely the responsibility of the authors and does not necessarily represent the official views of the National Institutes of Health.

Institutional Review Board Statement: Not applicable.

Informed Consent Statement: Not applicable.

Data Availability Statement: Not applicable.

Acknowledgments: This study is partially supported by Toffler neuroscience scholar award (to X.Y.).

Conflicts of Interest: The authors declare no conflict of interest.

References

1. van Niel, G.; D'Angelo, G.; Raposo, G. Shedding light on the cell biology of extracellular vesicles. *Nat. Rev. Mol. Cell Biol.* **2018**, *19*, 213–228. [CrossRef]
2. Battistelli, M.; Falcieri, E. Apoptotic bodies: Particular extracellular vesicles involved in intercellular communication. *Biology* **2020**, *9*, 21. [CrossRef] [PubMed]
3. Phan, T.K.; Ozkocak, D.C.; Poon, I.K.H. Unleashing the therapeutic potential of apoptotic bodies. *Biochem. Soc. Trans.* **2020**, *48*, 2079–2088. [CrossRef]
4. Meehan, B.; Rak, J.; Di Vizio, D. Oncosomes-large and small: What are they, where they came from? *J. Extracell Vesicles* **2016**, *5*, 33109. [CrossRef]
5. Rich, J.N. Cancer stem cells in radiation resistance. *Cancer Res.* **2007**, *67*, 8980–8984. [CrossRef]
6. Hauser, P.; Wang, S.; Didenko, V.V. Apoptotic bodies: Selective detection in extracellular vesicles. In *Signal Transduction Immunohistochemistry*; Springer: Berlin/Heidelberg, Germany, 2017; pp. 193–200.
7. Muralidharan-Chari, V.; Clancy, J.; Plou, C.; Romao, M.; Chavrier, P.; Raposo, G.; D'Souza-Schorey, C. ARF6-regulated shedding of tumor cell-derived plasma membrane microvesicles. *Curr. Biol.* **2009**, *19*, 1875–1885. [CrossRef] [PubMed]
8. Li, B.; Antonyak, M.A.; Zhang, J.; Cerione, R.A. RhoA triggers a specific signaling pathway that generates transforming microvesicles in cancer cells. *Oncogene* **2012**, *31*, 4740–4749. [CrossRef] [PubMed]
9. Akers, J.C.; Gonda, D.; Kim, R.; Carter, B.S.; Chen, C.C. Biogenesis of extracellular vesicles (EV): Exosomes, microvesicles, retrovirus-like vesicles, and apoptotic bodies. *J. Neuro-Oncol.* **2013**, *113*, 1–11. [CrossRef]
10. Schlienger, S.; Campbell, S.; Claing, A. ARF1 regulates the Rho/MLC pathway to control EGF-dependent breast cancer cell invasion. *Mol. Biol. Cell* **2014**, *25*, 17–29. [CrossRef]
11. Sedgwick, A.E.; Clancy, J.W.; Olivia Balmert, M.; D'Souza-Schorey, C. Extracellular microvesicles and invadopodia mediate non-overlapping modes of tumor cell invasion. *Sci. Rep.* **2015**, *5*, 14748. [CrossRef]
12. Pap, E.; Pallinger, E.; Pasztoi, M.; Falus, A. Highlights of a new type of intercellular communication: Microvesicle-based information transfer. *Inflamm. Res.* **2009**, *58*, 1–8. [CrossRef]
13. Bianco, F.; Perrotta, C.; Novellino, L.; Francolini, M.; Riganti, L.; Menna, E.; Saglietti, L.; Schuchman, E.H.; Furlan, R.; Clementi, E. Acid sphingomyelinase activity triggers microparticle release from glial cells. *EMBO J.* **2009**, *28*, 1043–1054. [CrossRef] [PubMed]
14. Wang, J.; Pendurthi, U.R.; Rao, L.V.M. Sphingomyelin encrypts tissue factor: ATP-induced activation of A-SMase leads to tissue factor decryption and microvesicle shedding. *Blood Adv.* **2017**, *1*, 849–862. [CrossRef]
15. Marrone, M.C.; Morabito, A.; Giustizieri, M.; Chiurchiù, V.; Leuti, A.; Mattioli, M.; Marinelli, S.; Riganti, L.; Lombardi, M.; Murana, E. TRPV1 channels are critical brain inflammation detectors and neuropathic pain biomarkers in mice. *Nat. Commun.* **2017**, *8*, 15292. [CrossRef] [PubMed]
16. Del Conde, I.; Shrimpton, C.N.; Thiagarajan, P.; López, J.A. Tissue-factor–bearing microvesicles arise from lipid rafts and fuse with activated platelets to initiate coagulation. *Blood* **2005**, *106*, 1604–1611. [CrossRef] [PubMed]
17. Lingwood, D.; Simons, K. Lipid rafts as a membrane-organizing principle. *Science* **2010**, *327*, 46–50. [CrossRef]
18. Théry, C.; Witwer, K.W.; Aikawa, E.; Alcaraz, M.J.; Anderson, J.D.; Andriantsitohaina, R.; Antoniou, A.; Arab, T.; Archer, F.; Atkin-Smith, G.K. Minimal information for studies of extracellular vesicles 2018 (MISEV2018): A position statement of the International Society for Extracellular Vesicles and update of the MISEV2014 guidelines. *J. Extracell. Vesicles* **2018**, *7*, 1535750. [CrossRef]
19. Raymond, C.K.; Howald-Stevenson, I.; Vater, C.; Stevens, T. Morphological classification of the yeast vacuolar protein sorting mutants: Evidence for a prevacuolar compartment in class E vps mutants. *Mol. Biol. Cell* **1992**, *3*, 1389–1402. [CrossRef]
20. McCullough, J.; Frost, A.; Sundquist, W.I. Structures, functions, and dynamics of ESCRT-III/Vps4 membrane remodeling and fission complexes. *Annu. Rev. Cell Dev. Biol.* **2018**, *34*, 85–109. [CrossRef] [PubMed]
21. Garrus, J.E.; von Schwedler, U.K.; Pornillos, O.W.; Morham, S.G.; Zavitz, K.H.; Wang, H.E.; Wettstein, D.A.; Stray, K.M.; Côté, M.; Rich, R.L. Tsg101 and the vacuolar protein sorting pathway are essential for HIV-1 budding. *Cell* **2001**, *107*, 55–65. [CrossRef]
22. Wang, T.; Gilkes, D.M.; Takano, N.; Xiang, L.; Luo, W.; Bishop, C.J.; Chaturvedi, P.; Green, J.J.; Semenza, G.L. Hypoxia-inducible factors and RAB22A mediate formation of microvesicles that stimulate breast cancer invasion and metastasis. *Proc. Natl. Acad. Sci. USA* **2014**, *111*, E3234–E3242. [CrossRef] [PubMed]
23. Nabhan, J.F.; Hu, R.; Oh, R.S.; Cohen, S.N.; Lu, Q. Formation and release of arrestin domain-containing protein 1-mediated microvesicles (ARMMs) at plasma membrane by recruitment of TSG101 protein. *Proc. Natl. Acad. Sci. USA* **2012**, *109*, 4146–4151. [CrossRef]
24. Brown, M.; Roulson, J.-A.; Hart, C.A.; Tawadros, T.; Clarke, N.W. Arachidonic acid induction of Rho-mediated transendothelial migration in prostate cancer. *Br. J. Cancer* **2014**, *110*, 2099–2108. [CrossRef] [PubMed]
25. Kim, J.; Morley, S.; Le, M.; Bedoret, D.; Umetsu, D.T.; Di Vizio, D.; Freeman, M.R. Enhanced shedding of extracellular vesicles from amoeboid prostate cancer cells: Potential effects on the tumor microenvironment. *Cancer Biol. Ther.* **2014**, *15*, 409–418. [CrossRef] [PubMed]
26. McConnell, R.E.; Higginbotham, J.N.; Shifrin, D.A., Jr.; Tabb, D.L.; Coffey, R.J.; Tyska, M.J. The enterocyte microvillus is a vesicle-generating organelle. *J. Cell Biol.* **2009**, *185*, 1285–1298. [CrossRef] [PubMed]

27. Raposo, G.; Stoorvogel, W. Extracellular vesicles: Exosomes, microvesicles, and friends. *J. Cell Biol.* **2013**, *200*, 373–383. [CrossRef] [PubMed]
28. Gurung, S.; Perocheau, D.; Touramanidou, L.; Baruteau, J. The exosome journey: From biogenesis to uptake and intracellular signalling. *Cell Commun. Signal.* **2021**, *19*, 47. [CrossRef] [PubMed]
29. Hurley, J.H. The ESCRT complexes. *Crit. Rev. Biochem. Mol. Biol.* **2010**, *45*, 463–487. [CrossRef] [PubMed]
30. Colombo, M.; Moita, C.; Van Niel, G.; Kowal, J.; Vigneron, J.; Benaroch, P.; Manel, N.; Moita, L.F.; Théry, C.; Raposo, G. Analysis of ESCRT functions in exosome biogenesis, composition and secretion highlights the heterogeneity of extracellular vesicles. *J. Cell Sci.* **2013**, *126*, 5553–5565. [CrossRef] [PubMed]
31. Wollert, T.; Hurley, J.H. Molecular mechanism of multivesicular body biogenesis by ESCRT complexes. *Nature* **2010**, *464*, 864–869. [CrossRef] [PubMed]
32. Takahashi, H.; Mayers, J.R.; Wang, L.; Edwardson, J.M.; Audhya, A. Hrs and STAM function synergistically to bind ubiquitin-modified cargoes in vitro. *Biophys. J.* **2015**, *108*, 76–84. [CrossRef] [PubMed]
33. Katzmann, D.J.; Babst, M.; Emr, S.D. Ubiquitin-dependent sorting into the multivesicular body pathway requires the function of a conserved endosomal protein sorting complex, ESCRT-I. *Cell* **2001**, *106*, 145–155. [CrossRef]
34. Bilodeau, P.S.; Winistorfer, S.C.; Kearney, W.R.; Robertson, A.D.; Piper, R.C. Vps27-Hse1 and ESCRT-I complexes cooperate to increase efficiency of sorting ubiquitinated proteins at the endosome. *J. Cell Biol.* **2003**, *163*, 237–243. [CrossRef]
35. Gill, D.J.; Teo, H.; Sun, J.; Perisic, O.; Veprintsev, D.B.; Emr, S.D.; Williams, R.L. Structural insight into the ESCRT-I/-II link and its role in MVB trafficking. *EMBO J.* **2007**, *26*, 600–612. [CrossRef] [PubMed]
36. Kostelansky, M.S.; Schluter, C.; Tam, Y.Y.C.; Lee, S.; Ghirlando, R.; Beach, B.; Conibear, E.; Hurley, J.H. Molecular architecture and functional model of the complete yeast ESCRT-I heterotetramer. *Cell* **2007**, *129*, 485–498. [CrossRef] [PubMed]
37. De Souza, R.F.; Aravind, L. UMA and MABP domains throw light on receptor endocytosis and selection of endosomal cargoes. *Bioinformatics* **2010**, *26*, 1477–1480. [CrossRef]
38. Im, Y.J.; Hurley, J.H. Integrated structural model and membrane targeting mechanism of the human ESCRT-II complex. *Dev. Cell* **2008**, *14*, 902–913. [CrossRef]
39. Hierro, A.; Sun, J.; Rusnak, A.S.; Kim, J.; Prag, G.; Emr, S.D.; Hurley, J.H. Structure of the ESCRT-II endosomal trafficking complex. *Nature* **2004**, *431*, 221–225. [CrossRef] [PubMed]
40. Teo, H.; Perisic, O.; González, B.; Williams, R.L. ESCRT-II, an endosome-associated complex required for protein sorting: Crystal structure and interactions with ESCRT-III and membranes. *Dev. Cell* **2004**, *7*, 559–569. [CrossRef]
41. Teis, D.; Saksena, S.; Emr, S.D. Ordered assembly of the ESCRT-III complex on endosomes is required to sequester cargo during MVB formation. *Dev. Cell* **2008**, *15*, 578–589. [CrossRef]
42. Wollert, T.; Wunder, C.; Lippincott-Schwartz, J.; Hurley, J.H. Membrane scission by the ESCRT-III complex. *Nature* **2009**, *458*, 172–177. [CrossRef] [PubMed]
43. Gonciarz, M.D.; Whitby, F.G.; Eckert, D.M.; Kieffer, C.; Heroux, A.; Sundquist, W.I.; Hill, C.P. Biochemical and structural studies of yeast Vps4 oligomerization. *J. Mol. Biol.* **2008**, *384*, 878–895. [CrossRef]
44. Yeo, S.C.; Xu, L.; Ren, J.; Boulton, V.J.; Wagle, M.D.; Liu, C.; Ren, G.; Wong, P.; Zahn, R.; Sasajala, P. Vps20p and Vta1p interact with Vps4p and function in multivesicular body sorting and endosomal transport in Saccharomyces cerevisiae. *J. Cell Sci.* **2003**, *116*, 3957–3970. [CrossRef]
45. Raiborg, C.; Bremnes, B.; Mehlum, A.; Gillooly, D.J.; D'Arrigo, A.; Stang, E.; Stenmark, H. FYVE and coiled-coil domains determine the specific localisation of Hrs to early endosomes. *J. Cell Sci.* **2001**, *114*, 2255–2263. [CrossRef]
46. Razi, M.; Futter, C. Distinct roles for Tsg101 and Hrs in multivesicular body formation and inward vesiculation. *Mol. Biol. Cell* **2006**, *17*, 3469–3483. [CrossRef]
47. Slagsvold, T.; Aasland, R.; Hirano, S.; Bache, K.G.; Raiborg, C.; Trambaiolo, D.; Wakatsuki, S.; Stenmark, H. Eap45 in Mammalian ESCRT-II Binds Ubiquitin via a Phosphoinositide-interacting GLUE Domain*♦. *J. Biol. Chem.* **2005**, *280*, 19600–19606. [CrossRef] [PubMed]
48. Böker, K.O.; Lemus-Diaz, N.; Ferreira, R.R.; Schiller, L.; Schneider, S.; Gruber, J. The impact of the CD9 tetraspanin on lentivirus infectivity and exosome secretion. *Mol. Ther.* **2018**, *26*, 634–647. [CrossRef]
49. Adell, M.A.Y.; Migliano, S.M.; Upadhyayula, S.; Bykov, Y.S.; Sprenger, S.; Pakdel, M.; Vogel, G.F.; Jih, G.; Skillern, W.; Behrouzi, R. Recruitment dynamics of ESCRT-III and Vps4 to endosomes and implications for reverse membrane budding. *eLife* **2017**, *6*, e31652. [CrossRef]
50. Chiaruttini, N.; Redondo-Morata, L.; Colom, A.; Humbert, F.; Lenz, M.; Scheuring, S.; Roux, A. Relaxation of loaded ESCRT-III spiral springs drives membrane deformation. *Cell* **2015**, *163*, 866–879. [CrossRef]
51. Coulter, M.E.; Dorobantu, C.M.; Lodewijk, G.A.; Delalande, F.; Cianférani, S.; Ganesh, V.S.; Smith, R.S.; Lim, E.T.; Xu, C.S.; Pang, S. The ESCRT-III protein CHMP1A mediates secretion of sonic hedgehog on a distinctive subtype of extracellular vesicles. *Cell Rep.* **2018**, *24*, 973–986.e8. [CrossRef]
52. Bishop, N.; Woodman, P. ATPase-defective mammalian VPS4 localizes to aberrant endosomes and impairs cholesterol trafficking. *Mol. Biol. Cell* **2000**, *11*, 227–239. [CrossRef]
53. Fujita, H.; Yamanaka, M.; Imamura, K.; Tanaka, Y.; Nara, A.; Yoshimori, T.; Yokota, S.; Himeno, M. A dominant negative form of the AAA ATPase SKD1/VPS4 impairs membrane trafficking out of endosomal/lysosomal compartments: Class E vps phenotype in mammalian cells. *J. Cell Sci.* **2003**, *116*, 401–414. [CrossRef]

54. Baietti, M.F.; Zhang, Z.; Mortier, E.; Melchior, A.; Degeest, G.; Geeraerts, A.; Ivarsson, Y.; Depoortere, F.; Coomans, C.; Vermeiren, E. Syndecan–syntenin–ALIX regulates the biogenesis of exosomes. *Nat. Cell Biol.* **2012**, *14*, 677–685. [CrossRef]
55. Roucourt, B.; Meeussen, S.; Bao, J.; Zimmermann, P.; David, G. Heparanase activates the syndecan-syntenin-ALIX exosome pathway. *Cell Res.* **2015**, *25*, 412–428. [CrossRef]
56. Laulagnier, K.; Grand, D.; Dujardin, A.; Hamdi, S.; Vincent-Schneider, H.; Lankar, D.; Salles, J.-P.; Bonnerot, C.; Perret, B.; Record, M. PLD2 is enriched on exosomes and its activity is correlated to the release of exosomes. *FEBS Lett.* **2004**, *572*, 11–14. [CrossRef] [PubMed]
57. Ghossoub, R.; Lembo, F.; Rubio, A.; Gaillard, C.B.; Bouchet, J.; Vitale, N.; Slavík, J.; Machala, M.; Zimmermann, P. Syntenin-ALIX exosome biogenesis and budding into multivesicular bodies are controlled by ARF6 and PLD2. *Nat. Commun.* **2014**, *5*, 3477. [CrossRef]
58. Stuffers, S.; Sem Wegner, C.; Stenmark, H.; Brech, A. Multivesicular endosome biogenesis in the absence of ESCRTs. *Traffic* **2009**, *10*, 925–937. [CrossRef] [PubMed]
59. Edgar, J.R.; Eden, E.R.; Futter, C.E. Hrs-and CD63-dependent competing mechanisms make different sized endosomal intraluminal vesicles. *Traffic* **2014**, *15*, 197–211. [CrossRef] [PubMed]
60. Trajkovic, K.; Hsu, C.; Chiantia, S.; Rajendran, L.; Wenzel, D.; Wieland, F.; Schwille, P.; Brugger, B.; Simons, M. Ceramide triggers budding of exosome vesicles into multivesicular endosomes. *Science* **2008**, *319*, 1244–1247. [CrossRef]
61. Chen, T.; Guo, J.; Yang, M.; Zhu, X.; Cao, X. Chemokine-containing exosomes are released from heat-stressed tumor cells via lipid raft-dependent pathway and act as efficient tumor vaccine. *J. Immunol.* **2011**, *186*, 2219–2228. [CrossRef]
62. Möbius, W.; Van Donselaar, E.; Ohno-Iwashita, Y.; Shimada, Y.; Heijnen, H.; Slot, J.; Geuze, H. Recycling compartments and the internal vesicles of multivesicular bodies harbor most of the cholesterol found in the endocytic pathway. *Traffic* **2003**, *4*, 222–231. [CrossRef]
63. Kajimoto, T.; Okada, T.; Miya, S.; Zhang, L.; Nakamura, S. Ongoing activation of sphingosine 1-phosphate receptors mediates maturation of exosomal multivesicular endosomes. *Nat. Commun.* **2013**, *4*, 2712. [CrossRef]
64. Van Niel, G.; Charrin, S.; Simoes, S.; Romao, M.; Rochin, L.; Saftig, P.; Marks, M.S.; Rubinstein, E.; Raposo, G. The tetraspanin CD63 regulates ESCRT-independent and-dependent endosomal sorting during melanogenesis. *Dev. Cell* **2011**, *21*, 708–721. [CrossRef]
65. Van Niel, G.; Bergam, P.; Di Cicco, A.; Hurbain, I.; Cicero, A.L.; Dingli, F.; Palmulli, R.; Fort, C.; Potier, M.C.; Schurgers, L.J. Apolipoprotein E regulates amyloid formation within endosomes of pigment cells. *Cell Rep.* **2015**, *13*, 43–51. [CrossRef]
66. Hurwitz, S.N.; Nkosi, D.; Conlon, M.M.; York, S.B.; Liu, X.; Tremblay, D.C.; Meckes, D.G., Jr. CD63 regulates epstein-barr virus LMP1 exosomal packaging, enhancement of vesicle production, and noncanonical NF-κB signaling. *J. Virol.* **2017**, *91*, e02251-16. [CrossRef] [PubMed]
67. Hurwitz, S.N.; Cheerathodi, M.R.; Nkosi, D.; York, S.B.; Meckes, D.G., Jr. Tetraspanin CD63 bridges autophagic and endosomal processes to regulate exosomal secretion and intracellular signaling of Epstein-Barr virus LMP1. *J. Virol.* **2018**, *92*, e01969-17. [CrossRef] [PubMed]
68. Gauthier, S.A.; Pérez-González, R.; Sharma, A.; Huang, F.-K.; Alldred, M.J.; Pawlik, M.; Kaur, G.; Ginsberg, S.D.; Neubert, T.A.; Levy, E. Enhanced exosome secretion in Down syndrome brain-a protective mechanism to alleviate neuronal endosomal abnormalities. *Acta Neuropathol. Commun.* **2017**, *5*, 1–13. [CrossRef]
69. Mazurov, D.; Barbashova, L.; Filatov, A. Tetraspanin protein CD 9 interacts with metalloprotease CD 10 and enhances its release via exosomes. *FEBS J.* **2013**, *280*, 1200–1213. [CrossRef]
70. Buschow, S.I.; Nolte-'T Hoen, E.N.; Van Niel, G.; Pols, M.S.; Ten Broeke, T.; Lauwen, M.; Ossendorp, F.; Melief, C.J.; Raposo, G.; Wubbolts, R. MHC II in dendritic cells is targeted to lysosomes or T cell-induced exosomes via distinct multivesicular body pathways. *Traffic* **2009**, *10*, 1528–1542. [CrossRef]
71. Chairoungdua, A.; Smith, D.L.; Pochard, P.; Hull, M.; Caplan, M.J. Exosome release of β-catenin: A novel mechanism that antagonizes Wnt signaling. *J. Cell Biol.* **2010**, *190*, 1079–1091. [CrossRef] [PubMed]
72. Guix, F.X.; Sannerud, R.; Berditchevski, F.; Arranz, A.M.; Horré, K.; Snellinx, A.; Thathiah, A.; Saido, T.; Saito, T.; Rajesh, S. Tetraspanin 6: A pivotal protein of the multiple vesicular body determining exosome release and lysosomal degradation of amyloid precursor protein fragments. *Mol. Neurodegener.* **2017**, *12*, 25. [CrossRef] [PubMed]
73. Nazarenko, I.; Rana, S.; Baumann, A.; McAlear, J.; Hellwig, A.; Trendelenburg, M.; Lochnit, G.; Preissner, K.T.; Zöller, M. Cell surface tetraspanin Tspan8 contributes to molecular pathways of exosome-induced endothelial cell activation. *Cancer Res.* **2010**, *70*, 1668–1678. [CrossRef]
74. Hirano, S.; Kawasaki, M.; Ura, H.; Kato, R.; Raiborg, C.; Stenmark, H.; Wakatsuki, S. Double-sided ubiquitin binding of Hrs-UIM in endosomal protein sorting. *Nat. Struct. Mol. Biol.* **2006**, *13*, 272–277. [CrossRef]
75. Ren, X.; Hurley, J.H. VHS domains of ESCRT-0 cooperate in high-avidity binding to polyubiquitinated cargo. *EMBO J.* **2010**, *29*, 1045–1054. [CrossRef] [PubMed]
76. Raiborg, C.; Wesche, J.; Malerød, L.; Stenmark, H. Flat clathrin coats on endosomes mediate degradative protein sorting by scaffolding Hrs in dynamic microdomains. *J. Cell Sci.* **2006**, *119*, 2414–2424. [CrossRef]
77. Shields, S.B.; Oestreich, A.J.; Winistorfer, S.; Nguyen, D.; Payne, J.A.; Katzmann, D.J.; Piper, R. ESCRT ubiquitin-binding domains function cooperatively during MVB cargo sorting. *J. Cell Biol.* **2009**, *185*, 213–224. [CrossRef]

78. Gangalum, R.K.; Atanasov, I.C.; Zhou, Z.H.; Bhat, S.P. αB-crystallin is found in detergent-resistant membrane microdomains and is secreted via exosomes from human retinal pigment epithelial cells. *J. Biol. Chem.* **2011**, *286*, 3261–3269. [CrossRef] [PubMed]

79. de Gassart, A.; Géminard, C.; Février, B.; Raposo, G.; Vidal, M. Lipid raft-associated protein sorting in exosomes. *Blood* **2003**, *102*, 4336–4344. [CrossRef] [PubMed]

80. Kosaka, N.; Iguchi, H.; Yoshioka, Y.; Takeshita, F.; Matsuki, Y.; Ochiya, T. Secretory mechanisms and intercellular transfer of microRNAs in living cells*♦. *J. Biol. Chem.* **2010**, *285*, 17442–17452. [CrossRef]

81. Mittelbrunn, M.; Gutiérrez-Vázquez, C.; Villarroya-Beltri, C.; González, S.; Sánchez-Cabo, F.; González, M.Á.; Bernad, A.; Sánchez-Madrid, F. Unidirectional transfer of microRNA-loaded exosomes from T cells to antigen-presenting cells. *Nat. Commun.* **2011**, *2*, 282. [CrossRef]

82. Bakhoum, S.F.; Ngo, B.; Laughney, A.M.; Cavallo, J.-A.; Murphy, C.J.; Ly, P.; Shah, P.; Sriram, R.K.; Watkins, T.B.; Taunk, N.K. Chromosomal instability drives metastasis through a cytosolic DNA response. *Nature* **2018**, *553*, 467–472. [CrossRef]

83. Shen, Y.J.; Le Bert, N.; Chitre, A.A.; Koo, C.X.E.; Nga, X.H.; Ho, S.S.; Khatoo, M.; Tan, N.Y.; Ishii, K.J.; Gasser, S. Genome-derived cytosolic DNA mediates type I interferon-dependent rejection of B cell lymphoma cells. *Cell Rep.* **2015**, *11*, 460–473. [CrossRef]

84. Takahashi, A.; Okada, R.; Nagao, K.; Kawamata, Y.; Hanyu, A.; Yoshimoto, S.; Takasugi, M.; Watanabe, S.; Kanemaki, M.T.; Obuse, C. Exosomes maintain cellular homeostasis by excreting harmful DNA from cells. *Nat. Commun.* **2017**, *8*, 15287. [CrossRef]

85. Vagner, T.; Spinelli, C.; Minciacchi, V.R.; Balaj, L.; Zandian, M.; Conley, A.; Zijlstra, A.; Freeman, M.R.; Demichelis, F.; De, S. Large extracellular vesicles carry most of the tumour DNA circulating in prostate cancer patient plasma. *J. Extracell. vesicles* **2018**, *7*, 1505403. [CrossRef]

86. Kalluri, R.; LeBleu, V.S. Discovery of double-stranded genomic DNA in circulating exosomes. *Cold Spring Harb. Symp. Quant. Biol.* **2016**, *81*, 275–280. [CrossRef] [PubMed]

87. Kitai, Y.; Kawasaki, T.; Sueyoshi, T.; Kobiyama, K.; Ishii, K.J.; Zou, J.; Akira, S.; Matsuda, T.; Kawai, T. DNA-containing exosomes derived from cancer cells treated with topotecan activate a STING-dependent pathway and reinforce antitumor immunity. *J. Immunol.* **2017**, *198*, 1649–1659. [CrossRef] [PubMed]

88. Lian, Q.; Xu, J.; Yan, S.; Huang, M.; Ding, H.; Sun, X.; Bi, A.; Ding, J.; Sun, B.; Geng, M. Chemotherapy-induced intestinal inflammatory responses are mediated by exosome secretion of double-strand DNA via AIM2 inflammasome activation. *Cell Res.* **2017**, *27*, 784–800. [CrossRef] [PubMed]

89. Lázaro-Ibáñez, E.; Sanz-Garcia, A.; Visakorpi, T.; Escobedo-Lucea, C.; Siljander, P.; Ayuso-Sacido, Á.; Yliperttula, M. Different gDNA content in the subpopulations of prostate cancer extracellular vesicles: Apoptotic bodies, microvesicles, and exosomes. *Prostate* **2014**, *74*, 1379–1390. [CrossRef]

90. Kahlert, C.; Melo, S.A.; Protopopov, A.; Tang, J.; Seth, S.; Koch, M.; Zhang, J.; Weitz, J.; Chin, L.; Futreal, A. Identification of double-stranded genomic DNA spanning all chromosomes with mutated KRAS and p53 DNA in the serum exosomes of patients with pancreatic cancer. *J. Biol. Chem.* **2014**, *289*, 3869–3875. [CrossRef]

91. Guduric-Fuchs, J.; O'Connor, A.; Camp, B.; O'Neill, C.L.; Medina, R.J.; Simpson, D.A. Selective extracellular vesicle-mediated export of an overlapping set of microRNAs from multiple cell types. *BMC Genom.* **2012**, *13*, 357. [CrossRef]

92. Xiao, D.; Ohlendorf, J.; Chen, Y.; Taylor, D.D.; Rai, S.N.; Waigel, S.; Zacharias, W.; Hao, H.; McMasters, K.M. Identifying mRNA, microRNA and protein profiles of melanoma exosomes. *PLoS ONE* **2012**, *7*, e46874. [CrossRef] [PubMed]

93. Huang, X.; Yuan, T.; Tschannen, M.; Sun, Z.; Jacob, H.; Du, M.; Liang, M.; Dittmar, R.L.; Liu, Y.; Liang, M. Characterization of human plasma-derived exosomal RNAs by deep sequencing. *BMC Genom.* **2013**, *14*, 319. [CrossRef] [PubMed]

94. Li, Y.; Zheng, Q.; Bao, C.; Li, S.; Guo, W.; Zhao, J.; Chen, D.; Gu, J.; He, X.; Huang, S. Circular RNA is enriched and stable in exosomes: A promising biomarker for cancer diagnosis. *Cell Res.* **2015**, *25*, 981–984. [CrossRef]

95. Goldie, B.J.; Dun, M.D.; Lin, M.; Smith, N.D.; Verrills, N.M.; Dayas, C.V.; Cairns, M.J. Activity-associated miRNA are packaged in Map1b-enriched exosomes released from depolarized neurons. *Nucleic Acids Res.* **2014**, *42*, 9195–9208. [CrossRef]

96. Squadrito, M.L.; Baer, C.; Burdet, F.; Maderna, C.; Gilfillan, G.D.; Lyle, R.; Ibberson, M.; De Palma, M. Endogenous RNAs modulate microRNA sorting to exosomes and transfer to acceptor cells. *Cell Rep.* **2014**, *8*, 1432–1446. [CrossRef]

97. Kosaka, N.; Iguchi, H.; Hagiwara, K.; Yoshioka, Y.; Takeshita, F.; Ochiya, T. Neutral sphingomyelinase 2 (nSMase2)-dependent exosomal transfer of angiogenic microRNAs regulate cancer cell metastasis. *J. Biol. Chem.* **2013**, *288*, 10849–10859. [CrossRef]

98. Villarroya-Beltri, C.; Gutiérrez-Vázquez, C.; Sánchez-Cabo, F.; Pérez-Hernández, D.; Vázquez, J.; Martin-Cofreces, N.; Martinez-Herrera, D.J.; Pascual-Montano, A.; Mittelbrunn, M.; Sánchez-Madrid, F. Sumoylated hnRNPA2B1 controls the sorting of miRNAs into exosomes through binding to specific motifs. *Nat. Commun.* **2013**, *4*, 2980. [CrossRef]

99. Santangelo, L.; Giurato, G.; Cicchini, C.; Montaldo, C.; Mancone, C.; Tarallo, R.; Battistelli, C.; Alonzi, T.; Weisz, A.; Tripodi, M. The RNA-binding protein SYNCRIP is a component of the hepatocyte exosomal machinery controlling microRNA sorting. *Cell Rep.* **2016**, *17*, 799–808. [CrossRef]

100. Shurtleff, M.J.; Temoche-Diaz, M.M.; Karfilis, K.V.; Ri, S.; Schekman, R. Y-box protein 1 is required to sort microRNAs into exosomes in cells and in a cell-free reaction. *eLife* **2016**, *5*, e19276. [CrossRef]

101. Shurtleff, M.J.; Yao, J.; Qin, Y.; Nottingham, R.M.; Temoche-Diaz, M.M.; Schekman, R.; Lambowitz, A.M. Broad role for YBX1 in defining the small noncoding RNA composition of exosomes. *Proc. Natl. Acad. Sci. USA* **2017**, *114*, E8987–E8995. [CrossRef]

102. Koppers-Lalic, D.; Hackenberg, M.; Bijnsdorp, I.V.; van Eijndhoven, M.A.; Sadek, P.; Sie, D.; Zini, N.; Middeldorp, J.M.; Ylstra, B.; de Menezes, R.X. Nontemplated nucleotide additions distinguish the small RNA composition in cells from exosomes. *Cell Rep.* **2014**, *8*, 1649–1658. [CrossRef] [PubMed]

103. Lee, Y.S.; Pressman, S.; Andress, A.P.; Kim, K.; White, J.L.; Cassidy, J.J.; Li, X.; Lubell, K.; Lim, D.H.; Cho, I.S. Silencing by small RNAs is linked to endosomal trafficking. *Nat. Cell Biol.* **2009**, *11*, 1150–1156. [CrossRef] [PubMed]

104. Gibbings, D.J.; Ciaudo, C.; Erhardt, M.; Voinnet, O. Multivesicular bodies associate with components of miRNA effector complexes and modulate miRNA activity. *Nat. Cell Biol.* **2009**, *11*, 1143–1149. [CrossRef] [PubMed]

105. Melo, S.A.; Sugimoto, H.; O'Connell, J.T.; Kato, N.; Villanueva, A.; Vidal, A.; Qiu, L.; Vitkin, E.; Perelman, L.T.; Melo, C.A. Cancer exosomes perform cell-independent microRNA biogenesis and promote tumorigenesis. *Cancer Cell* **2014**, *26*, 707–721. [CrossRef]

106. Shao, Y.; Pan, X.; Fu, R. Role and Function of T Cell-Derived Exosomes and Their Therapeutic Value. *Mediat. Inflamm.* **2021**, *2021*, 8481013. [CrossRef] [PubMed]

107. Raposo, G.; Nijman, H.W.; Stoorvogel, W.; Liejendekker, R.; Harding, C.V.; Melief, C.; Geuze, H.J. B lymphocytes secrete antigen-presenting vesicles. *J. Exp. Med.* **1996**, *183*, 1161–1172. [CrossRef]

108. Tkach, M.; Kowal, J.; Zucchetti, A.E.; Enserink, L.; Jouve, M.; Lankar, D.; Saitakis, M.; Martin-Jaular, L.; Théry, C. Qualitative differences in T-cell activation by dendritic cell-derived extracellular vesicle subtypes. *EMBO J.* **2017**, *36*, 3012–3028. [CrossRef]

109. Hemler, M.E. Tetraspanin proteins mediate cellular penetration, invasion, and fusion events and define a novel type of membrane microdomain. *Annu. Rev. Cell Dev. Biol.* **2003**, *19*, 397–422. [CrossRef]

110. Nolte-'t Hoen, E.N.; Buschow, S.I.; Anderton, S.M.; Stoorvogel, W.; Wauben, M.H. Activated T cells recruit exosomes secreted by dendritic cells via LFA-1. *Blood J. Am. Soc. Hematol.* **2009**, *113*, 1977–1981. [CrossRef]

111. Rana, S.; Yue, S.; Stadel, D.; Zöller, M. Toward tailored exosomes: The exosomal tetraspanin web contributes to target cell selection. *Int. J. Biochem. Cell Biol.* **2012**, *44*, 1574–1584. [CrossRef]

112. Klibi, J.; Niki, T.; Riedel, A.; Pioche-Durieu, C.; Souquere, S.; Rubinstein, E.; Le Moulec, S.; Guigay, J.; Hirashima, M.; Guemira, F. Blood diffusion and Th1-suppressive effects of galectin-9–containing exosomes released by Epstein-Barr virus–infected nasopharyngeal carcinoma cells. *Blood J. Am. Soc. Hematol.* **2009**, *113*, 1957–1966. [CrossRef] [PubMed]

113. Céspedes, P.F.; Jainarayanan, A.; Fernández-Messina, L.; Valvo, S.; Saliba, D.G.; Kurz, E.; Kvalvaag, A.; Chen, L.; Ganskow, C.; Colin-York, H. T-cell trans-synaptic vesicles are distinct and carry greater effector content than constitutive extracellular vesicles. *Nat. Commun.* **2022**, *13*, 3460. [CrossRef]

114. Kamerkar, S.; LeBleu, V.S.; Sugimoto, H.; Yang, S.; Ruivo, C.F.; Melo, S.A.; Lee, J.J.; Kalluri, R. Exosomes facilitate therapeutic targeting of oncogenic KRAS in pancreatic cancer. *Nature* **2017**, *546*, 498–503. [CrossRef]

115. Tian, T.; Zhu, Y.-L.; Zhou, Y.-Y.; Liang, G.-F.; Wang, Y.-Y.; Hu, F.-H.; Xiao, Z.-D. Exosome uptake through clathrin-mediated endocytosis and macropinocytosis and mediating miR-21 delivery. *J. Biol. Chem.* **2014**, *289*, 22258–22267. [CrossRef]

116. Feng, D.; Zhao, W.L.; Ye, Y.Y.; Bai, X.C.; Liu, R.Q.; Chang, L.F.; Zhou, Q.; Sui, S.F. Cellular internalization of exosomes occurs through phagocytosis. *Traffic* **2010**, *11*, 675–687. [CrossRef]

117. Laulagnier, K.; Javalet, C.; Hemming, F.J.; Chivet, M.; Lachenal, G.; Blot, B.; Chatellard, C.; Sadoul, R. Amyloid precursor protein products concentrate in a subset of exosomes specifically endocytosed by neurons. *Cell. Mol. Life Sci.* **2018**, *75*, 757–773. [CrossRef] [PubMed]

118. Parolini, I.; Federici, C.; Raggi, C.; Lugini, L.; Palleschi, S.; De Milito, A.; Coscia, C.; Iessi, E.; Logozzi, M.; Molinari, A. Microenvironmental pH is a key factor for exosome traffic in tumor cells. *J. Biol. Chem.* **2009**, *284*, 34211–34222. [CrossRef]

119. Record, M.; Carayon, K.; Poirot, M.; Silvente-Poirot, S. Exosomes as new vesicular lipid transporters involved in cell–cell communication and various pathophysiologies. *Biochim. Et Biophys. Acta (BBA)-Mol. Cell Biol. Lipids* **2014**, *1841*, 108–120. [CrossRef] [PubMed]

120. Coleman, B.M.; Hill, A.F. Extracellular vesicles–Their role in the packaging and spread of misfolded proteins associated with neurodegenerative diseases. *Semin. Cell Dev. Biol.* **2015**, *40*, 89–96. [CrossRef] [PubMed]

121. Van Dongen, H.M.; Masoumi, N.; Witwer, K.W.; Pegtel, D.M. Extracellular vesicles exploit viral entry routes for cargo delivery. *Microbiol. Mol. Biol. Rev.* **2016**, *80*, 369–386. [CrossRef] [PubMed]

122. Montecalvo, A.; Larregina, A.T.; Shufesky, W.J.; Beer Stolz, D.; Sullivan, M.L.; Karlsson, J.M.; Baty, C.J.; Gibson, G.A.; Erdos, G.; Wang, Z. Mechanism of transfer of functional microRNAs between mouse dendritic cells via exosomes. *Blood J. Am. Soc. Hematol.* **2012**, *119*, 756–766. [CrossRef]

123. Zhao, H.; Yang, L.; Baddour, J.; Achreja, A.; Bernard, V.; Moss, T.; Marini, J.C.; Tudawe, T.; Seviour, E.G.; San Lucas, F.A. Tumor microenvironment derived exosomes pleiotropically modulate cancer cell metabolism. *eLife* **2016**, *5*, e10250. [CrossRef] [PubMed]

124. Qiu, M.; Zhai, S.; Fu, Q.; Liu, D. Bone marrow mesenchymal stem cells-derived exosomal microRNA-150-3p promotes osteoblast proliferation and differentiation in osteoporosis. *Hum. Gene Ther.* **2021**, *32*, 717–729. [CrossRef] [PubMed]

125. Wang, X.; Omar, O.; Vazirisani, F.; Thomsen, P.; Ekström, K. Mesenchymal stem cell-derived exosomes have altered microRNA profiles and induce osteogenic differentiation depending on the stage of differentiation. *PLoS ONE* **2018**, *13*, e0193059. [CrossRef]

126. Mao, G.; Zhang, Z.; Hu, S.; Zhang, Z.; Chang, Z.; Huang, Z.; Liao, W.; Kang, Y. Exosomes derived from miR-92a-3p-overexpressing human mesenchymal stem cells enhance chondrogenesis and suppress cartilage degradation via targeting WNT5A. *Stem Cell Res. Ther.* **2018**, *9*, 247. [CrossRef] [PubMed]

127. Jiang, S.; Tian, G.; Yang, Z.; Gao, X.; Wang, F.; Li, J.; Tian, Z.; Huang, B.; Wei, F.; Sang, X. Enhancement of acellular cartilage matrix scaffold by Wharton's jelly mesenchymal stem cell-derived exosomes to promote osteochondral regeneration. *Bioact. Mater.* **2021**, *6*, 2711–2728. [CrossRef] [PubMed]

128. Jia, Y.; Ding, X.; Zhou, L.; Zhang, L.; Yang, X. Mesenchymal stem cells-derived exosomal microRNA-139-5p restrains tumorigenesis in bladder cancer by targeting PRC1. *Oncogene* **2021**, *40*, 246–261. [CrossRef]

129. Ma, Y.-S.; Liu, J.-B.; Lin, L.; Zhang, H.; Wu, J.-J.; Shi, Y.; Jia, C.-Y.; Zhang, D.-D.; Yu, F.; Wang, H.-M. Exosomal microRNA-15a from mesenchymal stem cells impedes hepatocellular carcinoma progression via downregulation of SALL4. *Cell Death Discov.* **2021**, *7*, 224. [CrossRef]

130. Jeong, K.; Yu, Y.J.; You, J.Y.; Rhee, W.J.; Kim, J.A. Exosome-mediated microRNA-497 delivery for anti-cancer therapy in a microfluidic 3D lung cancer model. *Lab Chip* **2020**, *20*, 548–557. [CrossRef]

131. Hou, C.; Sun, N.; Han, W.; Meng, Y.; Wang, C.; Zhu, Q.; Tang, Y.; Ye, J. Exosomal microRNA-23b-3p promotes tumor angiogenesis and metastasis by targeting PTEN in Salivary adenoid cystic carcinoma. *Carcinogenesis* **2022**, *43*, 682–692. [CrossRef]

132. Ma, Y.-S.; Wu, T.-M.; Ling, C.-C.; Yu, F.; Zhang, J.; Cao, P.-S.; Gu, L.-P.; Wang, H.-M.; Xu, H.; Li, L. M2 macrophage-derived exosomal microRNA-155-5p promotes the immune escape of colon cancer by downregulating ZC3H12B. *Mol. Ther.-Oncolytics* **2021**, *20*, 484–498. [CrossRef]

133. Yang, C.; Dou, R.; Wei, C.; Liu, K.; Shi, D.; Zhang, C.; Liu, Q.; Wang, S.; Xiong, B. Tumor-derived exosomal microRNA-106b-5p activates EMT-cancer cell and M2-subtype TAM interaction to facilitate CRC metastasis. *Mol. Ther.* **2021**, *29*, 2088–2107. [CrossRef] [PubMed]

134. Mao, G.; Liu, Y.; Fang, X.; Liu, Y.; Fang, L.; Lin, L.; Liu, X.; Wang, N. Tumor-derived microRNA-494 promotes angiogenesis in non-small cell lung cancer. *Angiogenesis* **2015**, *18*, 373–382. [CrossRef] [PubMed]

135. Chen, Y.-A.; Lu, C.-H.; Ke, C.-C.; Chiu, S.-J.; Jeng, F.-S.; Chang, C.-W.; Yang, B.-H.; Liu, R.-S. Mesenchymal stem cell-derived exosomes ameliorate Alzheimer's disease pathology and improve cognitive deficits. *Biomedicines* **2021**, *9*, 594. [CrossRef] [PubMed]

136. Müller, M.; Jäkel, L.; Bruinsma, I.B.; Claassen, J.A.; Kuiperij, H.B.; Verbeek, M.M. MicroRNA-29a is a candidate biomarker for Alzheimer's disease in cell-free cerebrospinal fluid. *Mol. Neurobiol.* **2016**, *53*, 2894–2899. [CrossRef]

137. Jahangard, Y.; Monfared, H.; Moradi, A.; Zare, M.; Mirnajafi-Zadeh, J.; Mowla, S.J. Therapeutic effects of transplanted exosomes containing miR-29b to a rat model of Alzheimer's disease. *Front. Neurosci.* **2020**, *14*, 564. [CrossRef] [PubMed]

138. Cui, G.H.; Wu, J.; Mou, F.F.; Xie, W.H.; Wang, F.B.; Wang, Q.L.; Fang, J.; Xu, Y.W.; Dong, Y.R.; Liu, J.R. Exosomes derived from hypoxia-preconditioned mesenchymal stromal cells ameliorate cognitive decline by rescuing synaptic dysfunction and regulating inflammatory responses in APP/PS1 mice. *FASEB J.* **2018**, *32*, 654–668. [CrossRef]

139. Yuan, X.; Sun, L.; Jeske, R.; Nkosi, D.; York, S.B.; Liu, Y.; Grant, S.C.; Meckes, D.G., Jr.; Li, Y. Engineering extracellular vesicles by three-dimensional dynamic culture of human mesenchymal stem cells. *J. Extracell. Vesicles* **2022**, *11*, e12235. [CrossRef] [PubMed]

140. Cone, A.S.; Yuan, X.; Sun, L.; Duke, L.C.; Vreones, M.P.; Carrier, A.N.; Kenyon, S.M.; Carver, S.R.; Benthem, S.D.; Stimmell, A.C. Mesenchymal stem cell-derived extracellular vesicles ameliorate Alzheimer's disease-like phenotypes in a preclinical mouse model. *Theranostics* **2021**, *11*, 8129. [CrossRef] [PubMed]

141. Pan, W.; Xu, X.; Zhang, M.; Song, X. Human urine-derived stem cell-derived exosomal miR-21-5p promotes neurogenesis to attenuate Rett syndrome via the EPha4/TEK axis. *Lab. Investig.* **2021**, *101*, 824–836. [CrossRef] [PubMed]

142. Huda, M.N.; Nurunnabi, M. Potential application of exosomes in vaccine development and delivery. *Pharm. Res.* **2022**, 1–37. [CrossRef]

143. Naseri, M.; Bozorgmehr, M.; Zoller, M.; Pirmardan, E.R.; Madjd, Z. Tumor-derived exosomes: The next generation of promising cell-free vaccines in cancer immunotherapy. *Oncoimmunology* **2020**, *9*, 1779991. [CrossRef] [PubMed]

144. Pan, D.; Liu, W.; Zhu, S.; Fan, B.; Yu, N.; Ning, G.; Feng, S. Potential of different cells-derived exosomal microRNA cargos for treating spinal cord injury. *J. Orthop. Transl.* **2021**, *31*, 33–40. [CrossRef] [PubMed]

145. Ashmwe, M.; Posa, K.; Rührnößl, A.; Heinzel, J.C.; Heimel, P.; Mock, M.; Schädl, B.; Keibl, C.; Couillard-Despres, S.; Redl, H. Effects of Extracorporeal Shockwave Therapy on Functional Recovery and Circulating miR-375 and miR-382-5p after Subacute and Chronic Spinal Cord Contusion Injury in Rats. *Biomedicines* **2022**, *10*, 1630. [CrossRef]

146. Kang, J.; Li, Z.; Zhi, Z.; Wang, S.; Xu, G. MiR-21 derived from the exosomes of MSCs regulates the death and differentiation of neurons in patients with spinal cord injury. *Gene Ther.* **2019**, *26*, 491–503. [CrossRef] [PubMed]

147. Hu, J.; Zeng, L.; Huang, J.; Wang, G.; Lu, H. miR-126 promotes angiogenesis and attenuates inflammation after contusion spinal cord injury in rats. *Brain Res.* **2015**, *1608*, 191–202. [CrossRef]

148. Xia, C.; Cai, Y.; Lin, Y.; Guan, R.; Xiao, G.; Yang, J. MiR-133b-5p regulates the expression of the heat shock protein 70 during rat neuronal cell apoptosis induced by the gp120 V3 loop peptide. *J. Med. Virol.* **2016**, *88*, 437–447. [CrossRef] [PubMed]

149. Liu, W.; Rong, Y.; Wang, J.; Zhou, Z.; Ge, X.; Ji, C.; Jiang, D.; Gong, F.; Li, L.; Chen, J. Exosome-shuttled miR-216a-5p from hypoxic preconditioned mesenchymal stem cells repair traumatic spinal cord injury by shifting microglial M1/M2 polarization. *J. Neuroinflamm.* **2020**, *17*, 47. [CrossRef] [PubMed]

150. Valadi, H.; Ekström, K.; Bossios, A.; Sjöstrand, M.; Lee, J.J.; Lötvall, J.O. Exosome-mediated transfer of mRNAs and microRNAs is a novel mechanism of genetic exchange between cells. *Nat. Cell Biol.* **2007**, *9*, 654–659. [CrossRef]

151. Jiang, D.; Gong, F.; Ge, X.; Lv, C.; Huang, C.; Feng, S.; Zhou, Z.; Rong, Y.; Wang, J.; Ji, C. Neuron-derived exosomes-transmitted miR-124-3p protect traumatically injured spinal cord by suppressing the activation of neurotoxic microglia and astrocytes. *J. Nanobiotechnol.* **2020**, *18*, 105. [CrossRef]

152. Zhang, G.; Yang, P. A novel cell-cell communication mechanism in the nervous system: Exosomes. *J. Neurosci. Res.* **2018**, *96*, 45–52. [CrossRef] [PubMed]

153. Fröhlich, D.; Kuo, W.P.; Frühbeis, C.; Sun, J.-J.; Zehendner, C.M.; Luhmann, H.J.; Pinto, S.; Toedling, J.; Trotter, J.; Krämer-Albers, E.-M. Multifaceted effects of oligodendroglial exosomes on neurons: Impact on neuronal firing rate, signal transduction and gene regulation. *Philos. Trans. R. Soc. B Biol. Sci.* **2014**, *369*, 20130510. [CrossRef] [PubMed]

154. Yu, B.; Xue, X.; Yin, Z.; Cao, L.; Li, M.; Huang, J. Engineered Cell Membrane-Derived Nanocarriers: The Enhanced Delivery System for Therapeutic Applications. *Front. Cell Dev. Biol.* **2022**, *10*, 844050. [CrossRef] [PubMed]

155. Deng, Y.; Chen, D.; Gao, F.; Lv, H.; Zhang, G.; Sun, X.; Liu, L.; Mo, D.; Ma, N.; Song, L. Exosomes derived from microRNA-138-5p-overexpressing bone marrow-derived mesenchymal stem cells confer neuroprotection to astrocytes following ischemic stroke via inhibition of LCN2. *J. Biol. Eng.* **2019**, *13*, 71. [CrossRef]

156. Ai, Z.; Cheng, C.; Zhou, L.; Yin, S.; Wang, L.; Liu, Y. Bone marrow mesenchymal stem cells-derived extracellular vesicles carrying microRNA-221-3p protect against ischemic stroke via ATF3. *Brain Res. Bull.* **2021**, *172*, 220–228. [CrossRef]

157. Xin, H.; Li, Y.; Liu, Z.; Wang, X.; Shang, X.; Cui, Y.; Zhang, Z.G.; Chopp, M. MiR-133b promotes neural plasticity and functional recovery after treatment of stroke with multipotent mesenchymal stromal cells in rats via transfer of exosome-enriched extracellular particles. *Stem Cells* **2013**, *31*, 2737–2746. [CrossRef]

158. Xin, H.; Li, Y.; Buller, B.; Katakowski, M.; Zhang, Y.; Wang, X.; Shang, X.; Zhang, Z.G.; Chopp, M. Exosome-mediated transfer of miR-133b from multipotent mesenchymal stromal cells to neural cells contributes to neurite outgrowth. *Stem Cells* **2012**, *30*, 1556–1564. [CrossRef]

159. Sun, X.-H.; Wang, X.; Zhang, Y.; Hui, J. Exosomes of bone-marrow stromal cells inhibit cardiomyocyte apoptosis under ischemic and hypoxic conditions via miR-486-5p targeting the PTEN/PI3K/AKT signaling pathway. *Thromb. Res.* **2019**, *177*, 23–32. [CrossRef]

160. Wang, W.; Zheng, Y.; Wang, M.; Yan, M.; Jiang, J.; Li, Z. Exosomes derived miR-126 attenuates oxidative stress and apoptosis from ischemia and reperfusion injury by targeting ERRFI1. *Gene* **2019**, *690*, 75–80. [CrossRef]

161. Li, Q.; Jin, Y.; Ye, X.; Wang, W.; Deng, G.; Zhang, X. Bone marrow mesenchymal stem cell-derived exosomal MicroRNA-133a restrains myocardial fibrosis and epithelial–mesenchymal transition in viral myocarditis rats through suppressing MAML1. *Nanoscale Res. Lett.* **2021**, *16*, 111. [CrossRef]

162. Pei, Y.; Xie, S.; Li, J.; Jia, B. Bone marrow-mesenchymal stem cell-derived exosomal microRNA-141 targets PTEN and activates β-catenin to alleviate myocardial injury in septic mice. *Immunopharmacol. Immunotoxicol.* **2021**, *43*, 584–593. [CrossRef] [PubMed]

163. Qian, C.; Xiao, Y.; Wang, J.; Li, Y.; Li, S.; Wei, B.; Du, W.; Feng, X.; Chen, P.; Liu, B.-F. Rapid exosomes concentration and in situ detection of exosomal microRNA on agarose-based microfluidic chip. *Sens. Actuators B Chem.* **2021**, *333*, 129559. [CrossRef]

164. Xiao, P.-P.; Wan, Q.-Q.; Liao, T.; Tu, J.-Y.; Zhang, G.-J.; Sun, Z.-Y. Peptide nucleic acid-functionalized nanochannel biosensor for the highly sensitive detection of tumor exosomal microRNA. *Anal. Chem.* **2021**, *93*, 10966–10973. [CrossRef] [PubMed]

165. Liu, N.; Lu, H.; Liu, L.; Ni, W.; Yao, Q.; Zhang, G.-J.; Yang, F. Ultrasensitive exosomal MicroRNA detection with a supercharged DNA framework nanolabel. *Anal. Chem.* **2021**, *93*, 5917–5923. [CrossRef]

166. Zhang, H.; Freitas, D.; Kim, H.S.; Fabijanic, K.; Li, Z.; Chen, H.; Mark, M.T.; Molina, H.; Martin, A.B.; Bojmar, L. Identification of distinct nanoparticles and subsets of extracellular vesicles by asymmetric flow field-flow fractionation. *Nat. Cell Biol.* **2018**, *20*, 332–343. [CrossRef]

167. Anand, S.; Samuel, M.; Mathivanan, S. Exomeres: A new member of extracellular vesicles family. In *New Frontiers: Extracellular Vesicles*; Springer: Berlin/Heidelberg, Germany, 2021; pp. 89–97.

168. Liang, Y.; Duan, L.; Lu, J.; Xia, J. Engineering exosomes for targeted drug delivery. *Theranostics* **2021**, *11*, 3183. [CrossRef]

169. Staufer, O.; Dietrich, F.; Rimal, R.; Schroter, M.; Fabritz, S.; Boehm, H.; Singh, S.; Moller, M.; Platzman, I.; Spatz, J.P. Bottom-up assembly of biomedical relevant fully synthetic extracellular vesicles. *Sci. Adv.* **2021**, *7*, eabg6666. [CrossRef]

170. Jeske, R.; Liu, C.; Duke, L.; Castro, M.L.C.; Muok, L.; Arthur, P.; Singh, M.; Jung, S.; Sun, L.; Li, Y. Upscaling human mesenchymal stromal cell production in a novel vertical-wheel bioreactor enhances extracellular vesicle secretion and cargo profile. *Bioact. Mater.* 2022; *in press.* [CrossRef]

171. Cao, J.; Wang, B.; Tang, T.; Lv, L.; Ding, Z.; Li, Z.; Hu, R.; Wei, Q.; Shen, A.; Fu, Y. Three-dimensional culture of MSCs produces exosomes with improved yield and enhanced therapeutic efficacy for cisplatin-induced acute kidney injury. *Stem Cell Res. Ther.* **2020**, *11*, 206. [CrossRef]

172. Zhang, Y.; Chen, J.; Fu, H.; Kuang, S.; He, F.; Zhang, M.; Shen, Z.; Qin, W.; Lin, Z.; Huang, S. Exosomes derived from 3D-cultured MSCs improve therapeutic effects in periodontitis and experimental colitis and restore the Th17 cell/Treg balance in inflamed periodontium. *Int. J. Oral Sci.* **2021**, *13*, 43. [CrossRef]

173. Kim, M.; Yun, H.-W.; Park, D.Y.; Choi, B.H.; Min, B.-H. Three-dimensional spheroid culture increases exosome secretion from mesenchymal stem cells. *Tissue Eng. Regen. Med.* **2018**, *15*, 427–436. [CrossRef]

174. Cha, J.M.; Shin, E.K.; Sung, J.H.; Moon, G.J.; Kim, E.H.; Cho, Y.H.; Park, H.D.; Bae, H.; Kim, J.; Bang, O.Y. Efficient scalable production of therapeutic microvesicles derived from human mesenchymal stem cells. *Sci. Rep.* **2018**, *8*, 1171. [CrossRef] [PubMed]

Article

Intrathecal Injection of the Secretome from ALS Motor Neurons Regulated for miR-124 Expression Prevents Disease Outcomes in SOD1-G93A Mice

Marta Barbosa [1], Marta Santos [1], Nídia de Sousa [2,3], Sara Duarte-Silva [2,3], Ana Rita Vaz [1,4], António J. Salgado [2,3] and Dora Brites [1,4,*]

[1] Faculty of Pharmacy, Research Institute for Medicines (iMed.ULisboa), Universidade de Lisboa, 1649-003 Lisbon, Portugal
[2] School of Medicine, Life and Health Sciences Research Institute (ICVS), University of Minho, 4710-057 Braga, Portugal
[3] ICVS/3B's Associate Lab, PT Government Associated Lab, 4806-909 Guimarães, Portugal
[4] Department of Pharmaceutical Sciences and Medicines, Faculty of Pharmacy, Universidade de Lisboa, 1649-003 Lisbon, Portugal
* Correspondence: dbrites@ff.ul.pt

Abstract: Amyotrophic lateral sclerosis (ALS) is a neurodegenerative disease with short life expectancy and no effective therapy. We previously identified upregulated miR-124 in NSC-34-motor neurons (MNs) expressing human SOD1-G93A (mSOD1) and established its implication in mSOD1 MN degeneration and glial cell activation. When anti-miR-124-treated mSOD1 MN (preconditioned) secretome was incubated in spinal cord organotypic cultures from symptomatic mSOD1 mice, the dysregulated homeostatic balance was circumvented. To decipher the therapeutic potential of such preconditioned secretome, we intrathecally injected it in mSOD1 mice at the early stage of the disease (12-week-old). Preconditioned secretome prevented motor impairment and was effective in counteracting muscle atrophy, glial reactivity/dysfunction, and the neurodegeneration of the symptomatic mSOD1 mice. Deficits in corticospinal function and gait abnormalities were precluded, and the loss of gastrocnemius muscle fiber area was avoided. At the molecular level, the preconditioned secretome enhanced NeuN mRNA/protein expression levels and the PSD-95/TREM2/IL-10/arginase 1/MBP/PLP genes, thus avoiding the neuronal/glial cell dysregulation that characterizes ALS mice. It also prevented upregulated GFAP/Cx43/S100B/vimentin and inflammatory-associated miRNAs, specifically miR-146a/miR-155/miR-21, which are displayed by symptomatic animals. Collectively, our study highlights the intrathecal administration of the secretome from anti-miR-124-treated mSOD1 MNs as a therapeutic strategy for halting/delaying disease progression in an ALS mouse model.

Keywords: ALS mouse model; anti-microRNA-124; intraspinal delivery route; neuroprotection; prevention of glial dysfunction; preservation of motor performance; secretome-based therapy; SOD1-G93A mutation

Citation: Barbosa, M.; Santos, M.; de Sousa, N.; Duarte-Silva, S.; Vaz, A.R.; Salgado, A.J.; Brites, D. Intrathecal Injection of the Secretome from ALS Motor Neurons Regulated for miR-124 Expression Prevents Disease Outcomes in SOD1-G93A Mice. *Biomedicines* **2022**, *10*, 2120. https://doi.org/10.3390/biomedicines10092120

Academic Editors: Milena Rizzo and Elena Levantini

Received: 30 June 2022
Accepted: 24 August 2022
Published: 29 August 2022

Publisher's Note: MDPI stays neutral with regard to jurisdictional claims in published maps and institutional affiliations.

1. Introduction

Amyotrophic lateral sclerosis (ALS) is a fast progressive disease that results from the degeneration of the upper and lower motor neurons (MNs) in the motor cortex, brainstem, and spinal cord (SC). Consequently, this neurodegeneration leads to voluntary muscle weakness and wasting, resulting in a diversity of symptoms, such as dysarthria, dysphagia, weakness, and atrophy of the limbs. The pathological mechanisms underlying the disease include misfolded protein aggregation, mitochondrial dysfunction, dysregulated axonal transport, altered synaptic dynamics, excitotoxicity, and neuroinflammation [1]. It is nowadays accepted that the dysregulation of glial cells also contributes to the pathogenesis and progression of the disease [2]. Lately, the single-cell RNAseq of ALS microglia evidenced a molecular signature of the disease driven by the triggering receptor expressed on myeloid cells 2 (TREM2), defined as disease-associated microglia (DAM) [3]. By identifying DAM profiles, it is possible to better understand the heterogeneous microglia cell responses observed along the disease development. Currently, there is no effective treatment to prevent disease progression or change the disease course. One of the most commonly used models to explore ALS pathophysiology are mice bearing mutations [4]. Among these, superoxide dismutase 1 mice carrying the glycine to alanine point mutation at residue 93 (SOD1-G93A, mSOD1) recapitulate the symptoms of ALS patients and some of the glial phenotypes associated with the disease [5,6].

microRNAs (miRNAs) have been defined as important regulatory molecules for gene expression, and their dysregulation in ALS can be modulated toward the recovery of cell function and an increase in mSOD1 mouse lifespan, as reviewed in [7]. Furthermore, these molecules can be released into the secretome as free species or encapsulated in small extracellular vesicles (sEVs) [8–11] and internalised by neighbouring cells, also having the ability to influence distant cells and the extracellular environment [11,12]. Since ALS is characterized as a neuroinflammatory and neurodegenerative disease, we have focused on the study of inflammatory-associated (inflamma)-miRNAs, including miR-124, miR-155, and miR-146a, which were shown to be differentially expressed at the cellular and regional levels [12,13]. In particular, miR-124 is one of the most abundant miRNAs in the central nervous system and is involved in several important neuronal functions, such as neuronal differentiation/maturation and the regulation of synaptic activity [14]. However, we showed that its upregulation in in vitro models of ALS MNs was associated with neurodegeneration and that miR-124 disseminates through sEVs and causes time-dependent alterations in recipient microglial cells [11]. miRNA mimics and inhibitors have been proposed as therapeutics to modulate dysregulated miRNAs in cancer and multiple sclerosis [15,16]. Recently, we demonstrated that the incubation of the secretome from anti-miR-124-treated mSOD1 MNs (preconditioned secretome) was able to counteract the increased levels of IL-1β, IL-18, HMGB1, arginase 1, and inducible nitric oxide synthase (iNOS) in microglia treated with the mSOD1 secretome [13]. Interestingly, our data evidenced that the targeting of normal miR-124 values in ALS MNs enriched in miR-124 abrogates miR-125b overexpression and causes miR-146a/miR-21 downregulation in both cells and the secretome, thus inhibiting a pathological inflamma-miRNA profile in mSOD1 MNs. Such a preconditioned secretome also showed similar benefits in the spinal cord organotypic cultures (SCOCs) from mSOD1 mice by preventing the dysregulation of inflammation-associated genes and cell demise. This is not without precedent, since the secretome from mesenchymal stem cells has also been shown to exert neuroprotective, immunomodulatory, and regenerative effects in retinal degeneration, Parkinson's disease, and SC injury, as examples [15–17]. Another study demonstrated the therapeutic potential of the administration of the secretome from adipose-derived stem cells in in vivo ALS mice by preventing MN loss and extending animal lifespan [18]. Importantly, conditioned media from human pluripotent stem cells showed neuroprotective effects on MNs, including those from ALS patients; improved neuromuscular junction; and delayed morbidity in mSOD1 mice [19]. The authors claimed that it has potential for autologous treatment in ALS. In addition, many studies have elucidated the therapeutic promise of miRNAs after their modulation, either by overexpression or suppression in mSOD1 mice and in astrocytes directly converted from patient somatic cells [8,20]. Some works have evidenced an increased lifespan of mice [20–23], followed by an improvement in MN survival [21,22] and muscle strength [21,23]. However, the benefits of the administration of the secretome derived from miRNA-modulated cells in in vivo ALS mice has never been explored.

For this reason, we decided to evaluate the therapeutic potential of anti-miR-124-treated mSOD1 MNs in preventing disabilities in mSOD1 mice. Thus, we performed an intrathecal injection of the preconditioned secretome in the mSOD1 mice at the early symptomatic disease stage. At the symptomatic stage, we evaluated the motor performance by testing the gait quality and the corticospinal function. Moreover, we collected the gastrocnemius muscle for muscle integrity evaluation and lumbar SC for neurodegeneration, neuroinflammation, and myelination assessments. Our results revealed that the preconditioned secretome prevented MN and glial pathological mechanisms and improved motor disabilities in the ALS mice. Therefore, this preconditioned secretome shows promise as a therapeutic tool to be tested in stratified patients with upregulated miR-124 levels, namely at symptom onset, contributing to the advance of precision medicine.

2. Materials and Methods

2.1. NSC-34 MN-like Cell Culture and Transfection Followed by sEV and Secretome Collection

NSC-34 MN cell line stably transfected with human WT SOD1 and human mSOD1 was grown in proliferation media (DMEM high-glucose with L-glutamine, no sodium pyruvate, supplemented with 10% FBS (Thermo Fisher Scientific, Waltham, MA, USA); 1% penicillin/streptomycin or pen/strep (Sigma-Aldrich, St. Louis, MO, USA); and 0.1% geneticin sulphate 418 (G418) for cell selection), as described in [13]. After 48 h of proliferation, differentiation was induced by changing the medium to DMEM-F12 plus 1% FBS; 1% nonessential amino acids (Merck, Darmstadt, Germany); 1% pen/strep; and 0.1% G418 [13]. After 24 h, mSOD1 MNs were transfected with 15 nM anti-miR-124 (Ambion® Anti-miR™ miRNA inhibitor, #AM10691) and mixed with X-tremeGENE™ HP DNA Transfection Reagent (Sigma-Aldrich, St. Louis, MO, USA) in a proportion of 2:1 and diluted in Opti-MEM™. Cells were left for 12 h, then fresh differentiation medium was added, and cells were maintained for an additional 48 h (4 DIV). This medium containing the factors secreted by mSOD1 MNs (secretome) was collected and centrifuged at $1000 \times g$ for 10 min to remove any cell debris. Since intrathecal injection requires very low volumes of the secretome [24], we concentrated it $100 \times$ using a Vivaspin™ 20 sample concentrator (5 kDa; GE Healthcare, Chicago, IL, USA). For this purpose, the secretome was subjected to a centrifugation of $4000 \times g$ at 4 °C for 180 min and then stored at -80 °C until being used in the intrathecal injection. In parallel and to confirm the successful entry of the injected secretome into the lumbar SC, we labelled the sEVs derived from the secretome of 4 DIV WT NSC-34 MNs and injected them intrathecally in the WT mice (as explained in Section 2.3). As previously described [12], we started to centrifugate the secretome at $1000 \times g$ for 10 min to remove cell debris. Then, we isolated the large EVs by centrifugation at $16,000 \times g$ for 1 h. The sEVs were further centrifuged in an Ultra L-XP100 centrifuge (Beckman Coulter Inc., Brea, CA, USA) at $100,000 \times g$ for 2 h. Finally, we labelled the sEVs with a PKH67 fluorescent linker kit (Sigma Aldrich, St. Louis, MO, USA) in accordance with the manufacturer's specifications and resuspended them in DMEM-F12 plus 1% FBS depleted from the sEVs.

2.2. Transgenic mSOD1 Mouse Model

Mice were purchased from The Jackson Laboratory (Bar Harbor, ME, USA), namely transgenic B6SJL-TgN mSOD1Gur/J males (no. 002726) and non-transgenic B6SJLF1/J wild-type (WT) females. They were housed at the animal facility of the Life and Health Science Research Institute (ICVS), University of Minho, where the colony was also established. They were maintained under standard conditions (12 h light/12 h dark cycles, room temperature (RT) at 22–24 °C, and 55% humidity) and provided with food and water ad libitum. The colony was maintained on a background B6SJL by breeding mSOD1 transgenic males with non-transgenic females. Transgenic mSOD1 mice were compared to aged-matched WT mice. The procedures performed were in accordance with the European Community guidelines (Directives 86/609/EU and 2010/63/EU, Recommendation 2007/526/CE, European Convention for the Protection of Vertebrate Animals used for Experimental or Other Scientific Purposes ETS 123 (https://rm.coe.int/168007a445 accessed on 1 June 2022) and the Portuguese Laws on Animal Care (Decreto-Lei 129/92, Portaria 1005/92, Portaria 466/95, Decreto-Lei 197/96, Portaria 1131/97). All the protocols used in this study were approved by the Portuguese National Authority (General Direction of Veterinary) and Ethical Subcommittee in Life and Health Sciences (SECVS; ID: 018/2019).

2.3. Intrathecal Injection of the sEVs and Secretome

The sequence of the experiments developed in this study is graphically summarized in Figure 1. The control groups used in the experiment were the WT and mSOD1 mice injected with the vehicle (differentiated NSC-34 cell media with 1% FBS and without sEVs), as used by others [25,26], with a total of 8 and 7 animals, respectively. The choice of such controls was based on our previous studies, whereby we evidenced that: (1) the secretome from cultured WT cells induced a decrease in several biomarkers associated with inflammatory status and phagocytic and synaptic genes in the SCOCs of early symptomatic mSOD1 mice (unpublished data); (2) the secretome from mSOD1 MNs caused cell demise in both WT and mSOD1 SCOCs [13]; and (3) the secretome from mSOD1 MNs activated the polarization of WT microglia (published in [13]). Therefore, we chose to use the MN cell culture media injection (vehicle) to not interfere with the natural life progression in the WT animals or the disease progression from the early stage to the symptomatic stage in the mSOD1 mice. By adopting this approach, we could document the behavioural and molecular disease progression in the transgenic mice relative to the WT animals. Such conditions are requisite for drawing conclusions on the potential benefits of using the preconditioned secretome from anti-miR-124-treated ALS MNs. Our final aim is the translation of such a strategy for autologous application in ALS patients showing upregulated levels of miR-124 after a stratification assessment. In parallel experiments, to assess the permanence of the secretome sEV components in the SC after their injection into the WT mice, we used the sEVs isolated from the secretome labelled with the PKH67 fluorescent cell linker kit, as described above.

Experiments to assess the therapeutic efficacy of the anti-miR-124-treated mSOD1 MNs were performed in the mSOD1 mice injected with the concentrated preconditioned secretome. The 12-week-old animals (early symptomatic stage of the disease) were anesthetized intraperitoneally (i.p.) with ketamine (75 mg/kg) plus medetomidine (0.5 mg/kg) [27]. Once anesthetized, we proceeded to the single intrathecal injection. WT and mSOD1 mice were firmly held by a pelvic girdle, which was in line with the sixth lumbar vertebral body (L6), as described previously [28]. Then, we identified the gap between the L4 and L5, located above the ileac crest, by palpation through the skin and inserted the needle at the midspinal line [28]. We used a 30-gauge needle Hamilton syringe (Hamilton, Bonaduz, Switzerland) to slowly inject the sEVs, vehicle, or secretome in a proportion of 1 μL per gram of animal weight [24]. The successful entry into the lumbar cistern was confirmed with a sudden tail flick after the needle insertion. Then, the animals were injected i.p. with atipamezole hydrochloride (1 mg/mL; Antisedan®, Pfizer, Inc., Brooklyn, NY, USA) for the reversal of the anesthesia [29]. The WT animals injected with sEVs derived from WT MNs were sacrificed at 8 h and 72 h post injection, based on our previous findings in microglia showing sEV engulfment after 24 h of incubation [11] and the results of other studies demonstrating the distribution of labelled sEVs in the SC 1 week after intrathecal injection [25]. Additional studies in other conditions are unanimous in considering the presence of sEVs at 72 h after their injection [26,30]. Then, we decided to explore the presence of labelled sEVs in the lumbar SC to determine their permanence using immunohistochemistry in the collected samples at 8 h and 72 h post injection, corresponding to acute and lasting distribution.

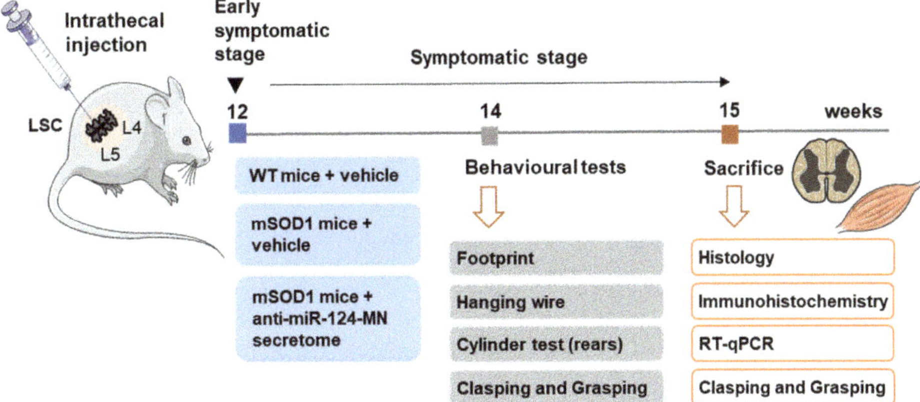

Figure 1. Experimental design. The wild-type (WT) and SOD1-G93A (mSOD1) mice were injected in the lumbar spinal cord [31] at the early symptomatic stage (12-week-old), either with the MN medium (vehicle, control group) or with the secretome from anti-miR-124-treated mSOD1 MNs (only the mSOD1 mice). Two weeks later, animals were behaviourally characterized through footprint, hanging wire, cylinder, clasping, and grasping tests. At 15 weeks of age, the animals were sacrificed, and the lumbar spinal cord [31] and gastrocnemius muscle were isolated for histological and immunohistological analysis, as well as for reverse transcription quantitative real-time polymerase chain reaction (RT-qPCR) and western blot evaluations. This Figure was partially created with Servier Medical Art (smart.servier.com).

Note that no statistical differences were found in the mouse weight (g) between WT and transgenic animals, before or after treatment (WT + vehicle, 25.9 ± 1.0 and 24.9 ± 1.4; mSOD1 + vehicle, 24.6 ± 0.9 and 25.2 ± 1.3; mSOD1 + secretome, 25.9 ± 0.5 and 26.8 ± 1.2).

2.4. Behavioural Tests

To determine the impact of the secretome from anti-miR-124-treated mSOD1 MNs on motor behaviour, we evaluated motor performance in the animals two weeks after the injection. We assessed the gait quality by the footprint test, the muscular strength by the hanging wire test, and the spontaneous activity by the cylinder test. Finally, the deficits in the corticospinal function were also examined using the limb clasping and grasping tests.

2.4.1. Footprint Test

The fore and hind paws of mice were painted with non-toxic dyes of different colours, and the mice were placed on absorbent paper in an inclined corridor so that they would walk up it since mice have the tendency to run upwards to escape [29]. We measured the stride length (in centimeters), which is the distance between the center of the fore-foot plantar and the center of the hind-foot plantar on the same side of the body, within the same stride [29]. A shorter stride length indicated abnormalities in the gait.

2.4.2. Hanging Wire Grid Test

The mice had to remain clinging to an inverted surface of a cage lid, demonstrating their grip strength [29]. The duration (in seconds) for which the mice remained grasping the cage was measured until it reached 120 s [29].

2.4.3. Cylinder Test

The mice were placed into a clear cylinder, and the number of times they reared up against the cylinder wall [32] was counted for 3 min. This test demonstrates animals' forelimb activity and their spontaneous exploration of the environment [32].

2.4.4. Clasping and Grasping Tests

The mice were suspended by holding their tails, and the position of their hindlimbs and toes was observed for several seconds. An abnormal phenotype is represented by the retraction of the hindlimbs toward the abdomen (clasping reflex) [29] and/or the curling of the toes (grasping reflex) during the suspension time, indicating the presence of motor dysfunctionality.

2.5. Homogenates and Tissue Slices

To avoid the animal suffering associated with the disease phenotype, we sacrificed the animals at 15 weeks of age (one week after the beginning of the behaviour tests). Mice were anesthetized i.p. with ketamine (75 mg/kg) plus medetomidine (0.5 mg/kg) and transcardially perfused with 0.1 M PBS at pH 7.4. Then, lumbar SC and left gastrocnemius muscle were dissected and rapidly frozen at $-80\ ^{\circ}$C for reverse transcription quantitative real-time polymerase chain reaction (RT-qPCR) and western blot (WB) assays. For histological and immunohistochemistry studies, animals were perfused with a fixative containing 4% paraformaldehyde in PBS, and the lumbar SC and left gastrocnemius muscle were removed. Then, the lumbar SC was post-fixed with the same fixative for 24 h at $4\ ^{\circ}$C and preserved in 30% sucrose solution. The gastrocnemius muscle was stretched over a notched wooden stick to maintain the in vivo length and to prevent over-contraction. Later, the muscle was frozen in isopentane cooled over liquid nitrogen and stored at $-80\ ^{\circ}$C [33].

For sectioning, the muscle was sliced directly in the cryostat, while the SCs were first embedded in Tissue-Tek® O.C.T. Compound (Sakura Finetek-USA, Torrance, CA, USA) before transversal 20 µm thick sections were serially cut using the cryostat Leica CM1900 (UV Leica; Wetzlar, Germany). The sections were collected on Superfrost Plus glass slides (Thermo Scientific, Waltham, MA, USA) and preserved at $-80\ ^{\circ}$C.

2.6. RT-qPCR

The lumbar SC and muscle were homogenized in TRIzol reagent (Invitrogen, Waltham, MA, USA) using a Pellet Mixer (VWR Life Science, EUA). The muscle homogenization in TRIzol reagent was performed by cutting the tissue into small pieces, followed by transferring the resulting mixture through a 1 mL syringe with a 20 G needle until a homogeneous mixture was obtained. The total RNA was extracted and quantified on a NanoDrop ND100 Spectrophotometer (NanoDrop Technologies, Wilmington, DE, USA) according to the standard procedure in our laboratory [12].

For gene expression, the total RNA was reverse transcribed into cDNA using the Xpert cDNA Synthesis Supermix kit (GRiSP, Porto, Portugal). RT-qPCR was performed using an Xpert Fast SYBR Mastermix BLUE kit (GRiSP) and the primer sequences listed in Table S1. The following conditions were used for each amplification product: $50\ ^{\circ}$C for 2 min, $95\ ^{\circ}$C for 2 min, followed by 40 amplification cycles at $95\ ^{\circ}$C for 5 s and $62\ ^{\circ}$C for 30 s. The amplification cycles for our samples were no higher than 30. The specificity of the amplified product was verified by a melting curve. The ribosomal protein L19 (RPL19) was used as the endogenous control to normalize each gene expression level.

For miRNA expression, cDNA was performed using a miRCURY LNA™ RT Kit (QIAGEN, Valencia, CA, USA) and RT-qPCR using PowerUp™ SYBR™ Green Master Mix (Applied Biosystems, Life Technologies, Waltham, MA, USA) and the primer sequences listed in Table S2. The operating conditions were $95\ ^{\circ}$C for 10 min, followed by 50 amplification cycles at $95\ ^{\circ}$C for 10 s and $60\ ^{\circ}$C for 1 min (ramp rate of 1.6°/s). Our results showed amplification cycles no higher than 40. At the end, a melting curve analysis was carried out to verify the specificity of the amplified product. SNORD110 was used as the endogenous control.

The expression of mRNA/miRNA was measured via the $2^{-\Delta\Delta Ct}$ method relative to that of the endogenous control. Each sample was measured in duplicate. The cDNA synthesis was performed in a thermocycler (Biometra®, Göttingen, Germany), and the RT-qPCR was run on a QuantStudio 7 Flex Real-Time PCR System (Applied Biosystems).

2.7. Histology and Immunohistochemistry

The gastrocnemius muscle was stained with hematoxylin and eosin using an AutoStainer XL (Leica). Briefly, the muscle sections were rinsed in distilled water and placed into Harris hematoxylin solution for 1 min, followed by a tap-water wash for 2 min. Then, differentiation with 0.5% ammonia water was performed for 10 s. After washing with tap water for 2 min, the sections were stained with eosin for 20 s; dehydrated using 70, 95, and absolute alcohol for 2 min each; and covered with xylol for 1 min. The samples were mounted with a glass coverslip using Entellan mounting media (Merck Millipore, Burlington, MA, USA). Photos were obtained using an Olympus widefield inverted microscope IX53 with 20× magnification. The images from three fields of the same muscle section per animal (from three animals per group) were analysed using Fiji software. We measured the mean area occupied by each fiber from the transversal section of the muscle.

To evaluate neurodegeneration, the lumbar SC was stained with Fluoro-Jade B (Chemicon International, Temecula, CA, USA). The samples were defrosted at RT for 10 min. Then, they were incubated in 0.06% potassium permanganate solution for 30 min with gentle agitation. After a 5 min PBS rinse, the samples were stained with a 0.001% solution of Fluoro-Jade B dissolved in 0.1% acetic acid vehicle for 30 min with mild agitation. The samples were then rinsed through three changes of PBS for 3 min each with agitation and mounted in Fluromount-G (Sigma-Aldrich) with a glass coverslip on the top.

To access the distribution of sEVs in the SC, we also stained the neurons and astrocytes with NeuN and glial fibrillary acidic protein (GFAP), respectively. The presence of glial cells in the lumbar SC was also evaluated through the quantification of ionized calcium-binding adaptor molecule 1 (Iba-1) for microglia and GFAP for astrocytes. The previously described protocol was followed with minor modifications [13]. After defrosting, the permeabilization/blocking of the sections was performed using Hank's balanced salt solution with 2% heat-inactivated horse serum, 10% FBS, 1% BSA, 0.25% Triton X-100, and 1 nM HEPES for 3 h at RT. Then, the sections were incubated for 48 h with the primary antibodies (indicated in Table S3) at 4 °C. Following 3 washes with 0.01% Triton-X in PBS for 20 min, the sections were incubated for 2 h at RT with secondary antibodies (indicated in Table S3). The cell nucleus was stained with 4′,6-diamidino-2-phenylindole (DAPI) for Neu N/GFAP staining and Hoechst dye for Iba-1/GFAP labelling. Both solutions were diluted in PBS (0.1 μg/mL) for 10 min. Finally, the sections were mounted in Flouromount-G with a glass coverslip.

The fluorescence images were obtained in a Leica DMi8-CS inverted microscope with Leica LAS X software, and the different z-stacks were merged and analysed with Fiji software. We measured the mean fluorescence of the ventral horn of one lumbar SC transversal section per animal (N = 3 per group) stained with Fluoro-Jade B and obtained with 20× magnification. Images stained with Iba-1 and GFAP were obtained with 40× magnification and analysed from five ventral horn fields per animal (from three animals per group). We measured the fraction of the area occupied by the GFAP- and Iba-1-positive cells.

2.8. Western Blot

Total protein was isolated from the organic phase of the TRIzol–chloroform from lumbar SC and muscle [34] and quantified using a Bradford protein assay kit. The protein expression was assessed by WB analysis according to the standard procedure in our laboratory [35]. Equivalent amounts of protein were separated on dodecyl sulfate polyacrylamide gel electrophoresis (SDS-PAGE) and transferred to a nitrocellulose membrane for 1 h 30. The membranes were blocked for 1 h with 5% non-fat milk in TBS-T (0.1% Tween-20), followed by overnight incubation with primary antibodies (listed in Table S3) and gentle agitation. Membranes were then washed 3 times, for 5 min, with TBS-T and incubated with secondary antibodies conjugated with horseradish peroxidase (listed in Table S3) for 1 h at RT with mild agitation. Immunoreactive bands were detected after the incubation with Western Bright™ Sirius (K-12043-D10, Advansta, Menlo Park, CA, USA) using an iBright™ FL1500 Imaging System. Bands were quantified using iBright™ analysis software. Results were normalized to the expression of β-actin and indicated as fold change.

2.9. Statistical Analysis

Non-parametric ANOVA (Kruskal–Wallis test) and unpaired and non-parametric Mann–Whitney tests were used when the data showed non-normal distribution. Non-continuous categorical variables were analysed using the chi-square (and Fisher's exact) test. For data with normal distribution (Shapiro–Wilk test $p > 0.05$), one-way ANOVA was applied, followed by multiple-comparisons Bonferroni post hoc correction, as well as unpaired and parametric Student's t-tests. Welch's t-test correction was applied when variances were different between the two groups. Values of $p < 0.05$ were considered statistically significant. Data were expressed as mean ± SEM values, except for the non-continuous categorical variables. Results were analysed using GraphPad Prism 8.0.1 (GraphPad Software, San Diego, CA, USA).

3. Results

3.1. sEVs Administered by Intrathecal Injection in B6SJLF1/J Wild-Type (WT) Mice Are Identified in Lumbar SC Slices after 8 and 72 h of Delivery in 12-Week-Old Animals

We previously demonstrated the upregulation of miR-124 in mSOD1 NSC-34 MNs and its recapitulation by the sEV-free secretome and sEVs, which caused alterations in the microglia phenotype once engulfed [11]. Increased miR-124 levels were also observed in the SC of mSOD1 mice at the symptomatic stage [36]. Considering that the regulation of miR-124 in MNs could have neuroprotective effects, we later demonstrated that the secretome from anti-miR-124-treated mSOD1 MNs prevented microglia activation and SC pathogenicity relative to the secretome from non-treated mSOD1 MNs [13]. Thus, in the present study, we tested whether the injection of the preconditioned secretome could halt or delay disease progression in an mSOD1 mouse model.

To gain an understanding of the permanence of the sEVs after being injected as components of the secretome, we labelled the sEVs isolated from the secretome of WT NSC-34 MNs with PKH67 fluorescent cell linker before their intrathecal injection in the 12-week-old WT mice, as explained in Section 2.3 of the Materials and Methods. We assessed their presence in the collected lumbar sections after short (8 h) and long (72 h) periods of time.

We were able to identify sEVs (in green) after the intrathecal injection for both periods of time (Figure 2A,B). To further assess their distribution among neurons and astrocytes, we performed immunohistochemical staining for NeuN (Figure 2C) and GFAP (Figure 2D), respectively, using DAPI for nuclei and PKH67 for sEV labelling. sEVs were identified in both cell images at 72 h after injection, demonstrating that they could be disseminated by the whole secretome and mediate lasting paracrine signalling with neural cells.

From here on we decided to use the whole secretome, and not only the injection of sEVs, considering its advantages for ALS therapeutics. On the one hand, storing and administering the secretome is easier than doing so for sEVs and facilitates their preservation. On the other hand, the secretome also contains additional bioactive molecules that may increase its neuroprotective potential, as a consequence of the effects of anti-miR-124 on MN survival [13]. When considering ALS patients, this study may also open up new possibilities regarding the manipulation and engineering of their cells toward the collection of the secretome for autologous transplantation.

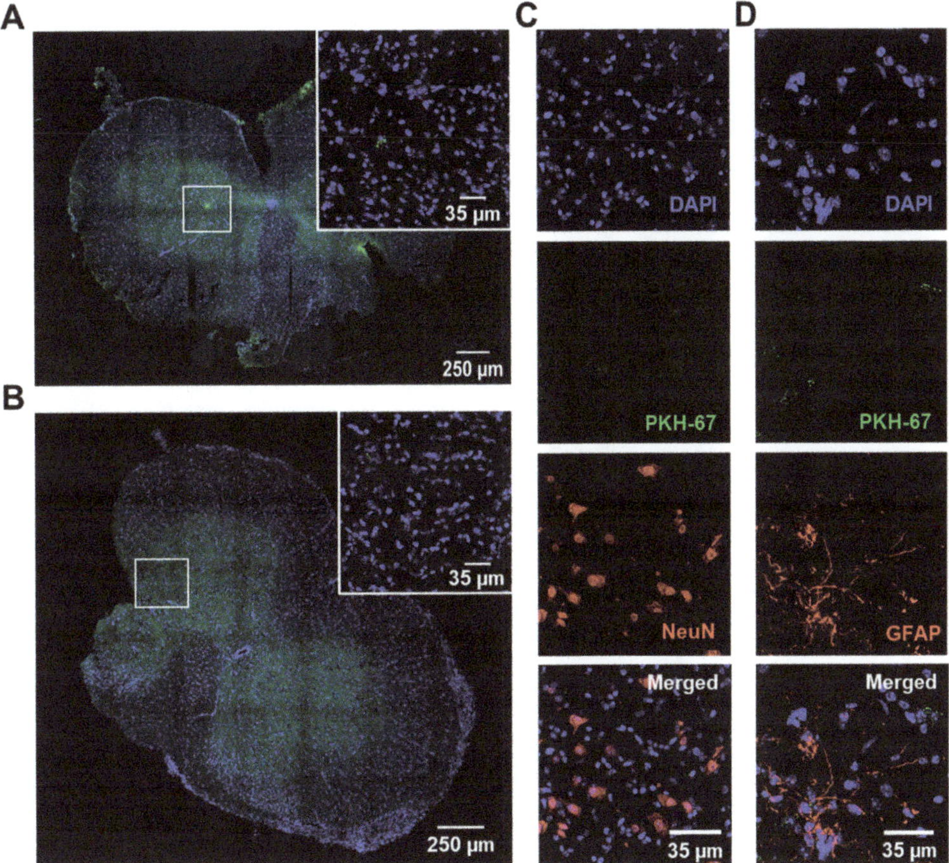

Figure 2. Intrathecal injection of labelled small extracellular vesicles (sEVs) in 12-week-old wild-type (WT) animals leads to their dissemination and interaction with nerve cells, as observed in the lumbar spinal cord (SC) sections at 8 h and 72 h after administration. sEVs were isolated by differential ultracentrifugation and labelled with PKH67 cell linker before injection in the WT mice, which were sacrificed at 8 h and 72 h thereafter. (**A,B**) Representative images of transversal SC slices with labelled sEVs (green) and respective insets at (**A**) 8 h and (**B**) 72 h post-injection. (**C,D**) Representative images of PKH67-labelled sEVs distributed among (**C**) NeuN-stained cells and (**D**) GFAP-stained cells from the lumbar SC of WT mice 72 h after injection. The nuclei were stained with DAPI (blue). DAPI, 4′,6-diamidino-2-phenylindole; GFAP, glial fibrillary acidic protein; NeuN, hexaribonucleotide binding protein 3.

3.2. Expression of miR-124 in the SC of mSOD1 Mice Is Downregulated after 3 Weeks of Intrathecal Injection of Secretome from Anti-miR-124-Treated mSOD1 MNs

In our prior studies, we demonstrated that miR-124 levels in the secretome recapitulate the levels observed in cells, either non-modulated or treated with its mimics and inhibitors [9–11,13]. Here, we intended to test whether the intrathecal injection of the secretome from anti-miR-124 mSOD1 MNs in pre-symptomatic mSOD1 mice was able to reduce the miR-124 expression in the lumbar SC of ALS mice when assessed at 15 weeks of age. This was important for inferring the potential beneficial effects that the approach may have.

As depicted in Figure 3A and Supplementary Table S4, the expression of miR-124 was found to be upregulated in the mSOD1 mice treated with the vehicle, as compared with the matched WT animals ($p < 0.01$). The injection of the secretome from anti-miR-124-treated MNs, under the same conditions, was successful in downregulating the expression of miR-124 (Figure 3B, $p < 0.05$ and Supplementary Table S4).

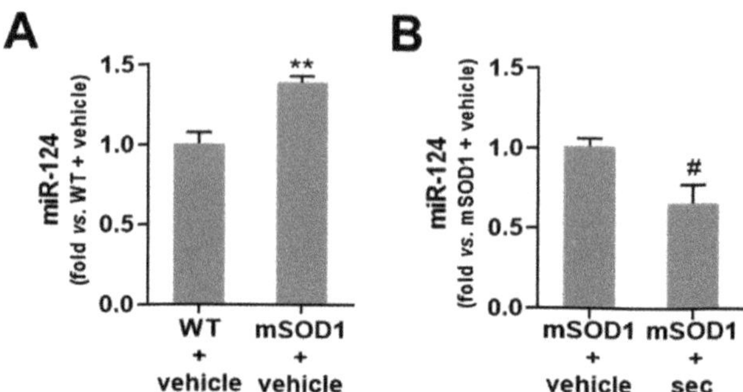

Figure 3. The secretome from mSOD1 MNs transfected with anti-miR-124 abolishes the upregulation of miR-124 in the spinal cord (SC) of ALS mice after 3 weeks of intrathecal injection. Expression of miRNA-124 in the lumbar SC of (**A**) SOD1-G93A (mSOD1) mice injected at 12 weeks of age with the vehicle (basal media of NSC-34 motor neurons (MNs)) in comparison with the respective wild-type (WT) animals, and (**B**) mSOD1 mice injected with the secretome derived from anti-miR-124-treated mSOD1 MNs (mSOD1 + sec) in comparison with those injected only with the vehicle. The results were obtained at 15 weeks of age and were normalized to SNORD110. Data are expressed as fold change vs. (**A**) WT + vehicle and (**B**) mSOD1 + vehicle (mean $\pm$ SEM) from at least 5 animals per group. ** $p < 0.01$ vs. WT + vehicle; # $p < 0.05$ vs. mSOD1 + vehicle, unpaired and parametric *t*-test (with Welch's correction when needed).

In the next sections, we will explore the efficacy of this novel strategy in inhibiting the progression of the disease in mSOD1 mice from the early symptomatic stage to the symptomatic stage by preventing motor disabilities, as well as neurodegeneration, neuroinflammation, and demyelination.

3.3. Motor Disabilities in Transgenic Mice at the Early Symptomatic Stage of ALS Are Reversed by Intrathecal Injection of Secretome from Anti-miR-124-Treated ALS MNs

As already mentioned, the downregulation of miR-124 in ALS MNs was shown to be neuroprotective and to regulate neuroinflammation in in vitro and ex vivo experimental ALS models [13]. However, this has never been tested in ALS in vivo models, such as mSOD1 mice. We started by assessing the behavioural alterations in the mSOD1 mice at the symptomatic stage (14-week-old mice), as the deterioration of motor performance at this stage of the disease was previously observed in an ALS mice model [37–44].

As indicated in Figure 4, we observed a decrease in the stride length, the delay before falling from the grid, and the number of times that the animal reared up and touched the cylinder for the vehicle-treated transgenic mice, in comparison with matched WT animals (Figure 4E–G, $p < 0.05$). These observations are associated with an abnormal gait quality, muscle weakness, and less spontaneous vertical activity, respectively. In addition, a large percentage of mSOD1 animals showed enhanced limb clasping (Figure 4H, $p < 0.01$ vs. WT + vehicle) and grasping (Figure 4I, $p < 0.0001$ vs. WT + vehicle) reflexes, in accordance with abnormalities in the corticospinal tract.

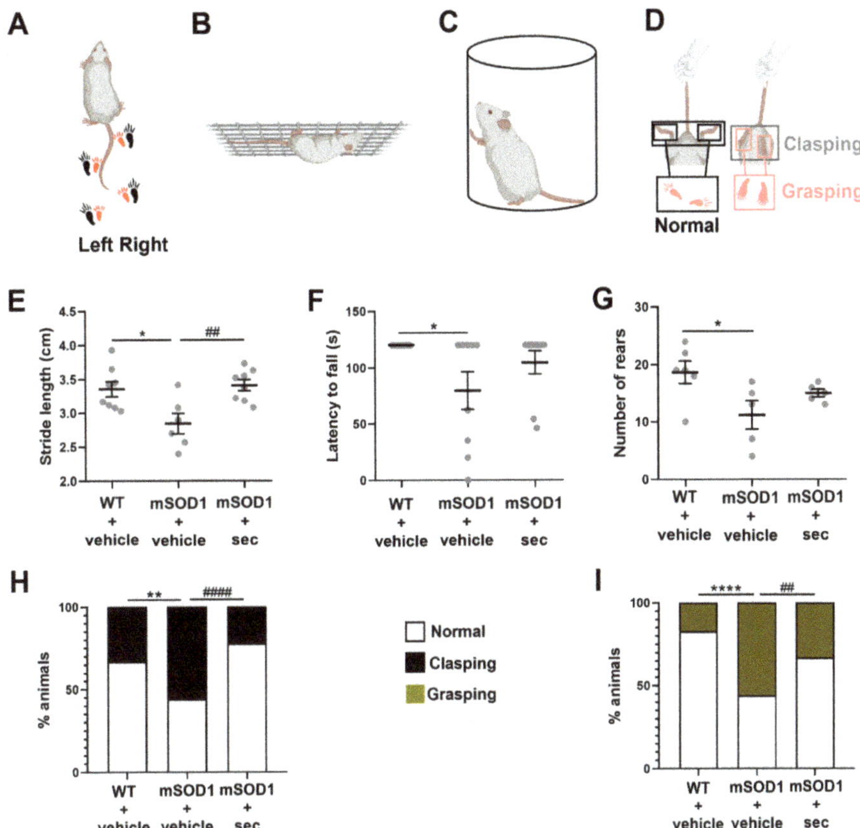

Figure 4. Motor performance, muscular strength, spontaneous activity, and corticospinal function in ALS mice are improved after 2 weeks of intrathecal injection of the secretome from anti-miR-124-treated mSOD1 motor neurons (MNs). Representative illustrations of (**A**) footprint test, (**B**) hanging wire test, (**C**) cylinder test, and (**D**) clasping/grasping reflexes test. Measurement of the (**E**) stride length (in centimeters, cm), (**F**) time holding onto the cage grid (in seconds, s), (**G**) number of times the mice reared up against the cylinder, and percentage (%) of animals showing the (**H**) clasping and (**I**) grasping reflexes in wild-type (WT)/SOD1-G93A (mSOD1) mice injected with the vehicle (basal media of NSC-34 MNs) and mSOD1 mice injected with the secretome derived from mSOD1 MNs modulated with anti-miR-124 (mSOD1 + sec). Data are expressed as mean ± SEM for (**E–G**) and percentage (%) for (**H,I**) from at least 5 animals per group. **** $p < 0.0001$, ** $p < 0.01$ and * $p < 0.05$ vs. WT + vehicle; #### $p < 0.0001$ and ## $p < 0.01$ vs. mSOD1 + vehicle. One-way ANOVA followed by multiple-comparisons Bonferroni post hoc correction was used for footprint and cylinder tests; unpaired and one-way non-parametric ANOVA (Kruskal–Wallis test) was used for hanging wire test; chi-square (and Fisher's exact) test was used for clasping and grasping tests. Panels (**A–D**) were partially created with Biorender (Biorender.com).

The secretome from anti-miR-124-treated mSOD1 MNs was able to inhibit such impairments. We noticed improvements in all the assessed behavioural tests (Figure 4E–G), though only significant for the stride length, i.e., the distance from the heel print of one foot to the heel print of the other foot during the walking test, evidencing the amelioration of the gait quality, muscle strength, and spontaneous activity. Moreover, the reduction in the percentage of animals with dystonic movements associated with clasping (Figure 4H, $p < 0.0001$ vs. mSOD1 + vehicle) and grasping reflexes (Figure 4I, $p < 0.01$ vs. mSOD1 + vehicle) was similarly clear 2 weeks after the treatment with the secretome from anti-miR-124-treated mSOD1 MNs.

3.4. Atrophy of Gastrocnemius Muscle in Early Symptomatic ALS Mice, as Well as Loss of Motor Neuron Viability and Synaptic Dynamics, Are Reversed by Intrathecal Injection of the Secretome from Anti-miR-124-mSOD1 MNs

Since we observed motor performance disabilities in the mSOD1 mice, a common symptom of ALS caused by muscle weakness [45], already described in SOD1 models of ALS [46], and their rescue by the intrathecal injection of the preconditioning secretome, we next investigated gastrocnemius muscle fibers in the absence of and after treatment using hematoxylin and eosin staining. The measurement of the cross-sectional areas revealed thinner muscle fibers in the transgenic than in the WT mice ($p < 0.0001$ vs. WT + vehicle, Figure 5A,B). We additionally identified the decreased gene expression of hexaribonucleotide binding protein 3 (NeuN), presynaptic synaptophysin, and postsynaptic density protein 95/PSD-95 (at least $p < 0.05$ vs. WT + vehicle, Figure 5C and Supplementary Table S4), suggesting the existence of neuronal dysfunction and the eventual compromise of neuromuscular transmission and muscle functionality.

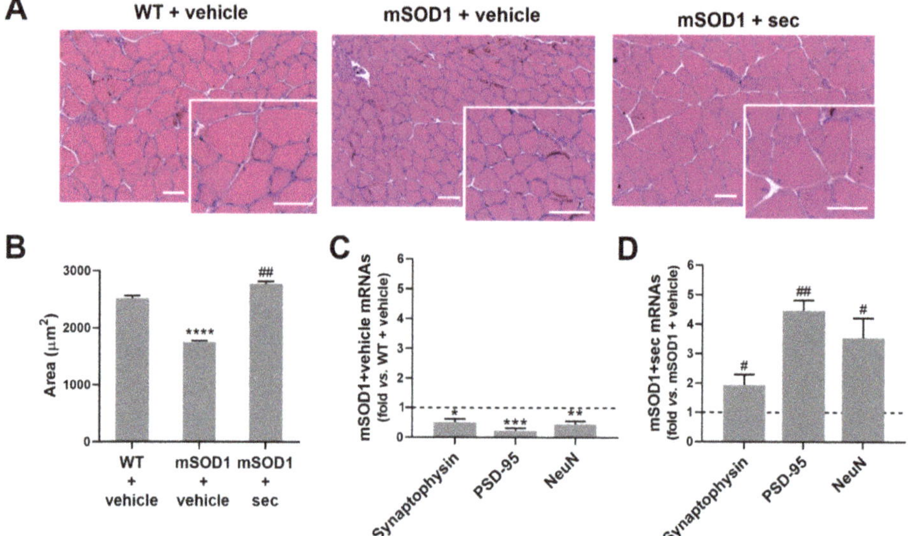

Figure 5. Intrathecal injection of anti-miR-124-treated ALS MN-derived secretome in 12-week-old mSOD1 mice prevents loss of muscle fiber area and the deregulation of genes that direct synaptic proteins at 3 weeks after treatment. (**A**) Representative images of transversal sections of gastrocnemius muscle from the wild-type (WT) and SOD1-G93A (mSOD1) mice injected with the vehicle (basal media of NSC-34 motor neurons (MNs)), as well as mSOD1 mice injected with the secretome derived from anti-miR-124-treated mSOD1 MNs (mSOD1 + sec), stained with hematoxylin–eosin. Scale bars: 50 μm. (**B**) Respective average muscle fiber area (in μm). (**C**) Gene expression of synaptophysin, PSD-95, and NeuN from mSOD1 mice injected with the vehicle in comparison with the respective WT mice, and (**D**) from mSOD1 mice injected with the secretome in comparison with those injected with the vehicle. Results are mean ($\pm$SEM) for (**B**) and expressed as fold change vs. WT + vehicle for (**C**) or fold change vs. mSOD1 + vehicle for (**D**). The images from three fields of the muscle per animal (from three animals per group) were used for histological analysis and five animals per group for RT-qPCR analysis. **** $p < 0.0001$, *** $p < 0.001$, ** $p < 0.01$, and * $p < 0.05$ vs. WT + vehicle; ## $p < 0.01$ and # $p < 0.05$ vs. mSOD1 + vehicle. One-way ANOVA followed by multiple-comparisons Bonferroni post hoc correction was used for (**B**) and unpaired and parametric *t*-test with Welch's correction for (**C,D**). NeuN, hexaribonucleotide binding protein 3; PSD-95, postsynaptic density protein 95.

Notably, 3 weeks after the intrathecal injection of the preconditioning secretome from anti-miR-124 mSOD1 MNs in the ALS mice, we noticed the reversal of the skeletal muscle atrophy ($p < 0.01$ vs. mSOD1 + vehicle, Figure 5A,B), as well as of neuronal demise and deregulated synaptic activity (at least $p < 0.05$ vs. mSOD1 + vehicle, Figure 5D and Supplementary Table S4).

The data highlighted the efficacy of the anti-miR-124 preconditioned secretome as a promising therapy for skeletal muscle remodelling in ALS with potential to be translated to patients.

3.5. Neurodegeneration in the Lumbar SC of Early-Symptomatic ALS Mice Is Prevented after Injection of Secretome from Anti-miR-124-Treated mSOD1 MNs

In previous studies, we observed the existence of neurodegeneration in the SC of mSOD1 mice [36] and that the downregulation of miR-124 restored mSOD1 MN viability toward control levels by preventing early apoptosis [13]. Based on the above results, we decided to explore the ability of the preconditioned secretome to prevent neurodegeneration when injected in 12-week-old mSOD1 mice. For this, we collected the lumbar SC at 3 weeks after the intrathecal administration and stained the transversal sections with Fluoro-Jade B. As expected, we found a higher fluorescence intensity in the SC from mSOD1 mice at 15 weeks of age than in the matched WT animals ($p < 0.01$, Figure 6A,B), demonstrating an increased number of degenerating neurons. In contrast, in the samples obtained from the mSOD1 mice treated at 12 weeks of age with the secretome from anti-miR-124-treated mSOD1 MNs, we noticed a decrease in Fluoro-Jade B-positive staining relative to the vehicle-treated mSOD1 mice ($p < 0.05$, Figure 6A,B), demonstrating the secretome's preventive effect against neurodegeneration in the ALS animals.

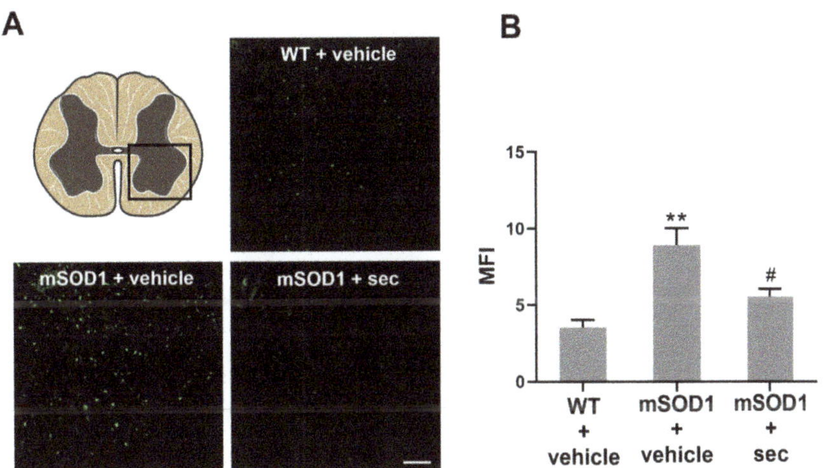

Figure 6. Intrathecal injection of the secretome from anti-miR-124-modulated mSOD1 motor neurons (MNs) in the lumbar spinal cord of mSOD1 mice at 12 weeks of age prevents age-associated neurodegeneration after 3 weeks of its administration. (**A**) Representative images of the ventral horn of the lumbar section grey matter stained by Fluoro-Jade B fluorescence (square) from 15-week-old wild-type (WT)/SOD1-G93A (mSOD1) mice injected with the vehicle (basal media of NSC-34 MNs) and mSOD1 mice injected with the secretome derived from mSOD1 MNs modulated with anti-miR-124 (mSOD1 + sec); (**B**) the respective quantification of mean fluorescence. Scale bar: 100 μm. Data from 3 animals per group are expressed as fold change (mean ± SEM) vs. WT + vehicle. ** $p < 0.01$ vs. WT + vehicle; # $p < 0.05$ vs. mSOD1 + vehicle, one-way ANOVA followed by multiple-comparisons Bonferroni post hoc correction. MFI, mean fluorescence intensity.

3.6. MN Loss and Dysfunction, as Well as Myelination Impairment, Are Averted in the Lumbar SC of ALS Mice Injected with the Secretome from Anti-miR-124-Treated mSOD1 MNs

Given our results suggesting the neuroprotective potential of the secretome from anti-miR-124-transfected mSOD1 MNs, we further explored the markers related to MN functionality. As shown in Figure 7A and Supplementary Table S4, we first observed the decreased gene expression of PSD-95, dynein, kinesin, and C-X3-C motif chemokine ligand 1 (CX3CL1, fractalkine), which accompanied that of NeuN (at least $p < 0.05$ vs. WT + vehicle) when assessed in the lumbar SC of the mSOD1 mice at 15-week-old. These results revealed an impairment of the postsynaptic dynamics, retrograde/anterograde axonal transport, and CX3CL1 that may have compromised paracrine signalling with the microglia receptor CX3CR1. Neurodegeneration was further confirmed by the decrease in the protein NeuN ($p < 0.01$ vs. WT + vehicle, Figure 7B and Supplementary Table S4). The data indicated that the postsynaptic deregulation and neuronal loss at the gastrocnemius muscle/neuromuscular junction was also present in the SC of the mSOD1 mice. It should also be noted that despite the alterations in PSD-95, we did not detect changes in the gene encoding for the presynaptic synaptophysin protein in the SC of the ALS animals.

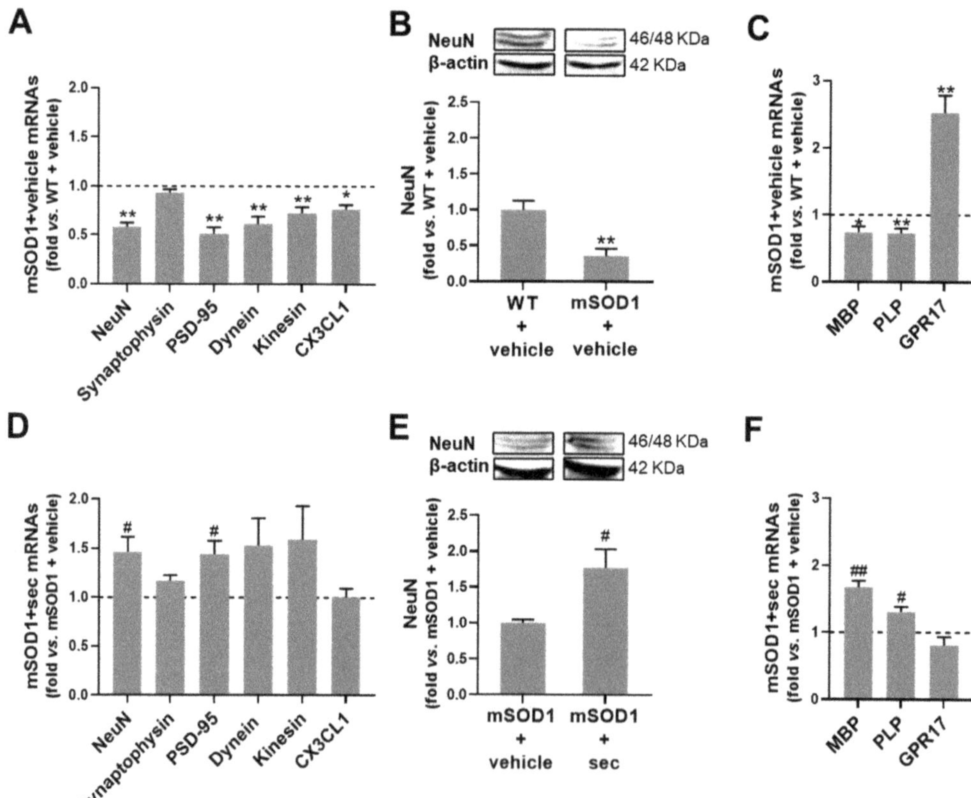

Figure 7. Neuronal demise and deficits in synaptic signalling, axonal transport, CX3CL1-CX3CR1 axis, and myelination in the lumbar spinal cord (SC) of 15-week-old mSOD1 mice are prevented by the intrathecal injection of the secretome from modulated mSOD1 motor neurons (MNs) in 12-week-old animals. (**A**) Gene expression of neuronal-related NeuN, synaptophysin, PSD-95, dynein, kinesin, and CX3CL1; (**B**) protein expression of NeuN; and (**C**) myelin-associated genes (MBP, PLP, and GPR17) in the SC of 15-week-old SOD1-G93A (mSOD1) mice that were treated with the vehicle (basal media of NSC-34 MNs) at 12-week-old, in comparison with the respective wild-type (WT)

values. (**D–F**) Data from matched experiments realized in mSOD1 mice injected with the secretome derived from anti-miR-124-treated mSOD1 MNs (mSOD1 + sec) in comparison to those treated with the vehicle. (**B,E**) Representative results from one blot. The results were normalized to RPL-19 for RT-qPCR and β-actin for western blot. Data are expressed as fold change vs. (**A–C**) WT + vehicle and (**D–F**) mSOD1 + vehicle (mean ± SEM) from at least 5 animals per group. ** $p < 0.01$ and * $p < 0.05$ vs. WT + vehicle; ## $p < 0.01$ and # $p < 0.05$ vs. mSOD1 + vehicle, unpaired and parametric t-test (with Welch's correction when needed). CX3CL1, C-X3-C motif chemokine ligand 1/fractalkine; GPR17, G-protein-coupled receptor 17; MBP, myelin basic protein; NeuN, hexaribonucleotide-binding protein 3; PLP, myelin proteolipid protein; PSD-95, postsynaptic density protein 95; RPL19, 60S ribosomal L19; RT-qPCR, reverse transcription quantitative real-time polymerase chain reaction.

Neuronal survival and axonal preservation depend on myelin integrity [47], an issue that has scarcely been assessed in ALS patients and experimental models. Here, we assessed the mRNA levels of the myelin basic protein (MBP) and the myelin proteolipid protein (PLP), which were revealed to be downregulated in the symptomatic mSOD1 SCs (at least $p < 0.05$ vs. WT + vehicle, Figure 7C and Supplementary Table S4). G protein-coupled receptor 17 (GPR17) mRNA, an important regulator of oligodendrocyte precursor (OPC) maturation, was previously found to be increased in mSOD1 mice [48]. We confirmed a 2.5-fold increase in its gene expression in these transgenic animals ($p < 0.01$).

The intrathecal injection of the preconditioned secretome exerted protective effects against most of the observed impairments. In fact, we observed an upregulation of PSD-95 and NeuN (including the protein expression) ($p < 0.05$ vs. mSOD1 + vehicle, Figure 7D,E and Supplementary Table S4), accompanied by an increase in MBP and PLP mRNA levels (at least $p < 0.05$ vs. mSOD1 + vehicle, Figure 7F and Supplementary Table S4), highlighting the benefits of the tested secretome-based strategy in avoiding demyelination.

These results call our attention to the efficacy of the conditioned secretome in halting disease progression in early symptomatic mSOD1 mice.

3.7. Intrathecal Injection of the Secretome from Anti-miR-124-Treated mSOD1 MNs in 12-Week-Old mSOD1 Mice Counteracts Lumbar SC Immune Deregulation at the Symptomatic Stage

Neuroinflammation is associated with ALS, and glial cell dysregulation is implicated in the initiation/propagation of neurodegeneration [49,50]. Here, based on the promising data obtained for neuronal function with the intrathecal injection of the preconditioned secretome in 12-week-old mSOD1 mice, we wondered whether such beneficial effects could be supported by the prevention of the homeostatic imbalance of microglia and astrocyte activation.

Concerning the categorization of microglia immunoreactivity, we found the downregulation of arginase 1 in the mSOD1 SC 3 weeks after vehicle injection in the early symptomatic mice (Figure 8A, $p < 0.05$ vs. WT + vehicle), known to exacerbate neuropathological and behavioural deficits [51], as well as the gene purinergic receptor p2y12/P2RY12 ($p < 0.01$ vs. WT + vehicle, Figure 8A and Supplementary Table S4), which may compromise microglial motility and migration [52]. In addition, we identified low levels of TREM2 ($p < 0.001$ vs. WT + vehicle) and the milk fat globule epidermal growth factor 8 (MFG-E8) ($p < 0.05$ vs. WT + vehicle), which have been shown to compromise phagocytic ability [3,53,54]. Notably, the significant changes observed in arginase 1 and MFG-E8 in mSOD1 SC + vehicle compared only with WT SC + vehicle were not observed when comparing the three conditions together (Supplementary Table S4). The induced regulation of microglial activation may have been determined by the upregulated levels of CX3CR1, TMEM119, and TIMP2 (at least $p < 0.05$ vs. WT + vehicle), which have been shown to regulate neuroinflammation [55–57]. An inverted signature was obtained when we assessed the SC parameters 3 weeks after the intrathecal injection of the preconditioned secretome, namely those related to microglia phagocytosis (e.g., TREM2, $p < 0.05$ vs. mSOD1 + vehicle,

Figure 8D and Supplementary Table S4) and arginase 1 ($p < 0.05$ vs. mSOD1 + vehicle), which are associated with a reparative phenotype.

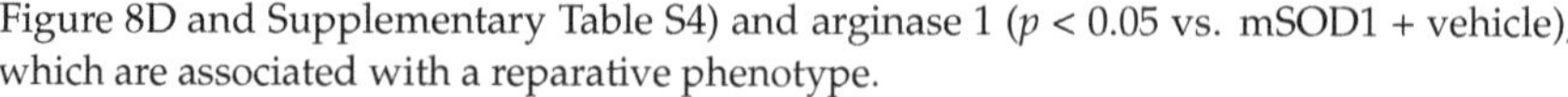

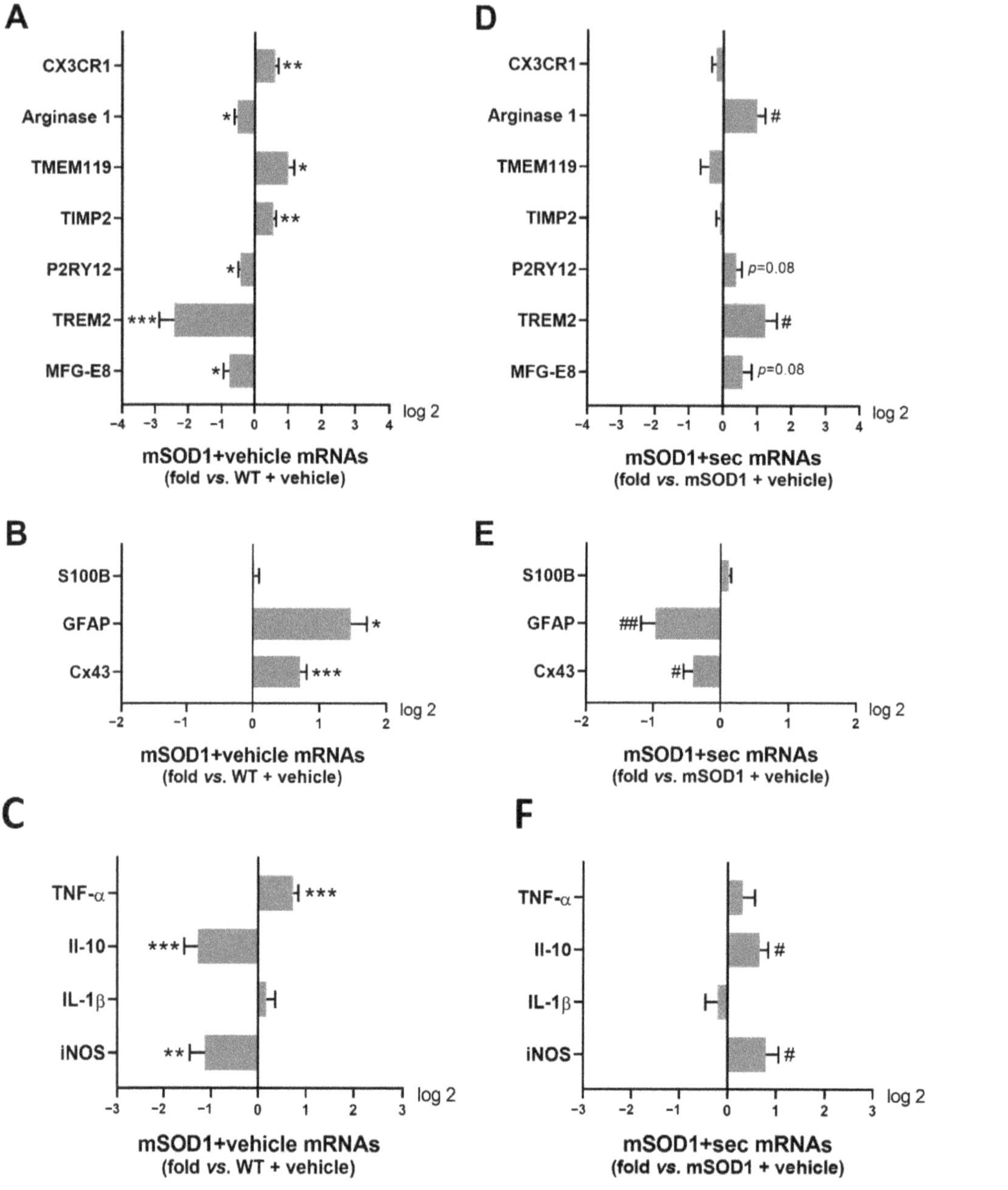

Figure 8. Intrathecal injection of the secretome from anti-miR-124-modulated mSOD1 motor neurons (MNs) in the lumbar spinal cord (SC) of mSOD1 mice at 12-week-old prevents microglia activation, astrocyte reactivity, TNF-α signalling, and inflammation associated with the symptomatic stage. Gene expression of (**A**) microglia-associated markers (MFG-E8, TREM 2, P2RY12, TIMP2, arginase 1, and CX3CR1); (**B**) astrocyte-related markers (Cx43, GFAP, and S100B); and (**C**) inflammatory-associated

markers (iNOS, IL-1β, IL-10, and TNF-α) in the SC of SOD1-G93A (mSOD1) mice injected with the vehicle (basal media of NSC-34 MNs) in comparison to those in wild-type (WT) animals; (**D–F**) after the administration of the secretome derived from anti-miR-124-treated mSOD1 MNs (mSOD1 + sec) as compared with mSOD1 mice treated with the vehicle. The results were normalized to RPL-19. Data are expressed as fold change vs. (**A–C**) WT + vehicle and (**D–F**) mSOD1 + vehicle (mean ± SEM) from at least 5 animals per group. *** $p < 0.001$, ** $p < 0.01$, and * $p < 0.05$ vs. WT + vehicle; ## $p < 0.01$ and # $p < 0.05$ vs. mSOD1 + vehicle, unpaired and parametric t-test (with Welch's correction when needed). CX3CR1, c-x3-c chemokine receptor 1; Cx43, connexin 43; GFAP, glial fibrillary acidic protein; IL-10, interleukin 10; IL-1β, interleukin 1β; iNOS, inducible nitric oxide synthase; MFG-E8, milk fat globule epidermal growth factor 8; P2RY12, purinergic receptor p2y12; RPL19, 60S ribosomal L19; S100B, S100 calcium-binding protein B; TIMP2, tissue inhibitor of metalloproteinases 2; TMEM119, transmembrane protein 119; TNF-α, tumour necrosis factor alpha; TREM2, triggering receptor expressed on myeloid cells 2.

Regarding the reactive profile of astrocytes, Figure 8B and Supplementary Table S4 show increased levels of GFAP ($p < 0.05$ vs. WT + vehicle, Figure 8B) and connexin-43/Cx43 ($p < 0.001$ vs. WT + vehicle, Figure 8B) in the lumbar SC of mSOD1 mice, pointing to the existence of reactive astrocytes.

Interestingly, the preconditioned secretome led to a significant reduction in these parameters (GFAP, $p < 0.01$ and Cx43, $p < 0.05$ vs. mSOD1 + vehicle). Note the non-existence of alterations in the gene expression of S100 calcium-binding protein B (S100B) in the absence or presence of the preconditioned secretome.

When we assessed the dysregulation of genes associated with inflammation, we observed an imbalance with decreased iNOS and interleukin (IL)-10, together with elevated levels of tumour necrosis α/TNF-α (at least $p < 0.01$ vs. WT + vehicle, Figure 8C and Supplementary Table S4), while no changes were observed for IL-1β. The secretome caused an elevation in IL-10 and iNOS ($p < 0.05$ vs. mSOD1 + vehicle), which may have compensated for each other to reach a steady state. Overall, our results highlighted that the preconditioned secretome regulated the lumbar SC immunoreactivity.

We also evaluated the microglial and astrocytic prevalence in the slices by assessing the area stained for Iba-1 and GFAP, respectively. An increase in the area occupied by GFAP- and Iba-1-positive cells was found in the mSOD1 SC ($p < 0.01$, Figure 9C,D, respectively) in comparison to the matched WT samples. The expression of proteins as assessed by WB analysis also indicated an elevation in GFAP and S100B (though not significant) and vimentin and Iba-1 in mSOD1 SC, in comparison with the WT samples (Figure 9E,F; $p < 0.05$ and Supplementary Table S4), corroborating the astrocyte reactivity and microglia activation. Interestingly, the modulated secretome not only decreased the area occupied by GFAP- and Iba-1-positive cells (at least $p < 0.05$ vs. mSOD1 + vehicle, Figure 9C,D), but also the protein levels of GFAP, S100B, and Iba-1 ($p < 0.05$, Figure 9G,H) in comparison with the SC from mSOD1 mice injected with the vehicle, supporting the immune remodelling ability of the preconditioned secretome.

3.8. The Secretome from Anti-miR-124-Treated mSOD1 MNs Counteracts the Upregulation of Inflamma-miRNAs in the Lumbar SC of mSOD1 Mice

Several miRNAs have been found to be dysregulated in ALS patients [58], as well as in cortical brain and SC samples from mSOD1 mice [36,59], including miR-155, miR-146a, miR-21, and miR-125b. Therefore, and based on the obtained data, we wondered whether the preconditioned secretome would be able to control such miRNAs, considering the reduction in anti-miR-124 levels observed in the SC of mSOD1 mice treated with the preconditioned secretome (Figure 3).

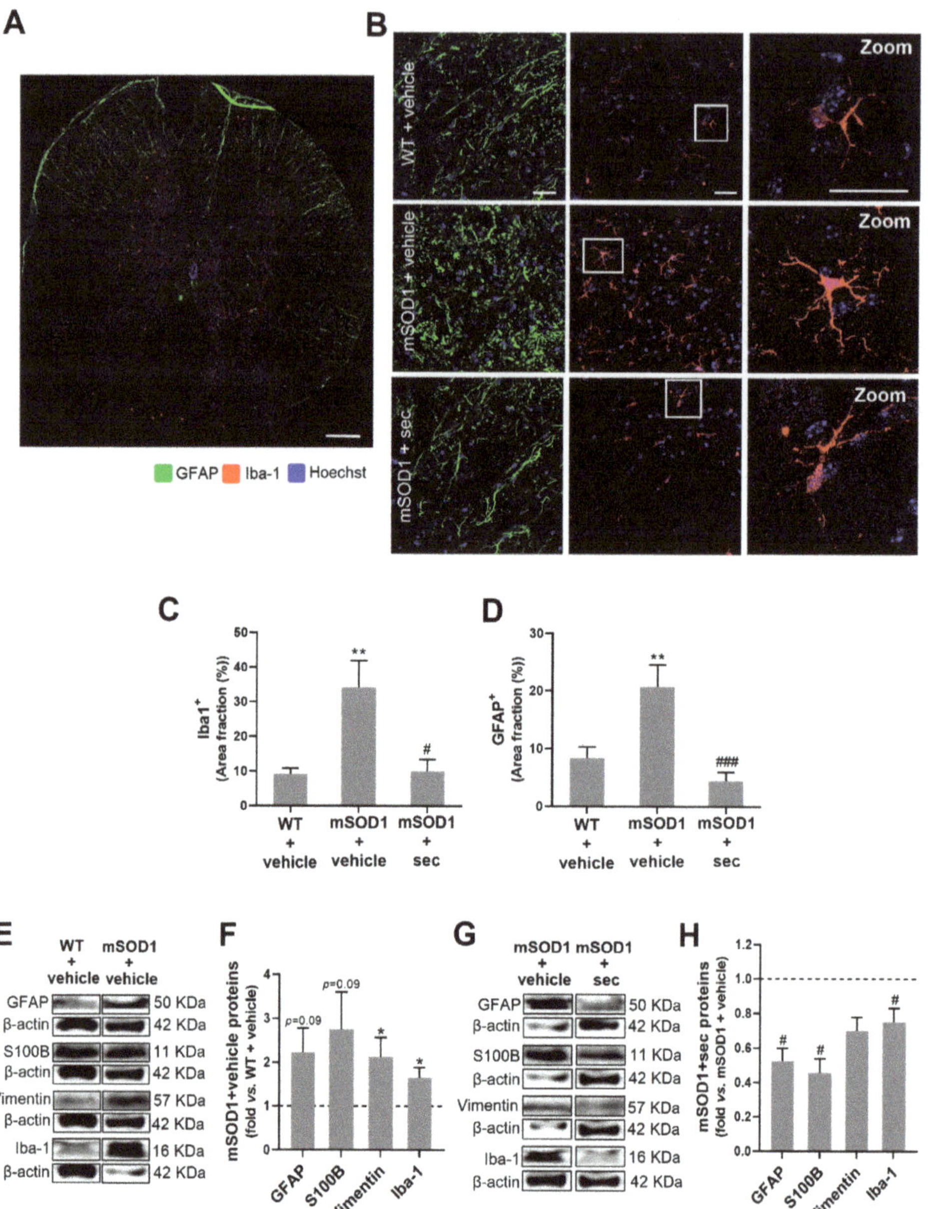

Figure 9. Intrathecal injection of the secretome from anti-miR-124-modulated mSOD1 motor neurons (MNs) in the lumbar spinal cord (SC) of mSOD1 mice at 12-week-old counteracts the upregulation of proteins linked to glia-driven immunoreactivity processes after 3 weeks of its administration. (**A**) Representative image of a transversal SC slice stained with Iba-1 (microglia-associated marker) and glial fibrillary acidic protein/GFAP (astrocyte-associated marker). (**B**) Representative images for GFAP- (green) and Iba-1- (red) positive cells from the lumbar SC of WT and SOD1-G93A (mSOD1)

mice injected with the vehicle (basal media of NSC-34 MNs) and mSOD1 mice injected with the secretome derived from anti-miR-124-treated mSOD1 MNs (mSOD1 + sec). A zoomed-in image of an Iba-1-positive cell is shown. Nuclei were stained with Hoechst (blue). Scale bar: (**A**) 300 μm and (**B**) 30 μm. (**C,D**) Area fraction (in percentage, %) occupied by GFAP- and Iba-1-positive cells, respectively. (**E,G**) Representative western blots (WB). (**F,H**) Data resulting from WB analysis of reactive astrocytic markers (GFAP, S100B, and vimentin) and microglial Iba-1. β-actin was used as a loading control for WB analysis. Results are mean (± SEM) for (**C,D**) and expressed as fold change vs. WT + vehicle for (**F**) or fold change vs. mSOD1 + vehicle for (**H**). The images were analysed from five ventral horn fields per animal (from three animals per group) for immunohistochemistry and five animals per group for WB analysis. ** $p < 0.01$ and * $p < 0.05$ vs. WT + vehicle; ### $p < 0.001$, and # $p < 0.05$ vs. mSOD1 + vehicle. One-way ANOVA followed by multiple-comparisons Bonferroni post hoc correction was used for (**C,D**), and unpaired and parametric *t*-test with Welch's correction for (**F,H**). GFAP, glial fibrillary acidic protein; Iba-1, ionized calcium-binding adaptor molecule 1; S100B, S100 calcium-binding protein B.

Firstly, we were interested in confirming that such miRNAs were upregulated in the mSOD1 mice treated with the vehicle at 12-week-old and analysed 3 weeks thereafter, the design we used to test the potential therapeutic benefits of the intrathecal injection of the preconditioned secretome from the anti-miR-124 treated mSOD1 MNs.

As expected, all the assayed miRNAs, except miR-125b, were found to be upregulated in the SC tissue collected from the mSOD1 mice at 15-week-old, corresponding to the symptomatic stage (Figure 10A, at least $p < 0.05$ vs. WT + vehicle and Supplementary Table S4).

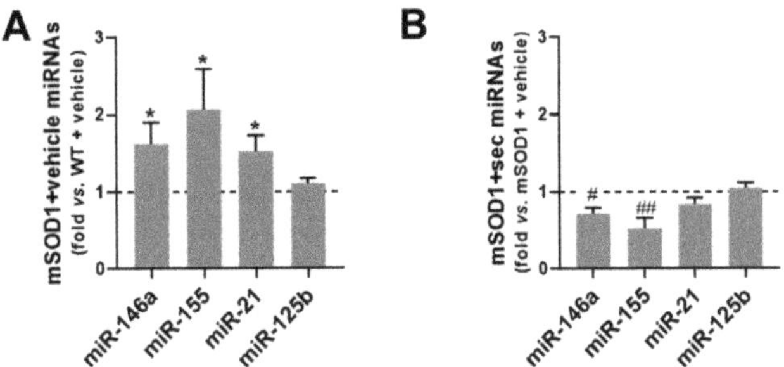

Figure 10. Upregulation of inflammatory-associated miRNA observed in the lumbar spinal cord (SC) of ALS mice is prevented by the secretome from the mSOD1 MNs engineered with anti-miR-124. Expression of inflammatory-associated micro(mi)RNAs (miR-146, miR-155, miR-21, miR-124, and miR-125b) in the SC of (**A**) SOD1-G93A (mSOD1) mice injected with the vehicle (basal media of NSC-34 motor neurons (MNs)) in comparison with those in wild-type (WT) animals, and (**B**) mSOD1 mice injected with the secretome derived from anti-miR-124-treated mSOD1 MNs (mSOD1 + sec) in comparison with those treated with the vehicle. The results were normalized to SNORD110. Data are expressed as fold change vs. (**A**) WT + vehicle and (**B**) mSOD1 + vehicle (mean ± SEM) from at least 5 animals per group. * $p < 0.05$ vs. WT + vehicle; ## $p < 0.01$ and # $p < 0.05$ vs. mSOD1 + vehicle, unpaired and parametric *t*-test (with Welch's correction when needed).

Interestingly, the preconditioned secretome was able to counteract the upregulation of the 3 main inflamma-miRNAs (Figure 10B and Supplementary Table S4), i.e., miR-146a ($p < 0.05$), miR-155 ($p < 0.01$), and miR-124 ($p < 0.05$), when compared with the mSOD1 mice injected with the vehicle (Figure 10A), attesting to its regulatory potential in the prevention of inflamma-miRNA deregulation.

In sum, our results highlight the therapeutic potential of the secretome from anti-miR-124-modulated mSOD1 MNs in preventing motor disabilities, muscular atrophy, MN degeneration/demise/dysfunctionality, glial reactivity, impaired phagocytosis, and inflammatory imbalance. We hope to translate this therapeutic strategy to ALS patients, mainly to those with upregulated miR-124, to slow down disease progression and prolong their lifetimes.

4. Discussion

In the present study, our main interest was to validate whether the benefits of regulating miR-124 in mSOD1 MNs using anti-miR-124 were translated to mSOD1 mice in terms of preventing disease progression. We showed that upregulated levels of miR-124 in mSOD1 MNs [11] were associated with neurodegeneration and that its secretome caused glial activation, findings that were preserved when we transfected mSOD1 MNs with anti-miR-124 [13].

miR-124 is the most abundant miRNA in the adult brain and has been described as a regulator of microglia-mediated inflammation in neurological diseases [10,60]. It is mainly expressed in neuronal cells, showing important functions in neuronal development, differentiation, synaptic plasticity, and even in memory signalling molecules [61]. The elevated expression of miR-124 in Alzheimer's disease (AD) was shown to benefit the neurite outgrowth, mitochondria activation, small Aβ oligomer reduction, and beta-site amyloid precursor protein cleaving enzyme 1 (BACE1) in experimental models and patients [9,10,62]. However, in ALS, the upregulation of miR-124 in mSOD1 MNs was shown to impair the neurite network, mitochondria dynamics, axonal transport, and synaptic signalling [13]. Moreover, the secretome from these ALS MNs also revealed elevated levels of miR-124 and induced pathological features in the SC organotypic cultures from early symptomatic ALS mice [11,13]. The beneficial or harmful effects of miR-124 may then depend on the cell type and the pathological condition.

miRNA mimics and inhibitors have been used to change the endogenous levels of certain miRNAs and regulate cell function [63,64], as in the case of miR-124 [65,66]. The overexpression of miR-124 was shown to act as a neuroprotective factor in Parkinson's disease [67,68] and in AD [62,69]. In human AD neuronal models with elevated levels of miR-124, we have shown by using an miR-124 mimic and inhibitor that an increase in miR-124 may be associated with mechanisms of defence and adaptation to the pathology [9]. miR-124 elevation was also observed in neuron-derived sEVs, and modulation was sensed in the cell secretome, recapitulating the intracellular levels. Interestingly, we also demonstrated that miR-124 was carried in sEVs from the donor neurons into the recipient microglial cells, leading to their reshaping and plasticity, with increased or decreased activation caused by the inhibitor and mimic, respectively [10]. In contrast with these data, our previous studies in mSOD1 MNs evidenced that when the miR-124 was restored to normal levels by transfecting ALS MNs with the miR-124 inhibitor, the secretome became neuroprotective [13]. Therefore, the beneficial or detrimental roles of miR-124 in AD and ALS, respectively, might be related to the type of neurons (neurons vs. MNs), the tested disease models, and the region of the central nervous system affected by the disorder.

Here, we tested the preconditioned secretome from anti-miR-124-treated mSOD1 MNs for its ability to halt motor disabilities, MN demise, neuron/glia deregulation, and muscle atrophy in mSOD1 mice. To achieve this, we performed an intrathecal injection of the preconditioned secretome in the lumbar section of the SC of mSOD1 mice at the early symptomatic stage (12-week-old) [70], and compared data with matched mSOD1 and WT controls injected with the vehicle (differentiated NSC-34 cell media with 1% FBS depleted in sEVs), as performed by others [25,26]. Such controls allowed us to evaluate the preventive effects of the preconditioned secretome injection on the pathological findings of mSOD1 mice at the symptomatic stage and to compare them with the normal behavioural and molecular status of the WT mice. By only injecting the vehicle, we were sure to not alter the course of the disease in the transgenic mice and to not cause any alteration in the WT animals, besides the eventual effect caused by the injection.

Cell secretomes are rich in soluble proteins and sEVs [71], which reach surrounding and distant cells and are engulfed by them [72]. sEVs and secretomes contain miRNAs [9–11,73] and therapeutic miRNA-enriched/depleted sEVs have emerged as novel therapeutics [17,74–76]. When administered in mice, the presence of sEVs/exosomes in several organs, including in the SC, was observed several hours after injection, until 72 h and disappearing at day 12 [25,30,77,78]. Since miR-124 elevation was observed in sEVs isolated from the secretome of mSOD1 MNs [11] and the sEVs also reflected the miR-124 modulation [13], we were interested in determining how long the labelled sEVs could be visualized in the SC, as we also had in mind their possible use in ALS therapeutics. Indeed, studies involving the intrathecal injection of PKH26-labelled sEVs from stem cells of different origins also identified their local distribution in the SC and their benefits in promoting recovery from spinal cord injuries and the inhibition of inflammation [25,79]. We observed that sEVs injected into the intrathecal space and labelled with the green probe PKH67 were visible in the SC at 8 and 72 h after administration, indicating that their dissemination and interaction with neural cells lasted for at least 3 days. This finding was important for inferring the benefits and lasting consequences of our preconditioned secretome administered by intrathecal injection in early symptomatic mSOD1 mice.

In previous studies, we found upregulated miR-124 in the SC of mSOD1 mice [36], which was also confirmed in the present study. Therefore, the next step was to validate that the preconditioned secretome from anti-miR-124-treated mSOD1 MNs was effective in regulating the expression of miR-124 toward normal values in the SC. In this study, we showed that the preconditioned secretome re-established the levels of miR-124 exhibited by the 15-week-old WT mice. The data were in conformity with our previous results demonstrating that such a secretome regulated the miR-124 levels in SC organotypic slices of mSOD1 mice [13]. This finding was important for us to proceed in our experimental design toward the establishment of the preconditioned secretome as a novel and promising therapeutic strategy for ALS.

Behavioural tests were performed on 14-week-old animals, and molecular assessments of the SC were carried out after 15-week-old mouse sacrifice, which preceded the advanced symptomatic stage that occurs at 16 weeks of age [70]. No approach for the modulation of miR-124 in SOD1-G93A transgenic mice, the best characterized model [80], has been performed before. The mice were phenotypically healthy until 80–90 days old, and then hindlimb tremors and weakness began, progressing to hyperreflexia and paralysis and culminating with death at around 120 days old [37,81]. Our results demonstrated that 14-week-old mSOD1 mice had several motor disabilities, including locomotor deficits that caused shorter steps, as previously demonstrated by other authors [38–40]. They also evidenced a lack of muscular strength, spending less time attached to the grid, as reported before [39,42–44]. Most of the ALS mice also evidenced an increase in spasticity by the retraction of the hindlimbs towards the abdomen and the closing of the fingers in a grasp, corresponding to the clasping and grasping reflexes, respectively. These disabilities associated with the corticospinal function might have been related to cerebellum dysfunction [82,83] and were observed previously in mSOD1 [41] and TDP-43 mouse models [84,85]. A less commonly used method, the cylinder test allowed us to evaluate the spontaneous activity and sensorimotor function of the mice, as described in [32,86]. We found significantly lower activity in the ALS mice compared to the WT mice, as demonstrated by the reduction in the number of times they reared up [44]. The injection of the preconditioned secretome prevented such motor disabilities from occurring, highlighting the importance of regulating miR-124 levels to preserve not only the MN function, but also the cell homeostasis in the SC [13] and, additionally, to maintain good coordination, balance, and motor functionality in mSOD1 mice.

The motor symptoms observed in ALS mice were unequivocally associated with muscular atrophy, the denervation of the neuromuscular junction, and the degeneration of the MNs. We observed a reduction in the area of gastrocnemius muscle fibers, together with MN loss and a lack of synaptic activity. Such a loss of muscle integrity is usually a result of MN demise [37,87]. However, other evidence points to defects in the muscle fibers that impair contractile function [46] and trigger MN degeneration [88]. These authors demonstrated severe skeletal muscle atrophy and the initiation of neuromuscular junction detachment from the muscle fibers caused by the expression of mSOD1 within skeletal muscle fibers. Moreover, Wong and Martin [88] showed that these mSOD1 mice developed weakness and abnormalities in motor function after performing hanging wire and clasping/grasping tests, corroborating our results. The loss of synaptic activity that we observed in the muscle contributed to the disassembly of the neuromuscular synapse, as previously demonstrated in ALS [89,90]. Again, the injection of the preconditioned secretome counteracted such alterations in the muscle and reinforced the significance of maintaining homeostatic miR-124 levels to support synaptic plasticity [91], as well as the myogenesis process, since the overexpression of miR-124 was shown to suppress myogenic differentiation [92].

We observed high levels of neurodegeneration and MN loss in the lumbar SC of symptomatic ALS mice, as previously published [36]. To contribute for these pathological features, the reduced expression of PSD-95 marked the disruption of postsynaptic signalling, as already described in mSOD1 mice [93] and evidenced in post mortem samples of ALS patients [94]. Thus, the retrograde and anterograde axonal transport systems seemed to be affected in the mSOD1 SC, with the decreased expression of dynein and kinesin, respectively. Such defects have been noticed in ALS mouse models and patients, as reviewed in [95]. Furthermore, we and others identified in the mSOD1 mouse model a reduction in kinesin-1 and dynein, together with significant spinal MN loss, reduced myelin fibers, and muscle pathology [96]. The axis between the neuronal fractalkine (CX3CL1) and its microglial receptor CX3CR1 is known to induce the production of soluble factors implicated in neuronal survival and microglia phagocytosis, thus exerting neuroprotective and immunoregulatory effects in mSOD1 mice [97]. Our results showed increased CX3CR1 but downregulated CX3CL1, as already described in [13,36], suggesting compromised CX3CL1/CX3CR1 neuronal–microglial communication that contributes to the progression of MN degeneration and pathophysiology in ALS [98].

Dysfunctional oligodendrocytes and myelin structures have been reported in the SC of ALS mouse models and patients, as reviewed in [99]. In accordance with published data [100], the reduced expression of MBP that we found in the SC of the ALS mice may have been implicated in the dysfunction of newly differentiated oligodendrocytes. Upregulated GPR17 levels additionally impact oligodendrocyte precursor cell (OPC) maturation due to its regulatory role [48]. Though important for the transition from OPCs to immature oligodendrocytes, GPR17 downregulation is mandatory to allow the maturation of OPCs, a process that is impaired in ALS [48]. Studies have demonstrated that toxic aggregates of SOD1 induce the demyelination of oligodendrocytes [101,102]. To compensate, OPCs increase their proliferation rate and differentiate into new mature oligodendrocytes. From this point of view, the upregulation of GPR17 is important for OPC maturation. However, the abnormal upregulation of this receptor in the SC of mSOD1 mice at the symptomatic stage precludes OPC terminal maturation [48]. For this reason, we suggest that upregulated GRP17 restrains oligodendrocyte maturation, which is further supported by the decreased gene levels of MBP and PLP that were observed. The preconditioned secretome from anti-miR-124-treated mSOD1 MNs prevented neurodegeneration occurring from weeks 12 to 15 in the mSOD1 mice, very likely by sustaining MN viability through the inhibition of early apoptosis [13]. Moreover, it also favoured postsynaptic signalling and myelin production, showing broad neuroprotection. The restoration of homeostatic miR-124 levels is crucial for optimal synaptic plasticity [91] and the prevention of demyelination [103]. Indeed, increased levels of miR-124 were shown to be associated with hippocampal demyelination and memory dysfunction, reinforcing the notion that its regulation is critical for the control of myelin processes [103].

Microglia activation and astrocyte reactivity contribute to the dissemination of neurodegeneration [49,50]. However, due to the glial cell heterogenous phenotypes, it has been hard to define the roles of microglia and astrocytes in pathological processes. Thus, our previous studies showed that several drivers of inflammation are upregulated in the SC of mSOD1 mice at the symptomatic stage [36] and that there is an immune imbalance in the

SCOCs of such mice [13,36]. The slightly decreased expression of MFG-E8, TREM2, and P2RY12 indicated the decreased phagocytic ability of the microglial cells. Reduced MFG-E8 was identified in this model at both the pre-symptomatic and symptomatic stages [36], and TREM2 was suggested to be important in ALS, but this has been scarcely explored [3]. Moreover, downregulated P2RY12 has also been associated with the loss of microglia's phagocytic abilities and cell migration [20]. Such results could be related to the upregulated miR-155 that we observed, considering the interference with microglia phagocytosis [20]. It should be noted that the preconditioned secretome tends to counteract the fingerprint of these parameters, indicating that they are characteristic of the symptomatic stage of mSOD1 mice and can be prevented. Such an effect may be related to the induced reduction we found for miR-155 together with the upregulation of arginase 1, supporting the existence of a more functional microglia after the intrathecal injection of the preconditioned secretome. Reduced levels of arginase 1, as we identified in the mSOD1 mice, have been shown to be linked to the existence of neuropathological mechanisms, neuroinflammation, and behavioural deficits in an AD mouse model [51] and to dysfunctional microglia in the SC of mSOD1 rats at disease end-stage [104]. The increased expression of Iba-1 microglia in mSOD1 mice is also associated with an activated microglia, and was found in our previous study [36]. Once more, the preconditioned secretome was effective in preventing such alterations from occurring in the mSOD1 mice at 15 weeks of age, thus supporting a more functional and reparative phenotype of microglia. Lately, the RNAseq analysis of microglia isolated from the SC of mSOD1 mice showed the existence of several activated phenotypes. A distinctive transcriptional signature, known as DAM, including the upregulation of TREM2 and TIMP2 and the downregulation of P2RY12, CX3CR1, and TMEM119, was identified [3,105,106]. The data we obtained for the dysregulation of P2RY12, CX3CR1, TMEM119, and TIMP2 gene expression levels suggested the existence of activated microglia and neuroinflammation [55–57], though we cannot discard the possibility that if evaluated for subtypes we could also identify the DAM microglia. However, with so many descriptions of microglia phenotypes in different diseases and through single-cell arrays, it was critical for us to investigate microglia functions instead of subtypes based on the gene signature. In summary, we observed activated microglia in the mSOD1 mice at the symptomatic stage with possible impairments in their phagocytic and migration abilities, which we believe to favour neurodegeneration. Remarkably, the preconditioned secretome was effective in preventing such pathological alterations in the microglia. When injected into 12-week-old pre-symptomatic mice, the secretome was shown to delay disease progression and sustain most of the markers associated with the neural cell steady state in 15-week-old animals.

We observed in a previous study that symptomatic mSOD1 mice also exhibited upregulated levels of miR-155 and miR-146a, alongside miR-124, in the SC, which are known to contribute to neuroinflammation [36]. Furthermore, another miRNA that was shown to regulate microglia activation and MN death in ALS was miR-125b [107]. Likewise, miR-21 was found to be upregulated in mSOD1 mice [108]. All of these, except for miR-125b, were upregulated in the SC of the symptomatic mSOD1 mice. Again, the preconditioned secretome prevented such dysregulation from occurring, clearly highlighting the anti-inflammatory efficacy of its intrathecal administration in the ALS mouse model. Ongoing studies are being performed with tricultures of spinal microglia, astrocytes, and MNs in microfluidic devices to better explore the signalling mechanisms and gain of function by the preconditioned secretome and miRNA-loaded sEVs, with relevant application to the development of effective target-driven therapies [109,110]. Our results also supported increased astrocyte reactivity and potential neurotoxic effects in the SC of mSOD1 mice, based on the increased levels of GFAP, Cx43, S100B, and vimentin, in agreement with other studies [35,36]. The preconditioned secretome completely averted the appearance of such a pathological astrocyte phenotype, highlighting its immune remodelling ability, if we also consider its preventive effects against TNF-α upregulation and IL-10 downregulation, which are known as inflammatory markers [13,36].

5. Conclusions

Overall, this study validated the therapeutic potential of the secretome derived from anti-miR-124-treated mSOD1 MNs in the prevention of pathological neural cell signatures in symptomatic mSOD1 mice and the subsequent dysregulation of the paracrine signalling implicated in disease dissemination. From a single intrathecal injection of the preconditioned secretome carrying basal levels of miR-124 at the early symptomatic stage of the mSOD1 mice, we were able to regulate miR-124 toward normal values. With such an approach, we sustained anti-inflammatory/phagocytic microglia, MN function, myelination, astrocyte neuroprotection, and immune regulation at 3 weeks after injection, i.e., at the symptomatic stage wherein all these mechanisms turn pathological. Importantly, the secretome also preserved synaptic signalling, functional skeletal muscle, and neuromuscular junction, which translated to the preservation of motor coordination, balance, corticospinal function, and spontaneous behaviour in the symptomatic mSOD1 mice. Our preconditioned secretome based on the strict regulation of miR-124 levels showed immunoregulatory and potential neuroregenerative properties, opening a new avenue for developing a novel effective therapeutic strategy that can be translated into clinical applications in the future. To achieve this, it will be crucial to stratify patients with upregulated levels of miR-124 based on MNs differentiated from induced pluripotent stem cells (iPSCs) or neural precursor cells generated from their somatic cells. In this sense, we propose the modulation of upregulated miR-124 expression levels in such MNs to produce the preconditioned personalized secretome. We can envisage, as was lately described for the secretome of iPSCs [19], that our preconditioned secretome might be used as an autologous treatment in judiciously selected patients to halt ALS pathogenicity or at least delay disease progression.

Supplementary Materials: The following supporting information can be downloaded at: https://www.mdpi.com/article/10.3390/biomedicines10092120/s1, Table S1: List of primer sequences used in RT-qPCR; Table S2: List of miRNA sequences used in RT-qPCR; Table S3: List of antibodies used for immunohistochemistry (IHC) or western blot (WB); Table S4: Expression of genes, miRNAs and proteins from the muscle and lumbar spinal cord of mSOD1 mice injected with the vehicle or the secretome derived from anti-miR-124-treated mSOD1 MNs (mSOD1 + sec).

Author Contributions: Conceptualization, M.B, A.R.V. and D.B.; methodology, M.B, A.R.V., N.d.S., S.D.-S., A.J.S. and D.B.; data analysis, M.B., M.S., A.R.V. and D.B.; investigation, M.B., M.S., N.d.S., A.R.V., A.J.S. and D.B.; writing—original draft preparation, M.B., M.S., A.R.V. and D.B.; writing—review and editing, A.R.V., A.J.S. and D.B.; supervision, A.R.V., A.J.S. and D.B.; funding acquisition, D.B. All authors have read and agreed to the published version of the manuscript.

Funding: This research was funded by Santa Casa da Misericórdia de Lisboa: ELA-2015-002 (to DB); Fundação para a Ciência e a Tecnologia (FCT): PTDC/MED-NEU/31395/2017 (to D.B.), UIDB/UIDP/04138/2020, and UID/DTP/04138/2019-2020 (to iMed.ULisboa); Programa Operacional Regional de Lisboa and the Programa Operacional Competitividade e Internacionalização LISBOA-01-0145-FEDER-031395 (to D.B.); La Caixa Foundation and Francisco Luzón Foundation through project HR21-00931 (to D.B.); and an individual fellowship from FCT: SFRH/BD/129586/2017 (to M.B.). This work was also funded by the ICVS Scientific Microscopy Platform, a member of the national infrastructure of PPBI—Portuguese Platform of Bioimaging (PPBI-POCI-01-0145-FEDER-022122).

Institutional Review Board Statement: The study was conducted in accordance with the European Community guidelines (Directives 86/609/EU and 2010/63/EU, Recommendation 2007/526/CE, European Convention for the Protection of Vertebrate Animals used for Experimental or Other Scientific Purposes ETS 123/Appendix A) and Portuguese Laws on Animal Care (Decreto-Lei 129/92, Portaria 1005/92, Portaria 466/95, Decreto-Lei 197/96, Portaria 1131/97). All the protocols used in this study were approved by the Portuguese National Authority (General Direction of Veterinary) and Ethical Subcommittee in Life and Health Sciences (SECVS; ID: 018/2019).

Informed Consent Statement: Not applicable.

Data Availability Statement: Data is contained within the article.

Acknowledgments: We would like to acknowledge Bárbara Pinheiro from the Life and Health Sciences Research Institute (ICVS), School of Medicine, University of Minho (Braga, Portugal) for helping to perform and analyse the motor behaviour tests. The Graphical Abstract and Figure 2A–D were created with BioRender.com. We are also grateful for the Servier medical art (https://smart.servier.com accessed on 1 June 2022), provided by Servier and licensed under a Creative Commons Attribution 3.0 unported license, which was used to partially generate Figure 1.

Conflicts of Interest: The authors declare no conflict of interest. The funders had no role in the design of the study; in the collection, analyses, or interpretation of data; in the writing of the manuscript; or in the decision to publish the results.

References

1. Hardiman, O.; Al-Chalabi, A.; Chio, A.; Corr, E.M.; Logroscino, G.; Robberecht, W.; Shaw, P.J.; Simmons, Z.; van den Berg, L.H. Amyotrophic lateral sclerosis. *Nat. Rev. Dis. Primers* **2017**, *3*, 17071. [CrossRef]
2. Vahsen, B.F.; Gray, E.; Thompson, A.G.; Ansorge, O.; Anthony, D.C.; Cowley, S.A.; Talbot, K.; Turner, M.R. Non-neuronal cells in amyotrophic lateral sclerosis—From pathogenesis to biomarkers. *Nat. Rev. Neurol.* **2021**, *17*, 333–348. [CrossRef]
3. Xie, M.; Zhao, S.; Bosco, D.B.; Nguyen, A.; Wu, L.J. Microglial TREM2 in amyotrophic lateral sclerosis. *Dev. Neurobiol.* **2022**, *82*, 125–137. [CrossRef]
4. Rosen, D.R.; Siddique, T.; Patterson, D.; Figlewicz, D.A.; Sapp, P.; Hentati, A.; Donaldson, D.; Goto, J.; O'Regan, J.P.; Deng, H.X.; et al. Mutations in Cu/Zn superoxide dismutase gene are associated with familial amyotrophic lateral sclerosis. *Nature* **1993**, *362*, 59–62. [CrossRef]
5. Alrafiah, A.R. From Mouse Models to Human Disease: An Approach for Amyotrophic Lateral Sclerosis. *In Vivo* **2018**, *32*, 983–998. [CrossRef]
6. Gurney, M.E.; Pu, H.; Chiu, A.Y.; Dal Canto, M.C.; Polchow, C.Y.; Alexander, D.D.; Caliendo, J.; Hentati, A.; Kwon, Y.W.; Deng, H.X.; et al. Motor neuron degeneration in mice that express a human Cu, Zn superoxide dismutase mutation. *Science* **1994**, *264*, 1772–1775. [CrossRef]
7. Martinez, B.; Peplow, P.V. MicroRNA expression in animal models of amyotrophic lateral sclerosis and potential therapeutic approaches. *Neural Regen. Res.* **2022**, *17*, 728–740. [CrossRef]
8. Gomes, C.; Sequeira, C.; Likhite, S.; Dennys, C.N.; Kolb, S.J.; Shaw, P.J.; Vaz, A.R.; Kaspar, B.K.; Meyer, K.; Brites, D. Neurotoxic Astrocytes Directly Converted from Sporadic and Familial ALS Patient Fibroblasts Reveal Signature Diversities and miR-146a Theragnostic Potential in Specific Subtypes. *Cells* **2022**, *11*, 1186. [CrossRef]
9. Garcia, G.; Pinto, S.; Cunha, M.; Fernandes, A.; Koistinaho, J.; Brites, D. Neuronal Dynamics and miRNA Signaling Differ between SH-SY5Y APPSwe and PSEN1 Mutant iPSC-Derived AD Models upon Modulation with miR-124 Mimic and Inhibitor. *Cells* **2021**, *10*, 2424. [CrossRef]
10. Garcia, G.; Fernandes, A.; Stein, F.; Brites, D. Protective Signature of IFNγ-Stimulated Microglia Relies on miR-124-3p Regulation From the Secretome Released by Mutant APP Swedish Neuronal Cells. *Front. Pharmacol.* **2022**, *13*, 833066. [CrossRef]
11. Pinto, S.; Cunha, C.; Barbosa, M.; Vaz, A.R.; Brites, D. Exosomes from NSC-34 Cells Transfected with hSOD1-G93A Are Enriched in miR-124 and Drive Alterations in Microglia Phenotype. *Front. Neurosci.* **2017**, *11*, 273. [CrossRef]
12. Barbosa, M.; Gomes, C.; Sequeira, C.; Goncalves-Ribeiro, J.; Pina, C.C.; Carvalho, L.A.; Moreira, R.; Vaz, S.H.; Vaz, A.R.; Brites, D. Recovery of Depleted miR-146a in ALS Cortical Astrocytes Reverts Cell Aberrancies and Prevents Paracrine Pathogenicity on Microglia and Motor Neurons. *Front. Cell Dev. Biol.* **2021**, *9*, 634355. [CrossRef]
13. Vaz, A.R.; Vizinha, D.; Morais, H.; Colaço, A.R.; Loch-Neckel, G.; Barbosa, M.; Brites, D. Overexpression of miR-124 in Motor Neurons Plays a Key Role in ALS Pathological Processes. *Int. J. Mol. Sci.* **2021**, *22*, 6128. [CrossRef]
14. Sun, A.X.; Crabtree, G.R.; Yoo, A.S. MicroRNAs: Regulators of neuronal fate. *Curr. Opin. Cell Biol.* **2013**, *25*, 215–221. [CrossRef]
15. Usategui-Martin, R.; Fernandez-Bueno, I. Neuroprotective therapy for retinal neurodegenerative diseases by stem cell secretome. *Neural Regen. Res.* **2021**, *16*, 117–118. [CrossRef]
16. Mendes-Pinheiro, B.; Anjo, S.I.; Manadas, B.; Da Silva, J.D.; Marote, A.; Behie, L.A.; Teixeira, F.G.; Salgado, A.J. Bone Marrow Mesenchymal Stem Cells' Secretome Exerts Neuroprotective Effects in a Parkinson's Disease Rat Model. *Front. Bioeng. Biotechnol.* **2019**, *7*, 294. [CrossRef]
17. Pinho, A.G.; Cibrao, J.R.; Silva, N.A.; Monteiro, S.; Salgado, A.J. Cell Secretome: Basic Insights and Therapeutic Opportunities for CNS Disorders. *Pharmaceuticals* **2020**, *13*, 31. [CrossRef]
18. Wang, J.; Zuzzio, K.; Walker, C.L. Systemic Dental Pulp Stem Cell Secretome Therapy in a Mouse Model of Amyotrophic Lateral Sclerosis. *Brain Sci.* **2019**, *9*, 165. [CrossRef]
19. Kim, D.; Kim, S.; Sung, A.; Patel, N.; Wong, N.; Conboy, M.J.; Conboy, I.M. Autologous treatment for ALS with implication for broad neuroprotection. *Transl. Neurodegener.* **2022**, *11*, 16. [CrossRef]
20. Butovsky, O.; Jedrychowski, M.P.; Cialic, R.; Krasemann, S.; Murugaiyan, G.; Fanek, Z.; Greco, D.J.; Wu, P.M.; Doykan, C.E.; Kiner, O.; et al. Targeting miR-155 restores abnormal microglia and attenuates disease in SOD1 mice. *Ann. Neurol.* **2015**, *77*, 75–99. [CrossRef]

21. Loffreda, A.; Nizzardo, M.; Arosio, A.; Ruepp, M.D.; Calogero, R.A.; Volinia, S.; Galasso, M.; Bendotti, C.; Ferrarese, C.; Lunetta, C.; et al. miR-129-5p: A key factor and therapeutic target in amyotrophic lateral sclerosis. *Prog. Neurobiol.* **2020**, *190*, 101803. [CrossRef] [PubMed]

22. Tung, Y.T.; Peng, K.C.; Chen, Y.C.; Yen, Y.P.; Chang, M.; Thams, S.; Chen, J.A. Mir-17 approximately 92 Confers Motor Neuron Subtype Differential Resistance to ALS-Associated Degeneration. *Cell Stem Cell* **2019**, *25*, 193–209.e7. [CrossRef] [PubMed]

23. Nolan, K.; Mitchem, M.R.; Jimenez-Mateos, E.M.; Henshall, D.C.; Concannon, C.G.; Prehn, J.H. Increased expression of microRNA-29a in ALS mice: Functional analysis of its inhibition. *J. Mol. Neurosci.* **2014**, *53*, 231–241. [CrossRef] [PubMed]

24. Hu, H.; Lin, H.; Duan, W.; Cui, C.; Li, Z.; Liu, Y.; Wang, W.; Wen, D.; Wang, Y.; Li, C. Intrathecal Injection of scAAV9-hIGF1 Prolongs the Survival of ALS Model Mice by Inhibiting the NF-kB Pathway. *Neuroscience* **2018**, *381*, 1–10. [CrossRef]

25. Noori, L.; Arabzadeh, S.; Mohamadi, Y.; Mojaverrostami, S.; Mokhtari, T.; Akbari, M.; Hassanzadeh, G. Intrathecal administration of the extracellular vesicles derived from human Wharton's jelly stem cells inhibit inflammation and attenuate the activity of inflammasome complexes after spinal cord injury in rats. *Neurosci. Res.* **2021**, *170*, 87–98. [CrossRef] [PubMed]

26. Zhang, J.; Ji, C.; Zhang, H.; Shi, H.; Mao, F.; Qian, H.; Xu, W.; Wang, D.; Pan, J.; Fang, X.; et al. Engineered neutrophil-derived exosome-like vesicles for targeted cancer therapy. *Sci. Adv.* **2022**, *8*, eabj8207. [CrossRef]

27. Mendes-Pinheiro, B.; Teixeira, F.G.; Anjo, S.I.; Manadas, B.; Behie, L.A.; Salgado, A.J. Secretome of Undifferentiated Neural Progenitor Cells Induces Histological and Motor Improvements in a Rat Model of Parkinson's Disease. *Stem Cells Transl. Med.* **2018**, *7*, 829–838. [CrossRef]

28. Li, D.; Liu, C.; Yang, C.; Wang, D.; Wu, D.; Qi, Y.; Su, Q.; Gao, G.; Xu, Z.; Guo, Y. Slow Intrathecal Injection of rAAVrh10 Enhances its Transduction of Spinal Cord and Therapeutic Efficacy in a Mutant SOD1 Model of ALS. *Neuroscience* **2017**, *365*, 192–205. [CrossRef]

29. Correia, J.S.; Neves-Carvalho, A.; Mendes-Pinheiro, B.; Pires, J.; Teixeira, F.G.; Lima, R.; Monteiro, S.; Silva, N.A.; Soares-Cunha, C.; Serra, S.C.; et al. Preclinical Assessment of Mesenchymal-Stem-Cell-Based Therapies in Spinocerebellar Ataxia Type 3. *Biomedicines* **2021**, *9*, 1754. [CrossRef]

30. Lee, B.R.; Kim, J.H.; Choi, E.S.; Cho, J.H.; Kim, E. Effect of young exosomes injected in aged mice. *Int. J. Nanomed.* **2018**, *13*, 5335–5345. [CrossRef]

31. Andersen, P.M.; Borasio, G.D.; Dengler, R.; Hardiman, O.; Kollewe, K.; Leigh, P.N.; Pradat, P.F.; Silani, V.; Tomik, B.; Group, E.W. Good practice in the management of amyotrophic lateral sclerosis: Clinical guidelines. An evidence-based review with good practice points. EALSC Working Group. *Amyotroph. Lateral. Scler.* **2007**, *8*, 195–213. [CrossRef] [PubMed]

32. Brooks, S.P.; Dunnett, S.B. Tests to assess motor phenotype in mice: A user's guide. *Nat. Rev. Neurosci.* **2009**, *10*, 519–529. [CrossRef] [PubMed]

33. Atkin, J.D.; Scott, R.L.; West, J.M.; Lopes, E.; Quah, A.K.; Cheema, S.S. Properties of slow- and fast-twitch muscle fibres in a mouse model of amyotrophic lateral sclerosis. *Neuromuscul. Disord.* **2005**, *15*, 377–388. [CrossRef]

34. Simões, A.E.; Pereira, D.M.; Amaral, J.D.; Nunes, A.F.; Gomes, S.E.; Rodrigues, P.M.; Lo, A.C.; D'Hooge, R.; Steer, C.J.; Thibodeau, S.N.; et al. Efficient recovery of proteins from multiple source samples after TRIzol((R)) or TRIzol((R))LS RNA extraction and long-term storage. *BMC Genom.* **2013**, *14*, 181. [CrossRef] [PubMed]

35. Gomes, C.; Sequeira, C.; Barbosa, M.; Cunha, C.; Vaz, A.R.; Brites, D. Astrocyte regional diversity in ALS includes distinct aberrant phenotypes with common and causal pathological processes. *Exp. Cell Res.* **2020**, *395*, 112209. [CrossRef] [PubMed]

36. Cunha, C.; Santos, C.; Gomes, C.; Fernandes, A.; Correia, A.M.; Sebastião, A.M.; Vaz, A.R.; Brites, D. Downregulated Glia Interplay and Increased miRNA-155 as Promising Markers to Track ALS at an Early Stage. *Mol. Neurobiol.* **2018**, *55*, 4207–4224. [CrossRef]

37. Gurney, M.E. Transgenic-mouse model of amyotrophic lateral sclerosis. *N. Engl. J. Med.* **1994**, *331*, 1721–1722. [CrossRef]

38. Knippenberg, S.; Thau, N.; Dengler, R.; Petri, S. Significance of behavioural tests in a transgenic mouse model of amyotrophic lateral sclerosis (ALS). *Behav. Brain Res.* **2010**, *213*, 82–87. [CrossRef]

39. Alves, C.J.; de Santana, L.P.; dos Santos, A.J.; de Oliveira, G.P.; Duobles, T.; Scorisa, J.M.; Martins, R.S.; Maximino, J.R.; Chadi, G. Early motor and electrophysiological changes in transgenic mouse model of amyotrophic lateral sclerosis and gender differences on clinical outcome. *Brain Res.* **2011**, *1394*, 90–104. [CrossRef]

40. Mead, R.J.; Bennett, E.J.; Kennerley, A.J.; Sharp, P.; Sunyach, C.; Kasher, P.; Berwick, J.; Pettmann, B.; Battaglia, G.; Azzouz, M.; et al. Optimised and rapid pre-clinical screening in the SOD1(G93A) transgenic mouse model of amyotrophic lateral sclerosis (ALS). *PLoS ONE* **2011**, *6*, e23244. [CrossRef]

41. Rodriguez-Cueto, C.; Santos-Garcia, I.; Garcia-Toscano, L.; Espejo-Porras, F.; Bellido, M.; Fernandez-Ruiz, J.; Munoz, E.; de Lago, E. Neuroprotective effects of the cannabigerol quinone derivative VCE-003.2 in SOD1(G93A) transgenic mice, an experimental model of amyotrophic lateral sclerosis. *Biochem. Pharmacol.* **2018**, *157*, 217–226. [CrossRef] [PubMed]

42. Miana-Mena, F.J.; Munoz, M.J.; Yague, G.; Mendez, M.; Moreno, M.; Ciriza, J.; Zaragoza, P.; Osta, R. Optimal methods to characterize the G93A mouse model of ALS. *Amyotroph. Lateral. Scler Other Motor Neuron Disord.* **2005**, *6*, 55–62. [CrossRef] [PubMed]

43. Olivan, S.; Calvo, A.C.; Rando, A.; Munoz, M.J.; Zaragoza, P.; Osta, R. Comparative study of behavioural tests in the SOD1G93A mouse model of amyotrophic lateral sclerosis. *Exp. Anim.* **2015**, *64*, 147–153. [CrossRef] [PubMed]

44. Trabjerg, M.S.; Andersen, D.C.; Huntjens, P.; Oklinski, K.E.; Bolther, L.; Hald, J.L.; Baisgaard, A.E.; Mork, K.; Warming, N.; Kullab, U.B.; et al. Downregulating carnitine palmitoyl transferase 1 affects disease progression in the SOD1 G93A mouse model of ALS. *Commun. Biol.* **2021**, *4*, 509. [CrossRef] [PubMed]

45. Lepore, E.; Casola, I.; Dobrowolny, G.; Musaro, A. Neuromuscular Junction as an Entity of Nerve-Muscle Communication. *Cells* **2019**, *8*, 906. [CrossRef]

46. Dobrowolny, G.; Aucello, M.; Rizzuto, E.; Beccafico, S.; Mammucari, C.; Boncompagni, S.; Belia, S.; Wannenes, F.; Nicoletti, C.; Del Prete, Z.; et al. Skeletal muscle is a primary target of SOD1G93A-mediated toxicity. *Cell Metab.* **2008**, *8*, 425–436. [CrossRef]

47. de Faria, O., Jr.; Gonsalvez, D.G.; Nicholson, M.; Xiao, J. Activity-dependent central nervous system myelination throughout life. *J. Neurochem.* **2019**, *148*, 447–461. [CrossRef]

48. Bonfanti, E.; Bonifacino, T.; Raffaele, S.; Milanese, M.; Morgante, E.; Bonanno, G.; Abbracchio, M.P.; Fumagalli, M. Abnormal Upregulation of GPR17 Receptor Contributes to Oligodendrocyte Dysfunction in SOD1 G93A Mice. *Int. J. Mol. Sci.* **2020**, *21*, 2395. [CrossRef]

49. Kwon, H.S.; Koh, S.H. Neuroinflammation in neurodegenerative disorders: The roles of microglia and astrocytes. *Transl. Neurodegener.* **2020**, *9*, 42. [CrossRef]

50. Garland, E.F.; Hartnell, I.J.; Boche, D. Microglia and Astrocyte Function and Communication: What Do We Know in Humans? *Front. Neurosci.* **2022**, *16*, 824888. [CrossRef]

51. Ma, C.; Hunt, J.B.; Selenica, M.B.; Sanneh, A.; Sandusky-Beltran, L.A.; Watler, M.; Daas, R.; Kovalenko, A.; Liang, H.; Placides, D.; et al. Arginase 1 Insufficiency Precipitates Amyloid-beta Deposition and Hastens Behavioral Impairment in a Mouse Model of Amyloidosis. *Front. Immunol.* **2020**, *11*, 582998. [CrossRef] [PubMed]

52. Lou, N.; Takano, T.; Pei, Y.; Xavier, A.L.; Goldman, S.A.; Nedergaard, M. Purinergic receptor P2RY12-dependent microglial closure of the injured blood-brain barrier. *Proc. Natl. Acad. Sci. USA* **2016**, *113*, 1074–1079. [CrossRef] [PubMed]

53. Akhter, R.; Shao, Y.; Formica, S.; Khrestian, M.; Bekris, L.M. TREM2 alters the phagocytic, apoptotic and inflammatory response to Abeta42 in HMC3 cells. *Mol. Immunol.* **2021**, *131*, 171–179. [CrossRef] [PubMed]

54. Fuller, A.D.; Van Eldik, L.J. MFG-E8 regulates microglial phagocytosis of apoptotic neurons. *J. Neuroimmune Pharmacol.* **2008**, *3*, 246–256. [CrossRef]

55. Hunter, M.; Spiller, K.J.; Dominique, M.A.; Xu, H.; Hunter, F.W.; Fang, T.C.; Canter, R.G.; Roberts, C.J.; Ransohoff, R.M.; Trojanowski, J.Q.; et al. Microglial transcriptome analysis in the rNLS8 mouse model of TDP-43 proteinopathy reveals discrete expression profiles associated with neurodegenerative progression and recovery. *Acta Neuropathol. Commun.* **2021**, *9*, 140. [CrossRef]

56. Lee, E.J.; Kim, H.S. The anti-inflammatory role of tissue inhibitor of metalloproteinase-2 in lipopolysaccharide-stimulated microglia. *J. Neuroinflamm.* **2014**, *11*, 116. [CrossRef]

57. Gonzalez-Prieto, M.; Gutierrez, I.L.; Garcia-Bueno, B.; Caso, J.R.; Leza, J.C.; Ortega-Hernandez, A.; Gomez-Garre, D.; Madrigal, J.L.M. Microglial CX3CR1 production increases in Alzheimer's disease and is regulated by noradrenaline. *Glia* **2021**, *69*, 73–90. [CrossRef] [PubMed]

58. Ricci, C.; Marzocchi, C.; Battistini, S. MicroRNAs as Biomarkers in Amyotrophic Lateral Sclerosis. *Cells* **2018**, *7*, 219. [CrossRef]

59. Gomes, C.; Cunha, C.; Nascimento, F.; Ribeiro, J.A.; Vaz, A.R.; Brites, D. Cortical Neurotoxic Astrocytes with Early ALS Pathology and miR-146a Deficit Replicate Gliosis Markers of Symptomatic SOD1G93A Mouse Model. *Mol. Neurobiol.* **2019**, *56*, 2137–2158. [CrossRef]

60. Zhao, A.D.; Qin, H.; Sun, M.L.; Ma, K.; Fu, X.B. Correction to: Efficient and rapid conversion of human astrocytes and ALS mouse model spinal cord astrocytes into motor neuron-like cells by defined small molecules. *Mil. Med. Res.* **2021**, *8*, 24. [CrossRef]

61. Han, D.; Dong, X.; Zheng, D.; Nao, J. MiR-124 and the Underlying Therapeutic Promise of Neurodegenerative Disorders. *Front. Pharmacol.* **2019**, *10*, 1555. [CrossRef] [PubMed]

62. An, F.; Gong, G.; Wang, Y.; Bian, M.; Yu, L.; Wei, C. MiR-124 acts as a target for Alzheimer's disease by regulating BACE1. *Oncotarget* **2017**, *8*, 114065–114071. [CrossRef] [PubMed]

63. He, P.Y.; Yip, W.K.; Jabar, M.F.; Mohtarrudin, N.; Dusa, N.M.; Seow, H.F. Effect of the miR-96-5p inhibitor and mimic on the migration and invasion of the SW480-7 colorectal cancer cell line. *Oncol. Lett.* **2019**, *18*, 1949–1960. [CrossRef] [PubMed]

64. Zhong, B.; Guo, S.; Zhang, W.; Zhang, C.; Wang, Y.; Zhang, C. Bioinformatics prediction of miR-30a targets and its inhibition of cell proliferation of osteosarcoma by up-regulating the expression of PTEN. *BMC Med. Genom.* **2017**, *10*, 64. [CrossRef]

65. Zhou, L.; Xu, Z.; Ren, X.; Chen, K.; Xin, S. MicroRNA-124 (MiR-124) Inhibits Cell Proliferation, Metastasis and Invasion in Colorectal Cancer by Downregulating Rho-Associated Protein Kinase 1(ROCK1). *Cell Physiol. Biochem.* **2016**, *38*, 1785–1795. [CrossRef]

66. Ohnuma, K.; Kasagi, S.; Uto, K.; Noguchi, Y.; Nakamachi, Y.; Saegusa, J.; Kawano, S. MicroRNA-124 inhibits TNF-alpha- and IL-6-induced osteoclastogenesis. *Rheumatol. Int.* **2019**, *39*, 689–695. [CrossRef]

67. Zhang, F.; Yao, Y.; Miao, N.; Wang, N.; Xu, X.; Yang, C. Neuroprotective effects of microRNA 124 in Parkinson's disease mice. *Arch. Gerontol. Geriatr* **2022**, *99*, 104588. [CrossRef]

68. Esteves, M.; Abreu, R.; Fernandes, H.; Serra-Almeida, C.; Martins, P.A.T.; Barao, M.; Cristovao, A.C.; Saraiva, C.; Ferreira, R.; Ferreira, L.; et al. MicroRNA-124-3p-enriched small extracellular vesicles as a therapeutic approach for Parkinson's disease. *Mol. Ther.* **2022**. [CrossRef]

69. Kong, Y.; Wu, J.; Zhang, D.; Wan, C.; Yuan, L. The Role of miR-124 in Drosophila Alzheimer's Disease Model by Targeting Delta in Notch Signaling Pathway. *Curr. Mol. Med.* **2015**, *15*, 980–989. [CrossRef]
70. Rubio, M.A.; Herrando-Grabulosa, M.; Gaja-Capdevila, N.; Vilches, J.J.; Navarro, X. Characterization of somatosensory neuron involvement in the SOD1(G93A) mouse model. *Sci. Rep.* **2022**, *12*, 7600. [CrossRef]
71. Driscoll, J.; Patel, T. The mesenchymal stem cell secretome as an acellular regenerative therapy for liver disease. *J. Gastroenterol.* **2019**, *54*, 763–773. [CrossRef] [PubMed]
72. Feng, J.; Zhang, Y.; Zhu, Z.; Gu, C.; Waqas, A.; Chen, L. Emerging Exosomes and Exosomal MiRNAs in Spinal Cord Injury. *Front. Cell Dev. Biol.* **2021**, *9*, 703989. [CrossRef] [PubMed]
73. Asgarpour, K.; Shojaei, Z.; Amiri, F.; Ai, J.; Mahjoubin-Tehran, M.; Ghasemi, F.; ArefNezhad, R.; Hamblin, M.R.; Mirzaei, H. Exosomal microRNAs derived from mesenchymal stem cells: Cell-to-cell messages. *Cell Commun. Signal.* **2020**, *18*, 149. [CrossRef] [PubMed]
74. Song, Y.; Kim, Y.; Ha, S.; Sheller-Miller, S.; Yoo, J.; Choi, C.; Park, C.H. The emerging role of exosomes as novel therapeutics: Biology, technologies, clinical applications, and the next. *Am. J. Reprod. Immunol.* **2021**, *85*, e13329. [CrossRef] [PubMed]
75. Munir, J.; Yoon, J.K.; Ryu, S. Therapeutic miRNA-Enriched Extracellular Vesicles: Current Approaches and Future Prospects. *Cells* **2020**, *9*, 2271. [CrossRef]
76. Naseri, Z.; Oskuee, R.K.; Jaafari, M.R.; Forouzandeh Moghadam, M. Exosome-mediated delivery of functionally active miRNA-142-3p inhibitor reduces tumorigenicity of breast cancer in vitro and in vivo. *Int. J. Nanomed.* **2018**, *13*, 7727–7747. [CrossRef]
77. Cao, H.; Yue, Z.; Gao, H.; Chen, C.; Cui, K.; Zhang, K.; Cheng, Y.; Shao, G.; Kong, D.; Li, Z.; et al. In Vivo Real-Time Imaging of Extracellular Vesicles in Liver Regeneration via Aggregation-Induced Emission Luminogens. *ACS Nano* **2019**, *13*, 3522–3533. [CrossRef]
78. Perets, N.; Betzer, O.; Shapira, R.; Brenstein, S.; Angel, A.; Sadan, T.; Ashery, U.; Popovtzer, R.; Offen, D. Golden Exosomes Selectively Target Brain Pathologies in Neurodegenerative and Neurodevelopmental Disorders. *Nano Lett.* **2019**, *19*, 3422–3431. [CrossRef]
79. Cao, Y.; Xu, Y.; Chen, C.; Xie, H.; Lu, H.; Hu, J. Local delivery of USC-derived exosomes harboring ANGPTL3 enhances spinal cord functional recovery after injury by promoting angiogenesis. *Stem. Cell Res. Ther.* **2021**, *12*, 20. [CrossRef]
80. Morrice, J.R.; Gregory-Evans, C.Y.; Shaw, C.A. Animal models of amyotrophic lateral sclerosis: A comparison of model validity. *Neural Regen. Res.* **2018**, *13*, 2050–2054. [CrossRef]
81. Turner, B.J.; Talbot, K. Transgenics, toxicity and therapeutics in rodent models of mutant SOD1-mediated familial ALS. *Prog. Neurobiol.* **2008**, *85*, 94–134. [CrossRef] [PubMed]
82. Nowak, D.A.; Topka, H.; Timmann, D.; Boecker, H.; Hermsdorfer, J. The role of the cerebellum for predictive control of grasping. *Cerebellum* **2007**, *6*, 7–17. [CrossRef] [PubMed]
83. Kandasamy, L.C.; Tsukamoto, M.; Banov, V.; Tsetsegee, S.; Nagasawa, Y.; Kato, M.; Matsumoto, N.; Takeda, J.; Itohara, S.; Ogawa, S.; et al. Limb-clasping, cognitive deficit and increased vulnerability to kainic acid-induced seizures in neuronal glycosylphosphatidylinositol deficiency mouse models. *Hum. Mol. Genet.* **2021**, *30*, 758–770. [CrossRef] [PubMed]
84. Alfieri, J.A.; Pino, N.S.; Igaz, L.M. Reversible behavioral phenotypes in a conditional mouse model of TDP-43 proteinopathies. *J. Neurosci.* **2014**, *34*, 15244–15259. [CrossRef] [PubMed]
85. Wu, L.S.; Cheng, W.C.; Shen, C.K. Targeted depletion of TDP-43 expression in the spinal cord motor neurons leads to the development of amyotrophic lateral sclerosis-like phenotypes in mice. *J. Biol. Chem.* **2012**, *287*, 27335–27344. [CrossRef]
86. Fleming, S.M.; Ekhator, O.R.; Ghisays, V. Assessment of sensorimotor function in mouse models of Parkinson's disease. *J. Vis. Exp.* **2013**, *76*, 50303. [CrossRef]
87. Hegedus, J.; Putman, C.T.; Tyreman, N.; Gordon, T. Preferential motor unit loss in the SOD1 G93A transgenic mouse model of amyotrophic lateral sclerosis. *J. Physiol.* **2008**, *586*, 3337–3351. [CrossRef]
88. Wong, M.; Martin, L.J. Skeletal muscle-restricted expression of human SOD1 causes motor neuron degeneration in transgenic mice. *Hum. Mol. Genet.* **2010**, *19*, 2284–2302. [CrossRef]
89. Vinsant, S.; Mansfield, C.; Jimenez-Moreno, R.; Del Gaizo Moore, V.; Yoshikawa, M.; Hampton, T.G.; Prevette, D.; Caress, J.; Oppenheim, R.W.; Milligan, C. Characterization of early pathogenesis in the SOD1(G93A) mouse model of ALS: Part II, results and discussion. *Brain Behav.* **2013**, *3*, 431–457. [CrossRef]
90. Campanari, M.L.; Garcia-Ayllon, M.S.; Ciura, S.; Saez-Valero, J.; Kabashi, E. Neuromuscular Junction Impairment in Amyotrophic Lateral Sclerosis: Reassessing the Role of Acetylcholinesterase. *Front. Mol. Neurosci.* **2016**, *9*, 160. [CrossRef]
91. Hou, Q.; Ruan, H.; Gilbert, J.; Wang, G.; Ma, Q.; Yao, W.D.; Man, H.Y. MicroRNA miR124 is required for the expression of homeostatic synaptic plasticity. *Nat. Commun.* **2015**, *6*, 10045. [CrossRef] [PubMed]
92. Qadir, A.S.; Woo, K.M.; Ryoo, H.M.; Yi, T.; Song, S.U.; Baek, J.H. MiR-124 inhibits myogenic differentiation of mesenchymal stem cells via targeting Dlx5. *J. Cell Biochem.* **2014**, *115*, 1572–1581. [CrossRef] [PubMed]
93. Broadhead, M.J.; Bonthron, C.; Waddington, J.; Smith, W.V.; Lopez, M.F.; Burley, S.; Valli, J.; Zhu, F.; Komiyama, N.H.; Smith, C.; et al. Selective vulnerability of tripartite synapses in amyotrophic lateral sclerosis. *Acta Neuropathol.* **2022**, *143*, 471–486. [CrossRef]
94. Genc, B.; Jara, J.H.; Lagrimas, A.K.; Pytel, P.; Roos, R.P.; Mesulam, M.M.; Geula, C.; Bigio, E.H.; Ozdinler, P.H. Apical dendrite degeneration, a novel cellular pathology for Betz cells in ALS. *Sci. Rep.* **2017**, *7*, 41765. [CrossRef] [PubMed]
95. De Vos, K.J.; Hafezparast, M. Neurobiology of axonal transport defects in motor neuron diseases: Opportunities for translational research? *Neurobiol. Dis.* **2017**, *105*, 283–299. [CrossRef] [PubMed]

96. Warita, H.; Itoyama, Y.; Abe, K. Selective impairment of fast anterograde axonal transport in the peripheral nerves of asymptomatic transgenic mice with a G93A mutant SOD1 gene. *Brain Res.* **1999**, *819*, 120–131. [CrossRef]

97. Liu, C.; Hong, K.; Chen, H.; Niu, Y.; Duan, W.; Liu, Y.; Ji, Y.; Deng, B.; Li, Y.; Li, Z.; et al. Evidence for a protective role of the CX3CL1/CX3CR1 axis in a model of amyotrophic lateral sclerosis. *Biol. Chem.* **2019**, *400*, 651–661. [CrossRef]

98. Zhang, J.; Liu, Y.; Liu, X.; Li, S.; Cheng, C.; Chen, S.; Le, W. Dynamic changes of CX3CL1/CX3CR1 axis during microglial activation and motor neuron loss in the spinal cord of ALS mouse model. *Transl. Neurodegener.* **2018**, *7*, 35. [CrossRef]

99. Raffaele, S.; Boccazzi, M.; Fumagalli, M. Oligodendrocyte Dysfunction in Amyotrophic Lateral Sclerosis: Mechanisms and Therapeutic Perspectives. *Cells* **2021**, *10*, 565. [CrossRef]

100. Philips, T.; Bento-Abreu, A.; Nonneman, A.; Haeck, W.; Staats, K.; Geelen, V.; Hersmus, N.; Kusters, B.; Van Den Bosch, L.; Van Damme, P.; et al. Oligodendrocyte dysfunction in the pathogenesis of amyotrophic lateral sclerosis. *Brain* **2013**, *136*, 471–482. [CrossRef]

101. Kang, S.H.; Li, Y.; Fukaya, M.; Lorenzini, I.; Cleveland, D.W.; Ostrow, L.W.; Rothstein, J.D.; Bergles, D.E. Degeneration and impaired regeneration of gray matter oligodendrocytes in amyotrophic lateral sclerosis. *Nat. Neurosci.* **2013**, *16*, 571–579. [CrossRef] [PubMed]

102. Kim, S.; Chung, A.Y.; Na, J.E.; Lee, S.J.; Jeong, S.H.; Kim, E.; Sun, W.; Rhyu, I.J.; Park, H.C. Myelin degeneration induced by mutant superoxide dismutase 1 accumulation promotes amyotrophic lateral sclerosis. *Glia* **2019**, *67*, 1910–1921. [CrossRef] [PubMed]

103. Dutta, R.; Chomyk, A.M.; Chang, A.; Ribaudo, M.V.; Deckard, S.A.; Doud, M.K.; Edberg, D.D.; Bai, B.; Li, M.; Baranzini, S.E.; et al. Hippocampal demyelination and memory dysfunction are associated with increased levels of the neuronal microRNA miR-124 and reduced AMPA receptors. *Ann. Neurol.* **2013**, *73*, 637–645. [CrossRef] [PubMed]

104. Nikodemova, M.; Small, A.L.; Smith, S.M.; Mitchell, G.S.; Watters, J.J. Spinal but not cortical microglia acquire an atypical phenotype with high VEGF, galectin-3 and osteopontin, and blunted inflammatory responses in ALS rats. *Neurobiol. Dis.* **2014**, *69*, 43–53. [CrossRef] [PubMed]

105. Keren-Shaul, H.; Spinrad, A.; Weiner, A.; Matcovitch-Natan, O.; Dvir-Szternfeld, R.; Ulland, T.K.; David, E.; Baruch, K.; Lara-Astaiso, D.; Toth, B.; et al. A Unique Microglia Type Associated with Restricting Development of Alzheimer's Disease. *Cell* **2017**, *169*, 1276–1290.e17. [CrossRef]

106. Krasemann, S.; Madore, C.; Cialic, R.; Baufeld, C.; Calcagno, N.; El Fatimy, R.; Beckers, L.; O'Loughlin, E.; Xu, Y.; Fanek, Z.; et al. The TREM2-APOE Pathway Drives the Transcriptional Phenotype of Dysfunctional Microglia in Neurodegenerative Diseases. *Immunity* **2017**, *47*, 566–581.e9. [CrossRef] [PubMed]

107. Parisi, C.; Napoli, G.; Amadio, S.; Spalloni, A.; Apolloni, S.; Longone, P.; Volonte, C. MicroRNA-125b regulates microglia activation and motor neuron death in ALS. *Cell Death Differ.* **2016**, *23*, 531–541. [CrossRef]

108. Parisi, C.; Arisi, I.; D'Ambrosi, N.; Storti, A.E.; Brandi, R.; D'Onofrio, M.; Volonte, C. Dysregulated microRNAs in amyotrophic lateral sclerosis microglia modulate genes linked to neuroinflammation. *Cell Death Dis* **2013**, *4*, e959. [CrossRef]

109. Masuda, T.; Sankowski, R.; Staszewski, O.; Prinz, M. Microglia Heterogeneity in the Single-Cell Era. *Cell Rep.* **2020**, *30*, 1271–1281. [CrossRef]

110. Luchena, C.; Zuazo-Ibarra, J.; Valero, J.; Matute, C.; Alberdi, E.; Capetillo-Zarate, E. A Neuron, Microglia, and Astrocyte Triple Co-culture Model to Study Alzheimer's Disease. *Front. Aging Neurosci.* **2022**, *14*, 844534. [CrossRef]

biomedicines

MDPI

Article

Targeting of microRNA-22 Suppresses Tumor Spread in a Mouse Model of Triple-Negative Breast Cancer

Riccardo Panella [1,2,3,*], Cody A. Cotton [2], Valerie A. Maymi [4], Sachem Best [4], Kelsey E. Berry [2], Samuel Lee [4], Felipe Batalini [1,5], Ioannis S. Vlachos [1,5], John G. Clohessy [1,4], Sakari Kauppinen [3] and Pier Paolo Pandolfi [1,6,7,*]

1   Cancer Research Institute, Beth Israel Deaconess Cancer Center, Departments of Medicine and Pathology, Beth Israel Deaconess Medical Center, Harvard Medical School, Boston, MA 02215, USA
2   Center for Genomic Medicine, Desert Research Institute, Reno, NV 89512, USA
3   Center for RNA Medicine, Department of Clinical Medicine, Aalborg University, 2450 Copenhagen, Denmark
4   Preclinical Murine Pharmacogenetics Facility and Mouse Hospital, Beth Israel Deaconess Medical Center, Harvard Medical School, Boston, MA 02215, USA
5   Broad Institute of MIT and Harvard, Cambridge, MA 02142, USA
6   Department of Molecular Biotechnology and Health Sciences, Molecular Biotechnology Center, University of Turin, 10154 Turin, Italy
7   Renown Institute for Cancer, Nevada System of Higher Education, Reno, NV 89502, USA
*   Correspondence: riccardop@dcm.aau.dk (R.P.); pierpaolo.pandolfiderinaldis@renown.org (P.P.P.)

Abstract: microRNA-22 (miR-22) is an oncogenic miRNA whose up-regulation promotes epithelial-mesenchymal transition (EMT), tumor invasion, and metastasis in hormone-responsive breast cancer. Here we show that miR-22 plays a key role in triple negative breast cancer (TNBC) by promoting EMT and aggressiveness in 2D and 3D cell models and a mouse xenograft model of human TNBC, respectively. Furthermore, we report that miR-22 inhibition using an LNA-modified antimiR-22 compound is effective in reducing EMT both in vitro and in vivo. Importantly, pharmacologic inhibition of miR-22 suppressed metastatic spread and markedly prolonged survival in mouse xenograft models of metastatic TNBC highlighting the potential of miR-22 silencing as a new therapeutic strategy for the treatment of TNBC.

Keywords: RNA medicine; microRNA; breast cancer

Citation: Panella, R.; Cotton, C.A.; Maymi, V.A.; Best, S.; Berry, K.E.; Lee, S.; Batalini, F.; Vlachos, I.S.; Clohessy, J.G.; Kauppinen, S.; et al. Targeting of microRNA-22 Suppresses Tumor Spread in a Mouse Model of Triple-Negative Breast Cancer. *Biomedicines* **2023**, *11*, 1470. https://doi.org/10.3390/biomedicines11051470

Academic Editors: Shaker A. Mousa and Albrecht Piiper

Received: 27 October 2022
Revised: 21 January 2023
Accepted: 9 May 2023
Published: 18 May 2023

1. Introduction

Breast cancer is the most widely diagnosed cancer in women and is one of the leading causes of cancer-related deaths after lung cancer [1–3]. Among the five different breast cancer subtypes, triple negative breast cancer (TNBC) is the most aggressive and lethal subtype [4]. TNBC is characterized by a very strong metastatic potential and its molecular complexity make it particularly hard to treat making the developing for new therapeutic options urgent to address the patient needs. Furthermore, the lack of receptor expression in these subtypes renders them entirely insensitive to endocrine and targeted therapies used for breast cancers [4–6]. At present, surgery and chemotherapy are the mainstay of treatment for TNBC patients with a limited number of adjuvant regimens recommended by the National Comprehensive Cancer Network [7]. When surgery with chemotherapy fails, there are no endocrine or targeted therapies available [6]; thus, new treatment strategies for TNBC are urgently needed.

microRNAs (miRNAs) are short non-coding RNAs of about 22 nucleotides in length that function as key post-transcriptional regulators of gene expression by guiding the RNA induced silencing complex (RISC) to partially complementary target sites located predominantly in the $3'$ untranslated regions (UTRs) of target mRNAs, resulting in translational repression and/or de-adenylation and degradation of the miRNA targets. Animal miRNAs have been implicated in the regulation of many biological processes, including developmental timing, apoptosis, differentiation, cell proliferation, and metabolism. Furthermore, miRNA dysregulation has been shown to be associated with a wide range of human diseases, such as CNS disorders, cardiovascular and metabolic diseases, and cancer, and thus, miRNAs have emerged as a new class of viable targets for miRNA-based therapeutics [7–28]. Currently, two strategies are deployed to therapeutically manipulate the activity of disease-associated miRNAs. The biological function of down-regulated or lost miRNAs can be restored using either synthetic double-stranded miRNAs or viral vector-based overexpression, whereas inhibition of overexpressed miRNAs can be achieved using single-stranded, chemically modified antisense oligonucleotides (ASOs) [9,10] called antimiRs. The use of oligonucleotides to manipulate miRNA levels has shown promise in the treatment of a variety of human diseases ranging from inflammation and viral infections to cancer and metabolism [11–14].

miR-22 is a 22-nucleotide-long microRNA encoded in the exon 2 of the miR-22 host gene (MIR22HG), located on the short arm of Chromosome 17 (GRCh38.p14) in a minimal loss of heterozygosity region. It is highly conserved across many vertebrate species, including chimp, mouse, rat, dog, and horse. This level of conservation suggests functional importance. We have previously shown that miR-22 functions as an oncogene increasing the metastatic potential of breast cancer through its ability to regulate TET2 and the miR-200 family [15–17], thereby promoting a strong activation of epithelial-mesenchymal transition (EMT) programs [16]. Specifically, we have previously shown that miR-22 contributes to the metastatic potential of breast cancer in an MCF7 xenograft model and a transgenic mouse model overexpressing miR-22 [15]. In the xenograft models, miR-22 overexpression markedly increased the proliferative rate of MCF7 (Ki67$^+$) cells, as well as their metastatic potential to the lung, compared to MCF7 control tumors. Similarly, in transgenic mice engineered to selectively overexpress miR-22 in mammary tissues, we observed a significant decrease in disease-free survival rate resulting from mammary cancer and spontaneous metastatic spread of the disease [15]. These findings prompted us to ask whether miR-22 plays a broader role in breast cancer. Here, we explore the role of miR-22 in TNBC and its impact on EMT and metastatic spread in preclinical models of TNBC.

2. Methods

2.1. Cell Culture

MDA-MB-231 (ATCC, ATCC HTB-26) cells were either transduced with lentivirus containing a plasmid designed to overexpress miR-22 and express RFP or a plasmid that only expressed RFP. Cells were sorted based on the presence of RFP. RFP-positive cells were collected and grown in high glucose DMEM (Thermo Fisher) containing 500 U/mL of penicillin-streptomycin (Thermo Fisher) and 20 mM L-glutamine (Thermo Fisher).

2.2. Mammospheres Preparation

MDA-MB-2312 cells overexpressing miR-22 and empty vector cells were seeded at a concentration of 4×10^3 cells/cm^2 in Mammocult media (StemCell Technologies) containing the Mammocult Proliferation Supplement (StemCell Technologies), 4 µg/mL heparin (StemCell Technologies), 0.48 µg/mL hydrocortisone (StemCell Technologies), and 500 U/mL Penicillin-Streptomycin (Thermo Fisher). Mammospheres grew in this media for 16 days, and the medium was changed every 2 days. Mammospheres were given 5 µL/mL doses of anti-miR-22 LNA, SCR-LNA for a final concentration of 500 nM, and VHL every 2 days when the medium was changed.

2.3. Synthesis of LNA Oligonucleotides

LNA oligonucleotides were obtained from Exiqon and resuspended in 0.9% NaCl to a concentration of 1 mg/mL in a sterile environment, and filter sterilized using a 0.22 µm vacuum filter.

2.4. RNA Extraction and Purification

At specific time points, cells/mammospheres were spun down at 1200 RPM for 5 min. The supernatant was aspirated; the cell/mammosphere pellet was resuspended in DPBS (Thermo Fisher), and the resuspended cells were spun down at 1200 RPM. The DPBS was aspirated, and 1 mL of TRIzol (Thermo Fisher) was used to lyse the pellet. The lysate was processed using the PureLink RNA Mini Kit (Thermo Fisher) using the instructions provided in the kit.

2.5. RT-qPCR

The purified RNA was subjected to Poly (A) tailing by using *E. coli* Poly (A) Polymerase and rATP from New England BioLabs following the manufacturer's instructions. Ten μL of the poly (A) tailed-RNA was reverse-transcribed using the SuperScript™ IV First-Strand Synthesis System (Thermo Fisher) following the manufacturer's instructions. PowerUp™ SYBR™ Green Master Mix (Thermo Fisher) was used in all qPCR reactions (Table 1).

Table 1. Primers for RT-qPCR.

Gene ID	Fwd 5′-3′	Rev 5′-3′
CDH1	tgctcttccaggaacctctg	gcggcattgtaggtgttc
CDH2	cctgaagccaaccttaactga	tggagggatgacccagtct
SOX9	tacccgcacttgcacaac	tctcgctctcgttcagaagtc
DSP	aaagaaaatgctgcctactttca	ggggtacttcttcctgatgga
TWIST	gggccggagacctagatg	tttccaagaaaatctttggcata
SNAIL	gcgagctgcaggactctaat	cggtggggttgaggatct
CD44	tgacacatattgcttcaatgctt	tgggcaggtctgtgactg
CD24	atgggcagagcaatggtg	ccagttgttgtttcactggaat
HUPO	gcttcctggagggtgtcc	ggactcgtttgtacccgttg

2.6. Protein Extraction and Quantification

RIPA buffer was prepared by adding 20 μL of $50\times$ EDTA-Free protease inhibitor cocktail (Sigma-Aldrich), and 100 μL PhosStop phosphatase inhibitor (Sigma-Aldrich) to 880 μL RIPA buffer (Boston Bioproducts). Mammospheres were pelleted using the same procedure described in the "RNA Extraction and Purification" section, and RIPA buffer plus protease and phosphatase inhibitor were used to resuspend and lyse the pellet. After incubating the cell lysate on ice for 30 min, the lysate was spun at $15,000\times g$ for 20 min; supernatants were collected, and the pellet discarded. The protein in the supernatant was quantified using the Bio-Rad Protein Assay Reagent and NanoDrop One© (Thermo Fisher).

2.7. Western Blot

3 to 5 μg of protein from each sample was loaded onto a 4–12% Bis-Tris NuPage Gel (Thermo Fisher), and after separation was complete, transferred onto a nitrocellulose membrane and immunoblotted for the proteins of interest.

2.8. Antibodies

The following antibodies were used from immunohistochemistry: anti-Ki67 (1:200, MA5-14520) from Invitrogen, and anti-AXL (1:2000, ab219651) from Abcam. The secondary antibodies used for immunohistochemistry staining were: Amersham ECL Mouse IgG (NA931-1ML) and Amersham ECL Rabbit IgG (NA9340-1mL) antibodies were used (GE Healthcare Life Sciences).

2.9. Immunohistochemistry

Immunohistochemistry staining was performed on the lungs and livers of the xenografted mice being studied. Slides were deparaffinized at 62 °C for 15 min, and then subjected to washes of xylene, 100% ethanol, and 95% ethanol. Antigen retrieval was performed using either a sodium citrate buffer or EDTA buffer based on the primary antibody being used, followed by chemical blocking using 30% hydrogen peroxide (Sigma-Aldrich), serum blocking with rabbit serum (MP Biomedicals), and an overnight incubation at 4 °C with the primary antibody was completed. The following day, the slides were washed with DPBS (Thermo Fisher), and then incubated in the secondary antibody at room temperature for 30 min. This was followed by an incubation in the working solution from the Vectastain ABC-HRP kit from Vector Labs for 30 min, diaminobenzidine (Sigma-Aldrich) staining, and counterstaining with hematoxylin (Fisher Scientific).

2.10. Hematoxylin and Eosin Staining

H&E staining was performed on the same samples as above. Slides were deparaffinized using multiple washes of xylene, 100% ethanol, and 95% ethanol. Slides were incubated in hematoxylin (Fisher Scientific), washed with 70% ethanol with 1% hydrochloric acid, followed by an incubation in eosin (Fisher Scientific). After final washes of 95% and 100% ethanol, the slides were ready to have a coverslip.

2.11. Mice

Animal experiments were performed in accordance with the guidelines of Beth Israel Deaconess Medical Center Institutional Animal Care and Use Committee. Nu/J mice were obtained from the Jackson Laboratory, and 5×10^6 of MDA-MB-231 overexpressing miR-22 cells were injected into 30 mice via tail vein, while MDA-MB-231 empty vector cells were injected into other 30 mice using the same delivery method. Tail vein injection was chose as transplant method to better mimic the process of metastasis formation and have the cells in the blood stream and to avoid forcing cell housing in a specific tissue of our choice. Each mouse within the two cohorts was separated into one of three treatment groups, for a total of 10 mice per experimental group, consisting of IP injections of either anti-miR-22 LNA, SCR-LNA, or VHL, given at a dose of 10 mg/Kg. Mice within ± 2.5 g of each other were given the same dosage volume.

2.12. Statistical Analysis

All the statistical analysis on qPCR data, staining, and animals was performed with the ANOVA tool (2-way ANOVA or 3-way ANOVA depending on the experiment) included in the Prism Graph Pad 8. P-values are represented as non-significant (ns) for $p > 0.1$; (*) for $p < 0.03$; (**) for $p < 0.002$; (***) for $p < 0.0002$ and (****) for $p < 0.0001$

2.13. Bioinformatic Analysis

The miRNA expression data from RNAseq (Illumina Hiseq) of samples with breast invasive carcinoma were downloaded from TCGA on 2 May 2020, from "https://tcga.xenahubs.net/download/TCGA.BRCA.sampleMap/miRNA_HiSeq_gene.gz". Comprehensive metadata were downloaded from https://tcga.xenahubs.net/download/TCGA.BRCA.sampleMap/miRNA_HiSeq_gene.json on 2 May 2020. Tumors with human epidermal growth factor receptor (HER2) were defined as HER2-positive (HER2+). Tumors that were HER2-negative (HER2-) but were positive for either estrogen receptor (ER) or progesterone receptor (PR) were defined as hormone receptor positive (HR+). Tumors that were negative for ER, PR, and HER2, were defined as triple-negative breast cancer (TNBC). Kruskal–Wallis rank sum test was used to detect differences of miR-22-3p expression levels across all breast cancer subtypes. For pair-wise comparisons, we used Wilcoxon rank sum test with continuity correction (Mann–Whitney U test). Survival analysis was performed using the Kaplan–Meier method. All analyses were done in R (version 3.6.1). Statistical analysis and graphing of all non-bioinformatical data were performed using GraphPad Prism 8.4.2.

3. Results

Molecularly, TNBC and HER2+ breast cancers share many similarities as both are enriched in loss-of-function mutations in TP53, PTEN, and RB1. HER2+ breast cancers are further characterized by *ERBB2* amplification. These features result in a higher histological grade and a more aggressive disease. In contrast, hormone-receptor positive (HR+) breast cancers typically share a luminal transcriptional profile, which is often estrogen-dependent and enriched in GATA3 mutations with lower mutation rates of TP53 [18]. In terms of prognosis and life expectancy, breast cancer is divided into four different stages, from I to IV, with stage IV being characterized by metastatic spread and considered incurable as per today. The vast majority of stage IV breast cancers are also TNBC [19]. Our in silico analyses suggest that miR-22 levels are elevated in stage IV breast cancer (Figure 1A) and this observation further prompted us to explore the role of miR-22 in this breast cancer subtype.

We first generated a cellular model from a human TNBC cell line, MDA-MB-231, in which we stably overexpressed miR-22 (Figure 2A). To achieve miR-22 overexpression in human TNBC cell lines, a plasmid containing an RFP reporter, designed to constitutively express miR-22, was transduced into MDA-MB-231 cells (referred to as LV-RFP miR-22). This line was compared to MDA-MB-231 cells transduced with an empty vector (referred to as LV-RFP Empty) containing the same reporter. We then isolated RFP-positive cells by sorting the top 10% of RFP-fluorescent cells. After cell recovery, equal numbers of each cell type were seeded into 10 cm dishes to determine whether overexpression of miR-22 resulted in increased proliferation compared to cells expressing basal levels of miR-22. As shown in Figure 1B, cells overexpressing miR-22 displayed a significant increase in the growth rate over control cells. Next, we asked whether miR-22 overexpression affects EMT in cultured MDA-MB-231 cells. To this end, we extracted RNA from both LV-RFP miR-22 and LV-RFP Empty cell lines and compared the expression levels of key EMT-related genes and mesenchymal markers such as SNAIL1, CDH2 (N-Cadherin), and TWIST1, as well as common epithelial structure markers such as DSP (Desmoplakin), and CDH1 (E-Cadherin). As shown in Figure 1C, we observed a significant increase in EMT-related genes and mesenchymal markers SNAIL1, TWIST1, and CDH2, as well as a significant decrease in the epithelial structural markers DSP and CDH1, implying that miR-22 impacts the EMT signature. To further assess the biological consequences of the aforementioned molecular perturbations, we established 3D cultures of the LV-RFP miR-22 and LV-RFP Empty cell lines by seeding the cells in a specialized medium under conditions (see Methods and Materials for details) that triggered the formation of mammospheres. The experiment was designed to compare the size and number of LV-RFP miR-22 and LV-RFP Empty mammospheres by seeding equal numbers of cells in ultra-low attachment dishes containing the specialized medium mentioned above. As shown in Figure 1D,E and Figure 2B, we observed a significant increase in the size and number of mammospheres in LV-RFP miR-22- seeded dishes compared to LV-RFP Empty controls, confirming the ability of miR-22 to confer stemness properties to human TNBC cells. To further corroborate the relevance of miR-22 in triggering EMT, we extracted RNA from both LV-RFP miR-22 and LV-RFP Empty cells and analyzed by qPCR the expression levels of the same key EMT-related genes as well as the mesenchymal and epithelial related markers listed above. The expression data demonstrate that the effect of miR-22 overexpression on EMT is maintained in mammospheres (Figure 1F).

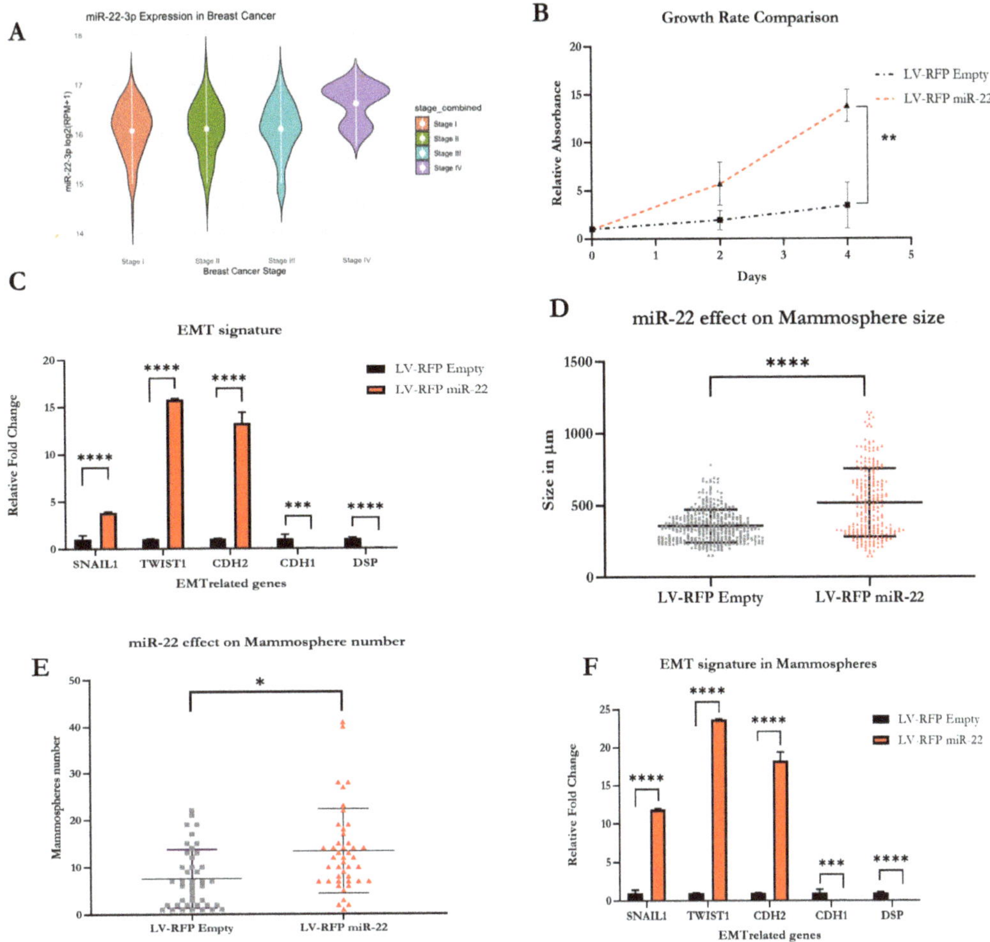

Figure 1. Overexpression of miR-22 promotes EMT in a 2D model of TNBC. Bioinformatic analysis of miR-22 expression levels at different stages of breast cancer (**A**). Overexpression of miR-22 causes a significant increase in the growth rate of a 2D human TNBC in-vitro model (**B**). Overexpression of miR-22 results in a significant upregulation of genes related to EMT, as well as downregulation of genetic markers related to genes found in epithelial cells (**C**) Overexpression of miR-22 in a 3D model of TNBC causes a significant increase in the size and number of mammospheres (**D,E**). Overexpression of miR-22 results in the significant increase in genes associated with mesenchymal cells, and a significant decrease in genes associated with epithelial cells in a 3D human TNBC model (**F**) [Statistical analysis are represented as (*) for $p < 0.03$; (**) for $p < 0.002$; (***) for $p < 0.0002$ and (****) for $p < 0.0001$].

To investigate whether inhibition of miR-22 function could revert the EMT signature and reduce stemness of cells in mammospheres, we seeded LV-RFP miR-22 cells or LV-RFP Empty cells in specialized medium in the presence of an antimiR-22 oligonucleotide (miR-22 LNA), VHL, or SCR-LNA control, respectively. As shown in Figure 3A–D, by day 8, inhibition of miR-22 reduced the size of mammospheres in both LV-RFP miR-22 and LV-RFP Empty cells, while the total number of mammospheres was only reduced in the LV-RFP Empty samples, probably due to the fact that the inhibition of miR-22 in a cell line not overexpressing the miRNA has faster kinetics. As shown in Figure 4B through

Figure 4E, by day 12, inhibition of miR-22 significantly slowed the growth and reduced the numbers of both LV-RFP miR-22 and LV-RFP Empty mammospheres, implying that the antimiR-22 treatment is effective at reducing the stemness of human TNBC cells exhibiting either overexpression or basal levels of miR-22. Indeed, the effect of miR-22 inhibition in mammosphere size could be easily observed under a brightfield microscope as shown in Figure 4F,G. Next, we determined whether the antimiR-22 treatment could revert the genetic signature of EMT in mammospheres. To this end, we obtained RNA from mammospheres treated as shown in Figure 4A. As shown in Figure 4H,I, we observed a partial reversion of EMT in human TNBC cells displaying basal levels of miR-22, and complete reversion of EMT in human TNBC cells overexpressing miR-22. Given the positive findings shown in Figure 1E, we decided to expand the panel of EMT genes. The data confirmed that inhibition of miR-22 function can revert EMT in a relevant 3D model of human TNBC and could represent a potential therapeutic strategy against the progression and metastatic spread of TNBC.

A

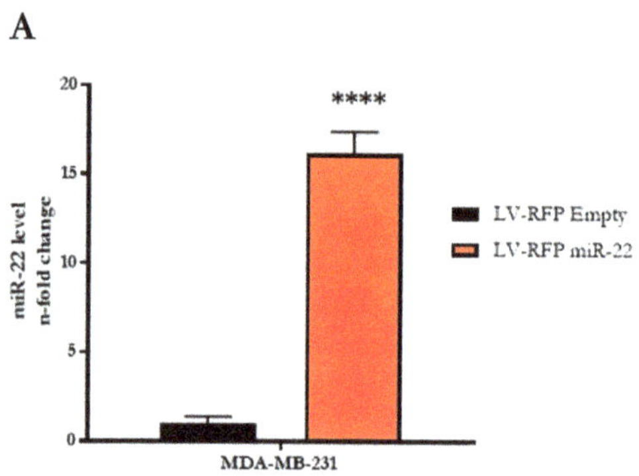

B

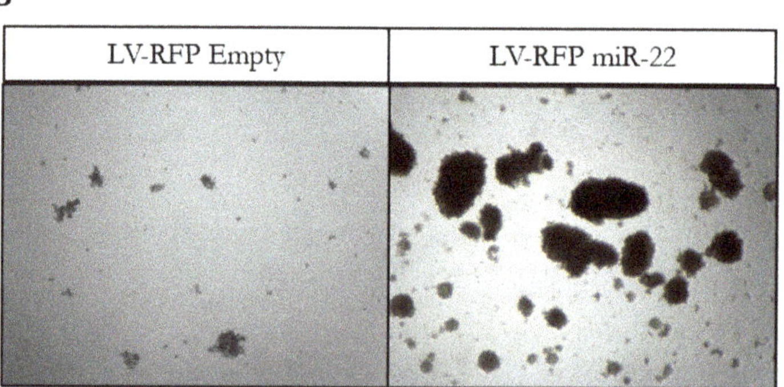

Figure 2. Validation that designed plasmids can upregulate the expression of miR-22 (in the case of LV-RFP miR-22) and maintain basal levels of miR-22 (in the case of LV-RFP Empty) (**A**). Brightfield images of MDA-MB-231 cultured in mammosphere medium in which overexpression of miR-22 results in an increase in the number and size of mammospheres (**B**). [Statistical analysis are represented as (****) for $p < 0.0001$].

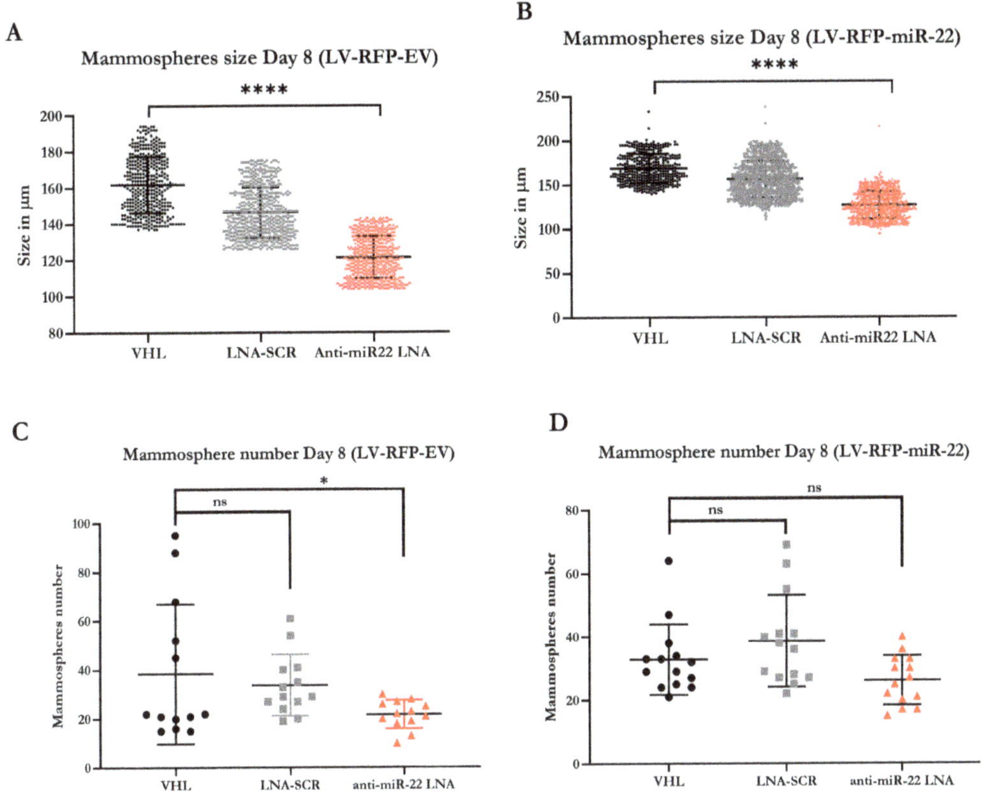

Figure 3. The antimiR-22 treatment is effective at decreasing the size of both LV-RFP miR-22 and LV-RFP Empty mammospheres by day 8, but is only able to significantly reduce the number of LV-RFP Empty mammospheres (**A–D**). [Statistical analysis are represented as non-significant (ns) for $p > 0.1$; (*) for $p < 0.03$ and (****) for $p < 0.0001$].

Finally, we investigated the effect of miR-22 overexpression on the metastatic potential and overall survival of human TNBC xenografts. As shown in Figure 5A,B, mice engrafted with human TNBC overexpressing miR-22 displayed a profound decrease in overall survival. This was caused by a much-increased metastatic spread of tumors to the lung and proliferation of the metastatic nodules as shown by H&E and IHC staining (Figure 5B), indicating that miR-22 overexpression impacts TNBC progression and outcome. We next assessed the in vivo efficacy of the antimiR-22 compound as shown in Figure 5C. Experimental cohorts were treated with either VHL, SCR-LNA, or anti-miR-22 LNA once weekly. Two weeks into the experiment, a postmortem census of both cohorts revealed that pharmacologic inhibition of miR-22 suppressed the metastatic spread and growth of tumors in both xenograft models as shown by H&E and IHC staining (Figure 5D,E). By the end of the experiment, the antimiR-22 treatment had significantly extended survival in both LV-RFP miR-22 and LV-RFP Empty mouse cohorts, as shown in Figure 5F,G. Collectively, these findings suggest that antimiR-22-mediated inhibition of miR-22 may constitute a promising therapeutic strategy for treatment of breast cancer of various molecular subtypes, irrespective of miR-22 expression levels.

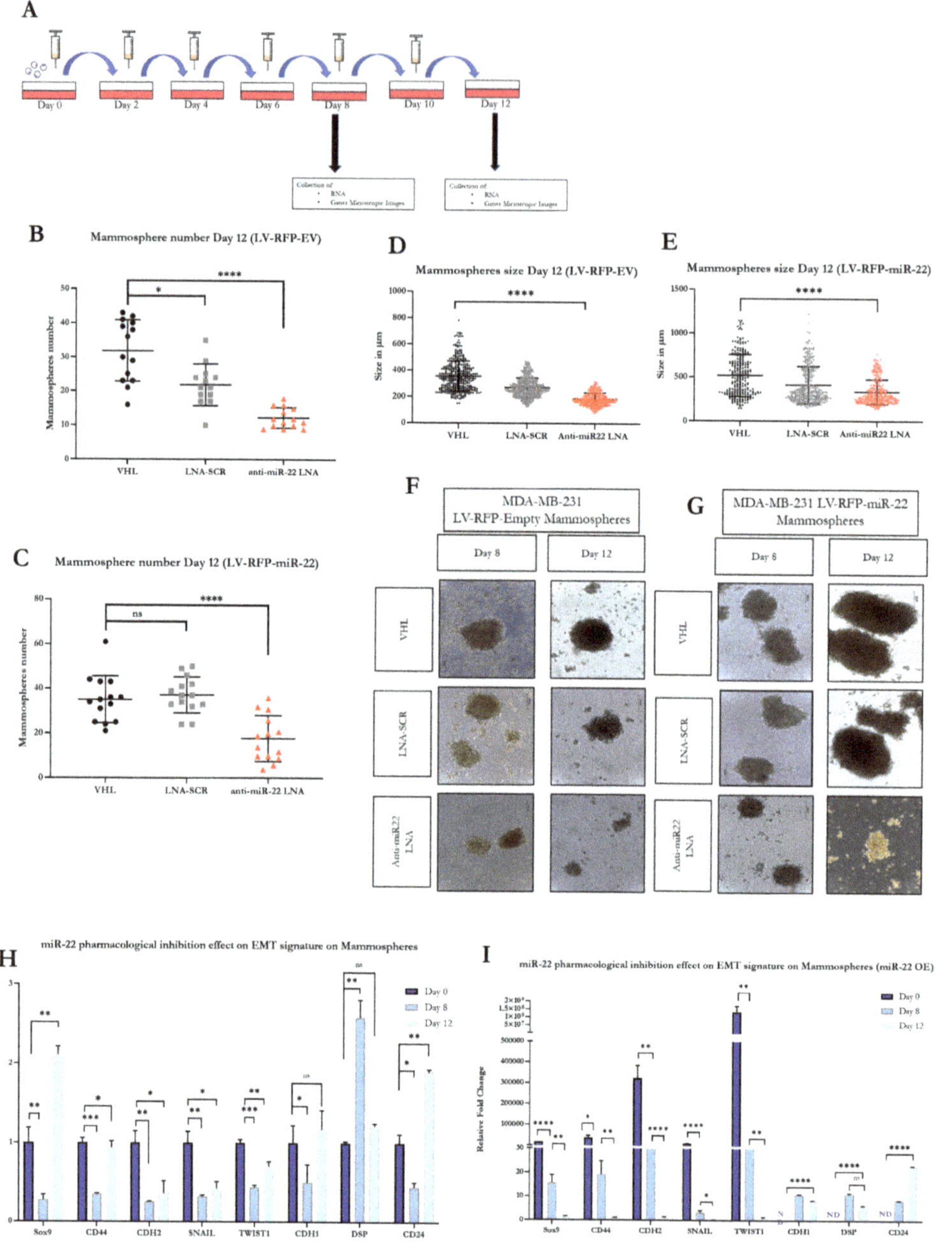

Figure 4. Inhibition of miR-22 function in TNBC mammospheres. A schematic of the in vitro experiments with the antimiR-22 oligonucleotide (**A**). Treatment of TNBC mammospheres with antimiR-22 significantly decreases the size and number of LV-RFP miR-22 and LV-RFP Empty mammospheres (**B–E**). Visual representations of the effect of antimiR-22 treatment on LV-RFP miR-22 and LV-RFP Empty mammospheres at days 8 and 12, respectively (**F,G**). Treatment with antimiR-22 LNA can partially revert the EMT signature of LV-RFP Empty mammospheres, and completely revert the EMT signature in LV-RFP miR-22 mammospheres (**H,I**). [Statistical analysis are represented as non-significant (ns) for $p > 0.1$; (*) for $p < 0.03$; (**) for $p < 0.002$; (***) for $p < 0.0002$ and (****) for $p < 0.0001$].

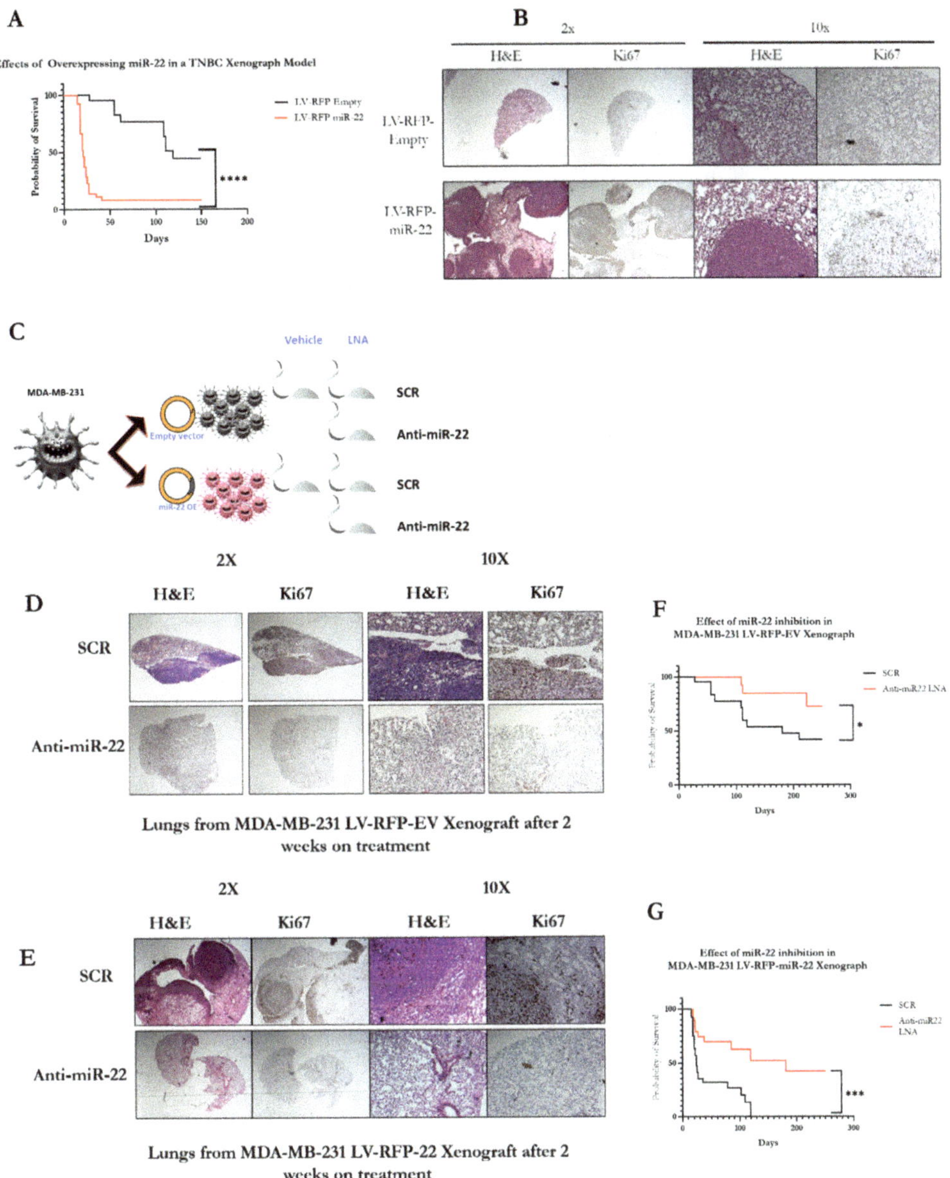

Lungs from MDA-MB-231 LV-RFP-EV Xenograft after 2 weeks on treatment

Lungs from MDA-MB-231 LV-RFP-22 Xenograft after 2 weeks on treatment

Figure 5. Therapeutic inhibition of miR-22 in TNBC xenograft mice. Mouse xenografts overexpressing miR-22 show a significant decrease in life expectancy compared to LV-RFP Empty xenografts, as well as larger tumors as shown by H&E and IHC (**A**,**B**). The animals were separated into three cohorts and treated with either vehicle (VHL), scramble control ASO (SCR), or the antimiR-22 compound (**C**). After two weeks, a census was conducted on the mice, and H&E showed smaller tumor sizes and a decrease in proliferation in the LV-RFP miR-22 and LV-RFP Empty xenograft cells treated with the antimiR-22 compound compared to mice treated with SCR LNA (**D**,**E**). Treatment with antimiR-22 resulted in a significant increase in the survival of LV-RFP miR-22 and LV-RFP Empty mice compared to treatment with vehicle or scramble control-treated mice (**F**,**G**). [Statistical analysis are represented as (*) for $p < 0.03$; (***) for $p < 0.0002$ and (****) for $p < 0.0001$].

4. Discussion

Although diagnosis of breast cancer cases has risen over the past decade, death rates have declined, suggesting that currently available therapeutics are effective in extending the lives of those diagnosed. Indeed, data from the NIH SEER database suggest that women diagnosed with local and regional breast cancers display 5-year survival rates of 99% and 86%, respectively [3,20]. However, once breast cancer progresses to present with distant metastatic spread, the outlook changes drastically and the prognosis becomes poor. By Stage III, breast cancer has spread to the lymph nodes surrounding the sternum, clavicles, and axillae, and the overall 5-year survival rate falls to 72%. Once breast cancer progresses to Stage IV and distant metastasis can be found in the lungs, liver, brain, and bones, the overall 5-year survival rate drops to 22%; a clear indication that a great deal of work is still to be done to make breast cancer fully curable [7,20–22].

Currently available treatments for TNBC are limited to standard chemotherapy, Poly ADP Ribose Polymerase (PARP) inhibitors for tumors harboring BRCA1/2 mutations, and PD-L1 checkpoint blockade for tumors with infiltrating lymphocytes, or tumors that are PD-L1 positive. The development of novel therapeutic strategies is therefore critically important to improve the survival rates of patients with breast cancer [23–25].

New promising therapeutic strategies have emerged over the past few years that focus on targeting of non-coding RNAs such as miRNAs and long noncoding RNAs (lncRNAs). For example, inhibition of the lncRNA MALAT1 triggered differentiation of breast cancer in animal models, suppressing tumorigenesis [26]. In contrast, MALAT1 overexpression was shown to repress the metastatic potential of TNBC cell lines and was proposed as a strategy for a novel anti-metastatic therapy [27]. These opposing functions of MALAT1, and potentially other ncRNAs, highlight the complexity of ncRNA biology. Hence, the development of new therapeutic approaches based on miRNAs and lncRNAs requires a deep understanding on the biological mechanisms of the targeted ncRNAs. We have previously shown that miR-22 contributes to the metastatic potential of breast cancer in an MCF7 xenograft model and a transgenic mouse model overexpressing miR-22 [15]. In the xenograft models, miR-22 overexpression markedly increased the proliferative rate of MCF7 (Ki67$^+$) cells, as well their metastatic potential to the lung, compared to MCF7 control tumors. Similarly, in transgenic mice engineered to selectively overexpress miR-22 in mammary tissues, we observed a significant decrease in disease-free survival rate resulting from mammary cancer and spontaneous metastatic spread of the disease [15]. These findings prompted us to ask whether miR-22 plays a broader role in breast cancer.

The well-known interaction of miR-22 with TET2 and the subsequent effect on miR-200 family, suggests that miR-22 is a central player in EMT. Through EMT, cancer cells acquire mesenchymal-like traits, and become more invasive, ultimately increasing their metastatic potential. Furthermore, remodeling of the extracellular matrix (ECM) by secretion of proteases allows tumor cells to move freely to their surroundings [28,29]. As tumor invasion progresses, cancer cells intravasate into the lymphatic and circulatory systems, granting them access to distant parts of the body and creating metastatic sites in various organs capable of sustaining tumor cell growth [30]. Thus, it is tempting to speculate that blocking EMT would slow the metastatic spread of cancer. We have previously demonstrated a correlation between metastatic potential of cancer and lipogenic switch [31], and more recently we showed that miR-22 acts as a positive master regulator of lipid metabolism and a key player in metabolic control (Panella et al. submitted). These findings provide further evidence of the crucial role of miR-22 in different and essential aspects of the metastatic process.

In conclusion, we demonstrated that overexpression of miR-22 triggers a more aggressive and mesenchymal-like phenotype in a TNBC model compared to cells expressing physiologic levels of miR-22. We observed a significant increase in the growth rate of the miR-22-overexpressing TNBC cells, increased expression of notable mesenchymal genes, and decreased expression of critical epithelial markers. Successful elucidation of the function of miR-22 in cultured cells and in vivo offers hope for an important and greatly needed treatment for TNBC. Interestingly, antimiR-22 treatment not only impacted the growth of TNBC cells overexpressing miR-22, but was also effective at slowing the growth of TNBC cells expressing basal amounts of miR-22. This finding suggests that an antimiR-22-based therapy could be effective in breast cancer patients irrespective of whether they overexpress miR-22 or exhibit normal expression levels of this miRNA. Additionally, overexpression of miR-22 dramatically reduced survival of TNBC xenograft mice. Given that the mature miR-22 sequence is 100% conserved between mouse and man, our data in mouse models of TNBC are indicative that an antimiR-22-based therapy could be further pursued in TNBC patients.

Author Contributions: R.P. conceived the project with P.P.P., R.P. designed and performed most experiments, interpreted the results. C.A.C. and K.E.B. executed experiments and edited the manuscript. F.B. and I.S.V. analyzed human genetic data and ran predictions on human breast cancer databases. V.A.M., S.B., S.L. and J.G.C. performed mouse treatment and sample collection. R.P. devised the experimental designs and interpreted results. S.K. provided reagents and interpreted results. P.P.P. supervised experimental designs and interpreted results. R.P. co-wrote the manuscript together with S.K. and P.P.P. All authors have read and agreed to the published version of the manuscript.

Funding: R.P. was partially supported by a postdoctoral fellowship from Breast Cancer Research Foundation and from Marie Curie iMOVE programme. This work was supported by Breast cancer research Foundation Grant to P.P.P.

Institutional Review Board Statement: Animal experiments were performed in accordance with the guidelines of Beth Israel Deaconess Medical Center Institutional Animal Care and Use Committee, Protocol number 076-2017.

Informed Consent Statement: Not applicable.

Data Availability Statement: The data presented in this study are available on request from the corresponding author. The data are not publicly available due to patent pending issues.

Acknowledgments: We thank the members of the P.P.P. laboratory for critical discussions and all the technical staff for their support.

Conflicts of Interest: RP is a founder, shareholder and member of the board of directors of Resalis Therapeutics Srl, a biotech company currently developing non-coding RNA targeted compounds in cancer and metabolic areas. SK is a founder and shareholder of Resalis Therapeutics Srl. RP, PPP and SK are inventors of patents and patent applications related to miR-22 in cancer and metabolism, owned by Beth Israel Deaconess Medical Center and Aalborg University and licensed to Resalis Therapeutics Srl.

References

1. DeSantis, C.E.; Ma, J.; Gaudet, M.M.; Newman, L.A.; Miller, K.D.; Goding Sauer, A.; Jemal, A.; Siegel, R.L. Breast cancer statistics, 2019. *CA Cancer J. Clin.* **2019**, *69*, 438–451. [CrossRef] [PubMed]
2. Valla, M.; Vatten, L.J.; Engstrøm, M.J.; Haugen, O.A.; Akslen, L.A.; Bjørngaard, J.H.; Hagen, A.I.; Ytterhus, B.; Bofin, A.M.; Opdahl, S. Molecular Subtypes of Breast Cancer: Long-term Incidence Trends and Prognostic Differences. *Cancer Epidemiol. Biomark. Prev.* **2016**, *25*, 1625–1634. [CrossRef] [PubMed]
3. NCI. *Cancer Stat Facts: Female Breast Cancer 2018*; NCI: Bethesda, MD, USA, 2018.
4. Bertucci, F.; Finetti, P.; Goncalves, A.; Birnbaum, D. The therapeutic response of ER+/HER2− breast cancers differs according to the molecular Basal or Luminal subtype. *npj Breast Cancer* **2020**, *6*, 8. [CrossRef] [PubMed]
5. Aysola, K.; Desai, A.; Welch, C.; Xu, J.; Qin, Y.; Reddy, V.; Matthews, R.; Owens, C.; Okoli, J.; Beech, D.J.; et al. Triple Negative Breast Cancer—An Overview. *Hered. Genet.* **2013**, *2013*. [CrossRef]
6. Yin, L.; Duan, J.-J.; Bian, X.-W.; Yu, S.-C. Triple-negative breast cancer molecular subtyping and treatment progress. *Breast Cancer Res.* **2020**, *22*, 61. [CrossRef]
7. Breastcancer.org. Triple-Negative Breast Cancer. (2020). Available online: https://www.breastcancer.org/types/triple-negative (accessed on 8 May 2023).
8. Gebert, L.F.R.; Macrae, I.J. Regulation of microRNA function in animals. *Nat. Rev. Mol. Cell Biol.* **2019**, *20*, 21–37. [CrossRef]
9. Exiqon. InVivo Guidelines. 2013.
10. Quemener, A.M.; Centomo, M.L.; Sax, S.L.; Panella, R. Small Drugs, Huge Impact: The Extraordinary Impact of Antisense Oligonucleotides in Research and Drug Development. *Molecules* **2022**, *27*, 536. [CrossRef]
11. Esau, C.; Davis, S.; Murray, S.F.; Yu, X.X.; Pandey, S.K.; Pear, M.; Watts, L.; Booten, S.L.; Graham, M.; McKay, R.; et al. miR-122 regulation of lipid metabolism revealed by in vivo antisense targeting. *Cell Metab.* **2006**, *3*, 87–98. [CrossRef]

12. Stenvang, J.; Petri, A.; Lindow, M.; Obad, S.; Kauppinen, S. Inhibition of microRNA function by antimiR oligonucleotides. *Silence* **2012**, *3*, 1. [CrossRef]
13. Straarup, E.M.; Fisker, N.; Hedtjärn, M.; Lindholm, M.W.; Rosenbohm, C.; Aarup, V.; Hansen, H.F.; Ørum, H.; Hansen, J.B.R.; Koch, T. Short locked nucleic acid antisense oligonucleotides potently reduce apolipoprotein B mRNA and serum cholesterol in mice and non-human primates. *Nucleic Acids Res.* **2010**, *38*, 7100–7111. [CrossRef]
14. Obad, S.; Dos Santos, C.O.; Petri, A.; Heidenblad, M.; Broom, O.; Ruse, C.; Fu, C.; Lindow, M.; Stenvang, J.; Straarup, E.M.; et al. Silencing of microRNA families by seed-targeting tiny LNAs. *Nat. Genet.* **2011**, *43*, 371–378. [CrossRef]
15. Song, S.J.; Poliseno, L.; Song, M.S.; Ala, U.; Webster, K.; Ng, C.; Beringer, G.; Brikbak, N.J.; Yuan, X.; Cantley, L.; et al. MicroRNA-Antagonism Regulates Breast Cancer Stemness and Metastasis via TET-Family-Dependent Chromatin Remodeling. *Cell* **2013**, *154*, 311–324. [CrossRef]
16. Song, S.J.; Pandolfi, P.P. miR-22 in tumorigenesis. *Cell Cycle* **2014**, *13*, 11–12. [CrossRef]
17. Song, S.J.; Ito, K.; Ala, U.; Kats, L.; Webster, K.; Sun, S.M.; Jongen-Lavrencic, M.; Manova-Todorova, K.; Teruya-Feldstein, J.; Avigan, D.E.; et al. The Oncogenic microRNA miR-22 Targets the TET2 Tumor Suppressor to Promote Hematopoietic Stem Cell Self-Renewal and Transformation. *Cell Stem Cell* **2013**, *13*, 87–101. [CrossRef]
18. Pondé, N.F.; Zardavas, D.; Piccart, M. Progress in adjuvant systemic therapy for breast cancer. *Nat. Rev. Clin. Oncol.* **2019**, *16*, 27–44. [CrossRef]
19. Chue, B.M.-F.; La Course, B.D. Can we cure stage IV triple-negative breast carcinoma? Another case report of long-term survival (7 years). *Medicine* **2019**, *98*, e17251. [CrossRef]
20. Breastcancer.org. Breast Cancer Stages. 2020. Available online: https://www.breastcancer.org/pathology-report/breast-cancer-stages (accessed on 8 May 2023).
21. NCI. *The Breast Cancer Risk Assessment Tool*; NCI: Bethesda, MD, USA, 2018.
22. Bergin, A.R.T.; Loi, S. Triple-negative breast cancer: Recent treatment advances. *F1000Research* **2019**, *8*, 1342. [CrossRef]
23. Marra, A.; Viale, G.; Curigliano, G. Recent advances in triple negative breast cancer: The immunotherapy era. *BMC Med.* **2019**, *17*, 90. [CrossRef]
24. Hill, B.S.; Sarnella, A.; D'avino, G.; Zannetti, A. Recruitment of stromal cells into tumour microenvironment promote the metastatic spread of breast cancer. *Semin. Cancer Biol.* **2020**, *60*, 202–213. [CrossRef]
25. Fedele, M.; Cerchia, L.; Chiappetta, G. The Epithelial-to-Mesenchymal Transition in Breast Cancer: Focus on Basal-Like Carcinomas. *Cancers* **2017**, *9*, 134. [CrossRef]
26. Arun, G.; Diermeier, S.; Akerman, M.; Chang, K.-C.; Wilkinson, J.E.; Hearn, S.; Kim, Y.; MacLeod, A.R.; Krainer, A.R.; Norton, L.; et al. Differentiation of mammary tumors and reduction in metastasis upon *Malat1* lncRNA loss. *Genes Dev.* **2016**, *30*, 34–51. [CrossRef] [PubMed]
27. Kim, J.; Piao, H.-L.; Kim, B.-J.; Yao, F.; Han, Z.; Wang, Y.; Xiao, Z.; Siverly, A.N.; Lawhon, S.E.; Ton, B.N.; et al. Long noncoding RNA MALAT1 suppresses breast cancer metastasis. *Nat. Genet.* **2018**, *50*, 1705–1715. [CrossRef] [PubMed]
28. Papadaki, M.A.; Stoupis, G.; Theodoropoulos, P.A.; Mavroudis, D.; Georgoulias, V.; Agelaki, S. Circulating Tumor Cells with Stemness and Epithelial-to-Mesenchymal Transition Features Are Chemoresistant and Predictive of Poor Outcome in Metastatic Breast Cancer. *Mol. Cancer Ther.* **2019**, *18*, 437–447. [CrossRef] [PubMed]
29. Ye, X.; Brabletz, T.; Kang, Y.; Longmore, G.D.; Nieto, M.A.; Stanger, B.Z.; Yang, J.; Weinberg, R.A. Upholding a role for EMT in breast cancer metastasis. *Nature* **2017**, *547*, E1–E3. [CrossRef]
30. Pastushenko, I.; Blanpain, C. EMT Transition States during Tumor Progression and Metastasis. *Trends Cell Biol.* **2019**, *29*, 212–226. [CrossRef]
31. Chen, M.; Zhang, J.; Sampieri, K.; Clohessy, J.G.; Mendez, L.; Gonzalez-Billalabeitia, E.; Liu, X.-S.; Lee, Y.-R.; Fung, J.; Katon, J.M.; et al. An aberrant SREBP-dependent lipogenic program promotes metastatic prostate cancer. *Nat. Genet.* **2018**, *50*, 206–218. [CrossRef]

Review

The Role of miR-29s in Human Cancers—An Update

Thuy T. P. Nguyen [1], Kamrul Hassan Suman [2], Thong Ba Nguyen [3], Ha Thi Nguyen [4,5,*] and Duy Ngoc Do [6,*]

[1] Division of Radiation and Genome Stability, Department of Radiation Oncology, Dana-Farber Cancer Institute, Boston, MA 02215, USA
[2] Department of Fisheries, Ministry of Fisheries and Livestock, Dhaka 1205, Bangladesh
[3] Department of Anatomy, Biochemistry, and Physiology, John A. Burns School of Medicine, University of Hawaii at Manoa, Honolulu, HI 96813, USA
[4] Institute of Research and Development, Duy Tan University, Danang 550000, Vietnam
[5] Center for Molecular Biology, College of Medicine and Pharmacy, Duy Tan University, Danang 550000, Vietnam
[6] Department of Animal Science and Aquaculture, Dalhousie University, Truro, NS B2N 5E3, Canada
* Correspondence: nguyenthiha23@duytan.edu.vn (H.T.N.); duy.do@dal.ca (D.N.D.)

Abstract: MicroRNAs (miRNAs) are small non-coding RNAs that directly bind to the 3′ untranslated region (3′-UTR) of the target mRNAs to inhibit their expression. The miRNA-29s (miR-29s) are suggested to be either tumor suppressors or oncogenic miRNAs that are strongly dysregulated in various types of cancer. Their dysregulation alters the expression of their target genes, thereby exerting influence on different cellular pathways including cell proliferation, apoptosis, migration, and invasion, thereby contributing to carcinogenesis. In the present review, we aimed to provide an overview of the current knowledge on the miR-29s biological network and its functions in cancer, as well as its current and potential applications as a diagnostic and prognostic biomarker and/or a therapeutic target in major types of human cancer.

Keywords: cancer; microRNA; miR-29; biomarker; therapeutic target

Citation: Nguyen, T.T.P.; Suman, K.H.; Nguyen, T.B.; Nguyen, H.T.; Do, D.N. The Role of miR-29s in Human Cancers—An Update. *Biomedicines* **2022**, *10*, 2121. https://doi.org/10.3390/biomedicines10092121

Academic Editors: Milena Rizzo and Elena Levantini

Received: 18 June 2022
Accepted: 24 August 2022
Published: 29 August 2022

Publisher's Note: MDPI stays neutral with regard to jurisdictional claims in published maps and institutional affiliations.

1. Introduction

MicroRNAs (miRNAs) are a class of small non-coding RNAs (ncRNAs) with approximately 18–24 nucleotides in length. The miRNAs control gene expression post-transcriptionally by binding to the 3′-untranslated region (3′-UTR) of its targeted messenger RNAs (mRNAs), resulting in mRNA cleavage or translation repression [1,2]. Noticeably, miRNAs have been indicated to be involved in all cancer stages from tumor initiation to metastasis [3], suggesting that miRNA could potentially be used as a diagnostic and/or prognostic biomarker or a therapeutic target in cancer treatments [4]. Among dozens of miRNAs that are abnormally expressed in cancer, miR-29s have been recognized as the critical one that acts as both oncogenic and tumor-suppressor regulators in cancer [5].

The miR-29 family consists of two clusters namely miR-29a/miR-29b-1 and miR-29c/miR-29b-2 located on chromosome 7q32.3 and 1q32.2, respectively [6]. Mature miR-29s in humans, mice, and rats share a common seed region [7] that plays a role in determining their target mRNAs. However, miR-29s exhibit differential expression and regulation in various cases. The miR-29a is the most abundantly expressed at all stages of the cell cycle; whereas, miR-29b exhibits low-level expression, rapid degradation, and becomes stable during mitosis; and miR-29c is undetectable [8]. Pulse-chase experiments have indicated that the miR-29a mimic has greater stability compared to miR-29b in Hela cells [9]. Besides, a deep sequencing miRNAs analysis revealed that miR-29s had distinct subcellular distributions [10]. While miR-29a is more prevalent in the cytoplasm, miR-29b is mostly localized in the nucleus [10]. The nuclear localization of miR-29b is mostly due to six nucleotides (nts) localized at the end of its sequences [8] (Figure 1A). Additionally, miR-29b-specific knock-out disrupts the tertiary structure of miRNA clusters and changes the sequence or structure of promoters, resulting in lower expression of miR-29a and miR-29c [11]. These results indicated that miR-29s may function differently in different conditions in both corporative and separate manner.

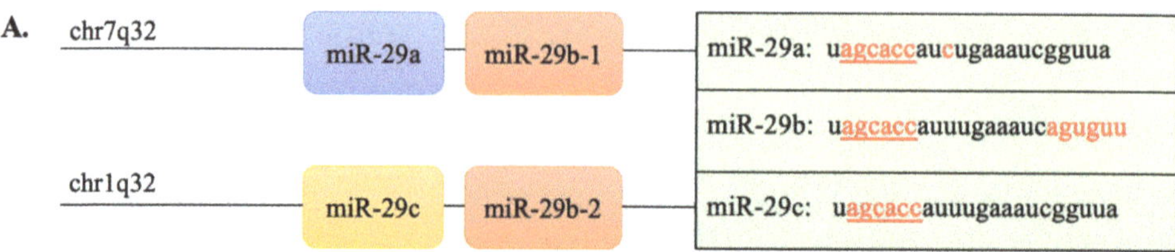

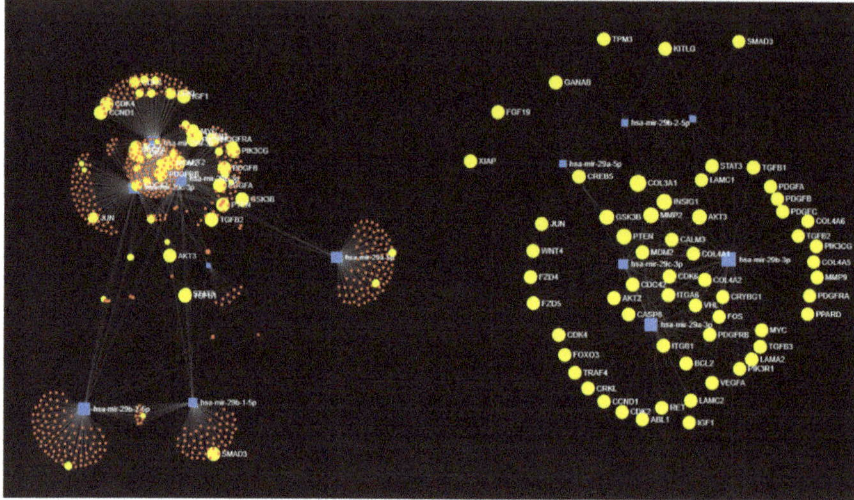

Figure 1. The miR-29 family members and their potential targets. (**A**) Mature sequences and chromosomal locations of miR-29s. MiR-29a/miR-29b-1 cluster is located in chr7q32, whereas miR-29b-2/miR-29c is in chr1q32. The mature miR-29s share a common seed region (agcacc) but are different at the cytosine position in miR-29a and a nucleotide sequence at the end of miR-29b (aguguu) which is known for nuclear localization. (**B**) The potential targets of miR-29s. The visualization is based on miR-29s' experimental targets of the miRNet platform (https://www.mirnet.ca, accessed on 24 July 2022). Each dot represents a gene. The yellow dots indicate the target genes involved in cancer pathways.

The miR-29s play an important role in a multitude of pathophysiological processes. According to the miRNet database, 677 human genes have been identified as potential targets of the hsa-miR-29 family (https://www.mirnet.ca; accessed on 27 November 2021), and many of them are involved in cancer pathways (Figure 1B). Several studies have revealed a strong antifibrotic property of miR-29s in multiple organs, such as heart [12], liver [13], lung [14], and kidney [15]. Specifically, the miR-29s negatively regulated multiple extracellular matrix (ECM) proteins [16–18], which are essential for matrix deposition, epithelial-mesenchymal transition (EMT) [14], and the progression of fibrosis. Additionally, dysregulated miR-29s were also identified in various conditions in liver fibrosis [13,19,20], cardiovascular diseases [12,21], and hepatitis C virus infection [13,19,22]. Intravenous injection of miR-29 mimics may reduce collagen biosynthesis and reverse pulmonary fibrosis [23].

Moreover, the expression of the miR-29 family was reduced in various types of cancer, suggesting their tumor-suppressing capacity as well as their potential role as a diagnostic and prognostic marker in these cancer types. Several studies have indicated that miR-29s can negatively regulate DNA methyl transferase proteins (DNMT3A/3B) in the lung, gastric, and liver cancer [24–26]. Additionally, miR-29s have also been found to suppress the expression of histone deacetylases (HDAC4) [27,28], thymine DNA glycosylase (TDG), and ten-eleven translocation 1 (TET1) [29]. Furthermore, miR-29s could act as pro-apoptotic, anti-proliferative, anti-metastatic/EMT, and immunomodulatory factors by directly binding to 3′-UTR of the target genes such as myeloid cell leukemia 1 (MCL1) [30], cyclin-dependent kinase 6 (CDK6) [31], cell division cycle 42 (CDC42) [32] and matrix metalloproteinase 2 (MMP2) [17] in human cancers. Therefore, miR-29-targeted interventions can be used as a therapeutic approach to inhibit tumorigenesis and invasion in different types of cancer. In contrast, miR-29s were also upregulated in several types of cancer. For example, miR-29a was upregulated in colorectal cancer metastasis [33], breast cancer [34], and in the urine of bladder cancer (BC) patients [35], while miR-29c-5p was significantly increased in the advanced stage of BC serum samples [36]. These studies suggested an oncogenic role of miR-29s and the potential link between the dysregulation of miR-29s and the carcinogenesis in these cancer types.

In the current review, we summarize and discuss the function of miR-29s across human cancers and the use of miR-29s as diagnostic, prognostic, and therapeutic biomarkers.

2. miR-29 Functions in Cancers

The miR-29s play a central role in the transcriptome networks, in which miR-29c was the most frequently reported one for its function in regulation of transcription factors. It has been reported that all three miR-29s were regulated by c-Myc, Yin and Yang 1 (YY1), and CCAAT/enhancer-binding protein-α (CEBPA). Particularly, c-Myc suppressed miR-29s transcription through a co-repressor complex with histone deacetylase 3 (HDAC3) and Enhancer of zeste homolog 2 (EZH2); and combined inhibition of HDAC3 and EZH2 restored miR-29s expression levels, which, in turn, caused lymphoma growth suppression [37]. Nuclear factor kappa B (NF-kB)-activated YY1 also inhibited miR-29s expression in myogenesis and rhabdomyosarcoma [38]. The CBEPA, on the other hand, selectively induced the transcription of miR-29a/b-1, but not miR-29b-2/c [39]. In addition, miR-29s have also been reported to be involved in the modulation of a set of transcription factors (Figure 2A), including tumor suppressors and oncogenic genes that are involved in different cancer biological pathways (Figure 2B).

The DNA methylation is a well-studied epigenetic gene silencing mechanism in mammalian cells and organisms [40]. Promoter hypermethylation of a tumor suppressor gene causes gene inactivation and thus may lead to cancer development. Numerous tumor suppressor genes were hypermethylated in various human cancers, such as BRCA1 in early breast cancer, MLH1 (mutL homolog (1) gene in colorectal cancer (CRC), and VHL (von Hippel–Lindau) gene in renal cell cancer [41]. The miR-29s directly suppress DNA methyltransferase enzymes and two other DNA methylation proteins, Thymine DNA Glycosylase (TDG) and Tet Methylcytosine Dioxygenase (TET1) [29,42]. According to Morita and colleagues, miR-29s protect cells from tumorigenesis by maintaining the existing DNA methylation profiles. In lung cancer, miR-29s induced silencing of DNMT3A/3B by binding to their 3′-UTR, thereby promoting tumor growth [24]. In multiple myeloma, miR-29b mimics reduced HDAC4 expression and myeloma cell migration, while increasing histone H4 acetylation and apoptosis [28].

Cyclin-dependent kinases (CDKs) were known for their central roles in cell cycle regulation. The CDK6 complexes promoted cancer cells to enter the S phase, thereby enhancing cell proliferation and growth. Numerous studies have reported that miR-29s suppressed the proliferation and invasion of cancer cells by inhibiting the expression of CDK6 in different types of malignancies, including osteosarcoma [31], and gastric carcinoma [43,44], and bladder cancer [45]. Additionally, miR-29s have also been shown to arrest the cell cycle at G0/G1 phases in gliomas [32] and breast cancer [46], and the G1-S phase in acute myeloid leukemia (AML) [47] by targeting cell division cycle 42 (CDC42) and cyclin D2 (CCND2), respectively; or to suppress tumor growth by repressing angiogenesis genes such as vascular endothelial growth factor (VEGF) [48,49] and insulin-like growth factor 1 (IGF-1) [50] in osteosarcoma and gastric cancer cells.

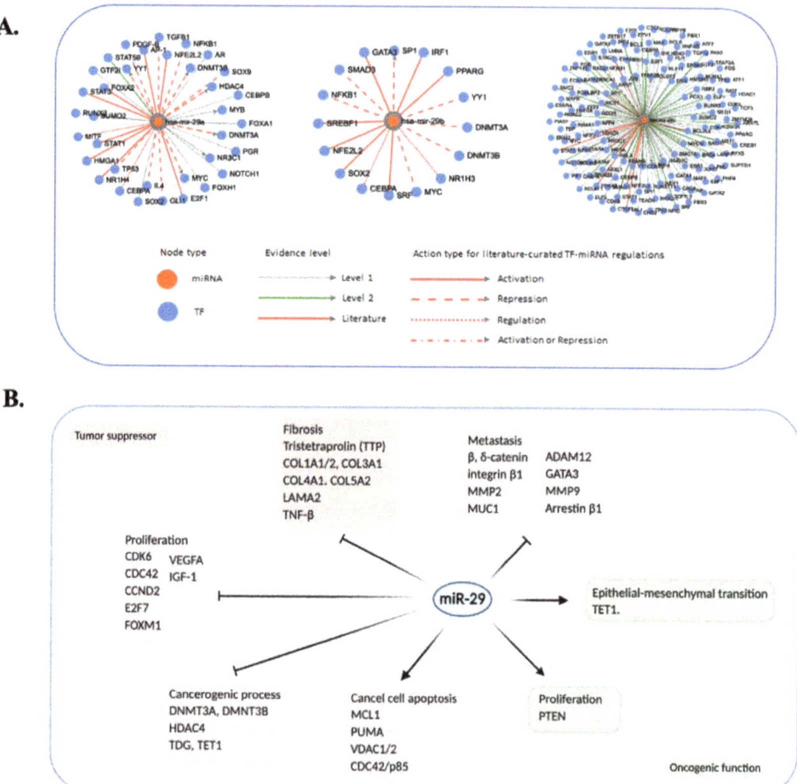

Figure 2. Roles of miR-29s in human cancers. (**A**) MiR-29s modulated different transcription factors. The visualization is based on miR-29s in transcriptome networks of the TransmiR v2.0 platform (cuilab/cn/transmir). The red dots represent has-mir-29a, -29b, and -29c from left to right, respectively, while the blue dots represent different cancer-related transcription factors. (**B**) cancer-related miR-29s targets. MiR-29s can act as both tumor suppressor and tumor inducer genes by contributing to different cancer pathways such as cancer cell apoptosis, cancerogenic process, proliferation, fibrosis, and metastasis.

The dysregulation of ECM remodeling proteins is a high-risk factor for cancer. Fibrosis is a complex process involved in the deposition and reorganization of the matrix, leading to the EMT and thus metastasis of cancer cells. Transforming growth factor beta (TGF-β) receptor binding induced the phosphorylation of the downstream transcription factors SMAD2/3 to stimulate fibrogenic gene expression, including COL1A1, COL1A2, and COL3A1 [51]. The miR-29s have been reported to be significantly downregulated by TGF-β/SMAD signaling in renal fibrosis [15,52]. Overexpression of miR-29s, however, can inhibit the expression of TGF-β1 and SMAD through a feedback loop, thus protecting cells from fibrosis development [53]. Besides, miR-29s also have negatively regulated other ECM-related genes such as laminins, integrins, MMPs, and ADAMs, strongly indicating its anti-fibrotic activity [16,17,54,55].

The miR-29s have also been linked to cancer metastasis, an indicator of poor prognosis. It has been reported that the expression of miR-29s were induced in chemo drugs-treated gastric cancer and that increased expression of miR-29c suppressed gastric cancer cell migration and invasion by negatively regulated δ-catenin [56]. Similarly, a strong downregulation of miR-29c has also been observed in pancreatic cancer, which was accompanied by hyperactivation of Wingless-related integration site (Wnt) signaling pathways [57]. Overexpression of miR-29c inhibited the Wnt/β-catenin signaling by down-regulating Wnt's upstream regulators, resulting in reduced invasion and metastasis in pancreatic cancer [57]. Membrane-bound mucin (MUC1), a stabilizer of β-catenin and Wnt/β-catenin signaling, has also been identified to be inhibited by miR-29a [58]. Another group of metastasis-induced proteins is the MMP family which has also been reported as a direct target of miR-29s. Particularly, miR-29b negatively regulated MMP2/9 by binding to its 3′-UTR, causing cell migration suppression in gastric cancer [17], or osteosarcoma cells [59].

Moreover, miR-29s induced apoptosis in cancer cells by negatively regulating anti-apoptotic proteins such as MCL-1 [30], VDAC1/2 [60], and CDC42/p85 complex [61]. Specifically, miR-29s elevated p53 levels and promoted the p53-dependent apoptotic pathway by directly suppressing two p53 inhibitors, p85 alpha and CDC42 [61]. Besides, by inhibiting the expression of MCL-1, an anti-apoptotic protein, and VDAC1/2 that is essential for the release of cytochrome C from mitochondria to the cytoplasm, miR-29a promoted apoptosis in cancer cells [30,60].

Lastly, some other studies have reported a contradictory role of miR-29s, which functioned as an oncogene in several types of human cancers. In osteosarcoma, for instance, miR-29s were shown to be an oncogenic factor where their downregulation resulted in significantly reduced cell growth and colony formation of osteosarcoma MG-63 cells, probably via miR-29/TGF-β1/PUMA (p53 upregulated modulator of apoptosis) axis [62]. Knockdown of PUMA in these cells, however, reversed miR-29s-induced cell growth suppression and apoptosis [62], due to its ability to induce mitochondrial translocation of Bax (Bcl-2 Associated X-protein) [63]. In addition, miR-29a was upregulated in estrogen receptor-negative (ER⁻) breast cancer that was strongly associated with tumor metastasis and shorter OS (overall survival) in patients with breast cancer [62]. The MiR-29a was proposed as a tumor activator that induces cell proliferation and migration by targeting and inhibiting TET1 [64]. Moreover, overexpression of miR-29s caused a reduction of Phosphatase and Tensin-Like Protein (PTEN), a tumor suppressor, resulting in a restoration of proliferation and migration in osteosarcoma cells [65]. In some cases, the functions of miR-29s have not been identified [66], suggesting that miR-29s can function as either an oncogene or a tumor suppressor depending on specific cellular contexts.

3. MiR-29s as Biomarkers

The miR-29s have been repeatedly reported for their abnormal expression across human cancers, suggesting their roles in cancer initiation and progression as well as their potential to be used as diagnostic and/or prognostic biomarkers in cancers. In this part, we summarized the potential use of miR-29s as biomarkers in major types of human cancer.

3.1. MiR-29s as Biomarkers in Colorectal Cancer

Integrative bioinformatics analysis has revealed the biological functions of the miR-29 family in CRC (colorectal cancer) occurrence and development [67]. Accordingly, pathway enrichment analysis indicated that the miR-29s-targeted genes were associated with the PI3K-AKT signaling pathway, p53-mediated apoptosis, cell cycle, FOXO (forkhead box transcription factors) signaling pathway, and miRNAs in cancer (Figure 3). Thus, miR-29s have been previously proposed to be used as potential biomarkers for CRC diagnosis and prognosis [68–72]. The testing samples ranged from serum, plasma, feces, and tissues were used to measure the levels of miR-29s in CRC. For quantitative measurement of the diagnosis accuracy, each study has calculated the area under the curve (AUC) of summary receiver operating characteristic (ROC), sensitivity, and specificity, which are listed in Table 1. Recently, a systemic meta-analysis based on a hundred single studies revealed that it was valuable to use miR-29s expression alone or in combination with other biomarkers to diagnose or prognoses CRC. Using the miR-29s alone method, however, had lower accuracy than combination methods, with AUC, sensitivity, and specificity of 0.82, 70%, 81%, and 0.86, 78%, and 91%, respectively [67]. The expression of miR-29s was mostly downregulated in CRC in all stages and higher in CRC patients with metastasis as compared to those without. In addition, CRC patients with higher miR-29s expression levels exhibited to have better survival outcomes with lower recurrence and metastasis rates [67]. Together, these results suggested the significant role of miR-29s as diagnostic and prognostic biomarkers in CRC.

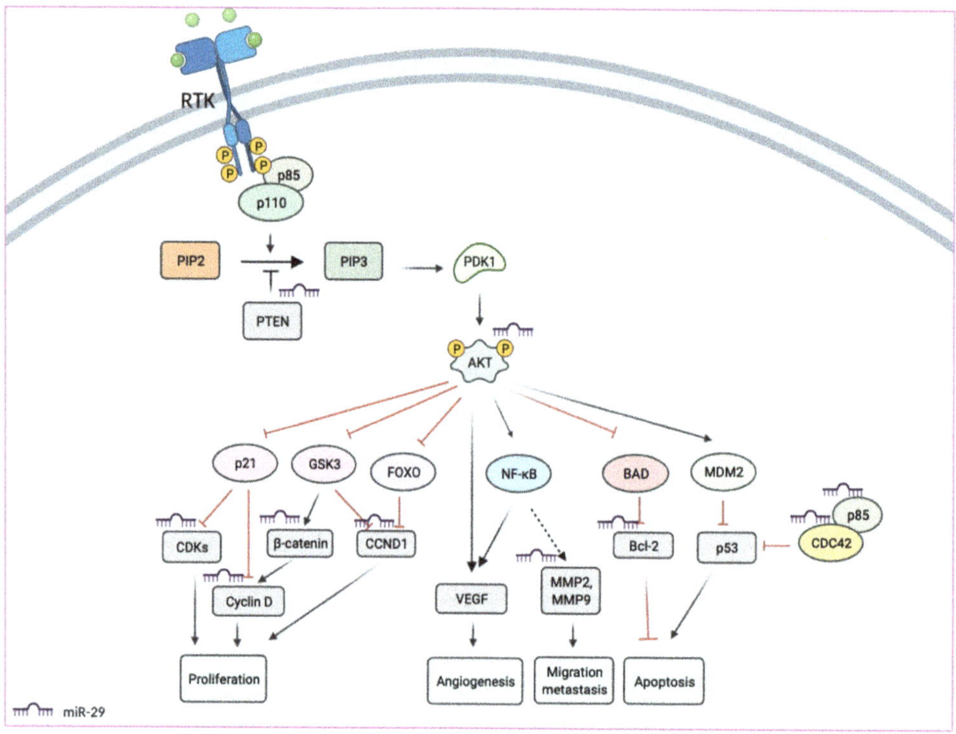

Figure 3. Potential functions of miR-29s in PI3K/AKT signaling pathways. MiR-29s directly inhibit the transcription of various PI3K/AKT downstream factors, resulting in the suppression of cell proliferation, angiogenesis, cell migration, and metastasis, while causing induction of cancer cell apoptosis.

Table 1. The potential use of miR-29s as biomarkers in colorectal cancer.

Sample	Sample Size	Outcome	Results	Ref.
Venous blood	114 CRC patients (58 patients with and 56 patients without metastasis)	MiR-29a was significantly increased in CRC patients with metastasis than in those without.	AUC: 80.3% Sensitivity: 75% Specificity: 75%	[73]
Serum	55 CRC patients and 55 normal controls	The serum level of miR-29b was lower in CRC as compared to the normal controls and inversely correlated with the advanced tumor stages.	AUC: 87% Sensitivity: 77% Specificity: 75%	[68]
Tissue Plasma	200 CRC patients and 400 normal controls	The level of miR-29b in plasma and tissue was highly correlated and significantly lower in CRC versus the normal controls.	Tissue: AUC: 88.3% Sensitivity: 81.6% Specificity: 84.9% Plasma: AUC: 74.3% Sensitivity: 61.4% Specificity: 72.5%	[69]

Table 1. *Cont.*

Sample	Sample Size	Outcome	Results	Ref.
Feces	80 CRC patients and 51 normal controls	The level of miR-29a in feces was significantly lower in CRC versus the normal controls.	AUC: 77.7% Sensitivity: 85% Specificity: 61%	[71]
Serum	160 colorectal neoplasms patients and 77 normal controls	The level of miR-29a in serum was significantly lower in colorectal neoplasms	AUC: 74.1%	[70]
Tissues	245 CRC patients (34 stages I, 63 stages II, 104 stage III, and 44 stages IV)	MiR-29b expression was significantly decreased in tumor versus normal tissues	Higher miR-29b is associated with higher 5-year DFS and OS.	[74]
Tissues	110 CRC patients (51 stages I and 59 stages II)	The level of miR-29a was a positive predictive factor for non-recurrence in stage II CRC.	Higher miR-29a is associated with longer DFS. Sensitivity: 67% Specificity: 88%	[75]

CRC: colorectal cancer; AUC: area under the curve; OS: overall survival; DFS: disease-free survival.

3.2. miR-29s as Biomarkers in Bladder Cancer

Several studies have suggested that urinary miRNAs could be used as potential biomarkers for the noninvasive diagnosis of BC [35,36,76]. Noticeably, the expression of the miR-29 family members was largely varied in BC. For example, miR-29c was significantly increased in the advanced stage of BC serum samples [36]; and miR-29a was up-regulated in urine samples of BC patients [35]. Additionally, after tumor removal, the level of miR-29a in urine samples was significantly decreased, suggesting a correlation between the level of miR-29a in urine and bladder tumor status [35]. Similarly, another study revealed that miR-29b-1 and miR-29c were upregulated in BC T24 cells as compared to normal cells; and knockdown of one of the two miR-29b-1/-29c caused growth suppression in T24 cells [77]. These data indicated the oncogenic role of miR-29s in this type of cancer.

The miR-29c, however, was significantly downregulated in BC samples [45,78–80]. Overexpression of miR-29c caused inhibition of cell growth, cell cycle, and cell mobility while induction of apoptosis in T24 cells [45,78]. Additionally, BC cells exposed to exosome-derived miR-29c are more likely to undergo apoptosis, which is achieved by inhibiting BCL-2 and MCL-1 [81]. These contradictory investigations on the roles of miR-29s in BC suggested their importance in biological pathways and their potential to be used as biomarkers for this type of cancer (Table 2). However, identifying whether they function as tumor suppressors or oncogenes in a typical condition of BC is necessary for better understanding their mechanism of action as well as their future applications in prognosis and diagnosis.

Table 2. miR-29s as biomarkers in bladder cancer.

Sample	Sample Size & Methods	Outcomes	Results	Ref.
Serum	- 392 BC samples and 100 normal controls - Bioinformatic analysis	MiR-29c was overexpressed in serum samples	MiR-29c was correlated to the advanced stage and OS time in BC patients.	[36]
Urine	- 276 BC samples: 276 normal controls - MiSeq and qRT-PCR	MiR-29a was upregulated in BC patients.	MiR-29a-3p in combination with six other miRNAs was used for the diagnosis of BC. AUC: 92.3% Sensitivity: 82% Specificity: 96%	[35]
Tissue	- 30 BC samples and 30 normal controls - qRT-PCR	MiR-29c was downregulated in BC. MiR-29c inhibited cell proliferation, migration, and cell cycle progression, and induce apoptosis through AKT signaling.	MiR-29c was inversely associated with bladder tumor stages.	[78]

Table 2. *Cont.*

Sample		Sample Size & Methods	Outcomes	Results	Ref.
Tissue	-	106 BC samples and 11 normal samples.	MiR-29b and miR-29c were downregulated in BC tumors	Higher miR-29c levels were correlated with longer DFS.	[79]
	-	Spotted locked nucleic acid-base oligonucleotide microarrays			
Specimen	-	108 bladder carcinomas and 29 carcinomas invading the bladder	MiR-29c was significantly under-expressed in progressed tumors.	High expression of miR-29c was associated with a better prognosis.	[80]
	-	Microarrays			

BC: bladder cancer; qRT-PCR: quantitative real-time PCR; AUC: area under the curve; OS: overall survival; DFS: disease-free survival.

3.3. miR-29s as Biomarkers in Hepatocellular Carcinoma

Hepatocellular carcinoma (HCC) is the most common cancer in the liver, with a high incidence and mortality rate [82]. The treatment strategy for HCC patients commonly depends on the tumor stage, but curative options are only available for patients with early stages of HCC [83]. Due to the limitation in early diagnosis, one-third of HCC patients cannot receive the appropriate therapy, and another one-third of those experience therapeutic delay, leading to significantly lower OS in HCC patients [84]. This fact suggests an urgent need for novel biomarkers for early and effective diagnosis and prognosis of HCC. As a tumor suppressor, miR-29s have been considered a potential diagnostic and prognostic biomarker for HCC (Table 3). The RNA from different sources such as serum and frozen tissues have been extracted and quantitatively measured by using the qRT-PCR method. Among miR-29 family members, miR-29a exhibited major functions in the liver as well as HCC tissues [85]. The miR-29a was significantly lower expressed in HCC tissues as compared to the controls and overexpression of miR-29a suppressed HCC cell growth by inhibiting the SPARC (Secreted protein acidic, rich in cysteine)-AKT pathway [85]. In hepatocytes, overexpression of miR-29a inhibited PTEN expression, leading to activation of the PI3K/AKT pathway that eventually induced cell migration [86]. The miRNA profile analysis of exosomes isolated from fast- and slow-migrated HCC patient-derived cells (PDCs) revealed a set of differentially expressed miRNAs that were further validated in HCC samples. The results showed a significant downregulation of miR-29b-3p gene in fast-growing PDCs as compared to slow-growing cells, suggesting its role as in metastasis and OS. Consequently, this cluster of miRNAs may serve as a biomarker for the proliferation of HCC cells [87]. Additionally, there is a statistically significant difference in the levels of miR-29c expression in HCC-derived exosomes amongst HCC, hepatitis B virus (HBV) infection, and cirrhosis patients [88]. It was indicated that TLR3 (Toll-like receptor 3) activated macrophages produced exosomes containing miRNA-29s that were proved to be able to prevent hepatitis C virus (HCV) replication in HCC cell line, suggesting the potential use of exosomes comprising miR-29 family members as a therapy to control HCV replication in infected hepatocytes [89]. Additionally, one other study reported that higher expression of miR-29a-3p was associated with a poorer prognosis, shorter OS, and disease-free survival (DFS) in HCC patients [90].

Table 3. MiR-29s as biomarkers in hepatocellular carcinoma.

Sample	Sample Size & Methods	Outcome	Results	Refs
Serum	- 58 NAFLD and 34 normal control - qRT-PCR	MiR-29a: lower in NAFLD patient MiR-29c: unchanged MiR-29b: undetectable	For miR-29a: AUC: 0.679 Sensitivity: 60.87% Specificity: 82.35%	[91]
Tissue	- 266 HCC - Taqman Low-Density Arrays qRT-PCR	MiR-29a-5p was associated with early HCC recurrence, resulting in lower OS	AUC: 0.708 Sensitivity: 74.2% Specificity: 68.2%	[92]
Venous blood	- 174 HCC - qRT-PCR	MiR-29a-3p was higher in both early and late stages of HCC	AUC: 0.71 (95%CI = 0.62–0.78)	[90]
Tissue	- 55 HCC and 55 normal control - qRT-PCR	MiR-29a was downregulated in HCC samples MiR-29a targeted SIRT1 and suppressed the HCC cell cycle and proliferation.	Lower miR-29a is associated with higher tumor size, vascular invasion, poor DFS	[93]
Specimen	- 110 HCC - qRT-PCR	MiR-29a was dramatically decreased in HCC tissues	miR-29a targeted to SPARC, downstream of AKT/mTOR to suppress cell growth.	[94]

HCC: hepatocellular carcinoma; NAFLD: Non-alcoholic fatty liver disease; AUC: area under the curve, qRT-PCR: quantitative real-time PCR; OS: overall survival; DFS: disease-free survival.

3.4. miR-29s as Biomarkers in Pancreatic Cancer

The miR-29a has been validated to be upregulated in tissue samples from patients with pancreatic ductal adenocarcinoma (PDAC) and was considered a potential diagnostic biomarker for this type of cancer [95]. A recent study comparing 38 patients with PDAC and 11 controls revealed that miR-29c-3p was typically downregulated in PDAC as compared to both normal pancreatic tissues and chronic pancreatitis [96]. Similarly, a study that employed high throughput screen figured out 42 candidate miRNAs that were significantly different between pancreatic cancer (PC) and healthy group, and, the miR-29b was noted to be downregulated 2.1 folds in PC samples [97]. Even though the fold-change and significant level of miR-29s were not high enough in both studies, miR-29s were not tested in the validated group and their role as crucial biomarkers in PC has not been confirmed. In one study, Humeau et al. used qRT-PCR to examine the change of 90 miRNAs in PC and identified four significant candidate miRNAs in saliva samples, including miR-29c [98]. However, the sample size (4 controls, 4 pancreatitis, and 7 PC samples) of the study was relatively small, making its finding of miR-29 as a biomarker for PC remains to be further confirmed.

Chemotherapy plays a significant role in the treatment of PC. Gemcitabine (GEM), an inhibitor of DNA synthesis and ribonucleotide reductase, has become a gold standard chemotherapeutic agent for PC [99,100]. Molecularly, miRNA-29a is involved in a PC cell's response to GEM by regulating the Wnt/β-catenin signaling pathway [101,102]. Wnt3a, an important ligand of the Wnt/β-catenin signaling pathway, has been shown to induce GEM-resistance in PC cells, probably by activating Wnt/β-catenin signaling in these cells [102]. Additionally, miR-29a was detected to be upregulated in PC tissues and cell lines, and its expression level was positively associated with metastasis [103]. Induced expression of miR-29a caused downregulation of tristetraprolin (TTP), thereby elevating the expression of pro-inflammatory factors and EMT markers. Ectopic overexpression of TTP decreased tumor growth and migration in vivo [103].

3.5. miR-29s as Diagnostic Biomarkers in Lung Cancer

Histologically, lung cancer is comprised of 85% of non-small-cell lung cancer (NSCLC) and 15% are small-cell lung cancer (SCLC) [104]. Despite recent advances in diagnosis, late diagnosis is still the main reason for poor prognosis and outcomes in lung cancer. It has been noted that miR-29c was overexpressed in the serum of NSCLC patients as compared to the normal controls [105,106]. Similarly, miR-29a was also found to be upregulated in peripheral blood of lung cancer patients as compared to the healthy control individuals [107]. These findings suggested that miR-29a and miR-29c could be used as potential diagnostic biomarkers for lung cancer. Additionally, Liu et al. reported that miR-29a was strongly downregulated in lung cancer tissues as compared to paired normal tissues and that induced expression of miR-29a suppressed cell proliferation and colony formation of lung cancer cells by targeting and negatively regulating the expression of NRAS (neuroblastoma ras viral oncogene homolog) oncogene. The study has also revealed that miR-29a increased the sensitivity of lung cancer cells to cisplatin treatment and that a combination of miR-29a and cisplatin-induced apoptosis in lung cancer cells, suggesting the potential role of miR-29a as a prognostic biomarker for lung cancer [108]. Recently, it was indicated that NSCLC generated exosomes that contain miR-29a. This miRNA can attach to TLRs in immune cells and elicit protumoral inflammation, hence increasing tumor growth and metastasis [109].

Lung adenocarcinoma (LAC) is a highly aggressive tumor though little is known about its underlying molecular mechanisms. Liu et al. discovered that downregulation of miR-29c was strongly correlated with unfavorable prognosis in stage IIIA LAC patients. The MiR-29c suppressed cell growth, migration, and invasion in human LAC cell lines by directly targeting vascular endothelial growth factor A (VEGFA) [110]. Therefore, miR-29c has been concluded as a tumor suppressor and may be considered a promising prognostic and therapeutic biomarker for LAC [110].

3.6. miR-29s as Biomarkers in Leukemia and Lymphoma

Serum miRNAs have been suggested as promising biomarkers for diffuse large B cell lymphoma [111]. Notably, miR-29a and miR-142-3p have been identified to be consistently under-expressed in AML and may act cooperatively in granulopoiesis and monopoiesis [112]. Therefore, dual evaluation of miR-29a and miR-142-3p is more effective for the diagnosis of AML. Additionally, downregulation of miR-29c was identified as a signature for chronic lymphocytic leukemia (CLL) [113], which was greatly correlated with disease progression in CLL patients harboring the 17p deletion [113]. Moreover, miR-29a was significantly downregulated in the bone marrow of pediatric AML patients as compared to the normal controls. The low expression of miR-29a was strongly correlated with shorter relapse-free and OS in these patients [114]. The study suggested that downregulation of miR-29a may be used as a prognostic marker in pediatric AML.

3.7. miR-29s as Biomarkers in Kidney Cancer

The miR-29s regulate genes that are closely related to the molecular pathogenesis of renal cell carcinoma (RCC). Serpin family H member 1 (SERPINH1), a direct target of miR-29, was noted to be overexpressed in RCC clinical samples and tyrosine kinase inhibitor failure autopsy specimens. Overexpression of SERPINH1 was significantly associated with advanced tumor stage, pathological grade, and poor prognosis, mostly due to its ability to induce cancer cell migration and invasion [115]. In addition, it was supported with evidence that miR-29b acted as an oncomiR and could be a potential prognostic marker for RCC. The miR-29b promoted proliferation and invasion in SN12-PM6 cells, which inhibited cell apoptosis by directly suppressing the expression of kinesin family member 1B, a tumor suppressor gene that induces cell apoptosis. Upregulation of miR-29b in both cell lines and clinical samples was significantly associated with tumor node metastasis and OS of RCC [116]. These studies suggested the clinical roles of miR-29s in RCC and its potential use as prognostic biomarkers in this type of cancer.

3.8. miR-29s as Biomarkers in Breast Cancer

Globally, breast cancer is regarded as one of the most diagnosed and deadly cancers, particularly in the case of women [117,118]. However, the prognosis of breast cancer is not satisfactory, and the 5-year survival rate is lower than 25% [119]. All these phenomena urge the discovery of novel biomarkers for the early diagnosis, and proper therapy of breast cancer. The miR-29s have been studied and suggested as a tumor suppressor in breast cancer [117,120,121]. Wu et al. illustrated that miR-29a was significantly downregulated in breast cancer cells, and its overexpression inhibited cancer cell growth which was achieved by repressing the expression of transcription factor B-Myb [117]. Additionally, overexpression of miR-29a resulted in cell cycle arrest at the G0/G1 phase. The findings denoted the partiality of miR-29a which exerts its tumor suppressor role in breast cancer cell lines by cessation of the cell cycle through negative regulation of CDC42 [46]. Later, Shinden and collaborators investigated the clinicopathological significance of miR-29b in breast cancer cases and illustrated that miR-29b acted as a tumor suppressive miRNA [121], suggesting it is a prominent biomarker for recurrence and metastasis in breast cancer patients. Moreover, overexpression of miR-29b-1/a significantly suppressed proliferation of Tamoxifen (TAM)-resistant breast cancer cells, indicating that miR-29b-1/a functions as a tumor suppressor in these cells [122]. Additionally, BRCA1 (Breast Cancer 1) was reported to bind to a specific region of the promoter and regulate the expression of miR-29b-1-5p. The higher significant level of miR-29b-1-5p as a prognostic marker than other widely used biomarkers signified the potential of this miRNA as a biomarker for BRCA1 deficiency and survival in breast cancer [123]. Consequently, miR-29s are significantly elevated in the whole blood, serum, and tissues samples from breast cancer patients (Table 4).

Furthermore, overexpressed miR-29s were testified both in tumor tissues and serum of breast cancer patients in comparison to that of healthy individuals [34]. Recently, it was reported that GATA binding protein 3 (GATA3), a transcription factor, elevated the miR-29b level in breast cancer whereas the destruction of miR-29b enhanced metastasis and accelerated EMT. Being a tumor-suppressor gene, the damage of GATA3 in breast cancer resulted in a poor prognosis [124]. A recent study revealed that miR-29a abated cell proliferation and promoted apoptosis in the MCF-7 (Michigan Cancer Foundation-7) cell line by negatively controlling NF-kB (nuclear factor-kappa B) and the levels of cyclinD1 and Bcl-2 proteins [125]. Additionally, the other study revealed that overexpression of miR-29a inhibited cell migration and invasion by negatively regulating Robo1 (Roundabout 1) in breast cancer cells, highlighting the significant role of miR-29a in carcinogenesis breast cancer [126]. Moreover, upregulation of miR-29a induced adriamycin resistance in MCF-7 breast cancer cells, possibly by inhibiting the PTEN/AKT/GSK3β pathway [127]. Treatment with progestin reduced migration and invasion in breast cancer cells, via the miR-29/ATP1B1(ATPase Na$^+$/K$^+$ transporting β1 polypeptide) axis [128].

Table 4. miR-29s as biomarkers in breast cancer.

Samples	Sample Size & Methods	Outcome	Results	Ref.
Blood samples	- 54 patients with Luminal A-like breast cancer and 56 healthy controls - qRT-PCR	MiR-29a was significantly down-regulated in the blood of patients with Luminal A-like breast tumors compared to healthy controls.	Combined miR-29a, miR-181a and miR-652 (AUC: 0.80, sensitivity: 77% and specificity: 74%)	[129]
Serum sample	- 76 breast cancer patients and 52 healthy controls. - SdM-qRT-PCR	MiR-2 was significantly higher in breast cancer patients compared to healthy controls.	MiR-29c AUC: 0.724 (95% CI 0.638–0.810)	[46]
Serum	- 20 breast cancer patients and 20 controls - SOLiD Sequencing (qRT-PCR)	MiR-29a was significantly elevated in the serum of breast cancer patients ($p < 0.05$).	MiR-29a was elevated more than 5-folds by SOLiD sequencing.	[130]
Tissue samples	- 15 breast cancer patients and 15 healthy controls - qRT-PCR	MiR-29a was significantly upregulated in breast cancer as compared with their respective healthy controls ($p < 0.001$).	MiR-29a (AUC:0.969, Sensitivity: 93.3%, specificity: 91.1%)	[131]

CI: confidence interval; AUC: area under the curve, qRT-PCR: quantitative real-time PCR.

4. Conclusions and Perspectives

The miR-29s are crucial regulators in numerous types of human cancer, which can act as either tumor suppressors or inducers. By regulating multiple target genes, they are indirectly involved in controlling different cellular pathways including cell proliferation, apoptosis, migration and invasion, and chemotherapeutic sensitivity, thereby contributing to cancer progression, metastasis, and drug resistance. The profound dysregulation of miR-29s in numerous types of cancer and their correlation to the patients' OS and metastasis have strongly signified them as potential diagnostic and prognostic biomarkers for specific types of cancer. However, due to its flexibility, the application of miR-29s as biomarkers and the development of miR-29s-based therapies need to be verified further for each type and stage of cancer specifically.

Fortunately, the recent advances in sequencing technologies (next generation of sequencing and long-read sequencing) and genome editing allows better validation of the target genes of miR-29s as well as an understanding of the roles of miR-29s in each cancer type. In addition, the rapid adoption of exosomes for the miRNA's delivery could also support the development of miR-29s for miR-29s-based therapies. In summary, exosomes have several desirable characteristics for delivering miRNAs including small sizes (30–200 nm), being able to cross the blood barriers, being specific to the target cells, and being relatively easy to be engineered. Consequently, the delivery by exosomes of miR-29s to unhealthy/abnormal cells will be adapted for a potential therapeutic approach.

Author Contributions: Conceptualization, D.N.D. and H.T.N.; methodology, T.T.P.N., K.H.S. and T.B.N.; investigation, T.T.P.N., K.H.S., D.N.D. and T.B.N.; resources, T.T.P.N., K.H.S., D.N.D. and H.T.N.; data curation, T.T.P.N., K.H.S., D.N.D., H.T.N. and T.B.N.; writing—original draft preparation, T.T.P.N. and K.H.S.; writing—review and editing, T.T.P.N., K.H.S., D.N.D. and H.T.N.; visualization, T.T.P.N.; supervision, D.N.D.; project administration, D.N.D. All authors have read and agreed to the published version of the manuscript.

Funding: This research received no external funding.

Institutional Review Board Statement: Not applicable.

Informed Consent Statement: Not applicable.

Data Availability Statement: No specific data used in the review article. All information has been provided in the manuscript.

Conflicts of Interest: The authors declare no conflict of interest.

References

1. Bartel, D.P. MicroRNAs: Genomics, biogenesis, mechanism, and function. *Cell* **2004**, *116*, 281–297. [CrossRef]
2. Bartel, D.P. MicroRNAs: Target recognition and regulatory functions. *Cell* **2009**, *136*, 215–233. [CrossRef] [PubMed]
3. Peng, Y.; Croce, C.M. The role of MicroRNAs in human cancer. *Signal Transduct. Target. Ther.* **2016**, *1*, 15004. [CrossRef] [PubMed]
4. Rupaimoole, R.; Slack, F.J. MicroRNA therapeutics: Towards a new era for the management of cancer and other diseases. *Nat. Rev. Drug Discov.* **2017**, *16*, 203–222. [CrossRef]
5. Kwon, J.J.; Factora, T.D.; Dey, S.; Kota, J. A Systematic Review of miR-29 in Cancer. *Mol. Ther. -Oncolytics* **2018**, *12*, 173–194. [CrossRef]
6. Chang, T.-C.; Yu, D.; Lee, Y.-S.; Wentzel, E.A.; Arking, D.E.; West, K.M.; Dang, C.V.; Thomas-Tikhonenko, A.; Mendell, J.T. Widespread microRNA repression by Myc contributes to tumorigenesis. *Nat. Genet.* **2008**, *40*, 43–50. [CrossRef]
7. Landgraf, P.; Rusu, M.; Sheridan, R.; Sewer, A.; Iovino, N.; Aravin, A.; Pfeffer, S.; Rice, A.; Kamphorst, A.O.; Landthaler, M.; et al. A Mammalian microRNA expression atlas based on small RNA library sequencing. *Cell* **2007**, *129*, 1401–1414. [CrossRef]
8. Hwang, H.-W.; Wentzel, E.A.; Mendell, J.T. A hexanucleotide element directs MicroRNA nuclear import. *Science* **2007**, *315*, 97–100. [CrossRef]
9. Zhang, Z.; Zou, J.; Wang, G.-K.; Zhang, J.-T.; Huang, S.; Qin, Y.-W.; Jing, Q. Uracils at nucleotide position 9–11 are required for the rapid turnover of miR-29 family. *Nucleic Acids Res.* **2011**, *39*, 4387–4395. [CrossRef]
10. Liao, J.-Y.; Ma, L.-M.; Guo, Y.-H.; Zhang, Y.-C.; Zhou, H.; Shao, P.; Chen, Y.-Q.; Qu, L.-H. Deep sequencing of human nuclear and cytoplasmic small RNAs reveals an unexpectedly complex subcellular distribution of miRNAs and tRNA 3′ trailers. *PLoS ONE* **2010**, *5*, e10563. [CrossRef]

11. Li, J.; Wang, L.; Hua, X.; Tang, H.; Chen, R.; Yang, T.; Das, S.; Xiao, J. CRISPR/Cas9-Mediated miR-29b Editing as a Treatment of Different Types of Muscle Atrophy in Mice. *Mol. Ther.* **2020**, *28*, 1359–1372. [CrossRef] [PubMed]

12. Sassi, Y.; Avramopoulos, P.; Ramanujam, D.P.; Grüter, L.; Werfel, S.; Giosele, S.; Brunner, A.-D.; Esfandyari, D.; Papadopoulou, A.-S.; De Strooper, B.; et al. Cardiac myocyte miR-29 promotes pathological remodeling of the heart by activating Wnt signaling. *Nat. Commun.* **2017**, *8*, 1614. [CrossRef] [PubMed]

13. Roderburg, C.; Urban, G.-W.; Bettermann, K.; Vucur, M.; Zimmermann, H.W.; Schmidt, S.; Janssen, J.; Koppe, C.; Knolle, P.; Castoldi, M.; et al. Micro-RNA Profiling Reveals a Role for miR-29 in Human and Murine Liver Fibrosis. *Hepatology* **2011**, *53*, 209–218. [CrossRef]

14. Sun, J.; Li, Q.; Lian, X.; Zhu, Z.; Chen, X.; Pei, W.; Li, S.; Abbas, A.; Wang, Y.; Tian, L. MicroRNA-29b mediates lung mesenchymal-epithelial transition and prevents lung fibrosis in the silicosis model. *Mol. Ther. -Nucleic Acids* **2019**, *14*, 20–31. [CrossRef]

15. Qin, W.; Chung, A.C.; Huang, X.R.; Meng, X.-M.; Hui, D.; Yu, C.-M.; Sung, J.J.Y.; Lan, H.Y. TGF-β/Smad3 signaling promotes renal fibrosis by inhibiting miR-29. *J. Am. Soc. Nephrol.* **2011**, *22*, 1462–1474. [CrossRef] [PubMed]

16. Nishikawa, R.; Goto, Y.; Kojima, S.; Enokida, H.; Chiyomaru, T.; Kinoshita, T.; Sakamoto, S.; Fuse, M.; Nakagawa, M.; Naya, Y.; et al. Tumor-suppressive microRNA-29s inhibit cancer cell migration and invasion via targeting LAMC1 in prostate cancer. *Int. J. Oncol.* **2014**, *45*, 401–410. [CrossRef]

17. Wang, T.; Hou, J.; Jian, S.; Luo, Q.; Wei, J.; Li, Z.; Wang, X.; Bai, P.; Duan, B.; Xing, J.; et al. miR-29b negatively regulates MMP2 to impact gastric cancer development by suppress gastric cancer cell migration and tumor growth. *J. Cancer* **2018**, *9*, 3776–3786. [CrossRef]

18. Duhachek-Muggy, S.; Zolkiewska, A. ADAM12-L is a direct target of the miR-29 and miR-200 families in breast cancer. *BMC Cancer* **2015**, *15*, 93. [CrossRef]

19. Kogure, T.; Costinean, S.; Yan, I.; Braconi, C.; Croce, C.; Patel, T. Hepatic miR-29ab1 expression modulates chronic hepatic injury. *J. Cell. Mol. Med.* **2012**, *16*, 2647–2654. [CrossRef]

20. Liang, C.; Bu, S.; Fan, X. Suppressive effect of microRNA-29b on hepatic stellate cell activation and its crosstalk with TGF-β1/Smad3. *Cell Biochem. Funct.* **2016**, *34*, 326–333. [CrossRef]

21. van Rooij, E.; Sutherland, L.B.; Thatcher, J.E.; DiMaio, J.M.; Naseem, R.H.; Marshall, W.S.; Hill, J.A.; Olson, E.N. Dysregulation of microRNAs after myocardial infarction reveals a role of miR-29 in cardiac fibrosis. *Proc. Natl. Acad. Sci. USA* **2008**, *105*, 13027. [CrossRef] [PubMed]

22. Bandyopadhyay, S.; Friedman, R.C.; Marquez, R.T.; Keck, K.; Kong, B.; Icardi, M.S.; Brown, K.E.; Burge, C.B.; Schmidt, W.N.; Wang, Y.; et al. Hepatitis C Virus Infection and Hepatic Stellate Cell Activation Downregulate miR-29: miR-29 Overexpression Reduces Hepatitis C Viral Abundance in Culture. *J. Infect. Dis.* **2011**, *203*, 1753–1762. [CrossRef] [PubMed]

23. Montgomery, R.L.; Yu, G.; Latimer, P.A.; Stack, C.; Robinson, K.; Dalby, C.M.; Kaminski, N.; van Rooij, E. MicroRNA mimicry blocks pulmonary fibrosis. *EMBO Mol. Med.* **2014**, *6*, 1347–1356. [CrossRef] [PubMed]

24. Fabbri, M.; Garzon, R.; Cimmino, A.; Liu, Z.; Zanesi, N.; Callegari, E.; Liu, S.; Alder, H.; Costinean, S.; Fernandez-Cymering, C.; et al. MicroRNA-29 family reverts aberrant methylation in lung cancer by targeting DNA methyltransferases 3A and 3B. *Proc. Natl. Acad. Sci. USA* **2007**, *104*, 15805–15810. [CrossRef]

25. Cui, H.; Wang, L.; Gong, P.; Zhao, C.; Zhang, S.; Zhang, K.; Zhou, R.; Zhao, Z.; Fan, H. Deregulation between miR-29b/c and DNMT3A is associated with epigenetic silencing of the CDH1 gene, affecting cell migration and invasion in gastric cancer. *PLoS ONE* **2015**, *10*, e0123926. [CrossRef]

26. Wu, H.; Zhang, W.; Wu, Z.; Liu, Y.; Shi, Y.; Gong, J.; Shen, W.; Liu, C. miR-29c-3p regulates DNMT3B and LATS1 methylation to inhibit tumor progression in hepatocellular carcinoma. *Cell Death Dis.* **2019**, *10*, 48. [CrossRef]

27. Gondaliya, P.; Dasare, A.P.; Jash, K.; Tekade, R.K.; Srivastava, A.; Kalia, K. miR-29b attenuates histone deacetylase-4 mediated podocyte dysfunction and renal fibrosis in diabetic nephropathy. *J. Diabetes Metab. Disord.* **2019**, *19*, 13–27. [CrossRef]

28. Amodio, N.; Stamato, M.A.; Gullà, A.M.; Morelli, E.; Romeo, E.; Raimondi, L.; Pitari, M.R.; Ferrandino, I.; Misso, G.; Caraglia, M.; et al. Therapeutic targeting of miR-29b/HDAC4 epigenetic loop in multiple myeloma. *Mol. Cancer Ther.* **2016**, *15*, 1364–1375. [CrossRef]

29. Zhang, P.; Huang, B.; Xu, X.; Sessa, W.C. Ten-eleven translocation (Tet) and thymine DNA glycosylase (TDG), components of the demethylation pathway, are direct targets of miRNA-29a. *Biochem. Biophys. Res. Commun.* **2013**, *437*, 368–373. [CrossRef]

30. Mott, J.L.; Kobayashi, S.; Bronk, S.F.; Gores, G.J. mir-29 regulates Mcl-1 protein expression and apoptosis. *Oncogene* **2007**, *26*, 6133–6140. [CrossRef]

31. Zhu, K.; Liu, L.; Zhang, J.; Wang, Y.; Liang, H.; Fan, G.; Jiang, Z.; Zhang, C.-Y.; Chen, X.; Zhou, G. MiR-29b suppresses the proliferation and migration of osteosarcoma cells by targeting CDK6. *Protein Cell* **2016**, *7*, 434–444. [CrossRef] [PubMed]

32. Shi, C.; Ren, L.; Sun, C.; Yu, L.; Bian, X.; Zhou, X.; Wen, Y.; Hua, D.; Zhao, S.; Luo, W.; et al. miR-29a/b/c function as invasion suppressors for gliomas by targeting CDC42 and predict the prognosis of patients. *Br. J. Cancer* **2017**, *117*, 1036–1047. [CrossRef] [PubMed]

33. Tang, W.; Zhu, Y.; Gao, J.; Fu, J.; Liu, C.; Liu, Y.; Song, C.; Zhu, S.; Leng, Y.; Wang, G.; et al. MicroRNA-29a promotes colorectal cancer metastasis by regulating matrix metalloproteinase 2 and E-cadherin via KLF4. *Br. J. Cancer* **2014**, *110*, 450–458. [CrossRef] [PubMed]

34. Wu, Q.; Wang, C.; Lu, Z.; Guo, L.; Ge, Q. Analysis of serum genome-wide microRNAs for breast cancer detection. *Clin. Chim. Acta* **2012**, *413*, 1058–1065. [CrossRef]

35. Du, L.; Jiang, X.; Duan, W.; Wang, R.; Wang, L.; Zheng, G.; Yan, K.; Wang, L.; Li, J.; Zhang, X.; et al. Cell-free microRNA expression signatures in urine serve as novel noninvasive biomarkers for diagnosis and recurrence prediction of bladder cancer. *Oncotarget* **2017**, *8*, 40832–40842. [CrossRef]

36. Lin, G.; Zhang, C.; Chen, X.; Wang, J.; Chen, S.; Tang, S.; Yu, T. Identification of circulating miRNAs as novel prognostic biomarkers for bladder cancer. *Math. Biosci. Eng.* **2019**, *17*, 834–844. [CrossRef]

37. Zhang, X.; Zhao, X.; Fiskus, W.; Lin, J.; Lwin, T.; Rao, R.; Zhang, Y.; Chan, J.C.; Fu, K.; Marquez, V.E.; et al. Coordinated silencing of MYC-Mediated miR-29 by HDAC3 and EZH2 as a therapeutic target of histone modification in aggressive B-Cell lymphomas. *Cancer Cell* **2012**, *22*, 506–523. [CrossRef]

38. Wang, H.; Garzon, R.; Sun, H.; Ladner, K.J.; Singh, R.; Dahlman, J.; Cheng, A.; Hall, B.M.; Qualman, S.J.; Chandler, D.S.; et al. NF-kappaB-YY1-miR-29 regulatory circuitry in skeletal myogenesis and rhabdomyosarcoma. *Cancer Cell* **2008**, *14*, 369–381. [CrossRef]

39. Eyholzer, M.; Schmid, S.; Wilkens, L.; Mueller, B.U.; Pabst, T. The tumour-suppressive miR-29a/b1 cluster is regulated by CEBPA and blocked in human AML. *Br. J. Cancer* **2010**, *103*, 275–284. [CrossRef]

40. Baylin, S.B. DNA methylation and gene silencing in cancer. *Nat. Clin. Pract. Oncol.* **2005**, *2*, S4–S11. [CrossRef]

41. Latif, F.; Tory, K.; Gnarra, J.; Yao, M.; Duh, F.-M.; Orcutt, M.L.; Stackhouse, T.; Kuzmin, I.; Modi, W.; Geil, L.; et al. Identification of the von Hippel-Lindau disease tumor suppressor gene. *Science* **1993**, *260*, 1317–1320. [CrossRef] [PubMed]

42. Morita, S.; Horii, T.; Kimura, M.; Ochiya, T.; Tajima, S.; Hatada, I. miR-29 represses the activities of DNA methyltransferases and DNA demethylases. *Int. J. Mol. Sci.* **2013**, *14*, 14647–14658. [CrossRef] [PubMed]

43. Jiang, H.; Liu, Z.-N.; Cheng, X.-H.; Zhang, Y.-F.; Dai, X.; Bao, G.-M.; Zhou, L.-B. MiR-29c suppresses cell invasion and migration by directly targeting CDK6 in gastric carcinoma. *Eur. Rev. Med. Pharmacol. Sci.* **2019**, *23*, 7920–7928. [PubMed]

44. Zhao, Z.; Wang, L.; Song, W.; Cui, H.; Chen, G.; Qiao, F.; Hu, J.; Zhou, R.; Fan, H. Reduced miR-29a-3p expression is linked to the cell proliferation and cell migration in gastric cancer. *World J. Surg. Oncol.* **2015**, *13*, 101. [CrossRef]

45. Zhao, X.; Li, J.; Huang, S.; Wan, X.; Luo, H.; Wu, D. MiRNA-29c regulates cell growth and invasion by targeting CDK6 in bladder cancer. *Am. J. Transl. Res.* **2015**, *7*, 1382.

46. Zhang, M.; Guo, W.; Qian, J.; Wang, B. Negative regulation of CDC42 expression and cell cycle progression by miR-29a in breast cancer. *Open Med.* **2016**, *11*, 78–82. [CrossRef]

47. Gong, J.-N.; Yu, J.; Lin, H.-S.; Zhang, X.-H.; Yin, X.-L.; Xiao, Z.; Wang, F.; Wang, X.-S.; Su, R.; Shen, C.; et al. The role, mechanism and potentially therapeutic application of microRNA-29 family in acute myeloid leukemia. *Cell Death Differ.* **2014**, *21*, 100–112. [CrossRef]

48. Zhang, K.; Zhang, C.; Liu, L.; Zhou, J. A key role of microRNA-29b in suppression of osteosarcoma cell proliferation and migration via modulation of VEGF. *Int. J. Clin. Exp. Pathol.* **2014**, *7*, 5701–5708.

49. Chen, L.; Xiao, H.; Wang, Z.-H.; Huang, Y.; Liu, Z.-P.; Ren, H.; Song, H. miR-29a suppresses growth and invasion of gastric cancer cells in vitro by targeting VEGF-A. *BMB Rep.* **2014**, *47*, 39–44. [CrossRef]

50. Zeng, Q.; Wang, Y.; Gao, J.; Yan, Z.; Li, Z.; Zou, X.; Li, Y.; Wang, J.; Guo, Y. miR-29b-3p regulated osteoblast differentiation via regulating IGF-1 secretion of mechanically stimulated osteocytes. *Cell. Mol. Biol. Lett.* **2019**, *24*, 11. [CrossRef]

51. Hu, H.-H.; Chen, D.-Q.; Wang, Y.-N.; Feng, Y.-L.; Cao, G.; Vaziri, N.D.; Zhao, Y.-Y. New insights into TGF-β/Smad signaling in tissue fibrosis. *Chem. Interact.* **2018**, *292*, 76–83. [CrossRef] [PubMed]

52. Wang, B.; Komers, R.; Carew, R.; Winbanks, C.E.; Xu, B.; Herman-Edelstein, M.; Koh, P.; Thomas, M.; Jandeleit-Dahm, K.; Gregorevic, P.; et al. Suppression of microRNA-29 expression by TGF-β1 promotes collagen expression and renal fibrosis. *J. Am. Soc. Nephrol.* **2012**, *23*, 252–265. [CrossRef] [PubMed]

53. Cushing, L.; Kuang, P.P.; Qian, J.; Shao, F.; Wu, J.; Little, F.; Thannickal, V.J.; Cardoso, W.V.; Lü, J. miR-29 is a major regulator of genes associated with pulmonary fibrosis. *Am. J. Respir. Cell Mol. Biol.* **2011**, *45*, 287–294. [CrossRef] [PubMed]

54. Koshizuka, K.; Kikkawa, N.; Hanazawa, T.; Yamada, Y.; Okato, A.; Arai, T.; Katada, K.; Okamoto, Y.; Seki, N. Inhibition of integrin β1-mediated oncogenic signalling by the antitumor *microRNA-29* family in head and neck squamous cell carcinoma. *Oncotarget* **2017**, *9*, 3663–3676. [CrossRef] [PubMed]

55. Tan, J.; Tong, B.-D.; Wu, Y.-J.; Xiong, W. MicroRNA-29 mediates TGFβ1-induced extracellular matrix synthesis by targeting wnt/β-catenin pathway in human orbital fibroblasts. *Int. J. Clin. Exp. Pathol.* **2014**, *7*, 7571–7577.

56. Wang, Y.; Liu, C.; Luo, M.; Zhang, Z.; Gong, J.; Li, J.; You, L.; Dong, L.; Su, R.; Lin, H. Chemotherapy-induced miRNA-29c/catenin-δ signaling suppresses metastasis in gastric cancer. *Cancer Res.* **2015**, *75*, 1332–1344. [CrossRef]

57. Jiang, J.; Yu, C.; Chen, M.; Zhang, H.; Tian, S.; Sun, C. Reduction of miR-29c enhances pancreatic cancer cell migration and stem cell-like phenotype. *Oncotarget* **2015**, *6*, 2767. [CrossRef]

58. Tréhoux, S.; Lahdaoui, F.; Delpu, Y.; Renaud, F.; Leteurtre, E.; Torrisani, J.; Jonckheere, N.; Van Seuningen, I. Micro-RNAs miR-29a and miR-330-5p function as tumor suppressors by targeting the MUC1 mucin in pancreatic cancer cells. *Biochim. Et Biophys. Acta* **2015**, *1853*, 2392–2403. [CrossRef]

59. Luo, D.-J.; Li, L.-J.; Huo, H.-F.; Liu, X.-Q.; Cui, H.-W.; Jiang, D.-M. MicroRNA-29b sensitizes osteosarcoma cells to doxorubicin by targeting matrix metalloproteinase 9 (MMP-9) in osteosarcoma. *Eur. Rev. Med. Pharmacol. Sci.* **2019**, *23*, 1434–1442.

60. Bargaje, R.; Gupta, S.; Sarkeshik, A.; Park, R.; Xu, T.; Sarkar, M.; Halimani, M.; Roy, S.S.; Yates, J.; Pillai, B. Identification of novel targets for miR-29a using miRNA proteomics. *PLoS ONE* **2012**, *7*, e43243. [CrossRef]

61. Park, S.-Y.; Lee, J.H.; Ha, M.; Nam, J.-W.; Kim, V.N. miR-29 miRNAs activate p53 by targeting p85α and CDC42. *Nat. Struct. Mol. Biol.* **2009**, *16*, 23–29. [CrossRef] [PubMed]

62. Wang, C.Y.; Ren, J.B.; Liu, M.; Yu, L. Targeting miR-29 induces apoptosis of osteosarcoma MG-63 cells via regulation of TGF-β1/PUMA signal. *Eur. Rev. Med. Pharmacol. Sci.* **2016**, *20*, 3552–3560. [PubMed]

63. Yu, J.; Wang, Z.; Kinzler, K.W.; Vogelstein, B.; Zhang, L. PUMA mediates the apoptotic response to p53 in colorectal cancer cells. *Proc. Natl. Acad. Sci. USA* **2003**, *100*, 1931–1936. [CrossRef]

64. Pei, Y.-F.; Lei, Y.; Liu, X.-Q. MiR-29a promotes cell proliferation and EMT in breast cancer by targeting ten eleven translocation 1. *Biochim. Biophys. Acta (BBA)-Mol. Basis Dis.* **2016**, *1862*, 2177–2185. [CrossRef] [PubMed]

65. Liu, Q.; Geng, P.; Shi, L.; Wang, Q.; Wang, P. miR-29 promotes osteosarcoma cell proliferation and migration by targeting PTEN. *Oncol. Lett.* **2019**, *17*, 883–890. [CrossRef]

66. Pekarsky, Y.; Croce, C.M. Is miR-29 an oncogene or tumor suppressor in CLL? *Oncotarget* **2010**, *1*, 224. [CrossRef]

67. Peng, Q.; Feng, Z.; Shen, Y.; Zhu, J.; Zou, L.; Shen, Y.; Zhu, Y. Integrated analyses of microRNA-29 family and the related combination biomarkers demonstrate their widespread influence on risk, recurrence, metastasis and survival outcome in colorectal cancer. *Cancer Cell Int.* **2019**, *19*, 181. [CrossRef]

68. Basati, G.; Razavi, A.E.; Pakzad, I.; Malayeri, F.A. Circulating levels of the miRNAs, miR-194, and miR-29b, as clinically useful biomarkers for colorectal cancer. *Tumor Biol.* **2016**, *37*, 1781–1788. [CrossRef]

69. Li, L.; Guo, Y.; Chen, Y.; Wang, J.; Zhen, L.; Guo, X.; Liu, J.; Jing, C. The diagnostic efficacy and biological effects of microRNA-29b for colon cancer. *Technol. Cancer Res. Treat.* **2016**, *15*, 772–779. [CrossRef]

70. Yamada, A.; Horimatsu, T.; Okugawa, Y.; Nishida, N.; Honjo, H.; Ida, H.; Kou, T.; Kusaka, T.; Sasaki, Y.; Yagi, M.; et al. Serum miR-21, miR-29a, and miR-125b are promising biomarkers for the early detection of colorectal neoplasia. *Clin. Cancer Res.* **2015**, *21*, 4234–4242. [CrossRef]

71. Zhu, Y.; Xu, A.; Li, J.; Fu, J.; Wang, G.; Yang, Y.; Cui, L.; Sun, J. Fecal miR-29a and miR-224 as the noninvasive biomarkers for colorectal cancer. *Cancer Biomark.* **2016**, *16*, 259–264. [CrossRef] [PubMed]

72. Ramzy, I.; Hasaballah, M.; Marzaban, R.; Shaker, O.; Soliman, Z.A. Evaluation of microRNAs-29a, 92a and 145 in colorectal carcinoma as candidate diagnostic markers: An Egyptian pilot study. *Clin. Res. Hepatol. Gastroenterol.* **2015**, *39*, 508–515. [CrossRef] [PubMed]

73. Wang, L.-G.; Gu, J. Serum microRNA-29a is a promising novel marker for early detection of colorectal liver metastasis. *Cancer Epidemiol.* **2012**, *36*, e61–e67. [CrossRef] [PubMed]

74. Inoue, A.; Yamamoto, H.; Uemura, M.; Nishimura, J.; Hata, T.; Takemasa, I.; Ikenaga, M.; Ikeda, M.; Murata, K.; Mizushima, T.; et al. MicroRNA-29b is a Novel Prognostic Marker in Colorectal Cancer. *Ann. Surg. Oncol.* **2015**, *22*, 1410–1418. [CrossRef]

75. Aharonov, R.; Weissmann-Brenner, A.; Kushnir, M.; Yanai, G.L.; Gibori, H.; Purim, O.; Kundel, Y.; Morgenstern, S.; Halperin, M.; Niv, Y.; et al. Tumor microRNA-29a expression and the risk of recurrence in stage II colon cancer. *Int. J. Oncol.* **2012**, *40*, 2097–2103. [CrossRef]

76. Wang, G.; Kwan, B.C.-H.; Lai, F.M.-M.; Chow, K.-M.; Li, P.K.-T.; Szeto, C.-C. Urinary miR-21, miR-29, and miR-93: Novel biomarkers of fibrosis. *Am. J. Nephrol.* **2012**, *36*, 412–418. [CrossRef]

77. Xu, F.; Zhang, Q.; Cheng, W.; Zhang, Z.; Wang, J.; Ge, J. Effect of miR-29b-1* and miR-29c knockdown on cell growth of the bladder cancer cell line T24. *J. Int. Med Res.* **2013**, *41*, 1803–1810. [CrossRef]

78. Fan, Y.; Song, X.; Du, H.; Luo, C.; Wang, X.; Yang, X.; Wang, Y.; Wu, X. Down-regulation of miR-29c in human bladder cancer and the inhibition of proliferation in T24 cell via PI3K-AKT pathway. *Med Oncol.* **2014**, *31*, 65. [CrossRef]

79. Dyrskjøt, L.; Ostenfeld, M.S.; Bramsen, J.B.; Silahtaroglu, A.N.; Lamy, P.; Ramanathan, R.; Fristrup, N.; Jensen, J.L.; Andersen, C.L.; Zieger, K.; et al. Genomic Profiling of MicroRNAs in Bladder Cancer: miR-129 Is Associated with Poor Outcome and Promotes Cell Death In vitro. *Cancer Res.* **2009**, *69*, 4851. [CrossRef]

80. Rosenberg, E.; Baniel, J.; Spector, Y.; Faerman, A.; Meiri, E.; Aharonov, R.; Margel, D.; Goren, Y.; Nativ, O. Predicting progression of bladder urothelial carcinoma using microRNA expression. *Br. J. Urol.* **2013**, *112*, 1027–1034. [CrossRef]

81. Xu, X.-D.; Wu, X.-H.; Fan, Y.-R.; Tan, B.; Quan, Z.; Luo, C.-L. Exosome-derived microRNA-29c induces apoptosis of BIU-87 cells by down regulating BCL-2 and MCL-1. *Asian Pac. J. Cancer Prev.* **2014**, *15*, 3471–3476. [CrossRef] [PubMed]

82. Llovet, J.M.; Kelley, R.K.; Villanueva, A.; Singal, A.G.; Pikarsky, E.; Roayaie, S.; Lencioni, R.; Koike, K.; Zucman-Rossi, J.; Finn, R.S. Hepatocellular carcinoma. *Nat. Rev. Dis. Primers* **2021**, *7*, 6. [CrossRef] [PubMed]

83. Lurje, I.; Czigany, Z.; Bednarsch, J.; Roderburg, C.; Isfort, P.; Neumann, U.P.; Lurje, G. Treatment Strategies for Hepatocellular Carcinoma—A Multidisciplinary Approach. *Int. J. Mol. Sci.* **2019**, *20*, 1465. [CrossRef] [PubMed]

84. Singal, A.G.; Waljee, A.K.; Patel, N.; Chen, E.Y.; Tiro, J.; Marrero, J.A.; Yopp, A.C. Therapeutic delays lead to worse survival among patients with hepatocellular carcinoma. *J. Natl. Compr. Cancer Netw.* **2013**, *11*, 1101–1108. [CrossRef]

85. Yang, Y.-L.; Chang, Y.-H.; Li, C.-J.; Huang, Y.-H.; Tsai, M.-C.; Chu, P.-Y.; Lin, H.-Y. New Insights into the Role of miR-29a in Hepatocellular Carcinoma: Implications in Mechanisms and Theragnostics. *J. Pers. Med.* **2021**, *11*, 219. [CrossRef]

86. Kong, G.; Zhang, J.; Zhang, S.; Shan, C.; Ye, L.; Zhang, X. Upregulated microRNA-29a by hepatitis B virus X protein enhances hepatoma cell migration by targeting PTEN in cell culture model. *PLoS ONE* **2011**, *6*, e19518. [CrossRef]

87. Yu, L.-X.; Zhang, B.-L.; Yang, Y.; Wang, M.-C.; Lei, G.-L.; Gao, Y.; Liu, H.; Xiao, C.-H.; Xu, J.-J.; Qin, H.; et al. Exosomal microRNAs as potential biomarkers for cancer cell migration and prognosis in hepatocellular carcinoma patient-derived cell models. *Oncol. Rep.* **2019**, *41*, 257–269. [CrossRef]

88. Lin, H.; Zhang, Z. Diagnostic value of a microRNA signature panel in exosomes for patients with hepatocellular carcinoma. *Int. J. Clin. Exp. Pathol.* **2019**, *12*, 1478.

89. Zhou, Y.; Wang, X.; Sun, L.; Zhou, L.; Ma, T.C.; Song, L.; Wu, J.G.; Li, J.L.; Ho, W.Z. Toll-like receptor 3-activated macrophages confer anti-HCV activity to hepatocytes through exosomes. *FASEB J.* **2016**, *30*, 4132–4140. [CrossRef]

90. Zhu, H.-T.; Hasan, A.M.E.; Liu, R.-B.; Zhang, Z.-C.; Zhang, X.; Wang, J.; Wang, H.-Y.; Wang, F.; Shao, J.-Y. Serum microRNA profiles as prognostic biomarkers for HBV-positive hepatocellular carcinoma. *Oncotarget* **2016**, *7*, 45637. [CrossRef]

91. Jampoka, K.; Muangpaisarn, P.; Khongnomnan, K.; Treeprasertsuk, S.; Tangkijvanich, P.; Payungporn, S. Serum miR-29a and miR-122 as potential biomarkers for non-alcoholic fatty liver disease (NAFLD). *MicroRNA* **2018**, *7*, 215–222. [CrossRef]

92. Zhu, H.-T.; Dong, Q.-Z.; Sheng, Y.-Y.; Wei, J.-W.; Wang, G.; Zhou, H.-J.; Ren, N.; Jia, H.-L.; Ye, Q.-H.; Qin, L.-X. MicroRNA-29a-5p is a novel predictor for early recurrence of hepatitis B virus-related hepatocellular carcinoma after surgical resection. *PLoS ONE* **2012**, *7*, e52393. [CrossRef]

93. Zhang, Y.; Yang, L.; Wang, S.; Liu, Z.; Xiu, M. MiR-29a suppresses cell proliferation by targeting SIRT1 in hepatocellular carcinoma. *Cancer Biomark.* **2018**, *22*, 151–159. [CrossRef]

94. Zhu, X.-C.; Dong, Q.-Z.; Zhang, X.-F.; Deng, B.; Jia, H.-L.; Ye, Q.-H.; Qin, L.-X.; Wu, X.-Z. microRNA-29a suppresses cell proliferation by targeting SPARC in hepatocellular carcinoma. *Int. J. Mol. Med.* **2012**, *30*, 1321–1326. [CrossRef]

95. Vila-Navarro, E.; Vila-Casadesús, M.; Moreira, L.; Duran-Sanchon, S.; Sinha, R.; Ginés, À.; Fernández-Esparrach, G.; Miquel, R.; Cuatrecasas, M.; Castells, A. MicroRNAs for detection of pancreatic neoplasia: Biomarker discovery by next-generation sequencing and validation in 2 independent cohorts. *Ann. Surg.* **2017**, *265*, 1226. [CrossRef]

96. Dobre, M.; Herlea, V.; Vlăduț, C.; Ciocîrlan, M.; Balaban, V.; Constantinescu, G.; Diculescu, M.; Milanesi, E. Dysregulation of miRNAs Targeting the IGF-1R Pathway in Pancreatic Ductal Adenocarcinoma. *Cells* **2021**, *10*, 1856. [CrossRef]

97. Ganepola, G.A.; Rutledge, J.R.; Suman, P.; Yiengpruksawan, A.; Chang, D.H. Novel blood-based microRNA biomarker panel for early diagnosis of pancreatic cancer. *World J. Gastrointest. Oncol.* **2014**, *6*, 22–33. [CrossRef]

98. Humeau, M.; Vignolle-Vidoni, A.; Sicard, F.; Martins, F.; Bournet, B.; Buscail, L.; Torrisani, J.; Cordelier, P. Salivary MicroRNA in Pancreatic Cancer Patients. *PLoS ONE* **2015**, *10*, e0130996. [CrossRef]

99. Lorber, G.; Benenson, S.; Rosenberg, S.; Gofrit, O.N.; Pode, D. A single dose of 240 mg gentamicin during transrectal prostate biopsy significantly reduces septic complications. *Urology* **2013**, *82*, 998–1003. [CrossRef]

100. Hsieh, T.-Y.; Wang, S.-C.; Kao, Y.-L.; Chen, W.-J. Adding gentamicin to fluoroquinolone-based antimicrobial prophylaxis reduces transrectal ultrasound-guided prostate biopsy-related infection rate. *Urol. Sci.* **2016**, *27*, 91–95. [CrossRef]

101. Makena, M.R.; Gatla, H.; Verlekar, D.; Sukhavasi, S.; Pandey, M.K.; Pramanik, K.C. Wnt/β-Catenin signaling: The culprit in pancreatic carcinogenesis and therapeutic resistance. *Int. J. Mol. Sci.* **2019**, *20*, 4242. [CrossRef]

102. Nagano, H.; Tomimaru, Y.; Eguchi, H.; Hama, N.; Wada, H.; Kawamoto, K.; Kobayashi, S.; Mori, M.; Doki, Y. MicroRNA-29a induces resistance to gemcitabine through the Wnt/β-catenin signaling pathway in pancreatic cancer cells. *Int. J. Oncol.* **2013**, *43*, 1066–1072. [CrossRef]

103. Sun, X.-J.; Liu, B.-Y.; Yan, S.; Jiang, T.-H.; Cheng, H.-Q.; Jiang, H.-S.; Cao, Y.; Mao, A.-W. MicroRNA-29a promotes pancreatic cancer growth by inhibiting tristetraprolin. *Cell. Physiol. Biochem.* **2015**, *37*, 707–718. [CrossRef]

104. Yang, S.; Zhang, Z.; Wang, Q. Emerging therapies for small cell lung cancer. *J. Hematol. Oncol.* **2019**, *12*, 45. [CrossRef]

105. Heegaard, N.H.H.; Schetter, A.J.; Welsh, J.A.; Yoneda, M.; Bowman, E.D.; Harris, C.C. Circulating micro-RNA expression profiles in early stage nonsmall cell lung cancer. *Int. J. Cancer* **2012**, *130*, 1378–1386. [CrossRef]

106. Yang, X.; Zhang, Q.; Zhang, M.; Su, W.; Wang, Z.; Li, Y.; Zhang, J.; Etherton-Beer, C.; Yang, S.; Chen, G. Serum microRNA signature is capable of early diagnosis for non-small cell lung cancer. *Int. J. Biol. Sci.* **2019**, *15*, 1712–1722. [CrossRef]

107. He, Q.; Fang, Y.; Lu, F.; Pan, J.; Wang, L.; Gong, W.; Fei, F.; Cui, J.; Zhong, J.; Hu, R.; et al. Analysis of differential expression profile of miRNA in peripheral blood of patients with lung cancer. *J. Clin. Lab. Anal.* **2019**, *33*, e23003. [CrossRef]

108. Liu, X.; Lv, X.; Yang, Q.; Jin, H.; Zhou, W.; Fan, Q. MicroRNA-29a functions as a tumor suppressor and increases cisplatin sensitivity by targeting NRAS in lung cancer. *Technol. Cancer Res. Treat.* **2018**, *17*, 1533033818758905. [CrossRef]

109. Fabbri, M.; Paone, A.; Calore, F.; Galli, R.; Gaudio, E.; Santhanam, R.; Lovat, F.; Fadda, P.; Mao, C.; Nuovo, G.J.; et al. MicroRNAs bind to Toll-like receptors to induce prometastatic inflammatory response. *Proc. Natl. Acad. Sci. USA* **2012**, *109*, E2110–E2116. [CrossRef]

110. Liu, L.; Bi, N.; Wu, L.; Ding, X.; Men, Y.; Zhou, W.; Li, L.; Zhang, W.; Shi, S.; Song, Y.; et al. MicroRNA-29c functions as a tumor suppressor by targeting VEGFA in lung adenocarcinoma. *Mol. Cancer* **2017**, *16*, 50. [CrossRef]

111. Fang, C.; Zhu, D.-X.; Dong, H.-J.; Zhou, Z.-J.; Wang, Y.-H.; Liu, L.; Fan, L.; Miao, K.-R.; Liu, P.; Xu, W.; et al. Serum microRNAs are promising novel biomarkers for diffuse large B cell lymphoma. *Ann. Hematol.* **2012**, *91*, 553–559. [CrossRef]

112. Wang, F.; Wang, X.-S.; Yang, G.-H.; Zhai, P.-F.; Xiao, Z.; Xia, L.-Y.; Chen, L.-R.; Wang, Y.; Bi, L.-X.; Liu, N.; et al. miR-29a and miR-142-3p downregulation and diagnostic implication in human acute myeloid leukemia. *Mol. Biol. Rep.* **2012**, *39*, 2713–2722. [CrossRef]

113. Visone, R.; Rassenti, L.Z.; Veronese, A.; Taccioli, C.; Costinean, S.; Aguda, B.D.; Volinia, S.; Ferracin, M.; Palatini, J.; Balatti, V.; et al. Karyotype-specific microRNA signature in chronic lymphocytic leukemia. *Blood* **2009**, *114*, 3872–3879. [CrossRef]

114. Zhu, C.; Wang, Y.; Kuai, W.; Sun, X.; Chen, H.; Hong, Z. Prognostic value of miR-29a expression in pediatric acute myeloid leukemia. *Clin. Biochem.* **2013**, *46*, 49–53. [CrossRef]

115. Yamada, Y.; Sugawara, S.; Arai, T.; Kojima, S.; Kato, M.; Okato, A.; Yamazaki, K.; Naya, Y.; Ichikawa, T.; Seki, N. Molecular pathogenesis of renal cell carcinoma: Impact of the anti-tumor miR-29 family on gene regulation. *Int. J. Urol.* **2018**, *25*, 953–965. [CrossRef]
116. Xu, Y.; Zhu, J.; Lei, Z.; Wan, L.; Zhu, X.; Ye, F.; Tong, Y. Expression and functional role of miR-29b in renal cell carcinoma. *Int. J. Clin. Exp. Pathol.* **2015**, *8*, 14161.
117. Wu, Z.; Huang, X.; Huang, X.; Zou, Q.; Guo, Y. The inhibitory role of Mir-29 in growth of breast cancer cells. *J. Exp. Clin. Cancer Res.* **2013**, *32*, 98. [CrossRef]
118. Sung, H.; Rosenberg, P.S.; Chen, W.-Q.; Hartman, M.; Lim, W.-Y.; Chia, K.S.; Mang, O.W.-K.; Chiang, C.-J.; Kang, D.; Ngan, R.K.-C.; et al. Female breast cancer incidence among asian and western populations: More similar than expected. *J. Natl. Cancer Inst.* **2015**, *107*, djv107. [CrossRef]
119. Jang, G.B.; Kim, J.Y.; Cho, S.D.; Park, K.S.; Jung, J.Y.; Lee, H.Y.; Hong, I.S.; Nam, J.S. Blockade of Wnt/β-catenin signaling suppresses breast cancer metastasis by inhibiting CSC-like phenotype. *Sci. Rep.* **2015**, *5*, srep12465. [CrossRef]
120. Rostas, J.W.; Pruitt, H.C.; Metge, B.J.; Mitra, A.; Bailey, S.K.; Bae, S.; Singh, K.P.; Devine, D.J.; Dyess, D.L.; Richards, W.O.; et al. microRNA-29 negatively regulates EMT regulator N-myc interactor in breast cancer. *Mol. Cancer* **2014**, *13*, 200. [CrossRef]
121. Shinden, Y.; Iguchi, T.; Akiyoshi, S.; Ueo, H.; Ueda, M.; Hirata, H.; Sakimura, S.; Uchi, R.; Takano, Y.; Eguchi, H.; et al. miR-29b is an indicator of prognosis in breast cancer patients. *Mol. Clin. Oncol.* **2015**, *3*, 919–923. [CrossRef] [PubMed]
122. Muluhngwi, P.; Alizadeh-Rad, N.; Vittitow, S.L.; Kalbfleisch, T.S.; Klinge, C.M. The miR-29 transcriptome in endocrine-sensitive and resistant breast cancer cells. *Sci. Rep.* **2017**, *7*, 5205. [CrossRef]
123. Milevskiy, M.J.; Sandhu, G.K.; Wronski, A.; Korbie, D.; Brewster, B.L.; Shewan, A.; Edwards, S.L.; French, J.D.; Brown, M.A. MiR-29b-1-5p is altered in BRCA1 mutant tumours and is a biomarker in basal-like breast cancer. *Oncotarget* **2018**, *9*, 33577. [CrossRef]
124. Chou, J.; Lin, J.H.; Brenot, A.; Kim, J.-W.; Provot, S.; Werb, Z. GATA3 suppresses metastasis and modulates the tumour microenvironment by regulating microRNA-29b expression. *Nat. Cell Biol.* **2013**, *15*, 201–213. [CrossRef]
125. Zhao, Y.; Yang, F.; Li, W.; Xu, C.; Li, L.; Chen, L.; Liu, Y.; Sun, P. miR-29a suppresses MCF-7 cell growth by downregulating tumor necrosis factor receptor 1. *Tumor Biol.* **2017**, *39*, 1010428317692264. [CrossRef]
126. Li, H.; Luo, J.; Xu, B.; Luo, K.; Hou, J. MicroRNA-29a inhibits cell migration and invasion by targeting Roundabout 1 in breast cancer cells. *Mol. Med. Rep.* **2015**, *12*, 3121–3126. [CrossRef]
127. Shen, H.; Li, L.; Yang, S.; Wang, D.; Zhong, S.; Zhao, J.; Tang, J. MicroRNA-29a contributes to drug-resistance of breast cancer cells to adriamycin through PTEN/AKT/GSK3β signaling pathway. *Gene* **2016**, *593*, 84–90. [CrossRef]
128. Cochrane, D.R.; Jacobsen, B.; Connaghan, K.D.; Howe, E.N.; Bain, D.L.; Richer, J.K. Progestin regulated miRNAs that mediate progesterone receptor action in breast cancer. *Mol. Cell. Endocrinol.* **2012**, *355*, 15–24. [CrossRef]
129. McDermott, A.M.; Miller, N.; Wall, D.; Martyn, L.M.; Ball, G.; Sweeney, K.J.; Kerin, M. Identification and Validation of Oncologic miRNA Biomarkers for Luminal A-like Breast Cancer. *PLoS ONE* **2014**, *9*, e87032. [CrossRef]
130. Wu, Q.; Lu, Z.; Li, H.; Lu, J.; Guo, L.; Ge, Q. Next-generation sequencing of microRNAs for breast cancer detection. *J. Biomed. Biotechnol.* **2011**, *2011*, 597145. [CrossRef]
131. Raeisi, F.; Mahmoudi, E.; Dehghani-Samani, M.; Hosseini, S.S.E.; Ghahfarrokhi, A.M.; Arshi, A.; Forghanparast, K.; Ghazanfari, S. Differential expression profile of miR-27b, miR-29a, and miR-155 in chronic lymphocytic leukemia and breast cancer patients. *Mol. Ther. -Oncolytics* **2020**, *16*, 230–237. [CrossRef] [PubMed]

Article

MicroRNAs Differentially Expressed in Actinic Keratosis and Healthy Skin Scrapings

Maria Vincenza Chiantore [1,*], Marco Iuliano [2], Roberta Maria Mongiovì [2,†], Fabiola Luzi [3], Giorgio Mangino [2], Lorenzo Grimaldi [2], Luisa Accardi [1], Gianna Fiorucci [1,4,‡], Giovanna Romeo [2] and Paola Di Bonito [1]

1   EVOR Unit, Department of Infectious Diseases, Istituto Superiore di Sanità, 00161 Rome, Italy; luisa.accardi@iss.it (L.A.); gianna.fiorucci@cnr.it (G.F.); paola.dibonito@iss.it (P.D.B.)
2   Department of Medico-Surgical Sciences and Biotechnologies, Sapienza University of Rome—Polo Pontino, 04100 Latina, Italy; marco.iuliano@uniroma1.it (M.I.); robertamaria.mongiovi@gmail.com (R.M.M.); giorgio.mangino@uniroma1.it (G.M.); lorenzo.grimaldi@uniroma1.it (L.G.); giovanna.romeo@uniroma1.it (G.R.)
3   Plastic and Reconstructive Surgery, San Gallicano Dermatologic Institute IRCCS, 00144 Rome, Italy; fabiolaluzi@yahoo.it
4   Institute of Molecular Biology and Pathology, CNR, 00185 Rome, Italy
*   Correspondence: mariavincenza.chiantore@iss.it
†   Current address: Department of Biomedical, Dental, Morphological and Functional Imaging Sciences, University of Messina, 98100 Messina, Italy.
‡   Retired.

Abstract: Actinic keratosis (AK) is a carcinoma in situ precursor of cutaneous squamous cell carcinoma (cSCC), the second most common cancer affecting the Caucasian population. AK is frequently present in the sun-exposed skin of the elderly population, UV radiation being the main cause of this cancer, and other risk factors contributing to AK incidence. The dysregulation of microRNAs (miRNAs) observed in different cancers leads to an improper expression of miRNA targets involved in several cellular pathways. The TaqMan Array Human MicroRNA Card assay for miRNA expression profiling was performed in pooled AK compared to healthy skin scraping samples from the same patients. Forty-three miRNAs were modulated in the AK samples. The expression of miR-19b ($p < 0.05$), -31, -34a ($p < 0.001$), -126, -146a ($p < 0.01$), -193b, and -222 ($p < 0.05$) was validated by RT-qPCR. The MirPath tool was used for MiRNA target prediction and enriched pathways. The top DIANA-mirPath pathways regulated by the targets of the 43 miRNAs are TGF-beta signaling, Proteoglycans in cancer, Pathways in cancer, and Adherens junction ($7.30 \times 10^{-10} < p < 1.84 \times 10^{-8}$). Selected genes regulating the KEGG pathways, i.e., *TP53*, *MDM2*, *CDKN1A*, *CDK6*, and *CCND1*, were analyzed. MiRNAs modulated in AK regulate different pathways involved in tumorigenesis, indicating miRNA regulation as a critical step in keratinocyte cancer.

Keywords: actinic keratosis; cSCC; microRNA; biomarkers

Citation: Chiantore, M.V.; Iuliano, M.; Mongiovì, R.M.; Luzi, F.; Mangino, G.; Grimaldi, L.; Accardi, L.; Fiorucci, G.; Romeo, G.; Di Bonito, P. MicroRNAs Differentially Expressed in Actinic Keratosis and Healthy Skin Scrapings. *Biomedicines* **2023**, *11*, 1719. https://doi.org/10.3390/biomedicines11061719

Academic Editors: Milena Rizzo and Elena Levantini

Received: 9 May 2023
Revised: 7 June 2023
Accepted: 13 June 2023
Published: 15 June 2023

1. Introduction

Actinic keratosis (AK) is currently recognized as a carcinoma in situ precursor of cutaneous squamous cell carcinoma (cSCC) [1], the second most common keratinocyte cancer affecting the White population worldwide, and its incidence has increased with the population aging [1–3]. Several risk factors influence this neoplasia of keratinocytes, such as genetic factors, family history for cSCC, immune system status, prolonged sun exposure, artificial UV tanning, male sex, fair skin, radiotherapy, and exposure to specific chemicals [4]. Recently, specific beta human papillomaviruses (HPV) have emerged as cofactors in the development of AK and cSCC [5], but cumulative UV exposure is the main risk factor for cSCC. In addition, immunosuppression (e.g., transplant recipients and patients with hematologic cancers such as chronic lymphocytic leukemia) is associated with a higher incidence and more aggressive course of cSCC. AK frequently occurs in older individuals; it can spontaneously regress or evolve into cSCC [1]. The epidemiology of AKs reflects their causation by cumulative sun exposure, with the highest prevalence seen in pale-skinned people living in low latitudes and in the most sun-exposed body sites, namely the hands, forearms, and face.

In recent years, significant progress has been made in the detection of specific biomarkers for accurate cancer diagnosis and prognosis in cancer, and to guide personalized therapy by estimating outcome risks. In this context, microRNAs (miRNAs) have been shown to be attractive candidates as biomarkers in cancer pathophysiology and diagnosis [6]. The role of miRNAs in controlling the activity of cutaneous stem cells and in driving skin development and regeneration has been demonstrated. Some miRNAs are involved in controlling the expression of key genes of stem cell activity in healthy skin and hair follicles; therefore, the research of cSCC biomarkers has focused on these miRNAs [7]. The majority of cSCC arise from AK: it has been reported that certain miRNAs are differentially expressed in AK and in cSCC [7–9] and that this difference could be correlated with the progression from AK to invasive cancer [10].

In this study, the miRNA profiles of pooled AK scraping samples were compared to those obtained with pooled HS skin scraping samples from the same patients. Among these miRNAs, the expression of miR-19b, -31, -34a, -126, -146a, -193b, and -222 was validated. The analysis of the pathways regulated by miRNA target genes was performed by the DIANA tool mirPath v.3 with all the miRNAs modulated in AK samples. In addition, the expression of selected miRNA target genes, i.e., *TP53*, *MDM2*, *CDKN1A*, *CDK6*, and *CCND1*, was analyzed.

2. Materials and Methods

2.1. Clinical Samples Collection

The collection of clinical samples for this study was carried out in conjunction with our previously published studies, approved by the Ethical Committees of NIHMP (2014) and the San Gallicano Dermatological Institute (CE943/17; RS/1090/18) [11,12]. The investigations were carried out following the rules of the Declaration of Helsinki of 1975 (https://www.wma.net/what-we-do/medical-ethics/declaration-of-helsinki/, accessed on 14 January 2019), revised in 2013. Samples were from 20 patients (median age of 72) with a single lesion eligible for laser surgery. The AK samples were collected by scraping the lesions with a sterile spatula without reaching the dermis. The HS samples were collected with a second sterile spatula from the glabellar area, a source of thick skin, of each patient. Both AK and HS samples were immediately stored at $-80\,^{\circ}$C until processing.

2.2. MicroRNA Extraction and TaqMan Array Human MicroRNA A Card Analysis

Scraping skin samples were pooled (two different pools of 10 samples for each experimental group) and lysed by the mirVana miRNA detection kit (Applied Biosystems, Waltham, MA, USA), following the manufacturer's procedure. The extracted RNA was retro-transcribed by using the TaqMan Micro-RNA Reverse Transcription Kit, and the Megaplex RT Primers and TaqMan Array Human MicroRNA A Card (Applied Biosystems, Waltham, MA, USA) was used to analyze the expression of multiple miRNA sequences in AK compared to HS samples. The obtained data were analyzed by using qPCR on Thermo Fisher Cloud (Thermo Fisher Scientific, Waltham, MA, USA) by setting global normalization, 35 as the maximum threshold cycle (Ct) and HS as reference.

2.3. Real Time RT-PCR

The results obtained by miRNA profiling were individually validated using specific TaqMan Small RNA assays following the manufacturer's instructions (Applied Biosystems, Waltham, MA, USA). Data were normalized using U6 as endogenous control and expressed using the $2^{-\Delta\Delta CT}$ method [13].

A total of 1 µg of extracted RNA of the AK and HS samples was retro-transcribed using the High-Capacity cDNA Reverse Transcription Kit (Applied Biosystem, Waltham, MA, USA), and cDNA products were analyzed by real-time RT-PCR using the SensiMix SYBR Hi-ROX Kit (Meridian Bioscience, Cincinnati, OH, USA). Data were normalized using as endogenous control HPRT-1 and expressed using the $2-\Delta\Delta CT$ method. Primers used are reported (Table 1).

Table 1. Primers used in Real Time RT-PCR.

Gene	FWD 5′–3′	REV 5′–3′	Ref.
TP53	CCTCAGCATCTTATCCGAGTGG	TGGATGGTGGTACAGTCAGAGC	[14]
MDM2	TGGGCAGCTTGAAGCAGTTG	CAGGCTGCCATGTGACCTAAGA	[15]
CDKN1A	GGCAGACCAGCATGACAGATT	GCGGATTAGGGCTTCCTCTT	[16]
CDK6	GCTGACCAGCAGTACGAATG	GCACACATCAAACAACCTGACC	[17]
CCND1	CCGTCCATGCGGAAGATC	GAAGACCTCCTCCTCGCACT	[18]

2.4. Principal Component Analysis (PCA)

PCA was performed using the ClustVis website and the "PCA method" of the R package (https://biit.cs.ut.ee/clustvis/#editions, accessed on 9 February 2023) [19]. Shortly, the analysis was conducted on ΔCt of all the miRNAs that were eligible for selection from the TaqMan Array Human MicroRNA A Cards analysis of AK samples. To allow for a comparison, the ΔCt from the analysis on miRNAs of HS samples were used.

2.5. MiRNA Target Prediction and Enriched Pathways

The mirPath tool (version 3.0) was used to predict target genes of the differentially regulated miRNAs using the microT-CDS v.5.0 database and to retrieve KEGG (i.e., Kyoto Encyclopedia of Genes and Genomes) molecular pathways, considering p values lower than 0.05 as significant for pathway enrichment.

2.6. Statistical Analysis

Two-tailed p value Mann-Whitney t test was performed to compare two sample groups. *: $p < 0.05$; **: $p < 0.01$; ***: $p < 0.001$. The presented data are mean $\pm$ standard deviation (SD).

3. Results

3.1. MicroRNA Profiling of AK Samples and Matched Healthy Skin

The skin scrapings were collected from patients with one AK lesion at an early stage. As only a small amount of RNA could be extracted from these samples, miRNA profiling was performed with a pool of scraping extracts. Real-time PCR array analysis of 384 miRNAs, mainly expressed in the human genome, was performed with two different pools of 10 samples collected from AK and HS, as described in Materials and Methods.

The principal component analysis (PCA) of the miRNA profiles was performed using the ClustVis website on ΔCt of all the miRNAs that were eligible for selection (Figure 1a). The miRNA dataset for each pool of samples was analyzed, and its variability was reduced and represented as a single dot on a first principal component PC1 vs. PC2 scatterplot. The PCA shows that the miRNA profiles of the two AK sample pools are different from the profiles of the two HS sample pools from the same donors. The PC1, which describes the majority of the variance, indicated that the miRNA datasets from the two AK pools were similar to each other but different from the miRNA datasets obtained from the HS pools. This suggested that the number of miRNAs with an altered expression in AK compared to HS was sufficient to allow for discrimination in PCA.

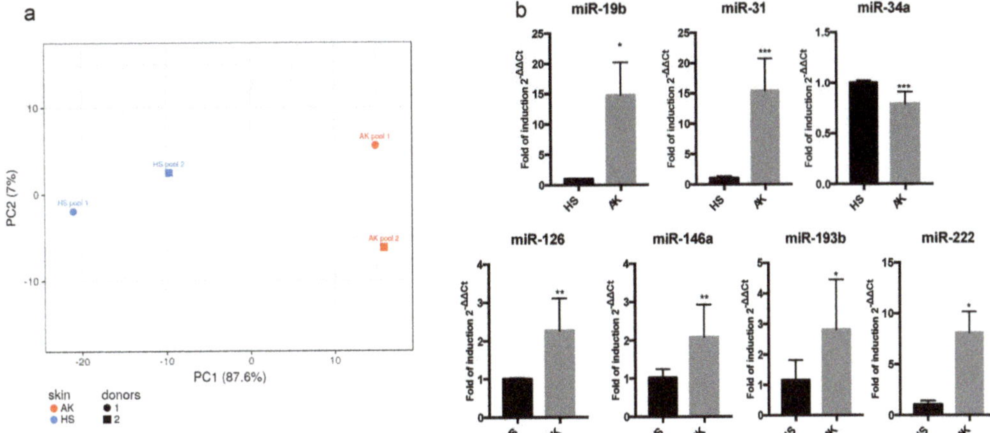

Figure 1. (**a**) Principal component analysis performed on ΔCts derived from the analysis of two couples of different pools of skin scrapings from AK and HS. Each dot represents the variance of the miRNA dataset for each sample type. PCA was performed by using ClustVis (https://biit.cs.ut.ee/clustvis/, accessed on 9 February 2023) algorithm. (**b**) MiRNA expression analysis performed in AK compared to HS. Total RNA from each sample was purified as described in Materials and Methods. Results were expressed as fold of induction or decrease using the $2^{-\Delta\Delta CT}$ method [13] using RNU6 as a calibrator and HS as a control. The reported values represent the mean of three independent experiments ± the SD. *p*-values (*p*) are indicated with asterisks. *** $p < 0.001$; ** $p < 0.01$, * $p < 0.05$.

The miRNA profiles of AK and HS pools were analyzed using the Thermofisher Cloud to determine the differentially expressed miRNAs. Table 2 shows the complete list of miRNAs dysregulated in AK compared to the HS samples and the relative quantification.

Table 2. List of dysregulated miRNAs in AK compared to HS, detected by TaqMan Array Human MicroRNA A Card analysis.

	MicroRNAs	Accession Number	Fold Change
1	hsa-miR-16-5p	MIMAT0000069	238.911
2	hsa-miR-150-5p	MIMAT0000451	177.837
3	hsa-miR-494-3p	MIMAT0002816	177.398
4	hsa-miR-31-5p	MIMAT0000089	177.057
5	hsa-miR-203a-5p	MIMAT0031890	160.851
6	hsa-miR-191-5p	MIMAT0000440	130.259
7	hsa-let-7e-5p	MIMAT0000066	110.853
8	hsa-miR-222-3p	MIMAT0000279	99.651
9	hsa-miR-146a-5p	MIMAT0000449	85.494
10	hsa-miR-19b-3p	MIMAT0000074	75.58
11	hsa-miR-200c-3p	MIMAT0000617	58.549
12	hsa-miR-320a	MIMAT0000510	56.054
13	hsa-let-7b-5p	MIMAT0000063	51.857
14	hsa-miR-24-3p	MIMAT0000080	48.505
15	hsa-miR-374-5p	MIMAT0000727	48.309
16	hsa-miR-126-3p	MIMAT0000445	47.816
17	hsa-miR-422a	MIMAT0001339	44.507
18	hsa-miR-146b-3p	MIMAT0004766	43.107
19	hsa-miR-486-5p	MIMAT0002177	40.989
20	hsa-miR-193b-3p	MIMAT0002819	40.612

Table 2. *Cont.*

	MicroRNAs	Accession Number	Fold Change
21	hsa-miR-454-3p	MIMAT0003885	34.13
22	hsa-miR-484	MIMAT0002174	24.427
23	hsa-miR-186-5p	MIMAT0000456	23.968
24	hsa-miR-504-5p	MIMAT0002875	20.504
25	hsa-miR-342-3p	MIMAT0000753	19.67
26	hsa-miR-487a-3p	MIMAT0002178	16.643
27	hsa-miR-195-5p	MIMAT0000461	16.126
28	hsa-miR-518f-3p	MIMAT0002842	15.495
29	hsa-miR-200a-3p	MIMAT0000682	15.161
30	hsa-miR-636	MIMAT0003306	13.205
31	hsa-miR-618	MIMAT0003287	12.183
32	hsa-miR-218-5p	MIMAT0000275	11.297
33	hsa-miR-302c-3p	MIMAT0000717	7.095
34	hsa-miR-214-3p	MIMAT0000271	3.675
35	hsa-miR-145-5p	MIMAT0000437	3.417
36	hsa-miR-519e-3p	MIMAT0002829	1.342
37	hsa-miR-182-5p	MIMAT0000259	1.296
38	hsa-miR-211-5p	MIMAT0000268	1.143
39	hsa-miR-518b	MIMAT0002844	1.023
40	hsa-miR-208b-3p	MIMAT0004960	0.272
41	hsa-miR-34a-5p	MIMAT0000255	0.173
42	hsa-miR-127-5p	MIMAT0004604	0.108
43	hsa-miR-370-3p	MIMAT0000722	0.083

Fold change: $2^{-\Delta\Delta CT}$.

Among the miRNAs differentially expressed in the AK samples, we selected seven miRNAs, i.e., miR-19b, -31, -34a, -126, -146a, -193b, and -222, based on their higher modulation with respect to the HS samples. Their altered expression was validated by specific TaqMan Small RNA assays. This analysis confirmed the results obtained by TaqMan Array Human MicroRNA A Card, showing that all these miRNAs were significantly upregulated, except for miR-34a, that was down-regulated in AK compared to HS (Figure 1b).

3.2. Pathways Regulated by Target Genes of miRNAs Modulated in AK Samples

In order to predict miRNA target and enriched pathways, the analysis with the DIANA tool mirPath v.3 was performed with all the 43 AK dysregulated miRNAs detected by the array. A total of 66 pathways were found to be regulated by target genes of these miRNAs, as reported in Table 3. Pathways are listed according to increasing *p*-value.

Table 3. MiRNA pathway analysis by DIANA mirPath v.3 on KEGG pathways linked to the 43 selected miRNAs.

	KEGG Pathway	*p*-Value	# Genes *	# miRNAs **
1.	TGF-beta signaling pathway (hsa04350)	7.30×10^{-10}	62	34
2.	Proteoglycans in cancer (hsa05205)	7.30×10^{-10}	136	36
3.	Pathways in cancer (hsa05200)	3.36×10^{-9}	251	37
4.	Adherens junction (hsa04520)	1.84×10^{-8}	58	35
5.	Axon guidance (hsa04360)	1.84×10^{-8}	91	37
6.	Hippo signaling pathway (hsa04390)	2.70×10^{-7}	99	37
7.	Fatty acid biosynthesis (hsa00061)	1.62×10^{-6}	7	12
8.	Ras signaling pathway (hsa04014)	4.68×10^{-6}	142	36
9.	ErbB signaling pathway (hsa04012)	1.32×10^{-5}	61	35
10.	Glioma (hsa05214)	1.65×10^{-5}	46	34

Table 3. *Cont.*

	KEGG Pathway	*p*-Value	# Genes *	# miRNAs **
11.	Wnt signaling pathway (hsa04310)	1.78×10^{-5}	95	35
12.	Melanoma (hsa05218)	3.69×10^{-5}	53	32
13.	Signaling pathways regulating pluripotency of stem cells (hsa04550)	3.69×10^{-5}	94	37
14.	Rap1 signaling pathway (hsa04015)	5.55×10^{-5}	134	35
15.	GABAergic synapse (hsa04727)	0.0001604280	54	35
16.	Thyroid hormone signaling pathway (hsa04919)	0.0001726085	77	36
17.	Estrogen signaling pathway (hsa04915)	0.0002032496	62	34
18.	MAPK signaling pathway (hsa04010)	0.0002242160	160	37
19.	Renal cell carcinoma (hsa05211)	0.0002295720	48	33
20.	Glycosaminoglycan biosynthesis—heparan sulfate/heparin (hsa00534)	0.0005046348	18	21
21.	Choline metabolism in cancer (hsa05231)	0.0005046348	69	34
22.	Neurotrophin signaling pathway (hsa04722)	0.0005046348	81	37
23.	Focal adhesion (hsa04510)	0.0007711823	129	36
24.	Glycosphingolipid biosynthesis—lacto and neolacto series (hsa00601)	0.0009080097	17	19
25.	FoxO signaling pathway (hsa04068)	0.0009080097	85	35
26.	Prostate cancer (hsa05215)	0.0010195428	60	33
27.	Glutamatergic synapse (hsa04724)	0.0020494968	71	34
28.	PI3K-Akt signaling pathway (hsa04151)	0.0020578098	199	37
29.	Regulation of actin cytoskeleton (hsa04810)	0.0032897905	131	35
30.	Amphetamine addiction (hsa05031)	0.0033806984	43	28
31.	Thyroid hormone synthesis (hsa04918)	0.0040209555	44	31
32.	Colorectal cancer (hsa05210)	0.0040209555	42	33
33.	Oocyte meiosis (hsa04114)	0.0040209555	72	35
34.	Prolactin signaling pathway (hsa04917)	0.0043891560	47	33
35.	Insulin secretion (hsa04911)	0.0052671844	56	35
36.	cAMP signaling pathway (hsa04024)	0.0059360368	120	35
37.	Chronic myeloid leukemia (hsa05220)	0.0060741359	48	35
38.	Pancreatic cancer (hsa05212)	0.0066389106	45	32
39.	Thyroid cancer (hsa05216)	0.0068643563	22	28
40.	Endometrial cancer (hsa05213)	0.0078889298	35	31
41.	Phosphatidylinositol signaling system (hsa04070)	0.0078889298	52	34
42.	Adrenergic signaling in cardiomyocytes (hsa04261)	0.0078889298	89	34
43.	Vasopressin-regulated water reabsorption (hsa04962)	0.0079518053	31	26
44.	Oxytocin signaling pathway (hsa04921)	0.0081386021	99	35
45.	Non-small-cell lung cancer (hsa05223)	0.0082341417	37	32
46.	cGMP-PKG signaling pathway (hsa04022)	0.0082341417	100	36
47.	Transcriptional misregulation in cancer (hsa05202)	0.0083247611	104	38
48.	Long-term potentiation (hsa04720)	0.0097964993	45	32
49.	mTOR signaling pathway (hsa04150)	0.0104774872	42	32
50.	Endocytosis (hsa04144)	0.0105456910	121	36
51.	Long-term depression (hsa04730)	0.011338320	39	32
52.	Viral carcinogenesis (hsa05203)	0.011338320	102	36
53.	Vascular smooth muscle contraction (hsa04270)	0.011568669	69	34
54.	Gap junction (hsa04540)	0.015258034	51	35
55.	Ubiquitin mediated proteolysis (hsa04120)	0.021158638	81	37
56.	Hepatitis B (hsa05161)	0.021158638	84	37
57.	Biotin metabolism (hsa00780)	0.022422627	2	4
58.	Morphine addiction (hsa05032)	0.027484035	53	36
59.	Inflammatory mediator regulation of TRP channels (hsa04750)	0.028782594	60	31
60.	N-Glycan biosynthesis (hsa00510)	0.032645322	27	24
61.	Nicotine addiction (hsa05033)	0.034469917	26	28
62.	Hedgehog signaling pathway (hsa04340)	0.035921144	34	24
63.	Gastric acid secretion (hsa04971)	0.036593705	48	34
64.	Melanogenesis (hsa04916)	0.038811067	63	33
65.	Tight junction (hsa04530)	0.038811067	81	37
66.	Bacterial invasion of epithelial cells (hsa05100)	0.043438632	47	32

represents the number of genes affected * and of miRNAs involved ** for each pathway.

The first two pathways regulated by target genes of the selected miRNAs are TGF-beta signaling and Proteoglycans in cancer, followed by Pathways in cancer and Adherens junction. Interestingly, among the regulated pathways there are also the Signaling pathways regulating pluripotency of stem cells and cAMP signaling pathway that, together with the TGF-beta signaling pathway, have been observed to be regulated by targets of miRNAs detected in peripheral blood samples of AK and cSCC patients by Dańczak-Pazdrowska et al. [10]. On the other hand, Proteoglycans in cancer is the second most regulated pathway by cSCC miRNAs according to this article and by AK miRNAs according to our work, and other pathways have been commonly found, such as Pathways in cancer, Fatty acid biosynthesis, MAPK signaling pathway, and mTOR signaling pathway.

Given the elevated number of the AK modulated miRNAs involved in all the KEGG pathways, the significant overlap among miRNAs was confirmed by the Venn diagram, which showed that 31 miRNAs are commonly present in the four top regulated pathways (Figure 2a and Supplementary Table S1). Therefore, a Venn diagram was constructed to analyze the overlaps among the genes regulating TGF-beta signaling, Proteoglycans in cancer, Pathways in cancer, and Adherens junction. There is one high overlap of 60 genes among the miRNA targets controlling Proteoglycans in cancer and Pathways in cancer, but only 3 genes commonly regulate the top four pathways (Figure 2b and Supplementary Table S2).

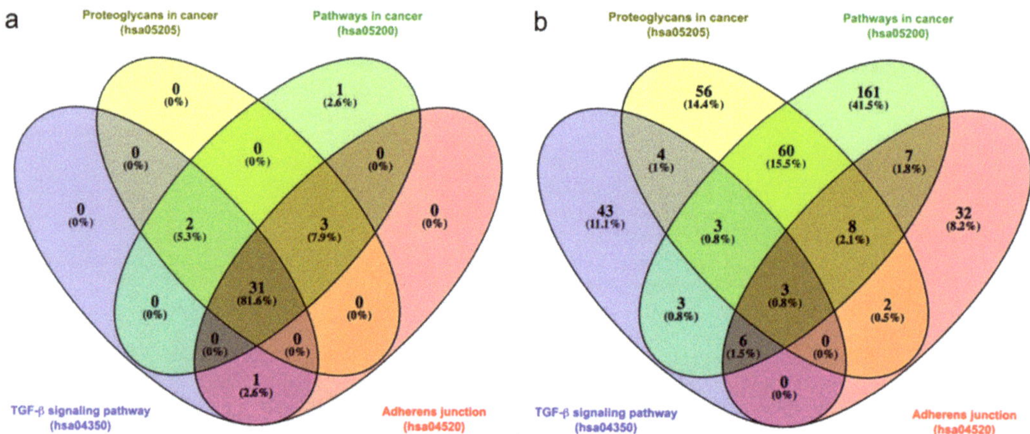

Figure 2. (**a**) Venn diagram representing the overlaps among the miRNAs regulating the first four KEGG pathways from Table 3. (**b**) Venn diagram representing the overlaps among all the target genes regulating the first four KEGG pathways from Table 3. Diagrams drawn using Venny 2.1 Internet site (https://bioinfogp.cnb.csic.es/tools/venny/, accessed on 23 May 2023). The list of the miRNAs and genes is reported in Supplementary Tables S1 and S2, respectively.

The expression of some targets controlling Proteoglycans in cancer and Pathways in cancer, key factors in apoptosis and cell cycle, was analyzed by realtime RT-PCR: *TP53*, *MDM2*, *CDKN1A*, *CDK6*, and *CCND1*. Interestingly, p53 is a regulator of miR34a, whereas *MDM2* and *CDKN1A* are common targets of miR-19b, -146a, and -193b. *CDK6* is shared by miR-19b, -34a, -186, -193b, and -214, and *CCND1* is a target of miR-19b, -31, -34a, and -193b. Except for MDM2, we observed the modulation of the mRNAs of these genes in AK compared to HS samples (Table 4).

Table 4. Gene expression analysis performed in AK compared to HS. Total RNA from each sample was purified as described in Materials and Methods. Results were expressed as fold of induction using the $2^{-\Delta\Delta CT}$ method using HPRT-1 as calibrator and HS as control.

Gene	Fold Change	SD	*p* Value
TP53	39.946	3.603	<0.001
MDM2	0.898	0.234	0.224
CDKN1A	161.751	0.518	<0.001
CDK6	3.371	1.417	<0.001
CCND1	204.732	33.863	<0.001

The reported values represent the mean of three independent experiments ± the SD. Fold change: $2^{-\Delta\Delta CT}$.

4. Discussion

A MiRNA profile of skin scraping samples was performed in RNA extracts obtained from pools of AK lesions and healthy-looking skin of the same individual. The results showed that 43 miRNAs are dysregulated in AK compared to HS samples, used as a control. The expression of seven miRNAs reported to play a role in skin cancer, namely miR-19b, -31, -34a, -126, -146a, -193b, and -222, was further analyzed by single qRT-PCR performed in the same AK and HS sample pools. Interestingly, a role in skin cancer has been reported for these miRNAs. MiR-19b belongs to the miR-17-92 cluster, known as oncomir since it is differentially expressed in different types of tumors, including cSCC [20]. MiR-31 and -222 are upregulated, whereas miR-34a is downregulated in cSCC [8,9,21]. MiR-19 and -126 seem to be early-stage specific markers of AK, while miR-193b is modulated throughout the malignant evolution of AK to cSCC [21]. MiR-146a is a modulator of inflammatory immune responses, and it has been reported to play a role in the development of both basal cell carcinoma and squamous cell carcinoma [22].

Among the other miRNAs differentially expressed in AK samples, miR-186, -203, and -214 also have been reported to be dysregulated in cSCC [23–25].

Our analysis with the DIANA tool mirPath v.3, performed with all the 43 miRNAs differentially expressed between the AK and HS samples, highlighted 66 pathways, most of which are regulated by targets of more than 30 miRNAs. Seventeen pathways are directly involved in cancer, because they are fundamental for the specific cancer progression or the principal pathways dysregulated in many cancers (i.e., Ras signaling, PI3K-Akt signaling, cAMP signaling), indicating their role in tumor development and confirming that AK could be a precursor of cSCC. In particular, the first two top regulated pathways are TGF-β signaling and Proteoglycans in cancer, followed by Pathways in cancer and Adherens junction. As expected, the miRNAs Venn diagram on these four pathways shows significant overlaps among miRNAs (31 common miRNAs), whereas the target genes Venn diagram highlights a significant overlap among the genes involved in the first two top pathways (60 genes) but only 3 genes commonly regulating all four pathways, i.e., SMAD2, RHOA, and MAPK1. The difference between the two Venn diagrams is due to the elevated number of target genes regulated by each miRNA.

TGF-β signaling pathway has a dual function, since its activation in healthy cells and in early-stage cancer induces cell-cycle arrest, whereas in late-stage cancer it can promote metastasis and chemoresistance. In this pathway, either SMAD or non-SMAD signaling can be activated. When activated, SMAD2 and SMAD3 are phosphorylated and form heterodimeric and trimeric complexes with SMAD4 that regulate the expression of target genes of the TGF-β signaling pathway [26]. As shown by the DIANA tool mirPath v.3 analysis, SMAD2 is a target of miR-186, SMAD3 of miR-214, and SMAD4 of both miR-34a and -146a. As reported by Cottonham et al. [27], TIAM1 (T lymphoma and metastasis gene 1) is a protein involved in the TGF-β signaling pathway, and it is a target of miR-31. Suppression of this protein leads to increased migration and invasion of a colon rectal cancer cell line, but its role in several types of cancer may be different. TIAM1 inhibits tumorigenesis in a Ras-induced skin cancer model, and miR-31 may be a negative regulator of metastasis development in breast cancer.

Our analysis highlighted that miR-19b, -146a, -186, -193b, and -203 share the target ErbB4, a member of the ErbB receptor family, also known as the EGF receptor family, which is involved in human cancer. ErbB receptors bind to many signaling proteins, inducing the activation of different signaling pathways, thus regulating several critical cellular processes, such as cell proliferation, differentiation, survival, metabolism, and migration. ErbB4 has both oncogenic and tumor suppressor functions even if in tumors it has been more frequently observed downregulated rather than upregulated [28]. ErbB4 is involved in the chondroitin sulfate (CS) and dermatan sulfate (DS) proteoglycan pathway. Both CS and DS regulate critical cellular processes, such as proliferation, apoptosis, migration, adhesion, and invasion. The CS/DS side chains of chondroitin sulfate proteoglycans take part in different interactions within the extracellular matrix and, therefore, have a key role in the regulation of proliferation, apoptosis, migration, adhesion, and invasion [29]. High levels of melanoma-associated chondroitin sulfate proteoglycans have been reported in melanoma, resulting in increased integrin function, activation of Erk1/2, cell growth, and motility [30]. ErbB2 is a well-established oncogene that is involved in different pathways in cancer [28], and the DIANA tool mirPath v.3 analysis showed that ErbB2 is a common target of miR-34a and -214, whereas miR-193b and -222 share the target PTEN, a known potent tumor suppressor, frequently mutated in human cancer [31].

During UV-induced DNA damage, the regulation of factors involved in apoptosis, cell growth, and cell cycle is pivotal to determine the fate of cells toward tumorigenesis. The DIANA tool mirPath v.3 analysis shows that in Pathways in cancer converge different pathways, among which are p53-signaling pathway and Cell cycle, regulated by common genes, such as *p53*, *MDM2*, *CDKN1A*, *CDK6*, and *CCND1*. Interestingly, these factors are targets of a group of miRNAs modulated in the AK samples. *MDM2* and *CDKN1A* are common targets of miR-19b, -146a, and -193b. *CDK6* is shared by miR-19b, -34a, -186, -193b, and -214. *CCND1* is a target of miR-19b, -31, -34a, and -193b. The tumor suppressor p53 is a regulator of miR34a that, in turn, regulates p53 through its target *SIRT1* [32]. Except for *MDM2*, the modulation of the mRNAs of p53, *CDKN1A*, *CDK6*, and *CCND1* we observed in the AK compared to HS samples.

It has been demonstrated that miR-34a regulates the activity of p53 in cells that undergo palmitate-induced lipoapoptosis [33]. After exposure to saturated fatty acids, such as palmitate acid, the expression of the transcription factor FOXO3 is upregulated, leading to an increment in miR-34a levels. This miRNA downregulates SIRT1, an enzyme that inactivates p53 via deacetylation, and cMET and KLF4, two anti-apoptotic factors. Consequently, the higher levels of active p53 could lead to apoptosis, thus blocking cancer progression [33]. Palmitate-induced apoptosis (PA) is a side process linked to the lipid biogenesis that can regulate miR-126 expression, whose target is the TNF receptor-associated factor, TRAF7. One of the main effects of PA on cells is the production of ROS and other factors that lead to apoptosis, namely through activation of the JNK and NF-κB pathways by TRAF7. MiR-126 blocks the translation of TRAF7 and prevents its recognition by TNF-α, but when PA is present, miR-126 is inhibited and the cell can undergo apoptosis [34]. In this study, we observed increased levels of miR-126 in AK-positive skin in comparison with healthy skin; therefore, it is possible to hypothesize that AK cells could regulate palmitic acid biosynthesis to prevent apoptosis and maintain tumor progression. Indeed, miR-126 is involved in fatty acid biosynthesis, regulating the expression of FASN enzymes such as ACSL1 and Insig1. In particular, it has been shown that in mammary luminal epithelial cells, inhibition of miR-126 leads to a decrease in both protein levels and in the number of cytoplasmic lipid droplets [35].

The biosynthesis of fatty acids may also be influenced by miR-222 and miR-31-3p [36], which share the target *ACOX1*, a catalytic rate-limiting gene in the β-oxidation of fatty acids from triglycerides, in particular miR-31 during oral squamous cell carcinoma (OSCC) and head and neck squamous cell carcinoma (HNSCC). Alterations in the lipidosome of OSCC caused by ACOX1 silencing enhance the motility and fitness of these tumor cells [37]. In our experiments, both miR-222 and miR-31 were upregulated in AK cells. Since AK could be considered a pre-malignant form of cSCC, it is possible to presume that the upregulation of these two miRNAs could indicate a transition from in situ carcinoma to full-blown epithelial cancer. Furthermore, considering that miR-34 and miR-126 are both involved in the silencing of genes that regulate apoptosis through fatty acid biosynthesis and catalysis [33,35], it is possible that the concertation between miR-222 and miR-31 to dysregulate palmitate and other free fatty acids could promote cancer progression.

Qian et al. [38] reported that exosomes from adipose mesenchymal cells are capable of paracrine gene regulation of cell renewal and proliferation during epithelial tissue healing. The mechanism behind this process has to do with the silencing of SOX9 by miR-19b. SOX9 is a crucial transcription factor that is involved in skin healing as it activates cell proliferation and differentiation through the Wnt/β-catenin pathway. The elevated levels of miR-19b that we observed in subjects with AK may be a physiological response to repair skin lesions inflicted by AK.

Epithelial cells exhibit several types of cell–cell junctions that play a key role in the maintenance of epithelial homeostasis. Dysregulation of molecules in the junctions promotes cell migration and tumor metastasis. In the adherens junctions, the cytoplasmic domain of E-cadherin forms a ternary complex with β-catenin and α-catenin, which in turn binds to F-actin, linking the catenin-based complexes to the actin cytoskeleton [39]. α-catenin (CTNNA)1, which is widely expressed in normal human tissues and in many malignancies, inhibits adhesion, invasion, and induces apoptosis of tumor cells by promoting or collaborating with E-cadherin. The expression of CTNNA1 is downregulated in different types of tumors, and it is often associated with reduced expression of other proteins of E-cadherin–catenin cell adhesion complex [40]. Among the top regulated pathways in our DIANA tool mirPath v.3 analysis is Adherens junction, where CTNNA1 and CTNNA2 are key factors. These two molecules are targets of miR-214 and -186, respectively, indicating that the upregulation of such miRNAs detected by the TaqMan Array can contribute to reduce the expression of catenin α and facilitate cell migration.

In skin cancer development, UV radiation is the main risk factor; however, virus infection, in particular human papillomavirus (HPV) infection, also seems to play a significant role. Progression of AK to cSCC is a multifactorial event, and many findings support the role of β-HPVs in the early steps of the carcinogenetic process, in cooperation with UV radiation [41]. We previously showed a large spectrum of β- and γ-HPVs in healthy-looking and lesion skin of patients affected by AK [11,12]. Interestingly, these samples belonged to the same cohort of patients as the present work; therefore, it is likely that in the AK and HS analyzed for miRNA expression, β- and γ-HPVs are abundantly present, contributing to modulate miRNA expression. Indeed, we showed that β-38 and -49 HPV E6 and E7 protein expression in primary human keratinocytes leads to the modulation of miRNAs involved in tumorigenesis [42,43].

5. Conclusions

Skin undergoes continuous renewal, and scrape cytology is a simple and cost-effective technique useful for the rapid diagnosis of some tumors [12]. In this study, scraping samples of actinic keratosis lesions proved to be useful in identifying different microRNAs that are modulated compared to healthy skin samples. However, further studies involving large numbers of samples are needed to confirm the hypothesis that the observed dysregulation of miRNAs in the AK samples may represent the combined result of UV exposure and human papillomavirus infection.

Supplementary Materials: The following are available online at https://www.mdpi.com/article/10.3390/biomedicines11061719/s1, Table S1: MiRNAs in Venn diagram; Table S2: Target genes in Venn diagram.

Author Contributions: Conceptualization, M.V.C., M.I., G.F. and P.D.B.; investigation, M.V.C., M.I., R.M.M., F.L., G.F. and P.D.B.; data curation, M.V.C., M.I., R.M.M., L.G., G.M. and P.D.B.; writing—original draft preparation, M.V.C., M.I. and P.D.B.; writing—review and editing, M.V.C.,

M.I., G.M., L.A., G.R. and P.D.B. All authors have read and agreed to the published version of the manuscript.

Funding: This research was funded in part by the Department of Infectious Diseases, Istituto Superiore di Sanità, and by Sapienza University of Rome ("Bando di Ateneo Sapienza-2020", Grant no. RM120172B8B7B997).

Institutional Review Board Statement: The study was conducted in accordance with the Declaration of Helsinki, and approved by the Ethics Committee of the NIHMP (2014) and the San Gallicano Dermatological Institute (CE943/17; RS/1090/18).

Informed Consent Statement: Informed consent was obtained from all subjects involved in the study. Samples were collected for a previous study [11] approved by the Ethics Committee of the NIHMP (2014) and the San Gallicano Dermatological Institute (CE943/17; RS/1090/18).

Data Availability Statement: The row data are available upon specific request.

Conflicts of Interest: The authors declare no conflict of interest.

References

1. Siegel, J.A.; Korgavkar, K.; Weinstock, M.A. Current perspective on actinic keratosis: A review. *Br. J. Dermatol.* **2017**, *177*, 350–358. [CrossRef]
2. Siegel, J.A.; Luber, A.J.; Weinstock, M.A.; Department of Veterans Affairs Topical Tretinoin Chemoprevention Trial and Department of Veterans Affairs Keratinocyte Carcinoma Chemoprevention Trial Groups. Predictors of actinic keratosis count in patients with multiple keratinocyte carcinomas: A cross-sectional study. *J. Am. Acad. Dermatol.* **2017**, *76*, 346–349. [CrossRef] [PubMed]
3. Yang, D.D.; Borsky, K.; Jani, C.; Crowley, C.; Rodrigues, J.N.; Matin, R.N.; Marshall, D.C.; Salciccioli, J.D.; Shalhoub, J.; Goodall, R. Trends in keratinocyte skin cancer incidence, mortality and burden of disease in 33 countries between 1990 and 2017. *Br. J. Dermatol.* **2023**, *188*, 237–246. [CrossRef] [PubMed]
4. De Oliveira, E.C.V.; da Motta, V.R.V.; Pantoja, P.C.; Ilha, C.S.O.; Magalhães, R.F.; Galadari, H.; Leonardi, G.R. Actinic keratosis—Review for clinical practice. *Int. J. Dermatol.* **2019**, *58*, 400–407. [CrossRef] [PubMed]
5. Rollison, D.E.; Amorrortu, R.P.; Zhao, Y.; Messina, J.L.; Schell, M.J.; Fenske, N.A.; Cherpelis, B.S.; Giuliano, A.R.; Sondak, V.K.; Pawlita, M.; et al. Cutaneous Human Papillomaviruses and the Risk of Keratinocyte Carcinomas. *Cancer Res.* **2021**, *81*, 4628–4638. [CrossRef] [PubMed]
6. Sohel, M.M.H. Circulating microRNAs as biomarkers in cancer diagnosis. *Life Sci.* **2020**, *248*, 117473. [CrossRef]
7. Konicke, K.; López-Luna, A.; Muñoz-Carrillo, J.L.; Servín-González, L.S.; Flores-de la Torre, A.; Olasz, E.; Lazarova, Z. The microRNA landscape of cutaneous squamous cell carcinoma. *Drug Discov. Today* **2018**, *23*, 864–870. [CrossRef]
8. García-Sancha, N.; Corchado-Cobos, R.; Pérez-Losada, J.; Cañueto, J. MicroRNA Dysregulation in Cutaneous Squamous Cell Carcinoma. *Int. J. Mol. Sci.* **2019**, *20*, 2181. [CrossRef]
9. Neagu, M.; Constantin, C.; Cretoiu, S.M.; Zurac, S. miRNAs in the Diagnosis and Prognosis of Skin Cancer. *Front. Cell Dev. Biol.* **2020**, *8*, 71. [CrossRef]
10. Dańczak-Pazdrowska, A.; Pazdrowski, J.; Polańska, A.; Basta, B.; Schneider, A.; Kowalczyk, M.J.; Golusiński, P.; Golusiński, W.; Adamski, Z.; Żaba, R.; et al. Profiling of microRNAs in actinic keratosis and cutaneous squamous cell carcinoma patients. *Arch. Dermatol. Res.* **2022**, *314*, 257–266. [CrossRef]
11. Donà, M.G.; Chiantore, M.V.; Gheit, T.; Fiorucci, G.; Vescio, M.F.; La Rosa, G.; Accardi, L.; Costanzo, G.; Giuliani, M.; Romeo, G.; et al. Comprehensive analysis of β- and γ-human papillomaviruses in actinic keratosis and apparently healthy skin of elderly patients. *Br. J. Dermatol.* **2019**, *181*, 620–622. [CrossRef]
12. Galati, L.; Brancaccio, R.N.; Robitaille, A.; Cuenin, C.; Luzi, F.; Fiorucci, G.; Chiantore, M.V.; Marascio, N.; Matera, G.; Liberto, M.C.; et al. Detection of human papillomaviruses in paired healthy skin and actinic keratosis by next generation sequencing. *Papillomavirus Res.* **2020**, *9*, 100196. [CrossRef]
13. Livak, K.J.; Schmittgen, T.D. Analysis of relative gene expression data using real-time quantitative PCR and the 2(-Delta Delta C(T)) Method. *Methods* **2001**, *25*, 402–408. [CrossRef] [PubMed]
14. Huang, X.; Wang, M.; Liu, Y.; Gui, Y. Synthesis of RNA-based gene regulatory devices for redirecting cellular signaling events mediated by p53. *Theranostics* **2021**, *11*, 4688–4698. [CrossRef] [PubMed]
15. Lezina, L.; Aksenova, V.; Fedorova, O.; Malikova, D.; Shuvalov, O.; Antonov, A.V.; Tentler, D.; Garabadgiu, A.V.; Melino, G.; Barlev, N.A. KMT Set7/9 affects genotoxic stress response via the Mdm2 axis. *Oncotarget* **2015**, *6*, 25843–25855. [CrossRef] [PubMed]
16. Gutekunst, M.; Oren, M.; Weilbacher, A.; Dengler, M.A.; Markwardt, C.; Thomale, J.; Aulitzky, W.E.; van der Kuip, H. p53 hypersensitivity is the predominant mechanism of the unique responsiveness of testicular germ cell tumor (TGCT) cells to cisplatin. *PLoS ONE* **2011**, *6*, e19198. [CrossRef]
17. Liu, G.; Zhao, H.; Ding, Q.; Li, H.; Liu, T.; Yang, H.; Liu, Y. CDK6 is stimulated by hyperthermia and protects gastric cancer cells from hyperthermia-induced damage. *Mol. Med. Rep.* **2021**, *23*, 1. [CrossRef]

18. Yang, P.; Chen, W.; Li, X.; Eilers, G.; He, Q.; Liu, L.; Wu, Y.; Yu, W.; Fletcher, J.A.; Ou, W.B. Downregulation of cyclin D1 sensitizes cancer cells to MDM2 antagonist Nutlin-3. *Oncotarget* **2016**, *7*, 32652–32663. [CrossRef]
19. Metsalu, T.; Vilo, J. ClustVis: A web tool for visualizing clustering of multivariate data using Principal Component Analysis and heatmap. *Nucleic Acids Res.* **2015**, *43*, W566–W570. [CrossRef]
20. Sand, M.; Hessam, S.; Amur, S.; Skrygan, M.; Bromba, M.; Stockfleth, E.; Gambichler, T.; Bechara, F.G. Expression of oncogenic miR-17-92 and tumor suppressive miR-143-145 clusters in basal cell carcinoma and cutaneous squamous cell carcinoma. *J. Dermatol. Sci.* **2017**, *86*, 142–148. [CrossRef]
21. Mizrahi, A.; Barzilai, A.; Gur-Wahnon, D.; Ben-Dov, I.Z.; Glassberg, S.; Meningher, T.; Elharar, E.; Masalha, M.; Jacob-Hirsch, J.; Tabibian-Keissar, H.; et al. Alterations of microRNAs throughout the malignant evolution of cutaneous squamous cell carcinoma: The role of miR-497 in epithelial to mesenchymal transition of keratinocytes. *Oncogene* **2018**, *37*, 218–230. [CrossRef]
22. Tamas, T.; Baciut, M.; Nutu, A.; Bran, S.; Armencea, G.; Stoia, S.; Manea, A.; Crisan, L.; Opris, H.; Onisor, F.; et al. Is miRNA Regulation the Key to Controlling Non-Melanoma Skin Cancer Evolution? *Genes* **2021**, *12*, 1929. [CrossRef] [PubMed]
23. Yu, X.; Li, Z. The role of miRNAs in cutaneous squamous cell carcinoma. *J. Cell Mol. Med.* **2016**, *20*, 3–9. [CrossRef] [PubMed]
24. Lohcharoenkal, W.; Harada, M.; Lovén, J.; Meisgen, F.; Landén, N.X.; Zhang, L.; Lapins, J.; Mahapatra, K.D.; Shi, H.; Nissinen, L.; et al. MicroRNA-203 Inversely Correlates with Differentiation Grade, Targets c-MYC, and Functions as a Tumor Suppressor in cSCC. *J. Investig. Dermatol.* **2016**, *136*, 2485–2494. [CrossRef]
25. Tian, J.; Shen, R.; Yan, Y.; Deng, L. miR-186 promotes tumor growth in cutaneous squamous cell carcinoma by inhibiting apoptotic protease activating factor-1. *Exp. Ther. Med.* **2018**, *16*, 4010–4018. [CrossRef]
26. Colak, S.; Ten Dijke, P. Targeting TGF-β Signaling in Cancer. *Trends Cancer* **2017**, *3*, 56–71. [CrossRef] [PubMed]
27. Cottonham, C.L.; Kaneko, S.; Xu, L. miR-21 and miR-31 converge on TIAM1 to regulate migration and invasion of colon carcinoma cells. *J. Biol. Chem.* **2010**, *285*, 35293–35302. [CrossRef]
28. Wang, Z. ErbB Receptors and Cancer. *Methods Mol. Biol.* **2017**, *1652*, 3–35. [CrossRef]
29. Mikami, T.; Kitagawa, H. Biosynthesis and function of chondroitin sulfate. *Biochim. Biophys. Acta* **2013**, *1830*, 4719–4733. [CrossRef]
30. Yang, J.; Price, M.A.; Li, G.Y.; Bar-Eli, M.; Salgia, R.; Jagedeeswaran, R.; Carlson, J.H.; Ferrone, S.; Turley, E.A.; McCarthy, J.B. Melanoma proteoglycan modifies gene expression to stimulate tumor cell motility, growth, and epithelial-to-mesenchymal transition. *Cancer Res.* **2009**, *69*, 7538–7547. [CrossRef]
31. Lee, Y.R.; Chen, M.; Pandolfi, P.P. The functions and regulation of the PTEN tumour suppressor: New modes and prospects. *Nat. Rev. Mol. Cell Biol.* **2018**, *19*, 547–562. [CrossRef]
32. Li, J.; Yu, L.; Shen, Z.; Li, Y.; Chen, B.; Wei, W.; Chen, X.; Wang, Q.; Tong, F.; Lou, H.; et al. miR-34a and its novel target, NLRC5, are associated with HPV16 persistence. *Infect. Genet. Evol.* **2016**, *44*, 293–299. [CrossRef] [PubMed]
33. Natarajan, S.K.; Stringham, B.A.; Mohr, A.M.; Wehrkamp, C.J.; Lu, S.; Phillippi, M.A.; Harrison-Findik, D.; Mott, J.L. FoxO3 increases miR-34a to cause palmitate-induced cholangiocyte lipoapoptosis. *J. Lipid Res.* **2017**, *58*, 866–875. [CrossRef]
34. Wang, Y.; Wang, F.; Wu, Y.; Zuo, L.; Zhang, S.; Zhou, Q.; Wei, W.; Zhu, H. MicroRNA-126 attenuates palmitate-induced apoptosis by targeting TRAF7 in HUVECs. *Mol. Cell. Biochem.* **2015**, *399*, 123–130. [CrossRef]
35. Chu, M.; Zhao, Y.; Feng, Y.; Zhang, H.; Liu, J.; Cheng, M.; Li, L.; Shen, W.; Cao, H.; Li, Q.; et al. MicroRNA-126 participates in lipid metabolism in mammary epithelial cells. *Mol. Cell. Endocrinol.* **2017**, *454*, 77–86. [CrossRef]
36. Wang, J.J.; Zhang, Y.T.; Tseng, Y.J.; Zhang, J. miR-222 targets ACOX1, promotes triglyceride accumulation in hepatocytes. *Hepatobiliary Pancreat. Dis. Int.* **2019**, *18*, 360–365. [CrossRef] [PubMed]
37. Lai, Y.H.; Liu, H.; Chiang, W.F.; Chen, T.W.; Chu, L.J.; Yu, J.S.; Chen, S.J.; Chen, H.C.; Tan, B.C. MiR-31-5p-ACOX1 Axis Enhances Tumorigenic Fitness in Oral Squamous Cell Carcinoma Via the Promigratory Prostaglandin E2. *Theranostics* **2018**, *8*, 486–504. [CrossRef]
38. Qian, L.; Pi, L.; Fang, B.R.; Meng, X.X. Adipose mesenchymal stem cell-derived exosomes accelerate skin wound healing via the lncRNA H19/miR-19b/SOX9 axis. *Lab. Investig.* **2021**, *101*, 1254–1266. [CrossRef] [PubMed]
39. Xu, Q.R.; Du, X.H.; Huang, T.T.; Zheng, Y.C.; Li, Y.L.; Huang, D.Y.; Dai, H.Q.; Li, E.M.; Fang, W.K. Role of Cell-Cell Junctions in Oesophageal Squamous Cell Carcinoma. *Biomolecules* **2022**, *12*, 1378. [CrossRef] [PubMed]
40. Huang, J.; Wang, H.; Xu, Y.; Li, C.; Lv, X.; Han, X.; Chen, X.; Chen, Y.; Yu, Z. The Role of CTNNA1 in Malignancies: An Updated Review. *J. Cancer* **2023**, *14*, 219–230. [CrossRef]
41. Tommasino, M. HPV and skin carcinogenesis. *Papillomavirus Res.* **2019**, *7*, 129–131. [CrossRef] [PubMed]
42. Chiantore, M.V.; Iuliano, M.; Mongiovì, R.M.; Dutta, S.; Tommasino, M.; Di Bonito, P.; Accardi, L.; Mangino, G.; Romeo, G. The E6 and E7 proteins of beta3 human papillomavirus 49 can deregulate both cellular and extracellular vesicles-carried microRNAs. *Infect. Agent. Cancer* **2022**, *17*, 29. [CrossRef] [PubMed]
43. Chiantore, M.V.; Mangino, G.; Iuliano, M.; Capriotti, L.; Di Bonito, P.; Fiorucci, G.; Romeo, G. Human Papillomavirus and carcinogenesis: Novel mechanisms of cell communication involving extracellular vesicles. *Cytokine Growth Factor. Rev.* **2020**, *51*, 92–98. [CrossRef] [PubMed]

 *biomedicines*

Review

Epigenetic Features in Uterine Leiomyosarcoma and Endometrial Stromal Sarcomas: An Overview of the Literature

Bruna Cristine de Almeida [1], Laura Gonzalez dos Anjos [1], Andrey Senos Dobroff [2,3], Edmund Chada Baracat [1], Qiwei Yang [4], Ayman Al-Hendy [4] and Katia Candido Carvalho [1,*]

[1] Laboratório de Ginecologia Estrutural e Molecular (LIM 58), Disciplina de Ginecologia, Departamento de Obstetricia e Ginecologia, Hospital das Clínicas da Faculdade de Medicina da Universidade de Sao Paulo (HCFMUSP), São Paulo 05403-010, Brazil
[2] UNM Comprehensive Cancer Center (UNMCCC), University of New Mexico, Albuquerque, NM 87131, USA
[3] Division of Molecular Medicine, Department of Internal Medicine, (UNM) School of Medicine, UNM Health Sciences Center, 1 University of New Mexico, Albuquerque, NM 87131, USA
[4] Department of Obstetrics and Gynecology, University of Chicago, Chicago, IL 60637, USA
* Correspondence: carvalhokc@gmail.com; Tel.: +55-011-3061-7486

Abstract: There is a consensus that epigenetic alterations play a key role in cancer initiation and its biology. Studies evaluating the modification in the DNA methylation and chromatin remodeling patterns, as well as gene regulation profile by non-coding RNAs (ncRNAs) have led to the development of novel therapeutic approaches to treat several tumor types. Indeed, despite clinical and translational challenges, combinatorial therapies employing agents targeting epigenetic modifications with conventional approaches have shown encouraging results. However, for rare neoplasia such as uterine leiomyosarcomas (LMS) and endometrial stromal sarcomas (ESS), treatment options are still limited. LMS has high chromosomal instability and molecular derangements, while ESS can present a specific gene fusion signature. Although they are the most frequent types of "pure" uterine sarcomas, these tumors are difficult to diagnose, have high rates of recurrence, and frequently develop resistance to current treatment options. The challenges involving the management of these tumors arise from the fact that the molecular mechanisms governing their progression have not been entirely elucidated. Hence, to fill this gap and highlight the importance of ongoing and future studies, we have cross-referenced the literature on uterine LMS and ESS and compiled the most relevant epigenetic studies, published between 2009 and 2022.

Keywords: uterine leiomyosarcoma; endometrial stromal sarcoma; epigenetics mechanisms; ncRNA; DNA methylation; histones modifications

Citation: de Almeida, B.C.; dos Anjos, L.G.; Dobroff, A.S.; Baracat, E.C.; Yang, Q.; Al-Hendy, A.; Carvalho, K.C. Epigenetic Features in Uterine Leiomyosarcoma and Endometrial Stromal Sarcomas: An Overview of the Literature. *Biomedicines* **2022**, *10*, 2567. https://doi.org/10.3390/biomedicines10102567

Academic Editors: Milena Rizzo and Elena Levantini

Received: 10 August 2022
Accepted: 6 October 2022
Published: 13 October 2022

1. Introduction

The body of the uterus is composed of a mucosa muscular interface derived from the Müllerian embryonic ducts and constituted of internal endometrium and external myometrium (MM) tissue layers [1–4]. The internal endometrium is composed of luminal epithelium, glandular epithelium, and endometrial stroma whereas the MM consists mainly of smooth muscle cells [2–4]. Cellular and molecular alterations in the endometrial stroma and smooth muscle cell layers can lead to uterine sarcoma (US) development [4–6].

US accounts for 3–9% of all uterine malignancies and shows high rates of recurrence and metastasis [7,8], occupying the second place among all gynecological tumors [7,9]. The American Cancer Society (ACS) registered a total of 66,570 new cases of uterine tumors with about 12,940 related deaths in 2021 [10] and estimates 65,950 new cases for 2022 (Figure 1) [11].

"Pure" sarcomas are composed exclusively of mesenchymal cells and include the leiomyosarcomas (LMS) and endometrial stromal sarcomas (ESS), which are morphologically classified mainly based on the tumor cells phenotype [12]. LMS arises from the smooth muscle compartment, while ESS arises from the stroma supporting the endometrial glands [8]. LMS and ESS are the most frequent uterine mesenchymal tumors in adult age [13].

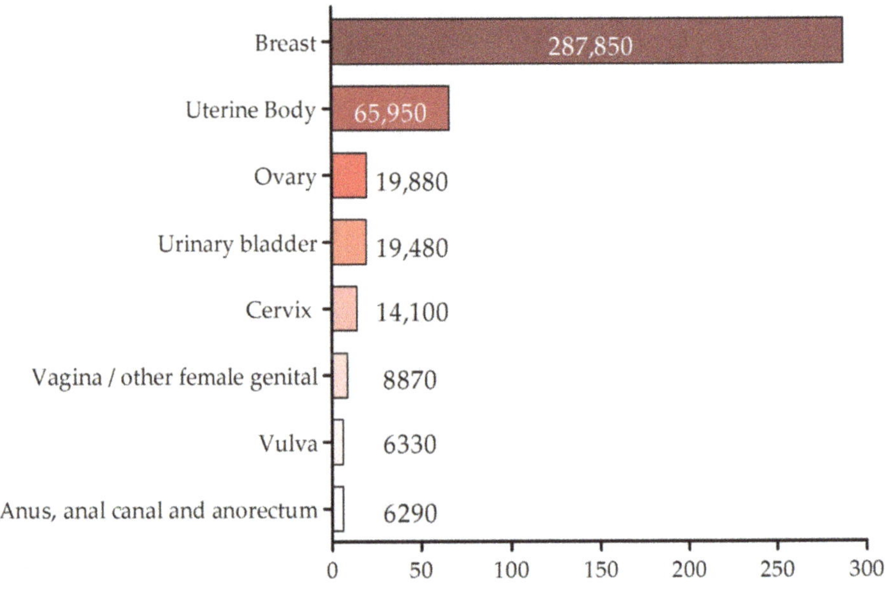

Figure 1. The estimated incidence of gynecological tumors for 2022 according to the ACS.

For LMS and ESS, the disease stage is the single most important prognostic factor [14]. The International Federation of Gynecology and Obstetrics (FIGO) classification and staging system has been specifically designed for these tumors [15]. In 2018, the ACS published the latest revision of the definitions and clinical staging of LMS and ESS (Table 1), based on the FIGO system and the American Joint Committee on Cancer (AJCC) TNM staging system [14,16–18].

It is well known that several molecular events may lead to tumor development. Among these, epigenetic mechanisms such as DNA methylation, post-translational modifications (PTMs), and non-coding RNA (ncRNA) regulation (e.g., microRNAs) can significantly affect the expression of relevant genes, leading to dramatic cell changes [19–21]. Epigenetic alterations are characterized by reversibility and susceptibility to external factors and are the main regulatory events governing the development and progression of uterine sarcomas [19,20,22]. Here, we reviewed and summarized the scientific and clinical reports from the past twelve years regarding epigenetic events and their role in the pathophysiology of the ESS and LMS. The most relevant articles written in English were meticulously reviewed and included in this review, and no restrictions for geographic location were applied. Articles without tumor type description or any identification as "uterine" were excluded.

1.1. LMS Etiology, Prognosis, and Treatment

LMS arises from the myometrium (MM) and often does not reach the endometrial cavity surface [9,23]. Its incidence is 0.36 per 100,000 women a year, affecting mainly women of ≥40 years of age, and representing approximately 70% of all US [24–29]. LMS is a very heterogeneous tumor and represents the most common sarcoma of the uterine body [14,17,25,26,30–32]. Its pathogenesis is poorly understood, but several studies focusing on tumor clonality indicate that many of these tumors are de novo entities [33–38]. Even though it is an extremely rare event [37], some authors defend the hypothesis that LMSs could arise from the malignant transformation of a pre-existing leiomyoma (LM) [15,30,35,39,40]. However, most of the patients do not exhibit predisposing factors such as prior radiation therapy to the pelvis (10–25%), tamoxifen use (1–2%), genetic syndromes (e.g., retinoblastoma and Li–Fraumeni syndrome), postmenopausal status, and ethnicity (African American) [41].

Table 1. Staging of LMS and ESS (FIGO [1] and AJCC [2]).

Stage	Features	Description
I	T1 N0 M0	Tumor limited to the uterus (T1).
IA	T1a N0 M0	Tumor restricted to the uterus (less than 5 cm) (T1a).
IB	T1b N0 M0	Tumor restricted to the uterus (more than 5 cm) (T1b).
II	T2 N0 M0	Tumor growing outside the uterus but is restricted to the pelvis (T2).
IIIA	T3a N0 M0	Tumor growing in a single tissue located in the abdomen (T3a).
IIIB	T3b N0 M0	Involvement of other extrauterine pelvic tissues, 2 or more sites (T3b).
IIIC	T1–T3 N1 M0	Tumor invades abdominal tissues (does not protrude from the abdomen) but does not grow into the bladder or rectum (T1 to T3). The cancer has spread to nearby lymph nodes (N1).
IVA	T4 Any N M0	Tumor spread to the rectum or urinary bladder (T4). It might or might not have spread to nearby lymph nodes (Any N).
IVB	Any T Any N M1	Tumor spread to distant sites (lungs, bones, or liver) (M1). It may or may not have grown into tissues in the pelvis and/or abdomen (any T) and it might or might not have spread to lymph nodes (Any N).

[1] FIGO: International Federation of Gynecology and Obstetrics classification (2009). [2] AJCC: American Joint Committee on Cancer TNM staging system (2018).

Clinically, LMS is associated with a poor prognosis even when diagnosed in the early stages, consequently leading to a significant increase in uterine cancer-associated deaths [13,17,26,28,29]. The recurrence rate of LMS reaches 53–75%, even at the initial stages of the disease, with locoregional or distant recurrence in the first two years after diagnosis [7,14,26,28,42–46]. The overall survival expectancy of LMS is 2.6 years, and the survival at 2, 5, and 10 years are approximately 57%, 24%, and 12%, respectively [26,28,42,44]. The survival rates for patients with LMS decrease as the disease progresses; thus, for localized disease (i.e., restricted to the uterus) the estimated survival rates are 64%, for regional disease (afflicting nearby and adjacent tissues, i.e., lymph nodes) the survival rates are 36%, and for disseminated disease [44,47–50] or metastatic disease (i.e., lungs and liver) the survival is 14% [51].

LMS-related symptoms are associated with vaginal bleeding in 56% of the cases, increased pelvic mass in 54% of the cases, and/or pelvic pain in 22% of the cases [14,15,17,37,39].

Typically, 75% of the patients present a large tumor mass with an average diameter of 10 cm at the time of diagnosis [14,17]. Although LMSs occur primarily in postmenopausal women [52], both progesterone receptor (PR) and estrogen receptor (ER) are found to be expressed in 40% and 70% of the cases, respectively [43,52–56]. A recent review has suggested that hormonal therapy applied to LMS expressing ER/PR is effective and presents favorable tolerance and reliability [57].

There are preoperative methods that allow the differential diagnosis of benign and malignant uterine disease. Magnetic resonance imaging (MRI) remains the optimal imaging modality to characterize pelvic masses originating from the uterus, but distinguishing LMS from LM remains a challenge [17].

LMS histopathological analysis is characterized by the presence of spindle cells, with ruptured nuclei, perinuclear vacuolization, and eosinophilic cytoplasm arranged in intersecting fascicles within the analyzed sample. Meeting the Stanford criteria, LMS should be deemed intrinsically as a high-grade tumor [58]. Cell atypia can vary from moderate to severe, while nuclear atypia is always severe, large areas of tumor cell necrosis with variable mitotic index and atypical mitosis are often observed [15,24,37,59,60]. There are two uncommon subtypes of uterine LMS: myxoid and epithelioid LMS. These present mild or focal nuclear pleomorphism and lower mitotic degree, compared to typical LMS. Diagnostic mistakes between these types of LMS and other smooth muscle tumors are often common [61,62].

Immunohistochemical co-expression of Desmin, h-Caldesmon, smooth muscle actin (SMA), and HDAC8 can assist with LMS diagnosis [39,40,56,63–65]. Several other potential biomarkers such as PDGFRA, WT1, GNRHR, BCL2, ESR, PGR, and LAMP2 have also been evaluated to distinguish LMS from other tumors, mainly from LM [63]. The cell proliferative index (determined by Ki-67 protein expression), the protein expression levels of the tumor suppressors p16 and p53, and the expression of several isoforms of the CD44 (hyaluronan receptor) are, however, routinely used [8,65]. Additionally, some patients show high amounts of lactate dehydrogenase (LDH) [12] and/or CA125 levels [13], but these markers are quite unspecific [17].

Most recently, gene expression profile analysis has enabled the classification of LMS into two subtypes according to their molecular signature. The subtype I recapitulated the low-grade LMS and was enriched for *LMOD1*, *SLMAP*, *MYLK*, and *MYH11*, all of them smooth muscle-specific markers. Subtype II of LMS included tumors with worse prognosis and expressed genes associated with cell cycle, proliferation, and tumorigenesis (*CDK6*, *BMP1*, *MAPK13*, *PDGFRL*, and *HOXA1*) [66,67].

The gold standard treatment for LMS is still the tumor surgical excision. Total hysterectomy and bilateral salpingo-oophorectomy are recommended for early-stage tumors [17, 32,46–48,56,58,68]. Adjuvant chemo and radiotherapies are indicated to avoid recurrences, or for early-stage disease, but their effectiveness is still unclear [17,44,49,69] and they do not offer a significant advantage to improve overall survival [22,43,49,55,70,71]. Recently, new chemotherapies, targeted therapies (pazopanib), and immunotherapies (nivolumab or pembrolizumab) seem to be promising new approaches to treat drug-resistant LMS [41].

1.2. ESS Etiology, Prognosis, and Treatment

ESS is the second most common type of US [72] and arises from the uterine stroma. It is composed of endometrial stromal cells reminiscent of proliferative phase endometrium [7,13,15,73]. It is predominantly intramural, showing both a myometrial invasion and myometrial lymphovascular space permeation [7]. ESS pathogenesis is unknown, but tamoxifen exposure and some medical conditions (e.g., polycystic ovary syndrome) may contribute to its development [73,74]. Although a rare tumor, representing less than 1% of all uterine tumors, ESS accounts for up to 25% of all uterine sarcomas [58]. Symptoms related to ESS development and progression include abnormal uterine bleeding (about 90% of patients), uterine enlargement (70% of cases), pelvic pain, and dysmenorrhea. In 25% of the cases, however, the patients can be asymptomatic [73,75].

The most recent World Health Organization (WHO) classification (2020) for ESS is based on both cytogenetic and molecular analyses (i.e., gene fusion or alterations [76]) where the tumors are divided into benign endometrial stromal nodules (ESN), low-grade endometrial stromal sarcoma (LG-ESS), high-grade endometrial stromal sarcoma (HG-ESS), and undifferentiated uterine sarcomas (UUS) (Table 2) [76–78]. Morphologically, ESN is differentiable from LG-ESS only for the absence of lymphovascular invasion and myometrial infiltration. LG-ESS is usually positive for CD10, ER, and PR and can express actin, keratins, and calretinin [79], which differ from the HG–ESS tumors carrying the *YWAE-NUTM2* fusion that do not express these markers. HG-ESS shows high expression of Cyclin-D1, c-KIT, and BCOR, and when *ZC3H7B-BCOR* fusion is present, CD10 and variable staining for ER and PR are also observed. Finally, UUS exhibits myometrial invasion, severe nuclear pleomorphism, high mitotic activity and/or necrosis, and loss of differentiation. This tumor, however, does not show a specific immunohistochemical profile, instead showing a diffused and atypical staining for CD10 as well as heterogeneous patterns of ER and PR staining [14,80].

HG-ESS and UUS can be difficult to diagnostically differentiate from LMS since the latter can mimic both ESS and UUS. In this case, the immunohistochemical markers and morphological features can be useful, but not accurate. Tumor location can also provide important information because LMS is exclusively related to MM, while UUS may also involve the endometrium. Furthermore, ESS is also diagnosed only post-surgery, but unlike LMS, they present an indolent course with relapses occurring up to 20 years after diagnosis [81].

Overall, patients with ESS have a better life expectancy than other sarcomas. Their five-year survival rates are higher than 80%. For disease stages I and II the five-year survival is approximately 90% whereas for stages III and IV (i.e., advanced disease) the survival rate for the same interval of time is significantly reduced, according to the FIGO stage system [75,82].

Treatment protocols are defined based on the grade and stage of the tumor at the time of diagnosis. Total hysterectomy with bilateral salpingo-oophorectomy remains the standard treatment for ESS, and lymphadenectomy does not appear to improve overall survival rates [83]. Adjuvant radiotherapy and hormone therapy are not well-established therapeutic options yet, even though some studies have shown that hormonal agents can be an alternative to the management of LG-ESS [82,84]. In contrast, HG-ESS is generally detected in advanced stages with no effective adjuvant therapy available. In this case, immunotherapy with adoptive T cells transfer, targeting tumor fusion proteins, can be useful. Such an approach has been proved to be efficient in inhibiting tumor recurrences in other cancer types, thus inducing long-term memory cells and the persistent presence of these cells in the patient's blood [85].

Table 2. Molecular features of endometrial stromal tumors (ESTs).

Category EST	Fusion/Gene Alteration [72,86–101]
Endometrial Stromal Nodule (ESN)	*JAZF1-SUZ12* [1] [86,87] *MEAF6-PHF1* [86,87]
Low-Grade Endometrial Stromal Sarcoma (LG-ESS)	*JAZF1-SUZ12* [1] [88] *JAZF1-PHF1* [88] *MEAF6-PHF1* [88] *EPC1-PHF1* [89] *MBTD1-EZHIP* [89] *JAZF1-BCORL1* [89] *MAGED2-PLAG1* [90] *MEAF6-SUZ12* [91] *EPC2-PHF1* [92] *BRD8-PHF1* [72] *EPC1-BCOR* [72] *EPC1-SUZ12* [72]

Table 2. *Cont.*

Category EST	Fusion/Gene Alteration [72,86–101]
High-Grade Endometrial Stromal Sarcoma (HG-ESS)	*YWHAE-NUTM2A/B* [1] [93] *BCOR*-rearrangement [94] *ZC3H7B-BCOR* [72,95] *EPC1-BCOR* [96] *EPC1-SUZ12* [96] *BCOR-ITD* [72] *LPP-BCOR* [72] *BRD8-PHF1* [97]
Undifferentiated Uterine Sarcoma (UUS)	*JAZF1-SUZ12* [97] *YWHAE-NUTM2* [97] *ZC3H7B-BCOR* [97] *YWHAE*-rearrangement [97] *HMGA-RAD51B* [98] *SMARCA4*-deficient [99]
NTRK-Rearranged Uterine Sarcomas (HG-ESS)	*RBPMS-NTRK3* [100,101] *TPR-NTRK1* [100,101] *LMNA-NTRK1* [100,101] *TPM3-NTRK1* [100,101] *EML4-NTRK3* [100,101] *STRN-NTRK3* [100,101]

[1] Most common alterations.

2. Genetics and Epigenetics Mechanisms in LMS and ESS

Genetic changes are related to alterations in the DNA sequences, whereas epigenetic modifications involve specific regulatory events apart from DNA codification [102–108]. Epigenetic events play an important role in several normal cellular processes, including embryonic development, genetic imprinting, and X-chromosome inactivation. When altered, epigenetic mechanisms may lead to several diseases, including cancer initiation and progression [106]. Epigenetic dysregulation affects gene functions by altering the gene expression mainly by (1) DNA methylation, (2) PTMs, and (3) RNA-mediated gene silencing by ncRNA (e.g., microRNA) (Figure 2) [102,109]. The main clinical and scientific interest in epigenetic events resides in the fact that they are reversible mechanisms [110,111].

2.1. DNA Methylation

DNA methylation is the most studied and understood epigenetic event described to date. Found in more than 70% of the human genome, DNA methylation is crucial for cellular differentiation and normal development [112]. It consists of the addition of a methyl radical (CH3) to the 5-carbon on cytosine residues (5mC) in CpG dinucleotides [103,104,108–114]. DNA methyltransferases (DNMTs)—enzymes responsible for DNA methylation—are known to act in cancer cells by either hypomethylation or hypermethylation of specific CpG regions in the DNA [114].

Global DNA hypomethylation or loss of methylation has been associated with genomic instability as well as aneuploidy, loss of imprinting, reactivation of transposable elements, and endogenous retrovirus (ERVs) [108,113,115]. In cancer, hypomethylation is commonly followed by hypermethylation of localized CpG islands at the promoter and regulatory regions of target genes, which remain unmethylated in normal cells [109,115,116]. Hypermethylation of regulatory regions leads to transcriptional silencing by directly blocking the transcription factors binding to the promoter region, or by the binding of proteins with a high affinity for methylated DNA that compete with the transcription factors binding sites (Figure 3) [108,117].

DNMTs are commonly found overexpressed in tumors, constituting an attractive target for specific therapy. The FDA has approved "epidrugs" such as 5-azacitidine (5-Aza), 5-aza-2-deoxycytidine (DAC), and the second-generation of the demethylation agent guadecitabine [116,118]. In LMS, Fischer et al. (2018) assessed the therapeutic potential of nucleoside analogs 5-Aza, DAC, and guadecitabine, using both in vitro and in vivo experiments. Their results show guadecitabine as a more effective inhibitor of both cell survival and colony formation in vitro. Additionally, animals who received this treatment showed a decrease in the tumor burden and increased survival [116].

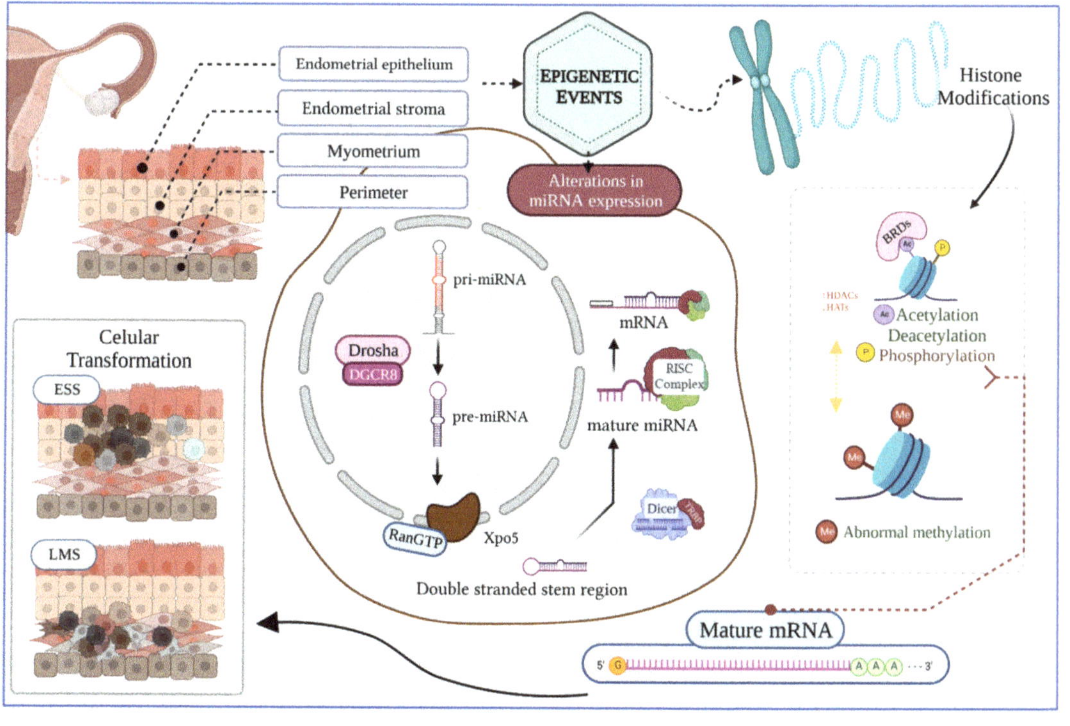

Figure 2. Graphical representation of epigenetic events potentially involved in the initiation and development of the tumors, including uterine LMS and ESS. The biogenesis of miRNAs starts in the nucleus and ends in the cytoplasm. This process includes the participation of several enzymes and protein complexes that regulate the production of mature molecules capable of regulating gene expression, both through induction of mRNA degradation and translational repression. Likewise, dynamic alterations of histone modifications, including acetylation, methylation, and phosphorylation, modify gene expression, thus affecting DNA replication and repair, chromatin compaction, and cell cycle control. In addition, histone modification readers such as BRDs can recognize modified histones, therefore altering gene expression and responding to different signals. Dysregulation in the epigenetic machinery leads to malignant transformation of cells culminating in the development of cancer. Created with BioRender.com.

Our group recently assessed the impact of DNMT inhibition on the Hedgehog (HH) signaling pathway with 5′-Aza-dc, in uterine LMS cells. We observed a reduction in the *GLI1* mRNA, and SMO and GLI1 protein in response to the treatment. Moreover, GLI1 and GLI2 nuclear translocation were also decreased while nuclear translocation of GLI3 was increased. Our data showed that DNMT inhibitor, alone or in combination with pharmacological treatment, was able to block the HH pathway and showed a high inhibitory effect on the LMS malignant cells phenotype [119].

MGMT silencing due to its promoter hypermethylation has been commonly observed in several malignancies, including uterine sarcomas [120,121]. The methylation of the *MGMT* promoter region, which contributes to genome instability and sensitizes the cells to alkylating agents (such as Temozolomide (TMZ)), has been correlated with improved prognosis and as a potential factor of response to TMZ-based therapy prediction [121].

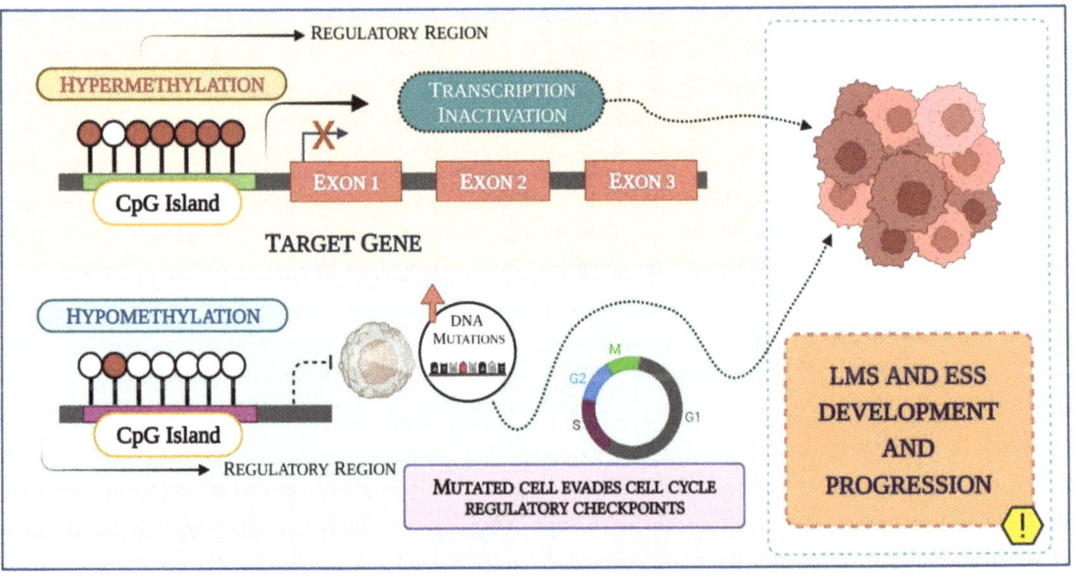

Figure 3. Schematic representation of the DNA methylation process in LMS and ESS during cancer development and progression. Methyl groups are added to the DNA molecule and change its activity. Promoter hypermethylation has been shown to silence tumor suppressor genes in cancer cells, leading to either dysregulation of cell growth or inducing resistance to cancer therapies. Hypomethylation promotes genomic instability causing missegregation of chromosomes during cell division. Created with BioRender.com.

Global DNA methylation studies have also found methylation patterns or signatures that have been associated with different cancer hallmarks [122]. Brraný et al. in 2019 observed differences in the methylation levels between MM and LM samples in the *KLF4* and *DLEC1* genes. Higher levels of methylation were found in LM compared to LMS cases, suggesting that methylation of *KLF4* and *DLEC1* are potential biomarkers to distinguish LM from LMS [123].

Hasan et al. (2021) identified differentially methylated and differentially expressed genes associated with LMS. Among the 77 hypermethylated genes, chromatin-modifying enzymes, including KAT6A, KMT2A and EZH2, and chromatin/DNA binding proteins such as CTNNB1, PBX3, SATB1, MEIS and COMMD1-BMI1 were observed. The findings indicate the possible involvement of chromatin modulation in regulating the DNA methylation of these genes [124].

A higher DNA damage response and hypomethylation of *estrogen receptor 1 (ESR1)* target genes were both observed when comparing uterine to extra uterine LMS [13].

Gene silencing through methylation can occur as frequently as mutations or deletions, leading to aberrant silencing of tumor suppressor genes [125]. In an LMS experimental model, the lack of *BRCA1* function was associated with tumor initiation and development. This protein expression was next investigated in human samples, and a loss of 29% was associated with promoter methylation [126]. Methylation in the *CDKN2A* gene, using uterine LMS samples with a rhabdomyosarcomatous component, has also been described [127]. The authors identified both methylated and unmethylated alleles, originating mainly from the smooth muscle component. Moreover, the loss of heterozygosity in the rhabdomyosarcomatous component has been described exclusively in the cells expressing p16 and p14 [127].

Hierarchical clustering based on the hypermethylation of ALX1, *CBLN1, CORIN, DUSP6, FOXP1, GATA2, IGLON5, NPTX2, NTRK2, STEAP4, PART1,* and *PRL* allowed differentiation among several uterine tumors with up to 70% accuracy [128,129]. Clusters of distinct DNA methylation patterns have also been useful in distinguishing tumor types regardless of the number of CpG sites. Thus, LM and LMS were separated into two different clusters each, while ESS samples (LG- and HG-ESS) were grouped into two subtypes with specific profiles. The HG-ESS cluster included *YWHAE* and *BCOR*-rearranged tumors, distinct from LG-ESS and LMS [128,129]. Although LMS and HG-ESS are morphologically similar, the results show that the DNA methylation profile may be useful to discriminate against these closely related tumors [128].

DNA methylation plays a key role in gene expression regulation, inducing functional changes in key genes that regulate endometrial homeostasis [113]. Li et al. (2017) showed that the *KLF4* promoter was hypermethylated in the ESS. They also found *PCDHGC5* was highly methylated in the ESS samples when compared to endometrioid and endometrial serous carcinoma. *sFRPs 1-5* are tumor suppressors that downregulate Wnt/β-catenin signaling [130–133]. Their consistent promoter hypermethylation and subsequent gene expression suppression were described in ESN, LG-ESS, and UUS.

Although the true role of DNA methylation patterns in the LMS and ESS initiation and progression is not completely understood, DNA methylation in normal endometrial stromal cells has been useful to identify signatures that indicate changes during decidualization or cell transformations [134–137]. Further studies to determine the precise methylation profile in pure mesenchymal tumors are certainly necessary to enable the characterization and differentiation of these tumors as well as to establish new therapeutic options.

2.2. Chromatin Remodeling

The chromatin is composed of DNA molecules tightly coiled around proteins called histones. This structure condensation degree is directly associated with greater or lesser RNA synthesis, with greater condensation (higher chromatin closing) being the state of more transcriptional repression [138]. The basic unit of chromatin, called nucleosome, is constituted by two copies of each core histone (H2A, H2B, H3, and H4) enveloped by DNA molecules [139]. Chromatin regulation occurs through PTMs of core histones and can involve phosphorylation, acetylation, methylation, ubiquitination, SUMOylation, and GlcNAcylation [139].

Currently, the two most studied and best understood mechanisms of chromatin regulation are histones acetylation and methylation. Such events are regulated by very specialized proteins called writers, erasers, and readers—which respectively add, remove, or recognize these PTMs [110–112,140–142]. Included in the writers' group are histone acetyltransferases (HATs), DNA methyltransferases (DNMTs), histone lysine methyltransferases (HKMTs), and histone methyltransferases (HMTs). The erasers' group includes histone deacetylase (HDACs) and histone demethylases (HDMs), ten eleven translocations (TETs), and histone lysine demethylases (HKDMs). In the reader's group are methyl-CpG-binding domains (MBDs) and bromodomains [143].

Histone modifications affect the chromatin structure providing binding sites for several transcriptional factors. Its modification has a direct influence on gene expression, DNA replication and repair, chromatin compaction, and cell cycle control. Thus, loss of regulation of the histone modifications can lead to cancer pathogenesis and several developmental defects (Figure 4) [143].

The lysine residues of histones H3 and H4 are targeted for methylation by site-specific enzymes, culminating in activation or repression of the gene expression. [144]. This molecular mechanism is uniquely able to originate three methylation levels: me1 (mono), me2 (di), and me3 (trimethylation). Lysine methylations may lead to both transcriptional activation and repression, depending on the lysine residue's location [143].

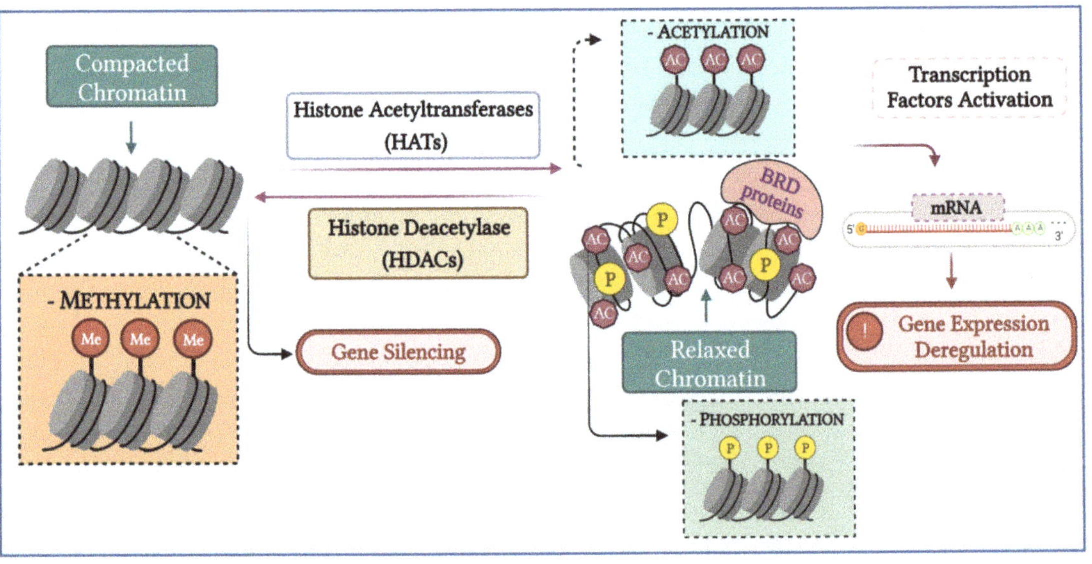

Figure 4. Chromatin remodeling process in LMS and ESS. Chromatin remodeling is an important mechanism of gene expression regulation. In the histone acetylation induced by HATs, the condensed chromatin is transformed into a more relaxed structure (euchromatin) that is associated with greater levels of gene transcription, while histone hypoacetylation induced by HDAC activity is associated with more condensed chromatin (heterochromatin), inducing gene silencing. Altered expression and mutations of genes that encode HDACs have been linked to tumor development. Created with BioRender.com.

In endometrial stromal cells, the transition from a proliferative to a decidual phenotype occurs due to the loss of the EZH2-dependent methyltransferase activity, which is part of the chromatin remodeling process [145]. The decidualization process in those cells downregulates EZH2, resulting in lower levels of H3K27me3 at the promoter region of *PRL* and *IGFBP1* (two decidual marker genes). The H3K27me3 loss, associated with acetylation enrichment in the same lysine residue, indicates the transition from a transcriptionally repressive chromatin form to a permissive one [145].

Little is known about the specific underlying mechanism of histone acetylation or methylation associated with the "pure" sarcoma pathobiology. A unique study available in the literature describes the fatty acid synthase (FASN)-enhanced expression inducing cell proliferation, migration, and invasion, in transfected cells of uterine LMS. It has been observed that FASN promotes H3K9me3 and H3K27ac by alteration in the HDAC, HDM, HMT, and HAT trimethylation activity. Thus, in the uterine LMS cells, the epigenome reprogramming by chromatin remodeling seems to induce a higher malignant phenotype [146].

The polycomb group (PcG) proteins are well-characterized transcriptional repressors that are essential for the regulation of physiological processes in several organisms. PcG proteins are known to form two distinct complexes with defined enzymatic activities: polycomb repressive complex 1 (*PRC1*), a histone ubiquitin ligase related to chromatin compaction; and *PRC2*, an *HMT* that mediates both H3K27me3 and target genes repression [147–150]. In several cancer types, the expression and function of PcG proteins are often found dysregulated [148,151–153], and their targeted deletions generally induce lethal phenotypes [153].

PRC1 catalytic core has two related E3 ubiquitin ligases, the *RING1* (RING1A) or *RNF2* (RING1B) that catalyze ubiquitination and BMI1 (polycomb ring finger oncogene), and one of six *PCGF* orthologues. The latter constitutes a *PRC1* variant containing

BCOR and *KDM2B* [154–157]. Translocations or chromosomal rearrangements, involving fusion proteins have also been implicated in PRCs mechanisms. Fusions such as *KDM2B-CREBBP* [158], *ZC3H7-BCOR* [79,155,156,159,160], *JAZF1-BCORL1* [79,91,161] *EPC1-BCOR* [72,95], *LPP-BCOR* [72] and *BCOR* internal tandem duplications (*BCOR-ITD*) are frequently found in ESS, and have recently been also found in LMS samples [80,160,162,163] Additionally, gene fusion such as *MBTD1-CXorf67* [79,91,151,164], *MBTD1-EZHIP* [90] and *MBTD1-PHF1* in ESS are found as products of PRC1-associated protein [152].

PRC2 has in its core subunits the *EZH2* (or its homolog *EZH1*), *EED*, *SUZ12*, and *RbAp46* (or *48*). *EZH1* and *EZH2* are responsible for the H3K27me3 generation. *EED* binds to H3K27me3, enhancing the *EZH2* catalytic activity, while *SUZ12* is essential for the *PRC2* activity, and bAp46/48 acts as a chaperone. H3K27 trimethylation leads to the suppression of several relevant genes, including tumor suppressors [165].

There are two well-characterized PcG proteins, the BMI1 and EZH2, which are required for the regulation of the PRC activity but are also known to display oncogenic functions in several cancers. BMI1 was the first identified PcG protein that was described as a proto-oncogene, and although there are no specific studies regarding the BMI1 role in uterine LMS [157,166]. Gao et al. (2021) have described one CD133 cell subpopulation that was derived from SK-UT-1 (uterine LMS cells) with enhanced levels of this protein. The authors found BMI1, among other CSCs-related (cancers stem cell) markers, up regulated in the CD133$^+$ cells when compared to the negative cell population [167].

Zhang et al. (2018) investigated both gene and protein expressions of the four PRC2 subunits (*EZH2*, *SUZ12*, *EED*, and *RbAp46*) in extra-uterine and uterine LMS samples. The authors observed 91% of sensitivity and 100% of specificity for *EZH2* positive staining in well-differentiated LMS, suggesting this expression is a specific marker for this tumor Furthermore, the increased expression of EZH2 was inversely correlated with *SUZ12* and *EED* expressions, leading to *PRC2* suppression and H3K27me3 decrease [168].

Chromosomal rearrangements in genes belonging to the PRC2 complex, or in proteins that interact with them, have previously been described in the ESS [131]. *JAZF1-SUZ12* (previously named *JAZF1-JJAZ1*) has been frequently reported as the most common feature of ESS [131,164,169–172]. Additionally, several other modifications such as *EPC1-PHF1* [158,172,173], *MEAF6-PHF1* [87,161,162,170], *EPC1-ZUZ12* [72,95], *MEAF6-SUZ12* [91,174], *BRD8-PHF1* [169], *JAZF1-PHF1* [88,158] and *YWHAE-NUTM2A/B* (previously known as *FAM22A/B*) [79,132,175] have also been reported. Panagopoulos et al. in 2012 observed that the rearrangement of genes involved in acetylation and methylation can be associated with ESS pathogenesis. LG-ESS harboring the *EPC1-PHF1* fusion gene has decreased levels of H3K27me3 and a concomitant increase of H3K36me3 [176]. *PHF1* acts in cell proliferation through the modulation of histone H3 methylation [171].

EZH2 can interact with HDAC1 and HDAC2, through the EED protein, suggesting that the transcriptional repression by the PRC2 complex may be mediated by *HDACs* [177] These enzymes act in the acetylation control of transcription factors [178], and their classification (Class I, IIa, IIb, III, and IV) is based on their activity, structural similarity, subcellular localization, and expression patterns [179]. In LMS patients, strong expression of HDACs 1, 2, 3, 4, 6, and 8 were associated with unfavorable prognosis, while HDACs 5, 7, or 9 weak expressions, together with p53 expression, were associated with favorable disease-free survival (DFS). HDACs 5, 7, and 9 were associated with better survival outcomes, whereas HDAC5 expression was an independent predictor for DFS in epithelioid subtype tumors [180]. In vitro analysis using SK-UT-1, SK-LMS-1, MES-SA, and DMR cell lines demonstrated that HDAC9 (Class IIa) transcription is under *MEF2D* direct control, and this axis sustains cell proliferation and survival through *FAS* repression [177].

In ESS, it has been observed that high expression of HDACs 1, 4, 6, 7, and 8 is associated with lower DFS [181] whereas, in UUS, distant tumor recurrence was associated with a strong expression of HDAC6 [140,182].

The increased HDAC activity often observed in cancers justifies the number of current studies investigating HDAC inhibitors as novel therapeutic agents [183,184]. These studies have shown promising results for metastatic LMS [140,180,181,185]. In this context, mocetinostat acts by turning on tumor suppressor genes, restoring their normal function, and reducing tumor growth [140,180]. Its use as mono- or combinatorial therapy has been evaluated in metastatic extra-uterine and uterine LMS with resistance to gemcitabine and found to induce regression of tumors with acquired chemoresistance. Romidepsin, LBH589, belinostat, SAHA, and valproate (other HDAC inhibitors) have shown good results alone or in combination with decitabine [180,186].

Combinatorial therapy using SAHA, LY294002 (PI3K inhibitor), and rapamycin (mTOR inhibitor) were tested in ESS cell culture [186]. The results show that SAHA combined with either LY294002 or rapamycin, or both, reduce specifically phospho-p70S6K and 4E-BP1 levels, inhibiting the tumor cell proliferation [186,187]. A strong reduction of mTOR and phospho-mTOR levels has been reported after treatment with either SAHA or rapamycin, by targeting phospho-S6rp, in ESS cells [188]. Fröhlich et al. (2014) showed the benefit of SAHA treatment associated with TRAIL/Apo-2L in two US cell lines [189].

Another study assessed the effects of combined therapy with valproic acid (VPA, a weak histone deacetylase inhibitor), bevacizumab (mAb against VEGF), gemcitabine, and docetaxel, for extra- and intrauterine unresectable or metastatic soft tissue sarcomas [184]. This study found partial response in one case of carcinosarcoma, two extrauterine LMSs, two undifferentiated pleomorphic sarcomas, and one uterine LMS patient. This pharmacological combination was well tolerated and overall safe, showing that the combination of traditional medication and "epidrugs" may truly represent a new treatment strategy for sarcoma [184].

New therapeutic strategies to specifically treat US, such as regional hyperthermia, combined with chemotherapy, radiotherapy, and/or immunotherapy have emerged in the last few years. Pazopanib (a multitargeted tyrosine kinase inhibitor with antiangiogenic effects) combined with hyperthermia has demonstrated synergistic effects mainly for LMS growth inhibition, in vitro and in vivo [185]. This approach induces *HAT1* downregulation by suppressing Clock which, in turn, is responsible for H3 and H4 acetylation [190].

Histone phosphorylation, which takes place predominantly but not exclusively on serine, threonine, and tyrosine residues at the histone tails [142], has gained considerable attention, especially regarding the histone H3, due to its close association with mitotic chromosome condensation in mammalian cells [191]. A preliminary study evaluating the mitotic index, based on H3 phosphorylation in LMS, found Phospho Histone H3 (PHH3) positive staining to be a promising mitosis-specific marker for this tumor [192].

Bromodomain-containing proteins (BRDs), as the "readers" of lysine acetylation are responsible for transducing regulatory signals carried by acetylated lysine residues into various biological phenotypes [193]. BRDs can exert a wide variety of functions via multiple gene regulatory mechanisms [194] and the deregulation of BRDs is involved in many diseases, including cancer [195–197]. BRD9 is a newly identified subunit of the noncanonical barrier-to-autointegration factor (ncBAF) complex and a member of the bromodomain family IV [198]. Studies have demonstrated that BRD9 plays an oncogenic role in multiple cancer types, by regulating tumor cell growth. Furthermore, the connection of BRD9 with the PI3K pathway [199], microRNAs [200], and STAT5 [201] is implicated in cancer progression. It has been shown that BRD9 is aberrantly overexpressed in uterine LMS tissues, compared to adjacent myometrium. [202]. In addition, BRD9 expression was upregulated in uterine LMS cell lines compared to benign LM and myometrium cell lines. Notably, targeting the BRD9 with a specific chemical inhibitor (TP-472) can suppress the LMS cell growth, concomitantly sculpting the transcriptome of uterine LMS cells, altering the important pathways, reprogramming the oncogenic epigenome, and inducing the miRNA-mediated gene regulation. These studies reveal that BRD9 constitutes a specific vulnerability in malignant LMS and that targeting non-bromodomain and extra-terminal BRDs in uterine LMS may provide a promising and novel strategy for treating patients with this aggressive uterine cancer [202].

In summary, histone modifications are frequently found in US, thus representing potential targets for new therapeutic strategies development. Several studies have demonstrated that HDAC inhibitors could modulate several signaling pathways, activating, or inhibiting numerous cascades that lead to an antitumor response. Moreover, combination with chemo- or targeted- therapies is likely to strengthen the activity of HDAC inhibitors. However, it is still necessary to further elucidate how the histone modifications are regulated, as well as to understand the mechanism of action of their available inhibitors in LMS and ESS. This information will enable more efficient clinical trials, which could lead to an improvement in patients' response to treatment and overall survival [140].

2.3. Non-Coding RNA (ncRNAs)

ncRNA is known to regulate gene expression both at transcriptional and post-transcriptional levels. ncRNAs play an important role in epigenetic processes, including modulation of heterochromatin, histone modification, DNA methylation, and gene silencing (Figure 5) [108]. These molecules can be divided into housekeeping and regulatory ncRNAs [110,203]. Regulatory ncRNAs are classified according to their size in small non-coding RNAs (sncRNAs), with approximately 19-200 nucleotides (nt), and long non-coding RNAs (lncRNAs), with more than 200 nt [105,108,113,204,205]. The sncRNAs have a wide range of structural and functional roles in gene expression regulation, RNA splicing, and chromatin structure [206,207]. sncRNAs includes four different categories: (1) small interfering RNA (siRNA), (2) microRNA (miRNA), (3) PIWI-interfering RNA (piRNA, with approximately 19-31 nt), and (4) small nucleolar RNA (snoRNA, with 60-300 nt) [111,203]. miRNA and piRNA are probably the most studied sncRNAs categories to date, and their functions are well established in the literature [206]. Due to miRNAs' broad roles, mainly at the post-transcriptional level, dysfunctions in their regulation have been associated with the development of several diseases, including cancer [205,206,208].

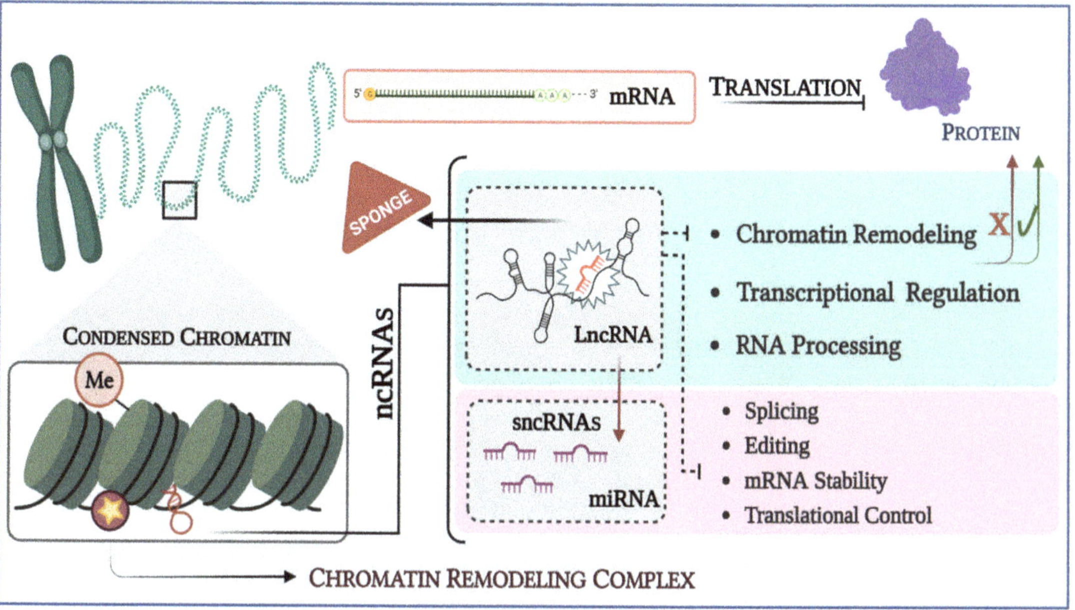

Figure 5. Graphical representation of the ncRNAs dysregulation in LMS and ESS. ncRNAs have been identified as oncogenic drivers or tumor suppressors in cancer. LncRNAs often affect the expression of their target genes by interacting with miRNAs, which are the main post-transcriptional regulation factors. Some lncRNAs act like sponges, thereby preventing miRNAs from binding to their target mRNAs. As lncRNAs work as decoys for miRNAs, oncogene mRNA translation is allowed, starting the LMS and ESS carcinogenesis. Created with BioRender.com.

To understand the complex biology of sarcomas, numerous correlative and functional studies aiming to integrate gene expression patterns and miRNAs have been carried out [209]. As mentioned above, the differential diagnosis of LMS is still a challenge, and studies focusing on new biomarkers to help distinguish uterine LMS from LM are extremely important [210–213]. Yokoi et al. (2019) demonstrate the feasibility of circulating serum miRNAs detection as a preoperative clinical assay to detect US. They identified two miRNA signatures (*miR-1246* and *miR-191-5p*) in uterine LMS (95% confidence interval of 0.91–1.00) [214]. Dvorská et al. (2019), and later, Wei et al. (2020) reviewed how liquid biopsies could increase the overall understanding of uterine LMS behavior and how its molecular profile could contribute to more accurate discrimination from LM [210,211].

Comparing LMS, LM, and MM, Anderson et al. (2014) found 37 miRNAs differentially expressed in uterine LMS. The lack of *miR-10b* in LMS samples was critical for tumor growth and metastasis. Indeed, rescuing *miR-10b* expression in the cell lines resulted in prominent inhibition of cell proliferation, migration, and invasion, and increased apoptosis. Similarly, stable *miR-10b* expression significantly reduced the number and size of tumor implants in vivo by reducing cell proliferation and increasing apoptosis [212].

Later, Schiavon et al. (2019) found that dysregulation of *miR-148a-3p, 27b-3p, 124-3p, 183-5p,* and *135b-5p* expression was associated with tumor relapse, increased metastasis, and poor survival rates in uterine LMS patients [213]. De Almeida et al. (2017) evaluated the miRNAs expression profile in cell lines of MM, LM, and LMS. Thirteen molecules presented differential expression profiles in LM and LMS, compared to normal tissue (MM). Additionally, the authors observed that *miR-1-3p, miR-130b-3p, miR-140-5p, miR-202, miR-205,* and *miR-7-5p* presented similar expression patterns between the cell lines and 16 patients' samples [215].

Zhang et al. (2014) demonstrated that miRNAs were significantly dysregulated among different types of uterine smooth muscle tumors (USMTs), including ordinary LM, mitotically active leiomyoma (MALM), cellular leiomyoma (CLM), atypical leiomyoma (ALM), uterine smooth muscle tumor of uncertain malignant potential (STUMP), and LMS samples. The miRNA expression profile showed that ALM and LMS shared similar signatures (including *miR-34a-5p, miR-10b-5p, miR-21-5p, miR-490-3p, miR-26a-5p* and *miR-650*). Unsupervised analysis divided the tumors into three clusters: LMS/ALM, LM/STUMP, and CLM/MM [216]. *miR-200c* was found to be significantly downregulated in LM, compared to MM [217], acting directly in *ZEB1/2, VEGFA, FBLNS,* and *TIMP2* regulation. Next, the authors observed a significant reduction of *miR-200c* in the SK-LMS-1 cells, compared to isolated LM cells, indicating this miRNA is an important marker for LM progression and malignance risk [4,218].

To date, the differential expression (i.e., up- or down-regulation) of several miRNAs has been directly correlated with US patients' prognosis. In 2018, Dos Anjos et al. analyzed the expression profile of 84 cancer-related miRNAs and associated their signatures with patients' clinical and pathological data. In LMS, specifically, the authors found an association between miRNA dysregulation and lower cancer-specific survival (CSS) and aggressive tumor phenotype. In ESS samples, alterations in miRNA regulation were related to both lower CSS and metastasis [219].

Shi et al. (2009) found a significant inverse correlation between endogenous *HMGA2* levels and *let-7* expression in uterine LMS. Their study revealed that the ectopic expression of *let-7a* inhibits LMS proliferation by *HMGA2* repression, suggesting that the *let-7* loss of expression can represent a worse prognostic factor [220]. Zavadil et al. (2010) identified the way in which let-7s is responsible for the direct regulation of PPP1R12B, *STARD13, TRIB1, BTG2, HMGA2,* and *ITGB3* genes (involved in the cell proliferation and extracellular matrix regulation) in LM samples. [221]. De Almeida et al. (2019), found that decreased expression of *let-7* family members was directly correlated with worse prognosis, affecting both the overall survival (OS) and the DFS rates of the LMS patients [22].

Dysregulation of some miRNAs has also been correlated with acquired chemoresistance in uterine neoplasm. For instance, the loss of *miR-34a* expression and its release from

LMS cells via exosomes contribute indirectly to the tumor doxorubicin chemoresistance. This mechanism seems to be mediated by *MELK* overexpression and the recruitment of M2 macrophages [216].

Although less studied than other ncRNAs, lncRNAs are known to interact with either DNA, RNA, or proteins, and play a significant regulatory function in several cellular processes [208]. lncRNAs are responsible for regulating transcription on three different levels: pre-transcriptional (chromatin remodeling), transcriptional, and post transcriptional [203,222]. Some similarities can be found between lncRNAs and mRNAs including size, transcription by RNA pol-II, 5′-capping, RNA splicing, and poly(A) tail (approximately 60% of all lncRNAs) [223]. LncRNAs can be stratified into five categories: (1) intergenic (present between two protein-coding genes), (2) intronic (between the introns of a protein-coding gene), (3) overlapping (a coding gene is located on the intron of a lncRNA), (4) antisense (the lncRNA is transcribed from the opposite strand of a protein coding gene), and (5) processed lncRNAs (lacks an open reading frame ORF) [208,224]. lncRNA can be expressed in distinguished cell regions and their functions are directly related to their sub-cellular location. However, these epigenetic regulators may suffer molecular alterations that affect their expression and, consequently, their physiological function. Accumulated evidence shows that several differentially expressed lncRNAs are related to cancer development, progression, and metastasis [204,222].

Unfortunately, in uterine LMS and ESS, the molecular role of lncRNAs and their regulation remains unclear. Yet, Guo et al. (2014) performed a microarray-based genome wide analysis of lncRNAs, including 35 LM and MM-matched samples. The authors showed, for the first time, the differential expression profile of the lncRNAs between these tissues. The expression pattern obtained was associated with the downregulation of the cytokine–cytokine receptor interaction pathway in large LM, and the upregulation of the fatty acid metabolism pathway in small LM. This study, although preliminary, sheds light on future studies that will attempt to elucidate the role of lncRNAs specifically in uterine mesenchymal tumors [225].

3. Conclusions

Uterine pure sarcomas constitute the most frequently diagnosed group of malignant neoplasms in the uterine body. LMS and ESS are distinct tumors with a variety of features similar to other uterine neoplasms. The high heterogeneity, morphological and molecular variations pose challenges to subtypes differentiation and diagnosis. The origin of these tumors remains unclear, as well as the molecular mechanisms that drive their clinical and biological behavior. However, genetic, and epigenetic mechanisms have been shown to directly and indirectly influence the USMT malignant transformation, but the high complexity of this group of tumors still represents a barrier to diagnosis and disease management. In this review, we provided insights into the most recent studies regarding epigenetic events in LMS and ESS, and their potential as novel biomarkers or for developing new therapeutic modalities to treat these tumors.

Author Contributions: B.C.d.A. and L.G.d.A. literature review, data collection and manuscript preparation; A.S.D., E.C.B., Q.Y. and A.A.-H. manuscript review, intellectual and technical support; K.C.C. idea conception, data collection and manuscript preparation, and review. All authors have read and agreed to the published version of the manuscript.

Funding: This research was funded by Fundação De Amparo à Pesquisa do Estado Do São Paulo (FAPESP 2019/01109-2) and Coordenação de Aperfeiçoamento de Pessoal De Nível Superior—Brasil (CAPES).

Institutional Review Board Statement: Not applicable.

Informed Consent Statement: Not applicable.

Data Availability Statement: Not applicable.

Acknowledgments: Not applicable.

Conflicts of Interest: The authors declare no conflict of interest.

References

1. Uduwela, A.S.; Perera, M.A.K.; Aiqing, L.; Fraser, I.S. Endometrial-myometrial interface: Relationship to adenomyosis and changes in pregnancy. *Obstet. Gynecol. Surv.* **2000**, *55*, 390–400. [CrossRef]
2. Commandeur, A.E.; Styer, A.K.; Teixeira, J.M. Epidemiological and genetic clues for molecular mechanisms involved in uterine leiomyoma development and growth. *Hum. Reprod. Update* **2015**, *21*, 593–615. [CrossRef] [PubMed]
3. Nothnick, W.B. Non-coding rnas in uterine development, function and disease. *Adv. Exp. Med. Biol.* **2016**, *886*, 171.
4. Brany, D.; Dvorska, D.; Nachajova, M.; Slavik, P.; Burjanivova, T. Malignant tumors of the uterine corpus: Molecular background of their origin. *Tumor Biol.* **2015**, *36*, 6615–6621. [CrossRef]
5. Kuperman, T.; Gavriel, M.; Gotlib, R.; Zhang, Y.; Jaffa, A.; Elad, D.; Grisaru, D. Tissue-engineered multi-cellular models of the uterine wall. *Biomech. Model. Mechanobiol.* **2020**, *19*, 1629–1639. [CrossRef] [PubMed]
6. De Almeida, T.G.; da Cunha, I.W.; Maciel, G.A.R.; Baracat, E.C.; Carvalho, K.C. Clinical and molecular features of uterine sarcomas. *Med. Exp.* **2014**, *1*, 291–297. [CrossRef]
7. Desar, I.M.E.; Ottevanger, P.B.; Benson, C.; van der Graaf, W.T.A. Systemic treatment in adult uterine sarcomas. *Crit. Rev. Oncol. Hematol.* **2018**, *122*, 10–20. [CrossRef]
8. Koivisto-Korander, R.; Butzow, R.; Koivisto, A.-M.; Leminen, A. Immunohistochemical studies on uterine carcinosarcoma, leiomyosarcoma, and endometrial stromal sarcoma: Expression and prognostic importance of ten different markers. *Tumour Biol.* **2011**, *32*, 451–459. [CrossRef] [PubMed]
9. American Cancer Society. What is Uterine Sarcoma? Available online: https://www.cancer.org/cancer/uterine-sarcoma/about/what-is-uterine-sarcoma.html (accessed on 21 April 2022).
10. Siegel, R.L.; Miller, K.D.; Fuchs, H.E.; Jemal, A. Cancer statistics, 2021. *CA Cancer J. Clin.* **2021**, *71*, 7–33. [CrossRef] [PubMed]
11. Analysis Tool American Cancer Society—Cancer Facts & Statistics. Available online: https://cancerstatisticscenter.cancer.org/#!/data-analysis/module/BmVYeqHT?type=barGraph (accessed on 15 April 2022).
12. Parra-Herran, C.; Howitt, B.E. Uterine mesenchymal tumors: Update on classification, staging, and molecular features. *Surg. Pathol. Clin.* **2019**, *12*, 363–396. [CrossRef]
13. Tsuyoshi, H.; Yoshida, Y. Molecular biomarkers for uterine leiomyosarcoma and endometrial stromal sarcoma. *Cancer Sci.* **2018**, *109*, 1743–1752. [CrossRef]
14. Mbatani, N.; Olawaiye, A.B.; Prat, J. Uterine sarcomas. *Int. J. Gynaecol. Obstet.* **2018**, *143*, 51–58. [CrossRef]
15. D'Angelo, E.; Prat, J. Uterine sarcomas: A review. *Gynecol. Oncol.* **2010**, *116*, 131–139. [CrossRef]
16. American Cancer Society. Uterine Sarcoma Stages. Available online: https://www.cancer.org/cancer/uterine-sarcoma/detection-diagnosis-staging/staging.html (accessed on 21 April 2022).
17. Roberts, M.E.; Aynardi, J.T.; Chu, C.S. Uterine leiomyosarcoma: A review of the literature and update on management options. *Gynecol. Oncol.* **2018**, *151*, 562–572. [CrossRef]
18. Sociedade Brasileira de Patologia. Útero Neoplasias Mesenquimais do Corpo Uterino. Available online: http://www.sbp.org.br/mdlhisto/utero-neoplasias-mesenquimais-corpo-uterino/ (accessed on 20 January 2022).
19. Laganà, A.S.; Vergara, D.; Favilli, A.; la Rosa, V.L.; Tinelli, A.; Gerli, S.; Noventa, M.; Vitagliano, A.; Triolo, O.; Rapisarda, A.M.C.; et al. Epigenetic and genetic landscape of uterine leiomyomas: A current view over a common gynecological disease. *Arch. Gynecol. Obstet.* **2017**, *296*, 855–867. [CrossRef] [PubMed]
20. Lu, Y.; Chan, Y.-T.; Tan, H.-Y.; Li, S.; Wang, N.; Feng, Y. Epigenetic regulation in human cancer: The potential role of epi-drug in cancer therapy. *Mol. Cancer* **2020**, *19*, 79. [CrossRef]
21. Navarro, A.; Yin, P.; Ono, M.; Monsivais, D.; Moravek, M.B.; Coon, J.S.V.; Dyson, M.T.; Wei, J.-J.; Bulun, S.E. 5-Hydroxymethylcytosine promotes proliferation of human uterine leiomyoma: A biological link to a new epigenetic modification in benign tumors. *J. Clin. Endocrinol. Metab.* **2014**, *99*, E2437. [CrossRef]
22. De Almeida, B.C.; dos Anjos, L.G.; Uno, M.; Cunha, I.W.; Soares, F.A.; Baiocchi, G.; Baracat, E.C.; Carvalho, K.C. Let-7 miRNA's expression profile and its potential prognostic role in uterine leiomyosarcoma. *Cells* **2019**, *8*, 1452. [CrossRef]
23. Tantitamit, T.; Huang, K.-G.; Manopunya, M.; Yen, C.-F. Outcome and management of uterine leiomyosarcoma treated following surgery for presumed benign disease: Review of literature. *Gynecol. Minim. Invasive Ther.* **2018**, *7*, 47–55. [CrossRef]
24. Benson, C.; Miah, A.B. Uterine sarcoma—Current perspectives. *Int. J. Womens Health* **2017**, *9*, 597–606. [CrossRef]
25. Seagle, B.L.L.; Sobecki-Rausch, J.; Strohl, A.E.; Shilpi, A.; Grace, A.; Shahabi, S. Prognosis and treatment of uterine leiomyosarcoma: A national cancer database study. *Gynecol. Oncol.* **2017**, *145*, 61–70. [CrossRef] [PubMed]
26. Bogani, G.; Ditto, A.; Martineli, F.; Signorelli, M.; Chiappa, V.; Fonatella, C.; Sanfilippo, R.; Leone Roberti Maggiore, U.; Ferrero, S.; Lorusso, D.; et al. Role of bevacizumab in uterine leiomyosarcoma. *Crit. Rev. Oncol. Hematol.* **2018**, *126*, 45–51. [CrossRef] [PubMed]
27. Juhasz-Böss, I.; Gabriel, L.; Bohle, R.M.; Horn, L.C.; Solomayer, E.F.; Breitbach, G.P. Uterine leiomyosarcoma. *Oncol. Res. Treat.* **2018**, *41*, 680–686. [CrossRef] [PubMed]

28. Ray-Coquard, I.; Serre, D.; Reichardt, P.; Martín-Broto, J.; Bauer, S. Options for treating different soft tissue sarcoma subtypes *Future Oncol.* **2018**, *14*, 25–49. [CrossRef] [PubMed]
29. Meng, Y.; Yang, Y.; Zhang, Y.; Li, X. Construction and validation of nomograms for predicting the prognosis of uterine leiomyosarcoma: A population-based study. *Med. Sci. Monit.* **2020**, *26*, e922739-1. [CrossRef]
30. Mittal, K.R.; Chen, F.; Wei, J.J.; Rijhvani, K.; Kurvathi, R.; Streck, D.; Dermody, J.; Toruner, G.A. Molecular and immunohistochemical evidence for the origin of uterine leiomyosarcomas from associated leiomyoma and symplastic leiomyoma-like areas. *Mod. Pathol.* **2009**, *22*, 1303–1311. [CrossRef] [PubMed]
31. Yasutake, N.; Ohishi, Y.; Taguchi, K.; Hiraki, Y.; Oya, M.; Oshiro, Y.; Mine, M.; Iwasaki, T.; Yamamoto, H.; Kohashi, K.; et al. Insulin-like growth factor II messenger RNA-binding protein-3 is an independent prognostic factor in uterine leiomyosarcoma. *Histopathology* **2017**, *12*, 3218–3221. [CrossRef]
32. Patel, D.; Handorf, E.; von Mehren, M.; Martin, L.; Movva, S. Adjuvant chemotherapy in uterine leiomyosarcoma: Trends and factors impacting usage. *Sarcoma* **2019**, *2019*, 3561501. [CrossRef] [PubMed]
33. Mittal, K.; Joutovsky, A. Areas with benign morphologic and immunohistochemical features are associated with some uterine leiomyosarcomas. *Gynecol. Oncol.* **2007**, *104*, 362–365. [CrossRef] [PubMed]
34. Banas, T.; Pitynski, K.; Okon, K.; Czerw, A. DNA fragmentation factors 40 and 45 (DFF40/DFF45) and B-cell lymphoma 2 (Bcl-2) protein are underexpressed in uterine leiomyosarcomas and may predict survival. *OncoTargets Ther.* **2017**, *10*, 4579–4589. [CrossRef]
35. Williams, A.R.W. Uterine fibroids—What's new? *F1000Research* **2017**, *6*, 2109. [CrossRef] [PubMed]
36. Zhang, P.; Zhang, C.; Hao, J.; Sung, C.J.; Quddus, M.R.; Steinhoff, M.M.; Lawrence, W.D. Use of X-chromosome inactivation pattern to determine the clonal origins of uterine leiomyoma and leiomyosarcoma. *Hum. Pathol.* **2006**, *37*, 1350–1356. [CrossRef] [PubMed]
37. Oh, J.; Park, S.B.; Park, H.J.; Lee, E.S. Ultrasound features of uterine sarcomas. *Ultrasound Q.* **2019**, *35*, 376–384. [CrossRef]
38. Chen, L.; Yang, B. Immunohistochemical analysis of P16, P53, and Ki-67 expression in uterine smooth muscle tumors. *Int. J. Gynecol. Pathol.* **2008**, *27*, 326–332. [CrossRef]
39. Loizzi, V.; Cormio, G.; Nestola, D.; Falagario, M.; Surgo, A.; Camporeale, A.; Putignano, G.; Selvaggi, L. Prognostic factors and outcomes in 28 cases of uterine leiomyosarcoma. *Oncology* **2011**, *81*, 91–97. [CrossRef]
40. Devereaux, K.A.; Schoolmeester, J.K. Smooth muscle tumors of the female genital tract. *Surg. Pathol. Clin.* **2019**, *12*, 397–455. [CrossRef]
41. Byar, K.L.; Fredericks, T. Uterine leiomyosarcoma. *J. Adv. Pract. Oncol.* **2022**, *13*, 70–76. [CrossRef]
42. Hoang, H.L.T.; Ensor, K.; Rosen, G.; Leon Pachter, H.; Raccuia, J.S. Prognostic factors and survival in patients treated surgically for recurrent metastatic uterine leiomyosarcoma. *Int. J. Surg. Oncol.* **2014**, *2014*, 919323. [CrossRef]
43. Friedman, C.F.; Hensley, M.L. Options for adjuvant therapy for uterine leiomyosarcoma. *Curr. Treat. Options Oncol.* **2018**, *19*, 7. [CrossRef]
44. Stope, M.B.; Cernat, V.; Kaul, A.; Diesing, K. Functionality of the tumor suppressor microrna-1 in malignant tissue and cell line cells of uterine leiomyosarcoma. *Anticancer Res.* **2018**, *1550*, 1547–1550. [CrossRef]
45. Rizzo, A.; Pantaleo, M.A.; Saponara, M.; Nannini, M. Current status of the adjuvant therapy in uterine sarcoma: A literature review. *World J. Clin. Cases* **2019**, *7*, 1753. [CrossRef] [PubMed]
46. Li, Y.; Ren, H.; Wang, J. Outcome of adjuvant radiotherapy after total hysterectomy in patients with uterine leiomyosarcoma or carcinosarcoma: A SEER-based study. *BMC Cancer* **2019**, *19*, 697. [CrossRef]
47. Amant, F.; Lorusso, D.; Mustea, A.; Duffaud, F.; Pautier, P. Management strategies in advanced uterine leiomyosarcoma: Focus on trabectedin. *Sarcoma* **2015**, *2015*, 704124. [CrossRef] [PubMed]
48. Barlin, J.N.; Zhou, Q.C.; Leitao, M.M.; Bisogna, M.; Olvera, N.; Shih, K.K.; Jacobsen, A.; Schultz, N.; Tap, W.D.; Hensley, M.L.; et al. Molecular subtypes of uterine leiomyosarcoma and correlation with clinical outcome. *Neoplasia* **2015**, *17*, 183–189. [CrossRef]
49. Ben-Ami, E.; Barysauskas, C.M.; Solomon, S.; Tahlil, K.; Malley, R.; Hohos, M.; Polson, K.; Loucks, M.; Severgnini, M.; Patel, T.; et al. Immunotherapy with single agent nivolumab for advanced leiomyosarcoma of the uterus: Results of a phase 2 study. *Cancer* **2017**, *123*, 3285–3290. [CrossRef] [PubMed]
50. Elvin, J.A.; Gay, L.M.; Ort, R.; Shuluk, J.; Long, J.; Shelley, L.; Lee, R.; Chalmers, Z.R.; Frampton, G.M.; Ali, S.M.; et al. Clinical benefit in response to palbociclib treatment in refractory uterine leiomyosarcomas with a common CDKN2A alteration. *Oncologist* **2017**, *22*, 416–421. [CrossRef]
51. American Cancer Society Survival. Rates for Uterine Sarcoma. Available online: https://www.cancer.org/cancer/uterine-sarcoma/detection-diagnosis-staging/survival-rates.html (accessed on 13 January 2022).
52. Kobayashi, H.; Uekuri, C.; Akasaka, J.; Ito, F.; Shigemitsu, A.; Koike, N.; Shigetomi, H. The biology of uterine sarcomas: A review and update. *Mol. Clin. Oncol.* **2013**, *1*, 599–609. [CrossRef] [PubMed]
53. Baek, M.-H.; Park, J.-Y.; Park, Y.; Kim, K.-R.; Kim, D.-Y.; Suh, D.-S.; Kim, J.-H.; Kim, Y.-M.; Kim, Y.-T.; Nam, J.-H. Androgen receptor as a prognostic biomarker and therapeutic target in uterine leiomyosarcoma. *J. Gynecol. Oncol.* **2018**, *29*, e30. [CrossRef]
54. Slomovitz, B.M.; Taub, M.C.; Huang, M.; Levenback, C.; Coleman, R.L. A randomized phase II study of letrozole vs. observation in patients with newly diagnosed uterine leiomyosarcoma (uLMS). *Gynecol. Oncol. Rep.* **2019**, *27*, 1. [CrossRef]
55. Chiang, S.; Oliva, E. Recent developments in uterine mesenchymal neoplasms. *Histopathology* **2013**, *62*, 124–137. [CrossRef]

6. Ducie, J.A.; Leitao, M.M., Jr. The role of adjuvant therapy in uterine leiomyosarcoma. *Expert Rev. Anticancer Ther.* **2016**, *16*, 45. [CrossRef] [PubMed]

7. Zang, Y.; Dong, M.; Zhang, K.; Gao, C.; Guo, F.; Wang, Y.; Xue, F. Hormonal therapy in uterine sarcomas. *Cancer Med.* **2019**, *8*, 1339–1349. [CrossRef]

8. Oliva, E.; Zaloudek, C.J.; Soslow, R.A. Mesenchymal tumors of the uterus. In *Blaustein's Pathology of the Female Genital Tract*; Springer International Publishing: Berlin/Heidelberg, Germany, 2019; pp. 535–647.

9. Gockley, A.A.; Rauh-Hain, J.A.; del Carmen, M.G. Uterine leiomyosarcoma: A review article. *Int. J. Gynecol. Cancer* **2014**, *24*, 1538–1542. [CrossRef] [PubMed]

10. Arend, R.C.; Toboni, M.D.; Montgomery, A.M.; Burger, R.A.; Olawaiye, A.B.; Monk, B.J.; Herzog, T.J. Systemic treatment of metastatic/recurrent uterine leiomyosarcoma: A changing paradigm. *Oncologist* **2018**, *23*, 1–13. [CrossRef] [PubMed]

11. Parra-Herran, C.; Schoolmeester, J.K.; Yuan, L.; Dal Cin, P.; Fletcher, C.D.; Quade, B.J.; Nucci, M.R. Myxoid leiomyosarcoma of the uterus: A clinicopathologic analysis of 30 cases and review of the literature with reappraisal of its distinction from other uterine myxoid mesenchymal neoplasms. *Am. J. Surg. Pathol.* **2016**, *40*, 285–301. [CrossRef]

12. Prayson, R.A.; Goldblum, J.R.; Hart, W.R. Epithelioid smooth-muscle tumors of the uterus: A clinicopathologic study of 18 patients. *Am. J. Surg. Pathol.* **1997**, *21*, 383–391. [CrossRef]

13. Guled, M.; Pazzaglia, L.; Borze, I.; Mosakhani, N.; Novello, C.; Benassi, M.S.; Knuutila, S. Differentiating soft tissue leiomyosarcoma and undifferentiated pleomorphic sarcoma: A miRNA analysis. *Genes Chromosomes Cancer* **2014**, *53*, 693–702. [CrossRef] [PubMed]

14. Ptáková, N.; Miesbauerová, M.; Kos, J.; Grossmann, P.; Henrieta, Š. Immunohistochemical and selected genetic reflex testing of all uterine leiomyosarcomas and STUMPs for ALK gene rearrangement may provide an effective screening tool in identifying uterine ALK-rearranged mesenchymal tumors. *Virchows Arch.* **2018**, *473*, 583–590. [CrossRef] [PubMed]

15. DeLair, D. Pathology of gynecologic cancer. In *Management of Gynecological Cancers in Older Women*; Springer: London, UK, 2013; pp. 21–38.

16. Dos Anjos, L.G.; da Cunha, I.W.; Baracat, E.C.; Carvalho, K.C. Genetic and epigenetic features in uterine smooth muscle tumors: An update. *Clin. Oncol.* **2019**, *4*, 1637.

17. An, Y.; Wang, S.; Li, S.; Zhang, L.; Wang, D.; Wang, H.; Zhu, S.; Zhu, W.; Li, Y.; Chen, W.; et al. Distinct molecular subtypes of uterine leiomyosarcoma respond differently to chemotherapy treatment. *BMC Cancer* **2017**, *17*, 17639. [CrossRef]

18. Cui, R.; Wright, J.; Hou, J. Uterine leiomyosarcoma: A review of recent advances in molecular biology, clinical management and outcome. *BJOG Int. J. Obstet. Gynaecol.* **2017**, *124*, 1028–1037. [CrossRef]

19. Kim, W.Y.; Chang, S.-J.; Chang, K.-H.; Yoon, J.-H.; Kim, J.H.; Kim, B.-G.; Bae, D.-S.; Ryu, H.-S. Uterine leiomyosarcoma: 14-year two-center experience of 31 cases. *Cancer Res. Treat.* **2009**, *41*, 24–28. [CrossRef] [PubMed]

20. Cordoba, A.; Prades, J.; Basson, L.; Robin, Y.M.; Taïeb, S.; Narducci, F.; Hudry, D.; Bresson, L.; Chevalier, A.; le Tinier, F.; et al. Adjuvant management of operated uterine sarcomas: A single institution experience. *Cancer Radiother.* **2019**, *23*, 401–407. [CrossRef] [PubMed]

21. Ricci, S.; Giuntoli, R.L.; Eisenhauer, E.; Lopez, M.A.; Krill, L.; Tanner, E.J.; Gehrig, P.A.; Havrilesky, L.J.; Secord, A.A.; Levinson, K.; et al. Does adjuvant chemotherapy improve survival for women with early-stage uterine leiomyosarcoma? *Gynecol. Oncol.* **2013**, *131*, 629–633. [CrossRef] [PubMed]

22. Micci, F.; Heim, S.; Panagopoulos, I. Molecular pathogenesis and prognostication of "low-grade" and 'high-grade' endometrial stromal sarcoma. *Genes Chromosomes Cancer* **2021**, *60*, 160–167. [CrossRef]

23. Puliyath, G.; Nair, M.K. Endometrial stromal sarcoma: A review of the literature. *Indian J. Med. Paediatr. Oncol.* **2012**, *33*, 1–6. [CrossRef]

24. Angelos, D.; Anastasios, L.; Dimosthenis, M.; Roxani, D.; Alexis, P.; Konstantinos, D. Endometrial stromal sarcoma presented as endometrial polyp: A rare case. *Gynecol. Surg.* **2020**, *17*, 1–3. [CrossRef]

25. Jabeen, S.; Anwar, S.; Fatima, N. Endometrial stromal sarcoma: A rare entity. *J. Coll. Physicians Surg. Pak.* **2015**, *25*, 216–217.

26. WHO Classification of Tumours Editorial Board. *Female Genital Tumours: WHO Classification of Tumours*, 5th ed.; IARC Publications: Lyon, France, 2020.

27. Zappacosta, R.; Fanfani, F.; Zappacosta, B.; Sablone, F.; Pansa, L.; Liberati, M.; Rosini, S. Uterine sarcomas: An updated overview part 2: Endometrial stromal tumor. In *Neoplasm*; IntechOpen: London, UK, 2018. [CrossRef]

28. Prat, J.; Mbatani, N. Uterine sarcomas. *Int. J. Gynaecol. Obstet.* **2015**, *131*, S105–S110. [CrossRef]

29. Hoang, L.; Chiang, S.; Lee, C.H. Endometrial stromal sarcomas and related neoplasms: New developments and diagnostic considerations. *Pathology* **2018**, *50*, 162–177. [CrossRef]

30. Subbaraya, S.; Murthy, S.S.; Devi, G.S. Immunohistochemical and molecular characterization of endometrial stromal sarcomas. *Clin. Pathol.* **2020**, *13*, 2632010X20916736. [CrossRef] [PubMed]

31. Conklin, C.M.J.; Longacre, T.A. Endometrial stromal tumors: The new who classification. *Adv. Anat. Pathol.* **2014**, *21*, 383–393. [CrossRef] [PubMed]

32. Gangireddy, M.; Chan Gomez, J.; Kanderi, T.; Joseph, M.; Kundoor, V. Recurrence of endometrial stromal sarcoma, two decades post-treatment. *Cureus* **2020**, *12*, e9249. [CrossRef]

33. DuPont, N.C.; Disaia, P.J. Recurrent endometrial stromal sarcoma: Treatment with a progestin and gonadotropin releasing hormone agonist. *Sarcoma* **2010**, *2010*, 353679. [CrossRef]

84. Serkies, K.; Pawłowska, E.; Jassem, J. Systemic therapy for endometrial stromal sarcomas: Current treatment options. *Ginekol. Pol.* **2016**, *87*, 594–597. [CrossRef]
85. Tuyaerts, S.; Amant, F. Endometrial stromal sarcomas: A Revision of their potential as targets for immunotherapy. *Vaccines* **2018**, *6*, 56. [CrossRef] [PubMed]
86. Nomura, Y.; Tamura, D. Detection of MEAF6-PHF1 translocation in an endometrial stromal nodule. *Genes Chromosomes Cancer.* **2020**, *59*, 702–708. [CrossRef]
87. Ali, R.H.; Rouzbahman, M. Endometrial stromal tumours revisited: An update based on the 2014 WHO classification. *J. Clin. Pathol.* **2015**, *68*, 325–332. [CrossRef]
88. Han, L.; Liu, Y.J.; Ricciotti, R.W.; Mantilla, J.G. A novel MBTD1-PHF1 gene fusion in endometrial stromal sarcoma: A case report and literature review. *Genes Chromosomes Cancer* **2020**, *59*, 428–432. [CrossRef]
89. Da Costa, L.T.; dos Anjos, L.G.; Kagohara, L.T.; Torrezan, G.T.; de Paula, C.A.A.; Baracat, E.C.; Carraro, D.M.; Carvalho, K.C. The mutational repertoire of uterine sarcomas and carcinosarcomas in a Brazilian cohort: A preliminary study. *Clinics* **2021**, *76*, 1–15. [CrossRef]
90. Mansor, S.; Kuick, C.H.; Lim-Tan, S.K.; Leong, M.Y.; Lim, T.Y.K.; Chang, K.T.E. 28. Novel fusion MAGED2-PLAG1 in endometrial stromal sarcoma. *Pathology* **2020**, *52*, S142–S143. [CrossRef]
91. Makise, N.; Sekimizu, M.; Kobayashi, E.; Yoshida, H.; Fukayama, M.; Kato, T.; Kawai, A.; Ichikawa, H.; Yoshida, A. Low-grade endometrial stromal sarcoma with a novel MEAF6-SUZ12 fusion. *Virchows Arch.* **2019**, *475*, 527–531. [CrossRef]
92. Brunetti, M.; Gorunova, L.; Davidson, B.; Heim, S.; Panagopoulos, I.; Micci, F. Identification of an EPC2-PHF1 fusion transcript in low-grade endometrial stromal sarcoma. *Oncotarget* **2018**, *9*, 19203–19208. [CrossRef]
93. Lee, C.H.; Nucci, M.R. Endometrial stromal sarcoma—The new genetic paradigm. *Histopathology* **2015**, *67*, 1–19. [CrossRef] [PubMed]
94. Chiang, S.; Lee, C.H.; Stewart, C.J.R.; Oliva, E.; Hoang, L.N.; Ali, R.H.; Hensley, M.L.; Arias-Stella, J.A., 3rd; Frosina, D.; Jungbluth, A.A.; et al. BCOR is a robust diagnostic immunohistochemical marker of genetically diverse high-grade endometrial stromal sarcoma, including tumors exhibiting variant morphology. *Mod. Pathol.* **2017**, *30*, 1251–1261. [CrossRef]
95. Akaev, I.; Yeoh, C.C.; Rahimi, S. Update on endometrial stromal tumours of the uterus. *Diagnostics* **2021**, *11*, 429. [CrossRef]
96. Dickson, B.C.; Lum, A.; Swanson, D.; Bernardini, M.Q.; Colgan, T.J.; Shaw, P.A.; Yip, S.; Lee, C.H. Novel EPC1 gene fusions in endometrial stromal sarcoma. *Genes Chromosomes Cancer* **2018**, *57*, 598–603. [CrossRef]
97. Cotzia, P.; Benayed, R.; Mullaney, K.; Oliva, E.; Felix, A.; Ferreira, J.; Soslow, R.A.; Antonescu, C.R.; Ladanyi, M.; Chiang, S. Undifferentiated uterine sarcomas represent under-recognized high-grade endometrial stromal sarcomas. *Am. J. Surg. Pathol.* **2019**, *43*, 662–669. [CrossRef]
98. Momeni-Boroujeni, A.; Mohammad, N.; Wolber, R.; Yip, S.; Köbel, M.; Dickson, B.C.; Hensley, M.L.; Leitao, M.M.; Antonescu, C.R.; Benayed, R.; et al. Targeted RNA expression profiling identifies high-grade endometrial stromal sarcoma as a clinically relevant molecular subtype of uterine sarcoma. *Mod. Pathol.* **2021**, *34*, 1008–1016. [CrossRef]
99. Murakami, I.; Tanaka, K.; Shiraishi, J.; Yamashita, H. SMARCA4-deficient undifferentiated uterine sarcoma: A case report *Gynecol. Obstet. Case Rep.* **2021**, *7*, 128.
100. Boyle, W.; Williams, A.; Sundar, S.; Yap, J.; Taniere, P.; Rehal, P.; Ganesan, R. TMP3-NTRK1 rearranged uterine sarcoma: A case report. *Case Rep. Womens Health* **2020**, *28*, e00246. [CrossRef] [PubMed]
101. Chiang, S.; Cotzia, P. NTRK fusions define a novel uterine sarcoma subtype with features of fibrosarcoma. *Am. J. Surg. Pathol.* **2018**, *42*, 791–798. [CrossRef] [PubMed]
102. Nebbioso, A.; Tambaro, F.P.; Dell'Aversana, C.; Altucci, L. Cancer epigenetics: Moving forward. *PLoS Genet.* **2018**, *14*, e1007362. [CrossRef]
103. Jones, P.A.; Laird, P.W. Cancer epigenetics comes of age. *Nat. Genet.* **1999**, *21*, 163–167. [CrossRef] [PubMed]
104. Cavalli, G.; Heard, E. Advances in epigenetics link genetics to the environment and disease. *Nature* **2019**, *571*, 489–499. [CrossRef]
105. Verma, M.; Rogers, S.; Divi, R.L.; Schully, S.D.; Nelson, S.; Su, L.J.; Ross, S.A.; Pilch, S.; Winn, D.M.; Khoury, M.J. Epigenetic research in cancer epidemiology: Trends, opportunities, and challenges. *Cancer Epidemiol. Biomark. Prev.* **2014**, *23*, 223–233 [CrossRef] [PubMed]
106. Herceg, Z.; Vaissière, T. Epigenetic mechanisms and cancer an interface between the environment and the genome. *Epigenetics* **2011**, *6*, 804–819. [CrossRef]
107. Werner, R.J.; Kelly, A.D.; Issa, J.P.J. Epigenetics and precision oncology. *Cancer J.* **2017**, *23*, 262–269. [CrossRef]
108. Kanwal, R.; Gupta, K.; Gupta, S. Cancer epigenetics: An introduction. *Methods Mol. Biol.* **2015**, *1238*, 3–25. [CrossRef]
109. Topper, M.J.; Vaz, M.; Marrone, K.A.; Brahmer, J.R.; Baylin, S.B. The emerging role of epigenetic therapeutics in immuno-oncology. *Nat. Rev. Clin. Oncol.* **2020**, *17*, 75–90. [CrossRef]
110. Wei, J.W.; Huang, K.; Yang, C.; Kang, C.S. Non-coding RNAs as regulators in epigenetics (Review). *Oncol. Rep.* **2017**, *37*, 3–9 [CrossRef]
111. Hosseini, A.; Minucci, S. Alterations of histone modifications in cancer. In *Epigenetics in Human Disease*; Elsevier: Amsterdam, The Netherlands, 2018; pp. 141–217. [CrossRef]
112. Zhao, Z.; Shilatifard, A. Epigenetic modifications of histones in cancer. *Genome Biol.* **2019**, *20*, 1–16. [CrossRef]
113. Caplakova, V.; Babusikova, E.; Blahovcova, E.; Balharek, T.; Zelieskova, M.; Hatok, J. DNA methylation machinery in the endometrium and endometrial cancer. *Anticancer Res.* **2016**, *36*, 4407–4420. [CrossRef]

Biomedicines **2022**, *10*, 2567

114. Kagohara, L.T.; Stein-O'Brien, G.L.; Kelley, D.; Flam, E.; Wick, H.C.; Danilova, L.V.; Easwaran, H.; Favorov, A.V.; Qian, J.; Gaykalova, D.A.; et al. Epigenetic regulation of gene expression in cancer: Techniques, resources and analysis. *Brief Funct. Genom.* **2018**, *17*, 49–63. [CrossRef] [PubMed]

115. Feinberg, A.P.; Vogelstein, B. Hypomethylation distinguishes genes of some human cancers from their normal counterparts. *Nature* **1983**, *301*, 89–92. [CrossRef]

116. Fischer, C.D.C.; Hu, Y.; Morreale, M.; Lin, W.Y.; Wali, A.; Thakar, M.; Karunasena, E.; Sen, R.; Cai, Y.; Murphy, L.; et al. Treatment with epigenetic agents profoundly inhibits tumor growth in leiomyosarcoma. *Oncotarget* **2018**, *9*, 19379–19395. [CrossRef]

117. Bird, A.P. DNA methylation–how important in gene control? *Nature* **1984**, *307*, 503–504. [CrossRef]

118. Gnyszka, A.; Jastrzebski, Z.; Flis, S. DNA methyltransferase inhibitors and their emerging role in epigenetic therapy of cancer. *Anticancer Res.* **2013**, *33*, 2989–2996.

119. Garcia, N.; Al-Hendy, A.; Baracat, E.C.; Carvalho, K.C.; Yang, Q. Targeting hedgehog pathway and DNA methyltransferases in uterine leiomyosarcoma cells. *Cells* **2020**, *10*, 53. [CrossRef]

120. Ferriss, J.S.; Atkins, K.A.; Lachance, J.A.; Modesitt, S.C.; Jazaeri, A.A. Temozolomide in advanced and recurrent uterine leiomyosarcoma and correlation with O6-methylguanine DNA methyltransferase expression a case series. *Int. J. Gynecol. Cancer* **2010**, *20*, 120–125. [CrossRef] [PubMed]

121. Bujko, M.; Kowalewska, M.; Danska-Bidzinska, A.; Bakula-Zalewska, E.; Siedecki, J.A.; Bidzinski, M. The promoter methylation and expression of the O6-methylguanine-DNA methyltransferase gene in uterine sarcoma and carcinosarcoma. *Oncol. Lett.* **2012**, *4*, 551–555. [CrossRef] [PubMed]

122. Romero-Garcia, S.; Prado-Garcia, H.; Carlos-Reyes, A. Role of DNA methylation in the resistance to therapy in solid tumors. *Front. Oncol.* **2020**, *10*, 1152. [CrossRef]

123. Brany, D.; Dvorská, D.; Grendár, M.; Ňachajová, M.; Szépe, P.; Lasabová, Z.; Žúbor, P.; Višňovský, J.; Halášová, E. Different methylation levels in the KLF4, ATF3 and DLEC1 genes in the myometrium and in corpus uteri mesenchymal tumours as assessed by MS-HRM. *Pathol. Res. Pract.* **2019**, *215*, 152465. [CrossRef]

124. Hasan, N.M.; Sharma, A.; Ruzgar, N.M.; Deshpande, H.; Olino, K.; Khan, S.; Ahuja, N. Epigenetic signatures differentiate uterine and soft tissue leiomyosarcoma. *Oncotarget* **2021**, *12*, 1566–1579. [CrossRef] [PubMed]

125. Baylin, S.B. DNA methylation and gene silencing in cancer. *Nat. Clin. Pract. Oncol.* **2005**, *2*, S4–S11. [CrossRef] [PubMed]

126. Xing, D.; Scangas, G.; Nitta, M.; He, L.; Xu, A.; Ioffe, Y.J.M.; Aspuria, P.J.; Hedvat, C.Y.; Anderson, M.L.; Oliva, E.; et al. A role for BRCA1 in uterine leiomyosarcoma. *Cancer Res.* **2009**, *69*, 8231–8235. [CrossRef]

127. Roncati, L.; Barbolini, G.; Sartori, G.; Siopis, E.; Pusiol, T.; Maiorana, A. Loss of CDKN2A promoter methylation coincides with the epigenetic transdifferentiation of uterine myosarcomatous cells. *Int. J. Gynecol. Pathol.* **2016**, *35*, 309–315. [CrossRef]

128. Kommoss, F.K.F.; Stichel, D.; Schrimpf, D.; Kriegsmann, M.; Tessier-Cloutier, B.; Talhouk, A.; McAlpine, J.N.; Chang, K.T.E.; Sturm, D.; Pfister, S.M.; et al. DNA methylation-based profiling of uterine neoplasms: A novel tool to improve gynecologic cancer diagnostics. *J. Cancer Res. Clin. Oncol.* **2020**, *146*, 97–104. [CrossRef]

129. Kommoss, F.K.; Chang, K.T.; Stichel, D.; Banito, A.; Jones, D.T.; Heilig, C.E.; Fröhling, S.; Sahm, F.; Stenzinger, A.; Hartmann, W.; et al. Endometrial stromal sarcomas with BCOR-rearrangement harbor MDM2 amplifications. *J. Pathol. Clin. Res.* **2020**, *6*, 178–184. [CrossRef]

130. Li, J.; Xing, X.; Li, D.; Zhang, B.; Mutch, D.G.; Hagemann, I.S.; Wang, T. Whole-Genome DNA methylation profiling identifies epigenetic signatures of uterine carcinosarcoma. *Neoplasia* **2017**, *19*, 100–111. [CrossRef]

131. Chiang, S.; Oliva, E. Cytogenetic and molecular aberrations in endometrial stromal tumors. *Hum. Pathol.* **2011**, *42*, 609–617. [CrossRef] [PubMed]

132. Kurihara, S.; Oda, Y.; Ohishi, Y.; Kaneki, E.; Kobayashi, H.; Wake, N.; Tsuneyoshi, M. Coincident expression of β-catenin and cyclin D1 in endometrial stromal tumors and related high-grade sarcomas. *Mod. Pathol.* **2010**, *23*, 225–234. [CrossRef]

133. Hrzenjak, A.; Dieber-Rotheneder, M.; Moinfar, F.; Petru, E.; Zatloukal, K. Molecular mechanisms of endometrial stromal sarcoma and undifferentiated endometrial sarcoma as premises for new therapeutic strategies. *Cancer Lett.* **2014**, *354*, 21–27. [CrossRef] [PubMed]

134. Matsukura, H.; Aisaki, K.I.; Igarashi, K.; Matsushima, Y.; Kanno, J.; Muramatsu, M.; Sudo, K.; Sato, N. Genistein promotes DNA demethylation of the steroidogenic factor 1 (SF-1) promoter in endometrial stromal cells. *Biochem. Biophys. Res. Commun.* **2011**, *412*, 366–372. [CrossRef] [PubMed]

135. Maekawa, R.; Tamura, I.; Shinagawa, M.; Mihara, Y.; Sato, S.; Okada, M.; Taketani, T.; Tamura, H.; Sugino, N. Genome-wide DNA methylation analysis revealed stable DNA methylation status during decidualization in human endometrial stromal cells. *BMC Genom.* **2019**, *20*, 324. [CrossRef]

136. Tamura, I.; Maekawa, R.; Jozaki, K.; Ohkawa, Y.; Takagi, H.; Doi-Tanaka, Y.; Shirafuta, Y.; Mihara, Y.; Taketani, T.; Sato, S.; et al. Transcription factor C/EBPβ induces genome-wide H3K27ac and upregulates gene expression during decidualization of human endometrial stromal cells. *Mol. Cell. Endocrinol.* **2021**, *520*, 111085. [CrossRef]

137. Nie, J.; Liu, X.; Sun-Wei, G. Promoter hypermethylation of progesterone receptor isoform B (PR-B) in adenomyosis and its rectification by a histone deacetylase inhibitor and a demethylation agent. *Reprod. Sci.* **2010**, *17*, 995–1005. [CrossRef]

138. Gillette, T.G.; Hill, J.A. Readers, writers, and erasers: Chromatin as the whiteboard of heart disease. *Circ. Res.* **2015**, *116*, 1245–1253. [CrossRef]

139. Luger, K.; Hansen, J.C. Nucleosome and chromatin fiber dynamics. *Curr. Opin. Struct. Biol.* **2005**, *5*, 188–196. [CrossRef]

140. Mastoraki, A.; Schizas, D.; Vlachou, P.; Melissaridou, N.M.; Charalampakis, N.; Fioretzaki, R.; Kole, C.; Savvidou, O.; Vassiliu, P.; Pikoulis, E. Assessment of synergistic contribution of histone deacetylases in prognosis and therapeutic management of sarcoma *Mol. Diagn. Ther.* **2020**, *24*, 557–569. [CrossRef]
141. Gujral, P.; Mahajan, V.; Lissaman, A.C.; Ponnampalam, A.P. Histone acetylation and the role of histone deacetylases in normal cyclic endometrium. *Reprod. Biol. Endocrinol.* **2020**, *18*, 1–11. [CrossRef] [PubMed]
142. Bannister, A.J.; Kouzarides, T. Regulation of chromatin by histone modifications. *Cell Res.* **2011**, *21*, 381–395. [CrossRef] [PubMed]
143. Biswas, S.; Rao, C.M. Epigenetic tools (the writers, the readers and the erasers) and their implications in cancer therapy. *Eur. J. Pharmacol.* **2018**, *837*, 8–24. [CrossRef] [PubMed]
144. Martin, C.; Zhang, Y. The diverse functions of histone lysine methylation. *Nat. Rev. Mol. Cell Biol.* **2005**, *6*, 838–849. [CrossRef] [PubMed]
145. Grimaldi, G.; Christian, M.; Steel, J.H.; Henriet, P.; Poutanen, M.; Brosens, J.J. Down-Regulation of the histone methyltransferase EZH2 contributes to the epigenetic programming of decidualizing human endometrial stromal cells. *Mol. Endocrinol.* **2011**, *25*, 1892. [CrossRef] [PubMed]
146. Guan, M.; Wu, X.; Chu, P.; Chow, W.A. Fatty acid synthase reprograms the epigenome in uterine leiomyosarcomas. *PLoS ONE* **2017**, *12*, e0179692. [CrossRef]
147. Dunican, D.S.; Mjoseng, H.K.; Duthie, L.; Flyamer, I.M.; Bickmore, W.A.; Meehan, R.R. Bivalent promoter hypermethylation in cancer is linked to the H327me3/H3K4me3 ratio in embryonic stem cells. *BMC Biol.* **2020**, *18*, 1–21. [CrossRef] [PubMed]
148. Piunti, A.; Shilatifard, A. The roles of polycomb repressive complexes in mammalian development and cancer. *Nat. Rev. Mol. Cell Biol.* **2021**, *22*, 326–345. [CrossRef]
149. Kim, H.; Ekram, M.B.; Bakshi, A.; Kim, J. AEBP2 as a transcriptional activator and its role in cell migration. *Genomics* **2015**, *105*, 108–115. [CrossRef]
150. Tan, J.Z.; Yan, Y.; Wang, X.X.; Jiang, Y.; Xu, H.E. EZH2: Biology, disease, and structure-based drug discovery. *Acta Pharmacol. Sin.* **2014**, *35*, 161–174. [CrossRef] [PubMed]
151. Piunti, A.; Smith, E.R.; Morgan, M.A.J.; Ugarenko, M.; Khaltyan, N.; Helmin, K.A.; Ryan, C.A.; Murray, D.C.; Rickels, R.A.; Yilmaz, B.D.; et al. CATACOMB: An endogenous inducible gene that antagonizes H3K27 methylation activity of Polycomb repressive complex 2 via an H3K27M-like mechanism. *Sci. Adv.* **2019**, *5*, 2887. [CrossRef] [PubMed]
152. Dewaele, B.; Przybyl, J.; Quattrone, A.; Ferreiro, J.F.; Vanspauwen, V.; Geerdens, E.; Gianfelici, V.; Kalender, Z.; Wozniak, A.; Moerman, P.; et al. Identification of a novel, recurrent MBTD1-CXorf67 fusion in low-grade endometrial stromal sarcoma. *Int. J. Cancer* **2014**, *134*, 1112–1122. [CrossRef] [PubMed]
153. Levine, S.S.; Weiss, A.; Erdjument-Bromage, H.; Shao, Z.; Tempst, P.; Kingston, R.E. The core of the polycomb repressive complex is compositionally and functionally conserved in flies and humans. *Mol. Cell. Biol.* **2002**, *22*, 6070. [CrossRef] [PubMed]
154. Oliviero, G.; Munawar, N.; Watson, A.; Streubel, G.; Manning, G.; Bardwell, V.; Bracken, A.P.; Cagney, G. The variant polycomb repressor complex 1 component PCGF1 interacts with a pluripotency sub-network that includes DPPA4, a regulator of embryogenesis. *Sci. Rep.* **2015**, *5*, 1–11. [CrossRef] [PubMed]
155. Astolfi, A.; Fiore, M.; Melchionda, F.; Indio, V.; Bertuccio, S.N.; Pession, A. BCOR involvement in cancer. *Epigenomics* **2019**, *11*, 835. [CrossRef]
156. Panagopoulos, I.; Thorsen, J.; Gorunova, L.; Haugom, L.; Bjerkehagen, B.; Davidson, B.; Heim, S.; Micci, F. Fusion of the ZC3H7B and BCOR genes in endometrial stromal sarcomas carrying an X;22-translocation. *Genes Chromosomes Cancer* **2013**, *52*, 610–618. [CrossRef]
157. Chase, A.; Cross, N.C. Aberrations of EZH2 in cancer. *Clin. Cancer Res.* **2011**, *17*, 2613–2618. [CrossRef]
158. Micci, F.; Gorunova, L.; Agostini, A.; Johannessen, L.E.; Brunetti, M.; Davidson, B.; Heim, S.; Panagopoulos, I. Cytogenetic and molecular profile of endometrial stromal sarcoma. *Genes Chromosomes Cancer* **2016**, *55*, 834–846. [CrossRef]
159. Aldera, A.P.; Govender, D. Gene of the month: BCOR. *J. Clin. Pathol.* **2020**, *73*, 314–317. [CrossRef]
160. Juckett, L.T.; Lin, D.I.; Madison, R.; Ross, J.S.; Schrock, A.B.; Ali, S. A pan-cancer landscape analysis reveals a subset of endometrial stromal and pediatric tumors defined by internal tandem duplications of BCOR. *Oncology* **2019**, *96*, 101–109. [CrossRef] [PubMed]
161. Allen, A.J.; Ali, S.M.; Gowen, K.; Elvin, J.A.; Pejovic, T. A recurrent endometrial stromal sarcoma harbors the novel fusion JAZF1-BCORL1. *Gynecol. Oncol. Rep.* **2017**, *20*, 51–53. [CrossRef] [PubMed]
162. Monaghan, L.; Massett, M.E.; Bunschoten, R.P.; Hoose, A.; Pirvan, P.-A.; Liskamp, R.M.J.; Jørgensen, H.G.; Huang, X. The emerging role of H3K9me3 as a potential therapeutic target in acute myeloid leukemia. *Front. Oncol.* **2019**, *9*, 705. [CrossRef]
163. Brahmi, M.; Franceschi, T.; Treilleux, I.; Pissaloux, D.; Ray-Coquard, I.; Dufresne, A.; Vanacker, H.; Carbonnaux, M.; Meeus, P.; Sunyach, M.P.; et al. Molecular classification of endometrial stromal sarcomas using RNA sequencing defines nosological and prognostic subgroups with different natural history. *Cancers* **2020**, *12*, 2604. [CrossRef]
164. Przybyl, J.; Kidzinski, L.; Hastie, T.; Debiec-Rychter, M.; Nusse, R.; van de Rijn, M. Gene expression profiling of low-grade endometrial stromal sarcoma indicates fusion protein-mediated activation of the Wnt signaling pathway. *Gynecol. Oncol.* **2018**, *149*, 388–393. [CrossRef]
165. Itoh, Y.; Takada, Y.; Yamashita, Y.; Suzuki, T. Recent progress on small molecules targeting epigenetic complexes. *Curr. Opin. Chem. Biol.* **2022**, *67*, 102130. [CrossRef] [PubMed]
166. Piunti, A.; Pasini, D. Epigenetic factors in cancer development: Polycomb group proteins. *Future Oncol.* **2011**, *7*, 57–75. [CrossRef]

67. Gao, J.; Yang, T.; Wang, X.; Zhang, Y.; Wang, J.; Zhang, B.; Tang, D.; Liu, Y.; Gao, T.; Lin, Q.; et al. Identification and characterization of a subpopulation of CD133+ cancer stem-like cells derived from SK-UT-1 cells. *Cancer Cell Int.* **2021**, *8*, 157. [CrossRef]
68. Zhang, N.; Zeng, Z.; Li, S.; Wang, F.; Huang, P. High expression of EZH2 as a marker for the differential diagnosis of malignant and benign myogenic tumors. *Sci. Rep.* **2018**, *8*, 12331. [CrossRef]
69. Micci, F.; Brunetti, M.; Cin, P.D.; Nucci, M.R.; Gorunova, L.; Heim, S.; Panagopoulos, I. Fusion of the genes BRD8 and PHF1 in endometrial stromal sarcoma. *Genes Chromosomes Cancer* **2017**, *56*, 841–845. [CrossRef] [PubMed]
70. Hrzenjak, A. JAZF1/SUZ12 gene fusion in endometrial stromal sarcomas. *Orphanet J. Rare Dis.* **2016**, *11*, 1–8. [CrossRef]
71. Hodge, J.C.; Bedroske, P.P.; Pearce, K.E.; Sukov, W.R. Molecular cytogenetic analysis of JAZF1, PHF1, and YWHAE in endometrial stromal tumors: Discovery of genetic complexity by fluorescence in situ Hybridization. *J. Mol. Diagn.* **2016**, *18*, 516–526. [CrossRef]
72. Tavares, M.; Khandelwal, G.; Mutter, J.; Viiri, K.; Beltran, M.; Brosens, J.J.; Jenner, R.G. JAZF1-SUZ12 dysregulates PRC2 function and gene expression during cell differentiation. *bioRxiv* **2021**, *15*, 440052. [CrossRef] [PubMed]
73. Sudarshan, D.; Avvakumov, N.; Lalonde, M.-E.; Alerasool, N.; Jacquet, K.; Mameri, A.; Rousseau, J.; Lambert, J.-P.; Paquet, E.; Setty, S.T.; et al. Recurrent chromosomal translocations in sarcomas create a mega-complex that mislocalizes NuA4/TIP60 to polycomb target loci. *bioRxiv* **2021**, *26*, 436670. [CrossRef]
74. Davidson, B.; Matias-Guiu, X.; Lax, S.F. The clinical, morphological, and genetic heterogeneity of endometrial stromal sarcoma. *Virchows Arch.* **2020**, *476*, 489–490. [CrossRef] [PubMed]
75. Lee, C.H.; Ou, W.B.; Mariño-Enriquez, A.; Zhu, M.; Mayeda, M.; Wang, Y.; Guo, X.; Brunner, A.L.; Amant, F.; French, C.A.; et al. 14-3-3 fusion oncogenes in high-grade endometrial stromal sarcoma. *Proc. Natl. Acad. Sci. USA* **2012**, *109*, 929–934. [CrossRef] [PubMed]
76. Panagopoulos, I.; Micci, F.; Thorsen, J.; Gorunova, L.; Eibak, A.M.; Bjerkehagen, B.; Davidson, B.; Heim, S. Novel fusion of MYST/Esa1-associated factor 6 and PHF1 in endometrial stromal sarcoma. *PLoS ONE* **2012**, *7*, e39354. [CrossRef]
77. Fiskus, W.; Pranpat, M.; Balasis, M.; Herger, B.; Rao, R.; Chinnaiyan, A.; Atadja, P.; Bhalla, K. Histone deacetylase inhibitors deplete enhancer of zeste 2 and associated polycomb repressive complex 2 proteins in human acute leukemia cells. *Mol. Cancer Ther.* **2006**, *5*, 3096–3104. [CrossRef]
78. Di Giorgio, E.; Dalla, E.; Franforte, E.; Paluvai, H.; Minisini, M.; Trevisanut, M.; Picco, R.; Brancolini, C. Different class IIa HDACs repressive complexes regulate specific epigenetic responses related to cell survival in leiomyosarcoma cells. *Nucleic Acids Res.* **2020**, *48*, 646–664. [CrossRef]
79. Park, S.Y.; Kim, J.S. A short guide to histone deacetylases including recent progress on class II enzymes. *Exp. Mol. Med.* **2020**, *52*, 204–212. [CrossRef]
80. Choy, E.; Ballman, K.; Chen, J.; Dickson, M.A.; Chugh, R.; George, S.; Okuno, S.; Pollock, R.; Patel, R.M.; Hoering, A.; et al. SARC018-SPORE02: Phase II study of mocetinostat administered with gemcitabine for patients with metastatic leiomyosarcoma with progression or relapse following prior treatment with gemcitabine-containing therapy. *Sarcoma* **2018**, *2018*, 2068517. [CrossRef] [PubMed]
81. Miller, H.; Ike, C.; Parma, J.; Masand, R.P.; Mach, C.M.; Anderson, M.L. Molecular targets and emerging therapeutic options for uterine leiomyosarcoma. *Sarcoma* **2016**, *2016*, 7018106. [CrossRef] [PubMed]
82. Baek, M.-H.; Park, J.-Y.; Rhim, C.C.; Kim, J.-H.; Park, Y.; Kim, K.-R.; Nam, J.-H. Investigation of new therapeutic targets in undifferentiated endometrial sarcoma. *Gynecol. Obstet. Investig.* **2017**, *82*, 329–339. [CrossRef] [PubMed]
83. Rafehi, H.; Karagiannis, T.C.; El-Osta, A. Pharmacological histone deacetylation distinguishes transcriptional regulators. *Curr. Top. Med. Chem.* **2017**, *17*, 1611–1622. [CrossRef] [PubMed]
84. Monga, V.; Swami, U.; Tanas, M.; Bossler, A.; Mott, S.L.; Smith, B.J.; Milhem, M. A phase I/II study targeting angiogenesis using bevacizumab combined with chemotherapy and a histone deacetylase inhibitor (Valproic acid) in advanced sarcomas. *Cancers* **2018**, *10*, 53. [CrossRef] [PubMed]
85. Yen, M.S.; Chen, J.R.; Wang, P.H.; Wen, K.C.; Chen, Y.J.; Ng, H.T. Uterine sarcoma part III—Targeted therapy: The taiwan association of gynecology (TAG) systematic review. *Taiwan. J. Obstet. Gynecol.* **2016**, *55*, 625–634. [CrossRef] [PubMed]
86. Quan, P.; Moinfar, F.; Kufferath, I.; Absenger, M.; Kueznik, T.; Denk, H.; Zatloukal, K.; Haybaeck, J. Effects of targeting endometrial stromal sarcoma cells via histone deacetylase and PI3K/AKT/mTOR signaling. *Anticancer Res.* **2014**, *34*, 2883–2897.
87. Tang, F.; Choy, E.; Tu, C.; Hornicek, F.; Duan, Z. Therapeutic applications of histone deacetylase inhibitors in sarcoma. *Cancer Treat. Rev.* **2017**, *59*, 33–45. [CrossRef]
88. Rikiishi, H. Autophagic and apoptotic effects of HDAC inhibitors on cancer cells. *J. Biomed. Biotechnol.* **2011**, *2011*, 830260. [CrossRef]
89. Fröhlich, L.F.; Mrakovcic, M.; Smole, C.; Lahiri, P.; Zatloukal, K. Epigenetic silencing of apoptosis-inducing gene expression can be efficiently overcome by combined saha and trail treatment in uterine sarcoma cells. *PLoS ONE* **2014**, *9*, e91558. [CrossRef]
90. Lin, C.Y.; Chao, A.; Wu, R.C.; Lee, L.Y.; Ueng, S.H.; Tsai, C.L.; Lee, Y.S.; Peng, M.T.; Yang, L.Y.; Huang, H.; et al. Synergistic effects of pazopanib and hyperthermia against uterine leiomyosarcoma growth mediated by downregulation of histone acetyltransferase 1. *J. Mol. Med.* **2020**, *98*, 1175–1188. [CrossRef]
91. Sawicka, A.; Seiser, C. Histone H3 phosphorylation—A versatile chromatin modification for different occasions. *Biochimie* **2012**, *94*, 2193. [CrossRef] [PubMed]
92. Veras, E.; Malpica, A.; Deavers, M.T.; Silva, E.G. Mitosis-specific marker phospho-histone h3 in the assessment of mitotic index in uterine smooth muscle tumors: A pilot study. *Int. J. Gynecol. Pathol.* **2009**, *28*, 316–321. [CrossRef] [PubMed]

193. Jain, A.K.; Barton, M.C. Bromodomain Histone Readers and Cancer. *J. Mol. Biol.* **2017**, *429*, 2003–2010. [CrossRef] [PubMed]
194. Fujisawa, T.; Filippakopoulos, P. Functions of bromodomain-containing proteins and their roles in homeostasis and cancer. *Nat. Rev. Mol. Cell Biol.* **2017**, *18*, 246–262. [CrossRef]
195. Sima, X.; He, J.; Peng, J.; Xu, Y.; Zhang, F.; Deng, L. The genetic alteration spectrum of the SWI/SNF complex: The oncogenic role of BRD9 and ACTL6A. *PLoS ONE* **2019**, *14*, e0222305. [CrossRef]
196. Barma, N.; Stone, T.C.; Carmona Echeverria, L.M.; Heavey, S. Exploring the Value of *BRD9* as a Biomarker, Therapeutic Target and Co-Target in Prostate Cancer. *Biomolecules* **2021**, *11*, 1794. [CrossRef]
197. Wang, X.; Wang, S.; Troisi, E.C.; Howard, T.P.; Haswell, J.R.; Wolf, B.K.; Hawk, W.H.; Ramos, P.; Oberlick, E.M.; Tzvetkov, E.P.; et al. BRD9 defines a SWI/SNF sub-complex and constitutes a specific vulnerability in malignant rhabdoid tumors. *Nat. Commun.* **2019**, *10*, 1881. [CrossRef]
198. Zhu, X.; Liao, Y.; Tang, L. Targeting BRD9 for Cancer Treatment: A New Strategy. *OncoTargets Ther.* **2020**, *13*, 13191–13200. [CrossRef]
199. Bell, C.M.; Raffeiner, P.; Hart, J.R.; Vogt, P.K. PIK3CA Cooperates with KRAS to Promote MYC Activity and Tumorigenesis via the Bromodomain Protein BRD9. *Cancers* **2019**, *11*, 1634. [CrossRef]
200. Huang, H.; Wang, Y.; Li, Q.; Fei, X.; Ma, H.; Hu, R. miR-140-3p functions as a tumor suppressor in squamous cell lung cancer by regulating BRD9. *Cancer Lett.* **2019**, *446*, 81–89. [CrossRef]
201. Del Gaudio, N.; Di Costanzo, A.; Liu, N.Q.; Conte, L.; Migliaccio, A.; Vermeulen, M.; Martens, J.H.A.; Stunnenberg, H.G.; Nebbioso, A.; Altucci, L. BRD9 binds cell type-specific chromatin regions regulating leukemic cell survival via STAT5 inhibition. *Cell Death Dis.* **2019**, *18*, 338. [CrossRef] [PubMed]
202. Yang, Q.; Bariani, M.V.; Falahati, A.; Khosh, A.; Lastra, R.R.; Siblini, H.; Boyer, T.G.; Al-Hendy, A. The Functional Role and Regulatory Mechanism of Bromodomain-Containing Protein 9 in Human Uterine Leiomyosarcoma. *Cells* **2022**, *11*, 2160. [CrossRef]
203. Vafadar, A.; Shabaninejad, Z.; Movahedpour, A.; Mohammadi, S.; Fathullahzadeh, S.; Mirzaei, H.R.; Namdar, A.; Savardashtaki, A.; Mirzaei, H. Long non-coding rnas as epigenetic regulators in cancer. *Curr. Pharm. Des.* **2019**, *25*, 3563–3577. [CrossRef]
204. Hanly, D.J.; Esteller, M.; Berdasco, M. Interplay between long non-coding RNAs and epigenetic machinery: Emerging targets in cancer? *Philos. Trans. R. Soc. Lond. B Biol. Sci.* **2018**, *373*, 20170074. [CrossRef] [PubMed]
205. Romano, G.; Veneziano, D.; Acunzo, M.; Croce, C.M. Small non-coding RNA and cancer. *Carcinogenesis* **2017**, *38*, 485–491. [CrossRef] [PubMed]
206. Guisier, F.; Barros-Filho, M.C.; Rock, L.D.; Constantino, F.B.; Minatel, B.C.; Sage, A.P.; Marshall, E.A.; Martinez, V.D.; Lam, W.L. Small noncoding rna expression in cancer. In *Gene Expression Profiling in Cancer*; IntechOpen: London, UK, 2019.
207. Eddy, S.R. Noncoding RNA genes. *Curr. Opin. Genet. Dev.* **1999**, *9*, 695–699. [CrossRef]
208. Lu, J.; Getz, G.; Miska, E.A.; Alvarez-Saavedra, E.; Lamb, J.; Peck, D.; Sweet-Cordero, A.; Ebert, B.L.; Mak, R.H.; Ferrando, A.A.; et al. MicroRNA expression profiles classify human cancers. *Nature* **2005**, *435*, 834–838. [CrossRef] [PubMed]
209. Subramanian, S.; Kartha, R.V. MicroRNA-mediated gene regulations in human sarcomas. *Cell. Mol. Life Sci.* **2012**, *69*, 3571–3585. [CrossRef]
210. Wei, J.; Liu, X.; Li, T.; Xing, P.; Zhang, C.; Yang, J. The new horizon of liquid biopsy in sarcoma: The potential utility of circulating tumor nucleic acids. *J. Cancer* **2020**, *11*, 5293–5308. [CrossRef]
211. Dvorská, D.; Škovierová, H.; Brány, D.; Halašová, E.; Danková, Z. Liquid biopsy as a tool for differentiation of leiomyomas and sarcomas of corpus uteri. *Int. J. Mol. Sci.* **2019**, *20*, 3825. [CrossRef] [PubMed]
212. Anderson, M.L.; Kim, G.E.; Mach, C.M.; Creighton, C.J.; Lev, D. Abstract 5242: miR-10b functions as a novel tumor suppressor in uterine leiomyosarcoma by promoting overexpression of SDC1. *Cancer Res.* **2014**, *74*, 5242. [CrossRef]
213. Schiavon, B.N.; Carvalho, K.C.; Coutinho-Camillo, C.M.; Baiocchi, G.; Valieris, R.; Drummond, R.; da Silva, I.T.; De Brot, L.; Soares, F.A.; da Cunha, I.W. MiRNAs 144-3p, 34a-5p, and 206 are a useful signature for distinguishing uterine leiomyosarcoma from other smooth muscle tumors. *Surg. Exp. Pathol.* **2019**, *2*, 1–8. [CrossRef]
214. Yokoi, A.; Matsuzaki, J.; Yamamoto, Y.; Tate, K.; Yoneoka, Y.; Shimizu, H.; Uehara, T.; Ishikawa, M.; Takizawa, S.; Aoki, Y.; et al. Serum microRNA profile enables preoperative diagnosis of uterine leiomyosarcoma. *Cancer Sci.* **2019**, *110*, 3718. [CrossRef] [PubMed]
215. De Almeida, B.; Garcia, N.; Maffazioli, G.; Gonzalez dos Anjos, L.; Chada Baracat, E.; Candido Carvalho, K. Oncomirs expression profiling in uterine leiomyosarcoma cells. *Int. J. Mol. Sci.* **2017**, *19*, 52. [CrossRef]
216. Zhang, Q.; Ubago, J.; Li, L.; Guo, H.; Liu, Y.; Qiang, W.; Kim, J.J.; Kong, B.; Wei, J.-J. Molecular analyses of 6 different types of uterine smooth muscle tumors: Emphasis in atypical leiomyoma. *Cancer* **2014**, *120*, 3165–3177. [CrossRef] [PubMed]
217. Chuang, T.-D.; Panda, H.; Luo, X.; Chegini, N. miR-200c is aberrantly expressed in leiomyomas in an ethnic-dependent manner and targets ZEBs, VEGFA, TIMP2, and FBLN5. *Endocr. Relat. Cancer* **2012**, *19*, 541. [CrossRef]
218. Chuang, T.-D.; Ho, M.; Khorram, O. The regulatory function of mir-200c on inflammatory and cell-cycle associated genes in SK-LMS-1, a leiomyosarcoma cell line. *Reprod. Sci.* **2015**, *22*, 563–571. [CrossRef] [PubMed]
219. Dos Anjos, L.G.; de Almeida, B.C.; de Almeida, T.G.; Rocha, A.M.L.; Maffazioli, G.D.N.; Soares, F.A.; da Cunha, I.W.; Baracat, E.C.; Carvalho, K.C. Could miRNA signatures be useful for predicting uterine sarcoma and carcinosarcoma prognosis and treatment? *Cancers* **2018**, *10*, 315. [CrossRef]

220. Shi, G.; Perle, M.A.; Mittal, K.; Chen, H.; Zou, X.; Narita, M.; Hernando, E.; Lee, P.; Wei, J.J. Let-7 repression leads to HMGA2 overexpression in uterine leiomyosarcoma. *J. Cell Mol. Med.* **2009**, *13*, 3898–3905. [CrossRef] [PubMed]
221. Zavadil, J.; Ye, H.; Liu, Z.; Wu, J.; Lee, P.; Hernando, E.; Soteropoulos, P.; Toruner, G.A.; Wei, J.-J. Profiling and functional analyses of microRNAs and their target gene products in human uterine leiomyomas. *PLoS ONE* **2010**, *5*, e12362. [CrossRef] [PubMed]
222. Ayub, A.L.P.; D'Angelo Papaiz, D.; da Silva Soares, R.; Jasiulionis, M.G. The function of lncrnas as epigenetic regulators. In *Non-Coding RNAs*; IntechOpen: Rijeka, Croatia, 2020. [CrossRef]
223. Grillone, K.; Riillo, C.; Riillo, C.; Scionti, F.; Rocca, R.; Rocca, R.; Tradigo, G.; Guzzi, P.H.; Alcaro, S.; Alcaro, S.; et al. Non-coding RNAs in cancer: Platforms and strategies for investigating the genomic "dark matter". *J. Exp. Clin. Cancer Res.* **2020**, *39*, 1–19. [CrossRef] [PubMed]
224. Rasool, M.; Malik, A.; Zahid, S.; Basit Ashraf, M.A.; Qazi, M.H.; Asif, M.; Zaheer, A.; Arshad, M.; Raza, A.; Jamal, M.S. Non-coding RNAs in cancer diagnosis and therapy. *Non-Coding RNA Res.* **2016**, *1*, 69–76. [CrossRef] [PubMed]
225. Guo, H.; Zhang, X.; Dong, R.; Liu, X.; Li, Y.; Lu, S.; Xu, L.; Wang, Y.; Wang, X.; Hou, D.; et al. Integrated analysis of long noncoding RNAs and mRNAs reveals their potential roles in the pathogenesis of uterine leiomyomas. *Oncotarget* **2014**, *5*, 8625. [CrossRef] [PubMed]

biomedicines

Review

An Update on Circular RNA in Pediatric Cancers

Angela Galardi [1,†], Marta Colletti [1,†], Alessandro Palma [2] and Angela Di Giannatale [1,*]

1   Department of Pediatric Hemato-Oncology and Cell and Gene Therapy, IRCCS, Bambino Gesù Children's Hospital, Viale San Paolo 15, 00146 Rome, Italy
2   Translational Cytogenomics Research Unit, IRCCS, Bambino Gesù Children's Hospital, Viale San Paolo 15, 00146 Rome, Italy
*   Correspondence: angela.digiannatale@opbg.net
†   These authors contributed equally to this work.

Abstract: Circular RNAs (circRNAs) are a class of single-stranded closed noncoding RNA molecules which are formed as a result of reverse splicing of mRNAs. Despite their relative abundance, only recently there appeared an increased interest in the understanding of their regulatory importance. Among their most relevant characteristics are high stability, abundance and evolutionary conservation among species. CircRNAs are implicated in several cellular functions, ranging from miRNA and protein sponges to transcriptional modulation and splicing. Additionally, circRNAs' aberrant expression in pathological conditions is bringing to light their possible use as diagnostic and prognostic biomarkers. Their use as indicator molecules of pathological changes is also supported by their peculiar covalent closed cyclic structure which bestows resistance to RNases. Their regulatory role in cancer pathogenesis and metastasis is supported by studies involving human tumors that have investigated different expression profiles of these molecules. As endogenous competitive RNA, circRNAs can regulate tumor proliferation and invasion and they arouse great consideration as potential therapeutic biomarkers and targets for cancer. In this review, we describe the most recent findings on circRNAs in the most common pediatric solid cancers (such as brain tumors, neuroblastomas, and sarcomas) and in more rare ones (such as Wilms tumors, hepatoblastomas, and retinoblastomas).

Keywords: circRNA; pediatric cancer; biomarker

Citation: Galardi, A.; Colletti, M.; Palma, A.; Di Giannatale, A. An Update on Circular RNA in Pediatric Cancers. *Biomedicines* **2023**, *11*, 36. https://doi.org/10.3390/biomedicines11010036

Academic Editors: Milena Rizzo and Elena Levantini

Received: 21 November 2022
Revised: 15 December 2022
Accepted: 20 December 2022
Published: 23 December 2022

1. Introduction

Circular RNAs (circRNAs) are single-stranded covalently closed RNAs that have been identified in several species such as viruses, prokaryotes, unicellular eucaryotes, and mammals [1–4]. Thanks to the advancements in sequencing techniques and the bioinformatics approach, it has been recognized that circRNAs are highly enriched in the human transcriptome and may play different functions. Indeed, they can act as protein scaffolds or miRNA sponges and can be translated in polypeptides. These molecules are characterized by high structural stability and are mainly localized in the cytoplasm of several cell types [5], with tissue- and developmental stage-specific expression [6]. Furthermore, several reports showed that circRNAs may play a key role in different pathological conditions, including neurological disorders [7], diabetes mellitus, cardiovascular diseases, inflammatory diseases, and tumors [8,9].

In particular, emerging body of evidence has shown that abnormal expression of circRNAs is tightly related to tumorigenesis [10]. In this context, the elucidation of circRNAs' functional role and the identification of pathways they regulate in cancer can be fundamental for the discovery of new biomarkers and for the development of new targeted therapies. In this review, we reported the most recent findings on the role of circRNAs in solid pediatric tumors (Summarized in Table 1).

Biogenesis, Regulation, and Functions of CircRNAs

Splicing is the process by which introns are removed from pre-mRNA to connect exons together to produce mature mRNA for the expression of most eukaryotic genes. The different combination of exons in mRNA producing diversified mature mRNA is called alternative splicing, which is a highly regulated mechanism depending on many trans-acting factors and cis-acting elements. RNA splicing occurs within the nucleus and is operated by an enzyme protein complex called spliceosome which recognizes the junction of introns and exons. The spliceosome consists of several small nuclear ribonucleoprotein particles (snRNPs) and numerous protein factors which recognize specific sequences at the 5′ and 3′ splice sites. The specific association of snRNPs to certain splice sites at the junctions of exons and introns initiates a series of reactions which lead to the removal of the intron and the union of two exons between which the intron was located to form linear mRNA (canonical splicing, Figure 1). Alternative splicing is a ubiquitous regulatory mechanism of gene expression that allows generation of more than one unique mRNA species from a single gene. Alternative splicing can generate mRNAs that differ in untranslated regions (UTR) or the coding sequence, with mechanisms including intron retention, exon skipping (removal of specific exons, referred to as cassette exons, that can be either included or excluded from the mature transcript depending on alternative splicing decisions), or the usage of alternative splice sites (that affects the boundaries between the introns and exons involved in this alternative splicing event contributing to transcript diversity) and intron retention (alternative splicing, Figure 1).

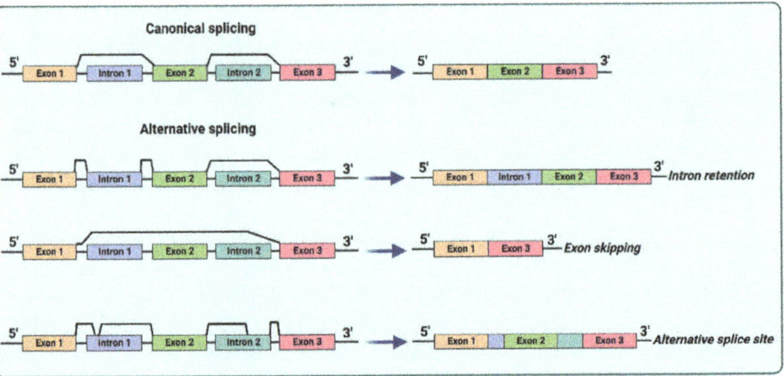

Figure 1. Schematic diagram depicting canonical splicing and the various types of alternative splicing (intron retention, exon skipping, and exon rearrangement following the use of alternative 3′ and 5′ splice sites).

Circular RNAs (circRNAs) include a large class of noncoding RNAs derived from a noncanonical splicing event named back splicing where a downstream splice donor site is covalently linked to an upstream splice acceptor site [8]. CircRNAs are composed of 1–5 exons and the introns flanking the exons are up to three times as long as their linear counterparts [11]. Several studies emphasized the existence of many complementary inverter Alu repeats in intron segments, suggesting that such conditions facilitate and promote circularization [12]. Alu elements, generally located in introns, are abundant and constitute 11% of the reference human genome [13], and a recent analysis has indicated that repetitive elements, such as Alu repeats flanking exons, are responsible for RNA circularization and circRNA formation [14]. In fact, it has been reported that pairs of Alu repeats are enriched in the introns flanking exons that produce circRNAs [12,15], which is further confirmed by the fact that the majority of circRNA sequence is complementary to Alu sequences [16]. This was also proved in vitro by mutagenized plasmids that express different human circular RNAs, where the authors found that splicing sites and short inverted repeated sequences were sufficient to circularize RNA and that mutating some nucleotides inside the Alu elements also prevented circularization [17]. Zhang et al. also confirmed that the complementarity and pairing between Alu elements with reverse orientation regulates the expression of circular RNAs [16,18]. In particular, the adenosine-to-inosine editing is facilitated by double-stranded Alu sequences, thus promoting RNA circularization [16,19].

CircRNAs are subdivided into three classes depending on their synthesis: exonic (EciRNAs), intronic (CiRNAs), and exon–intron circRNAs (EIciRNAs) [20]. The biogenesis of circRNAs is closely linked to the alignment and the scrambling of exon and intron parties [21]. EciRNAs are synthesized via back splicing of the $5'$ splice site to a $3'$ splice site. CiRNAs are formed from intronic lariat precursors escaping from the debranching step of canonical linear splicing while CiRNAs synthesis is different as it depends on GU-rich sequences near the $5'$ splice site and C-rich sequences near the branch point. To form mature circRNA, during the back splicing event, first, the two segments bind to form a circular structure followed by the incision of exonic and intronic sequences in the binding portion of the spliceosome, and after that, the residual introns are stitched together [22]. CircRNA biogenesis relies on canonical splicing machinery, and their expression level can be regulated in cells at three levels: precursor RNA transcription, post- or co-transcriptional processing, and turnover. If canonical splicing happens first, it generates a linear RNA with skipped exons and a long intron lariat containing these skipped exons, which are then back-spliced to generate circRNA (exon skipping or lariat intermediate model). If back splicing happens first, it directly generates circRNA together with an exon–intron(s)-exon intermediate that can meet two fates: be further processed to produce linear RNA with skipped exons or be degraded (direct back splicing model). In both models, the expression of circRNA is associated with an alternatively spliced linear RNA with exon exclusion, and the way in which the spliceosome chooses either canonical splicing or back splicing to start the production of circRNA is still unclear (Figure 2A). These two steps may happen stochastically or synergistically. To bring the downstream donor and upstream acceptor sites close together to promote back splicing, both cis-elements and trans-factors are required because back splicing is unfavorable for spliceosome assembly and is less efficiently catalyzed. The majority of circRNAs are formed from one or more exons of known protein-coding genes. Products resulting from all of the basic types of alternative splicing of linear RNA can be found within circRNAs, and some circRNAs contain exons that are not included in linear transcripts. Despite the lack of polyadenylation (poly(A)) and capping, circRNAs generally localize to the cytoplasm. However, internal intron retention may lead to the production of circRNAs that contain sequences derived from both exons and introns [23]. Although they are expressed at low levels, recent discoveries have revealed a wide range of regulatory functions performed by these circular molecules. CircRNA are involved in gene expression regulation through distinct mechanisms: the processing of circRNAs affects splicing of their linear mRNA counterparts, can regulate transcription of their parental genes or splicing of their linear cognate. Some circRNAs containing the internal ribosome entry site (IRES) can be translated into proteins through a process that involves the N^6-methyladenosine (m^6A) RNA base modification, eukaryotic translation initiation factor 4 gamma (eIF4G2), methyltransferase-like 3 (METTL3)/methyltransferase-like 14 (METTL14), and the association of the ribosome. Furthermore, circRNAs can act as endogenous competitive RNAs (ceRNAs) defined as miRNA sponges: they bind to the corresponding miRNAs via their complementary sequences, thereby regulating downstream target gene expression which is inhibited by miRNAs. CircRNAs can also interact with different proteins to form specific circRNPs that subsequently influence modes of action of the associated proteins (Figure 2B).

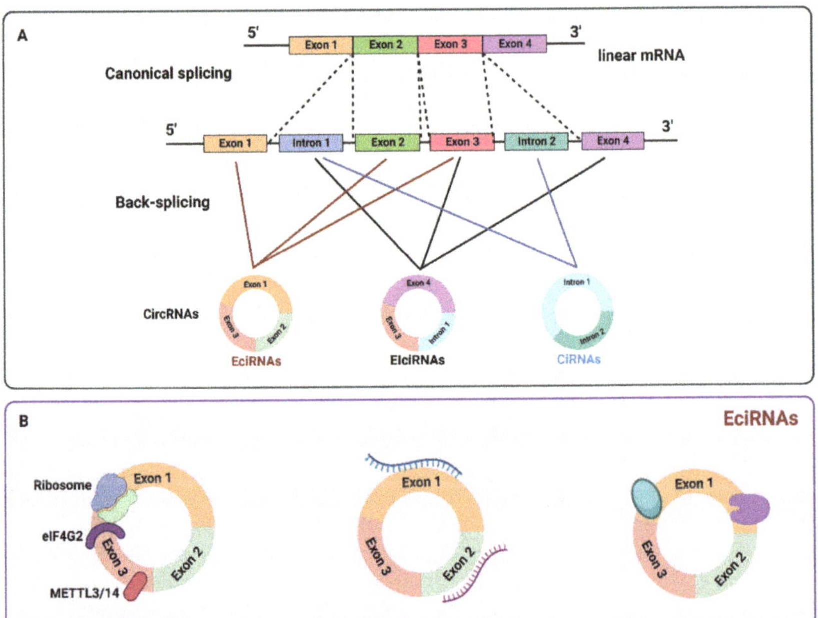

Figure 2. (**A**) Biogenesis of circRNAs: circRNAs are created by noncanonical splicing process known as back splicing. A downstream splice donor is joined to an upstream splice acceptor. Exonic circRNAs (EciRNAs) are formed via back splicing (red lines) with multiple exons or a single exon, the major forms of circRNAs. Exon–intron circRNAs (ElciRNAs) are circularized with the retained intronic sequences between the circularized exons, predominantly in the nucleus (black lines). Intronic circRNAs (ciRNAs) are formed from intronic lariat precursors escaping from the debranching step of canonical linear splicing, abundant in the nucleus (blu lines). (**B**) CircRNA are involved in gene expression regulation through distinct mechanisms: the processing of circRNAs affects splicing of their linear mRNA counterparts, can regulate transcription and translation of their parental genes. The association of translation initiation factor eIF4G2 with circRNA triggers its translation, and this process can be enhanced by METTL3/14. Furthermore, circRNAs can act as sponges for miRNAs through their binding sites or interact with different proteins to form specific circRNPs that subsequently influence modes of action of the associated proteins.

2. CircRNAs in Cancer

CircRNAs have been found to be dysregulated in cancer and involved as oncogenes or tumor suppressors in tumor development and progression. Since they have a different expression in tumors compared to normal tissues [21–23], they represent potential diagnostic/prognostic biomarkers and interesting targets for the development of new therapeutic approaches. CircRNAs may affect tumor progression acting in different ways: as miRNA sponges by sequestering specific miRNAs to regulate their downstream targets, thereby suppressing or promoting tumor progression [24,25], or interacting directly with specific proteins to regulate their or their targets' function/translation. This mechanism allows the activation or inhibition of downstream pathways to affect malignant behavior of tumors [26,27]. Furthermore, it has been suggested that circRNA could be selectively translated from the same linear RNA, though only few circRNA translation products have been actually detected [28,29].

2.1. Pediatric Brain Tumors

2.1.1. Medulloblastoma

Medulloblastoma (MB) is one of the most common malignant tumors affecting the central nervous system (CNS) in children. Although considerable progress has been achieved with the improvement of classic therapeutic approaches such as surgery, radiotherapy, and chemotherapy, the prognosis of some patients with MB remains very poor. Therefore, highly specific molecular targeted treatment, which can improve therapeutic efficacy and reduce side effects in patients, has become a research hotspot. Among the molecules with a potential prognostic/therapeutic value, circRNAs are also emerging in MB. The first evidence of circRNA involvement in the regulation of proliferation and growth of MB was reported in 2018 by Lv et al. With a next-generation sequencing approach on RNAs extracted from four human MBs and four human normal cerebellum tissues, 44,184 distinct circRNAs were identified, and among those, two were expressed differently between the two groups: circ-SKA3 and circ-DTL. Silencing studies on DAOY cells using si-circ-SKA3 and si-circ-DTL plasmids demonstrated that the downregulation of expression of these two circRNAs suppressed cell proliferation, migration, and invasion in human MB cell lines [30]. In order to deepen the role of circSKA3 in MB, Wang et al. confirmed a higher level of these molecules in MB tissues compared to the adjacent normal tissues and demonstrated the relationship between circSKA3 and its predicted target miR-383-5p, whose expression is reduced in MB tissues. In particular, the authors showed that circSKA3 facilitated MB progression through the suppression of miR-383-5p by acting as a sponge for this miRNA. This action leads to the regulation of FOXM1 expression and accelerates MB progression. In conclusion, these data suggest that circSKA3 depletion blocks MB progression by regulating FOXM1 expression via miR-383-5p [31]. Last year, it was demonstrated that circ-SKA3 decoyed miR-326 to increase the ID3 expression leading to cell proliferation, migration, and invasion in MB [32]. Moreover, it was observed that circSKA3 expression in MB tissues was negatively correlated with the miR-520 expression and positively correlated with the CDK6 expression, suggesting that circSKA3 facilitates MB progression through the miR-520 h/CDK6 axis [33].

2.1.2. Ependymoma

Ependymomas represent one of the most common CNS tumors among children, even though they are rare. This malignancy occurs predominantly intracranially (supratentorial brain and posterior fossa) in children between 0 and 4 years of age, but it can also arise in older children and adults. Patients with intracranial ependymomas have high morbidity and mortality rates, and clinical management is challenging. A profiling of circRNAs was performed in a cohort of 10 pediatric patients diagnosed with ependymoma and three healthy controls in 2018. The authors found a significant downregulation of circRNAs in the ependymoma tissues compared to the healthy control samples, with 1167 circRNAs that showed a high abundance, of which 263 were detected in the ependymomas and 1126—in the controls. Following bioinformatic analysis, five circRNAs were found to be upregulated: circRMST, circLRBA circWDR78, circDRC1, and circBBS9 [34].

2.2. Neuroblastoma

Neuroblastoma (NB) is the most common extracranial solid cancer in childhood originating from primitive neuronal crest cells of the sympathetic nervous system and typically develops in the adrenal medulla or the paraspinal ganglia. NB is a complex tumor capable of rapid progression or spontaneous resolution, the latter usually occurring in under-one children. In 25% of the cases, a solitary mass that may be cured by surgery is present, whereas about 60% of the patients present disseminated disease at diagnosis, involving mainly bone, bone marrow, lymph nodes, and liver. In patients with metastasis, survival is poor, with a high mortality rate [35]. Despite scientific advances contributing to an increased understanding of genetic, molecular, and cellular mechanisms in NB, the prognosis of children with regional or distant metastasis disease remains still poor. Thus, efforts are being made to unveil the mechanisms underlying NB progression in order to identify new therapeutic targets. Even though in the past decade some evidence showed that noncoding RNAs have a role in the regulation of different cellular processes such as chromatin remodelling, transcription, transduction, and post-transcriptional modification, the role of circular RNAs in NB is just starting to emerge. In 2019, Chen et al. identified one circRNA, circAGO2 (has_circ_0135889), that was upregulated in several cancer tissues, including NB. In particular, they observed that the overexpression of circAGO2 is associated with the promotion of growth and invasion of cancer cell lines in vitro and poor prognosis of patients. The promotion of growth and invasiveness are related to the interaction between circAGO2 and the human antigen R (HuR) protein which implies the competitive enrichment on the 3′ untranslated region (UTR) of target genes, resulting in the reduction of AGO2 binding, repression of AGO2/miRNA-mediated gene silencing, and promotion of tumorigenesis and aggressiveness [36]. Shortly after this observation, CUX1-generated intron-containing circular RNA (circ-CUX1) was identified by Li et al. as a circRNA with oncogenic properties in NB cell lines. In particular, elevated circ-CUX1 promotes aerobic glycolysis, growth, and aggressiveness of NB cells by binding to EWS RNA-binding protein 1 (EWSR1) and facilitating its interaction with the MYC-associated zinc finger protein (MAZ); this results in transactivation and transcriptional alteration of CUX1 [37]. CUX1 is a transcriptional factor regulating glycolytic genes such as ENO1, GPI, and PGK1 in NB, and its elevated expression has been reported to be associated with poor prognosis in several tumors [38]. Zhang et al. observed that miR-16-5p is downregulated in NB cells and tissues, and it is a direct target of circ-CUX1. Thus, circ-CUX1 acts as a sponge for miR-16-5p, which, in turn, no longer inhibits the messenger of DMRT2, which is upregulated, accelerating proliferation, migration, invasion, and glucose uptake in NB cells [39]. The target prediction approach revealed miR-388-3p as another possible target for circCUX1. In 2021, Wang et al. confirmed the role of circCUX1 in triggering the progression and glycolysis of NB by serving as competing endogenous RNA (ceRNA) for miR-388-3p and upregulating plant homeodomain finger protein 20 (PHF20) [40]. Another circRNA upregulated in NB tissues and associated with a poor prognosis is circDGKB (has_circ_0133622). Overexpression of this molecule promotes proliferation, migration, and invasion of SK-N-SH cells, inhibits apoptosis, and induces S phase arrest, while its downregulation reverts these effects. In particular, circDGKB targets miR-873, which is an oncosuppressor in several cancers. The loss of the inhibitory effect exerted by miR-873 on its target GLI1 (glioma-associated oncogene 1) leads to increased proliferation, migration, and invasion in NB cells [41]. Zhang et al. collected RNA sequencing data of 39 NB cell lines and two normal cell lines (RPE-1 and fetal brain) as the control and identified 39,022 functional circRNAs [42]. From their analysis, it emerged that circRNAs located within the hotspot at 2p originated from the genes adjacent to MYCN, probably deriving from *MYCN* amplification. In particular, they found 29 circRNAs dysregulated in NB; those located within the amplified regions of MYCN were upregulated in the cell lines with MYCN amplification and a high activation level of MYCN targets. Furthermore, they identified has_circ_0002343, which is implicated in the regulation of PI3K/Akt/mTOR signaling, and has_cir_0001361 was associated with the genes implicated in epithelial-to-mesenchymal transition (EMT), such as NOTCH2, SERPINHI, LAMCI [43]. In 2021, Lin et al. performed high-throughput RNA sequencing of five paired NB tumors and adjacent normal adrenal medulla samples and identified 4704 differentially expressed circRNAs (2462 up- and 2242 downregulated). Among the differentially expressed circRNAs, 10 highly downregulated circRNAs were confirmed to be less expressed in 20 NB specimens compared to the adjacent normal adrenal gland samples. Moreover, the authors chose to extend the analysis to seven other circRNAs that target miR-21, which plays an important role in the regulation of proliferation and apoptosis in NB [44]. Three miR-21-related circRNAs (circRNA-TBC1D4, circRNA-NAALAD2, circRNA-TGFBR3) were significantly downregulated in NB samples. In particular, circRNA-TBC1D4 was associated with the MYCN status, circRNA-TGFBR3—with the histological classification. In order to investigate the biological functions of these circRNAs, the authors overexpressed circRNA TBC1D4 in the SH-SY5Y NB cell lines and observed a significant decrease in the migratory properties of the cells [45]. Two papers published in 2021 illustrated that nuclear protein 4 like (NOL4L) was regulated by microRNAs, which were regulated, in turn, by circular RNAs. In the first work, miR-362-5p, which has suppressive functions in NB progression was sponged by circRNA phosphodiesterase 5 A (circPDE5A, has_circ_0002474), which, in turn, was upregulated in NB cells and tissues. Among the functional targets of miR-362-5p the authors identified NOL4L, whose expression correlates with the circPDE5A levels in NB [46]. In the second work, NOL4L was negatively regulated by miR-432-5p and positively regulated by circ_0132817. Inhibition of circ_0132817 restrained tumor growth by upregulating miR-432-5p and downregulating NOL4L, suggesting that circ_0132817 knock down inhibits the progression of NB through the modulation of the miR-432-5p/NOL4L axis [47]. In 2022, Tang et al. performed a high-throughput microarray for circRNAs on three NB and three gangliocytoma tissues [48]. From their analysis, circ0125803 emerged as the most upregulated circRNA. Circ0125803 interacts with miR-197-5p, which has E2F1 as the target, and the upregulation of this gene promotes NB progression. Yang et al observed that in NB tumor specimens and cells, circ_0135889 and neuronal differentiation 1 (NEUROD1) were upregulated while miR-127-5p was down-regulated. Cell proliferation, migration, and invasion were suppressed following circ_0135889 silencing. Circ_0135889 acts as a sponge for miR-127-5p, and inhibition of miR-127-5p counteracts the inhibitory impact of circ_0135889 knockdown on the malignant behavior of NB cells. Moreover NEUROD1 was a direct target of miR-127-5p, and miR-127-5p exerted antitumor activity in NB cells by targeting NEUROD1 [49]. The expression level of circular RNA (circRNA kinesin superfamily protein 2A (circKIF2A, also known as hsa_circ_0129276) has been reported to be upregulated in NB tissues, SK-N-AS and LAN-6 cell lines. Silencing of circKIF2A inhibits cell proliferation, migration, invasion, and glycolysis, boosts apoptosis in NB cells in vitro, and blocks the growth in vivo. CircKI2A acts as a sponge of miR-377-3p, targeting phosphoribosyl pyrophosphate synthetase 1 (PRPS1), leading to an increase in its expression that determines a decreased aggressivity of NB cells [50]. Karami Fath et al. recently published a review where they reported the potential role of circRNA in NB progression [51]. This body of evidence underlines the potential of these molecules as possible biomarkers and therapeutic targets in NB, even though extensive validation is needed, in particular in relation to the individual characteristics of patients.

2.3. Sarcomas

2.3.1. Rhabdomyosarcoma

Rhabdomyosarcoma (RMS) is the most-common soft tissue malignancy in children and adolescents, accounting for up to 3–4% of childhood cancer cases and approximately 50% of all sarcomas [52,53]. Among the main RMS subtypes, embryonal (ERMS) and alveolar sarcomas (ARMS) account for 60% and 20% of all RMS cases, respectively. ARMS is associated to specific genetic alterations and generally has a worse prognosis due to its low response to treatment [54]. In 2019, Rossi et al. reported that circ-ZNF609 is expressed in growing myoblasts, showing that its depletion leads to cell proliferation modulation Indeed, circ-ZNF609 is upregulated in RMS cell lines, in particular in ARMS, and has a role as a positive regulator of cell proliferation pathways. Knockdown of circ-ZNF609 induces the block of the G1–S transition, specifically in ERMS cells, with a decrease in the p-AKT protein levels in both ERMS and ARMS cell types. Moreover, circ-ZNF609 has been described to be overexpressed in RMS primary tissues [55]. In 2021, the same research group described a high expression of circVAMP3 in ARMS cells lines and demonstrated its involvement in cell cycle progression through the alteration of AKT-related pathways. In particular, downregulation of circVAMP3 leads to the upregulation of CDKN1A and WEE1, which directly regulate the CCNB1/CDK1 complex, controlling the G2/M checkpoint, as well as the downregulation of AKT and ERK1, therefore leading to an accumulation of cells in the G2 phase [56].

2.3.2. Osteosarcoma

Osteosarcoma (OS) accounts for approximately 35% of primary malignant bone tumors; its metastatic form presents the lowest survival rate out of all pediatric cancers despite the use of multiple chemotherapeutic regimens [57]. The MNAT1 protein (menage a trois 1, MAT1) is upregulated in different sarcomas (Ewing's sarcoma, synovial sarcoma), but in particular in OS tissues compared to normal bone tissue. Among the four miR-NAs (miRNA-26a-5p, miRNA-26b-5p, miRNA-200a-3p, miRNA-141-3p) that are aberrantly expressed in OS and are predicted to target MNAT1, miRNA-26a-5p expression shows negative correlation with MNAT1 expression. This effect is due to the action of has_circ-0001146 that promotes MNAT1 expression by sponging miR-26a-5p and promoting proliferation and invasion ability [58]. Zhang et al. reported that in 10 pediatric OS tissue samples, has_circ-0005909 was upregulated in respect to the adjacent normal tissue. This circRNA was confirmed to also be upregulated in OS cell lines MG63, HOS, 143B, and U2OS. Knockdown of circ_0005909 in OS cells demonstrated that this circRNA represses proliferation by targeting miR-338-3p that could bring the 3′UTR of HMGA1 (high mobility group A1) [59].

2.4. Wilms' Tumor

Wilms' tumor (WT, nephroblastoma) develops before the age of 3 years and accounts for 6% of childhood tumors and 95% of pediatric kidney tumors. In recent years, with the improvement of therapeutic strategy, around 85% of the patients affected by this tumor are cured, with a few children presenting relapse, metastasis, and chemoresistance who have a poor prognosis. Thus, the identification of new molecules such as circRNAs would be fundamental to improving the survival of patients with high-risk disease. In 2021, Cao et al. conducted high-throughput microarray sequencing to screen circRNAs in Wilms' tumor cells and tissues [60]. Wilms' tumor samples and the adjacent normal kidney samples were collected and analyzed for circRNA expression. Circ0093740 was identified as upregulated in the tumor samples in respect to the normal kidney samples and the Wilms' tumor cell lines. Silencing of circ0093740 inhibited the proliferation and migration ability of the Wilms' tumor cells, and this mechanism is mediated by the sponging of miR-136/145 operated by circ0093740 that leads to the upregulation of DNA (cytosine-5-)-methyltransferase 3 alpha (DNMT3A), transcript variant 3. An opposite effect was observed for has_circ_0008285 (circCDYL) which was downregulated in WT tissues compared to normal adjacent tissues, and its overexpression suppresses proliferation, migration, and invasion of WT cells in vitro and in vivo [61]. These effects are due to the upregulation of the tight junction protein 1 (TPJI) expression following the ability to circCDYL to sponge miR-145-5p. In 2022. Jiaju et al. observed that circSLC7A6 was upregulated in WT tumor samples and cells in respect to the controls. Cell apoptosis was increased while cell viability, migration, and invasion were repressed by circSLC7A6 silencing. Furthermore, miR-107 was a direct target of this circRNA, and circSLC7A6 could upregulate ABL2 expression by serving as a ceRNA of miR-107 [62].

2.5. Hepatoblastoma

Hepatoblastoma (HB), one of the major liver cancers in infants, accounts for about 1% of all pediatric cancers [63]. To date, surgical resection, adjuvant chemotherapy, and liver transplantation are the methods mainly used for HB treatment. However, a high percentage of patients have a high risk of relapse or metastasis, and the mortality rate in those advanced cases is still over 35% [64]. Furthermore, in the field of HB, considerable effort has been devoted to the identification of circRNAs. In 2018, Liu et al. compared the HB expression profile with the paired adjacent normal tissue samples and identified 869 differentially expressed circRNAs. Among these molecules, has_circ_0015756 was the most significantly upregulated both in tissues and in metastatic HB cell lines. A reduction of viability, proliferation, and invasion ability in SMMC-7221 and HepG2 HB cells was observed following the silencing of this circRNA. Predictive bioinformatic analyses suggested miR-6134, miR-7854-3p, miR-4778-3p, miR-1250-3p, and miR-4659a-3p as a putative miRNA target on circ_0015756. Functionality experiments suggested that miR-1250-3p directly regulates circ_0015756, and this was confirmed by the reduction of viability, proliferation, and invasion following miR-1250-3p mimic transfection in HuH-6 cells [65]. In 2019, two papers were published on dysregulated circRNAs in HB. Song et al. demonstrated by using qRT-PCR and in situ hybridization that has_circ_0000594 is upregulated in HB samples collected during surgery. This circRNA directly regulates miR-217, which is a well-recognized tumor suppressor miRNA, to act on SIRT1 (Sirtuin1) expression, this latter being an oncogene implicated in the promotion of the malignant phenotype [66]. Zhen et al. investigated by means of sequencing analyses the expression profile of circRNAs in HB tissues and discovered that circHMGCS1 (has_circ_0072391) is strongly upregulated in patients' tissues and is negatively correlated with the prognosis. Furthermore, they found that this circRNA exerts its oncogenic role via sponging tumor suppressor miR-503-5p to upregulate the IGF–PI3K–Akt signaling and regulating glutamine metabolism [67]. Using circBase, Liu et al. looked at the circRNAs which are derived from the STAT3 gene sequence and that were expressed in HepG2 and HuH-6 cells compared to a normal liver cell line, THLE-3, and identified 14 sequences [68,69]. Among these sequences, circ_0043800 and circ_0043804 were highly expressed in cancer cells in respect to the control cell line and were also detected in HB and the adjacent nontumor tissues. Only circ_0043800 was significantly upregulated in cancer tissues, and the authors renamed it as circ-STAT3. The silencing of circ-STAT3 led to the inhibition of HB (HepG2 and HuH-6) cell growth and migration and a stem cell phenotype. Circ_0043800, that is predominantly located in the cytoplasm, was found to upregulate STAT-3 and Gli2 (GLI family zinc finger 2) via sponging miR-29a/b/c-3p, and this leads to the promotion of HB tumor growth [69]. In 2021, Chen et al. explored the role of circRNAs in stemness maintenance of HB cells. CircRNA CDR1 was highly expressed in HB cancer stem cells (CSC)-enriched population, and its knockdown decreased the proportion of stem cells. This effect was due to its sponge activity on miR-7-5p, which leads to the increase in Kruppel-like factor 4 (KLF-4) expression [70]. The expression of circSETD3 was investigated by Li et al. who observed a significant reduction in HB tissues and cell lines (HepG2 and HuH-6) compared to normal tissues and a liver cell line (THLE-3). Furthermore, low expression of circSETD3 has been shown in advanced stages of disease and to be associated with poor survival in HB patients. These results indicate that circSETD3 acts as a tumor suppressor in HB, and functional experiments on this tumor demonstrated that its expression inhibits cell proliferation, migration, epithelial-to-mesenchymal transition (EMT) and induces apoptosis via sponging miR-423-3p to promote Bim expression [71].

2.6. Retinoblastoma

Emerging evidence suggests that abnormal expression of circRNA is very closely related to the development and progression of various ocular disorders, including retinoblastoma (RB). RB, a malignancy originating from embryonal retinal cells, is the most common eye cancer occurring in the pediatric population. Several approaches have been employed to treat RB, with survival rates currently exceeding 95%. However, in the presence of resistant retinal, vitreous, or humor aqueous disease, the management of local tumor control and the visual outcome are still challenging. Therefore, the understanding of molecular events underlying the initiation and the progression of RB remains crucial for the treatment of this cancer. First evidence of the involvement of circRNAs in RB progression was provided in 2018 by Xing et al. with a work illustrating the clinical value and biological functions of has_circ_0001649. Has_circ_0001649 expression levels were significantly downregulated in RB cell lines in respect to retinal pigment epithelial (RPE) cell line ARPE-19 and in RB tissues compared with normal retinal tissues. Low expression of has_circ_0001649 was associated with an aggressive phenotype of RB and predicted a poor prognosis for RB patients after surgery [72]. Further functional studies demonstrated the ability of has_circ_0001649 to induce changes in proliferation and apoptosis in RB cells through the regulation of AKT/mTOR signaling. In 2019, Jiao et al. investigated the expression of circRNA in primary RB tissue samples compared to healthy retinas by means of RNA sequencing. They identified 557 differentially expressed circRNAs, with 550 circRNAs downregulated and seven circRNAs upregulated in RB tissues compared to the corresponding normal retinas. Gene Ontology (GO) enrichment analyses showed that the host genes of differentially expressed circRNAs were associated predominantly with chromatin modification and phosphorylation. In particular, among circRNAs whose host genes participated in chromatin modification, TET1-has_circ_0093996 was significantly downregulated in the RB samples. Prediction using TargetScat reveals that cluster miR-96-182-186 is sponged by TET1-has_circ_0093996 and that programmed cell death 4 (PDCD4) is a potential candidate gene impacted in RB tissue as the target of miR-183 [73]. In another work, Fu et al. observed that among the potential targets of circTET1, there are miR-492 and miR-494-3p in RB. The ability to sponge these miRNAs by circTET1 has as a consequence the inhibition of the Wnt/β-catenin signaling pathway and the inhibition of tumor progression [74]. In 2020, Wenbo et al. provided the evidence of the potential role of circ_0075804 in regulating RB cell proliferation by binding to heterogeneous nuclear ribonucleoprotein K (HNRNPK) and enhancing the stability of E2F3 mRNA [75]. Du Shanshan et al. identified that circ_ODC1 is implicated in the regulation of RB proliferation through a mechanism that involves miR-422a and S phase kinase-associated protein 2 (SKP2). In particular, they observed that circ_ODC1 sponging miR-422a, which targets SKP2 mRNA, indirectly increases the SKP2 protein to promote cell growth in RB [76]. Circ_0000527 has been demonstrated to be expressed at a high level in RB tissues and cells implicated in the promotion of RB progression. Specifically, circ_0000527, by sponging miR-646, increases the expression of its target BCL-2 through the promotion of malignant phenotypes on RB cells [77]. Among the messengers targeted by miR-646, there is also LDL receptor-related protein (LRP)-6 mRNA. Zhang et al. demonstrated that upregulation of miR-646 and knockdown of circ_0000527 induce a decrease in LRP6 expression in RB cells; on the other hand, inhibition of miR-646 or overexpression of circ_0000527 enhance LRP6 expression, promoting cell proliferation and migration [78]. Another mechanism by which circ_0000527 facilitates RB progression is through the regulation of the miR-98-5p/X-linked inhibitor of apoptosis (XIAP) axis. Indeed, the silencing of circ_0000527 suppresses proliferation, migration, and invasion of RB cells and promotes apoptosis by increasing expression of miR-98-5p that targets XIAP [79]. Recently, two papers have highlighted the role of circular RNA circ_0000034 in the development of RB. Liu et al. showed that the level of circ_0000034 and syntaxin (STX17) was increased in RB tissues and cells, while miR-361-3p was decreased, suggesting a role of the miR-361-3p/STX17 axis in promoting RB cell growth [80]. Sun et al. found that has_0000034 expression in RB tissues was high compared to normal retinal tissues and correlated with advanced disease in RB patients. Thus, they suggested that has_circ_0000034 may promote RB aggressiveness by sponging miR-361-3p [81]. Le et al. evaluated the expression level of circMKLN1 in 55 RB tissue samples, showing its downregulation compared to retinal tissue [82]. To evaluate the role of this circRNA in RB development, they transfected into Y79 and WERI-RB1 cells a vector overexpressed circMKLN1, showing inhibition of its target miR-425-5p and suppression of proliferation and aggressiveness of RB cells. This anticancer effect is due to the upregulation of oncosuppressor gene PDCD4 following the inhibition of miR-425-5p carried out by circMKLN1. Conversely, it was observed that circ-FAM158A was overexpressed in RB tissues and cells, and its knockdown inhibited RB cell growth, migration, and invasion. Moreover, in RB specimens and cells, miR-138-5p was lowly expressed, while solute carrier family 7 member 5 (SLC7A5) was highly expressed, suggesting a possible relationship between these two molecules. In fact, it was reported in RB that circ-FAM158A may act as an oncogene by sponging miR-138-5p to regulate *SLC7A5* expression [83]. In 2022, Guangwei analyzed the expression of four different circRNAs identified with bioinformatics analysis via the circBank database (circ_0084811, circ_0137212, circ_0137213, circ_0137214) in RB cell lines. Among the four, only circ_0084811 was expressed at a high level in HXO-Rb44,

Y79, SO-Rb50, and WERI-Rb-1 in respect to the human retinal pigment epithelial ARPE-19 cell line. Functional assays demonstrated that circ_0084811 facilitated cell proliferation but inhibited cell apoptosis. Moreover, circ_0084811 regulated its host gene E2F transcription factor 5 (E2F5) whose expression at the protein and mRNA levels was reduced following circ_0084811 silencing. Furthermore, circ_ 0084811 deficiency induced cell apoptosis, and this effect could be partially countervailed by the knockdown of miR-18a-5p or miR-18b-5p while it could be greatly counteracted by E2F5 upregulation. These observations suggested that circ_0084811 regulates E2F5 expression via sponging miR-18a-5p and miR-18b-5p [84].

Table 1. CircRNAs involved in various pediatric cancers.

Tumor	CircRNA	Related Network	Function	Reference
MB	circ-DTL		Promotion of proliferation	[30]
MB	circSKA3	miR-383-5p	Promotion of proliferation	[30]
			Promotion of cancer progression	[31]
		miR-326	Increased proliferation	[32]
		miR-520 h	Promotion of cancer progression	[33]
EP	circRMST, circLRBA circWDR78, circDRC1, and circBBS9			[34]
NB	circAGO2 (hsa_circ_0135889)		Promotion of tumorigenesis and aggressiveness	[36]
NB	circCUX1	EWS RNA-binding protein	Promotion of glycolysis, growth, and aggressiveness	[37]
NB	circCUX1	miR-16-5p/DMTR2	Promotion of proliferation, migration, invasion, and glucose uptake	[39]
NB	circCUX1	miR-388-3p/PHF20	Promotion of cancer progression	[40]
NB	circDGKB (has_circ_0133622)	miR-873/GLI1	Increased proliferation, migration, and invasion	[41]
NB	has_circ_0002343	PI3K/Akt/mTOR signaling	Survival	[42]
NB	has_circ_0001361	NOTCH2, SERPINH1, LAMC1	Epithelial-to-mesenchymal transition	[44]
NB	circTBC1D4, circNAALAD2, circTGFBR3	miR-21	Decreased migratory properties	[45]
NB	circPDE5A (has_circ_0002474)	miR-362-5p/NOL4L	Promotion of proliferation and migration	[46]
NB	circ_0132817	miR-432-5p/NOL4L	Promotion of tumor progression	[45]
NB	circ0125803	miR-197-5p/E2F1	Promotion of tumor progression	[48]
NB	circ_0135889	miR-127-5p/NEUROD1	Promotion of proliferation and tumorigenicity	[49]
NB	circKIF2A (hsa_circ_0129276)	miR-377-3p/PRPS1	Promotion of cell proliferation, migration, invasion, and glycolysis	[50]
ERMS	circZNF609	AKT	Regulation of cell proliferation	[55]
ARMS	circVAMP3	AKT	Regulation of cell proliferation	[56]
OS	has_circ_0001146	miR-26a-5p/MNAT1	Promotion of proliferation and invasiveness	[58]
OS	has_circ_0005909	miR-338-3p/HMGA1	Promotion of cancer development	[59]
WT	circ0093740	miR-136/145/DNMT3A	Promotion of proliferation and migration ability	[60]

Table 1. *Cont.*

Tumor	CircRNA	Related Network	Function	Reference
WT	circCDYL	miR-145-5p/TJP1	Reduction of cell proliferation, migration, and invasion	[61]
WT	circSLC7A6	miR-107/ABL2	Cancer promotion	[62]
HB	circ_0015756	miR-1250-3p	Increased viability, proliferation, and invasion	[63]
HB	has_circ_0000594	miR-217/SIRT1	Promotion of proliferation, viability, and migration	[66]
HB	circHMGCS1 (has_circ_0072391)	miR-503-5p/IGF-PI3K-Akt signaling	Regulation of glutamine metabolism	[67]
HB	circ_0043800	miR-29a/b/c-3p/STAT3/GLI1	Promotion of tumor growth	[69]
HB	CDR1as	miR-7-5p/KLF4	Promotion of proliferation and stemness	[70]
HB	circSETD3	miR-423-3p/Bim	Inhibition of proliferation, migration, and EMT and induction of apoptosis	[71]
RB	has_circ_0001649	AKT/mTOR	Regulation of proliferation and apoptosis	[72]
RB	TET1-has_circ_0093996	miR-183/PDCD4 miR-494/miR-494-3p	Chromatin modification Inhibition of tumor progression	[73] [74]
RB	circ_0075804	HNRNPK	Regulation of proliferation	[75]
RB	circ_ODC1	miR-422a/SKP2	Regulation of cell growth	[76]
RB	circ_0000527	miR-646/BCL-2	Promotion of RB progression	[77]
		miR-646/LRP6	Promotion of cell proliferation and migration	[78]
		miR-98-5p/XIAP	Promotion of cell proliferation and migration	[79]
RB	has_circ_0000034	miR-361-3p/STX17	Promotion of cell growth	[80,81]
RB	circMKLN1	miR-425-5p/PDCD4	Inhibition of tumor progression	[82]
RB	circ-FAM158A	miR-138-5p/SLC7A5	Regulation of cell growth and migration	[83]
RB	circ_0084811	miR-18a-5p/miR-18b-5p/E2F5	Promotion of proliferation and inhibition of apoptosis	[84]

MB, medulloblastoma; EP, ependymoma; NB, neuroblastoma; OS, osteosarcoma; ERMS, embryonal rhabdomyosarcoma; ARMS, alveolar rhabdomyosarcoma; WT, Wilms' tumor; HB, hepatoblastoma; RB, retinoblastoma.

3. Brief View on circRNAs in Blood Cancers

Although we focused our attention on the latest scientific evidence regarding the identification and possible role of circRNAs in different solid pediatric tumors, a brief consideration of their involvement in blood cancers must be made. Indeed, circRNAs could be applied as new clinical biomarkers for blood cancers due to their close connections with multiple biological functions in the disease, as well as their elevated levels of stability and abundance in bone marrow and body fluids [85]. Several circRNAs have been used as diagnostic and prognostic biomarkers in blood malignancies, such as multiple myeloma (MM), acute lymphocytic leukemia (ALL), and juvenile myelomonocytic leukemia (JMML). For example, in MM, circRNAs involved in cell proliferation and tumor progression (such as circ-SMARCA5, hsa_circ_0007841, hsa_circ_0005273, hsa_circ_0007146, hsa_circ_0001947, hsa_circ_0001910, circ-CDYL, circ-MYBL2, circ-ITCH, hsa_circ_0087776, circMYC, circ-G042080, circ-ATP10A) [86] or in chemoresistance (such as circRNA_101237, circ_0007841, circITCH, circ-CCT3, ciRS-7, circPVT1) were identified [87]. In ALL, circPVT1 was proposed to promote ALL leukemogenesis [88], while circPVT1, circHIPK3, and circPAX5 were proposed to interfere with B cell maturation and promote disease progression [89]. In 2021, Liu et al. identified circRNF220, which was specifically abundant and accumulated in the peripheral blood and bone marrow of pediatric patients with AML. Interestingly, they suggested the use of this circRNA as a prognostic biomarker because its elevated expression correlated with relapse. Moreover, they found that circRNF220 knockdown specifically inhibited proliferation and promoted apoptosis in AML cell lines and primary cells [90]. In JMML, CircMCTP, circLYN, and circAFF2 were highly upregulated, suggesting their involvement in this tumor's pathogenesis [91]. As in solid pediatric tumors, the search for circRNAs in blood malignancies is still in its infancy and is an important resource for the identification of biomarkers on which the spotlight is turned on.

4. Bioinformatic Tools and Resources for circRNAs

Identification of circRNAs in RNA sequencing data remains a major challenge, especially for their detection within pooled heterogeneous RNAs. In fact, circRNAs generally lack poly(A) tails [19], even though some A-rich stretches can still be found within their sequences. Random priming can be used in place of oligo(dT) priming to enrich circRNA libraries. Moreover, despite circRNAs are retained in rRNA-depleted libraries, they could require the use of RNase R to digest linear RNA. In addition, adaptor ligation could represent an artifact in circRNA enrichment as two different cDNAs may be ligated in a noncanonical order during the step of adaptor ligation. Another problem is template switching that is promoted in case of long homologous sequences, such as those found in the genes coding for multiple isoforms that share identical constitutive exons [92,93]. All of these experimental artifacts should be taken into account by the algorithms that are used to identify circRNAs. To date, different algorithms have been developed to resolve these issues [94], each one with its own methodologies. The most common algorithms are listed in Table 2 [93]. These are used for single-end or paired-end data, with the latter being more sensitive, especially when dealing with high read coverage [95]. Reads that are not contiguously aligned to the reference genome and that align to back-splice junctions are further processed by distinct algorithms in order to identify and count circRNAs. A list of bioinformatic resources was developed to allow not only the discovery, but also the study of the mechanism of function of circRNAs. These tools comprise both those for circRNA identification and quantification and resources for circRNA annotation. Indeed, there are several databases that collect circRNA information related to many aspects of this RNA's biology, such as interaction with other biological entities as well as expression data across cells and tissues and tools for experiment design. Such resources were comprehensively reviewed by Chen et al. [96].

Table 2. Tools for circRNA identification and annotation.

Algorithm	Aligner	Reference
find_circ	Bowtie2	[6]
MapSplice	Bowtie	[97]
CIRCexplorer	TopHat	[16]
circRNA_finder	STAR	[98]
CIRI	BWA	[99]
Salzman 2012	Bowtie	[100]
Salzman 2013	Bowtie2	[101]

Table 2. *Cont.*

Algorithm	Aligner	Reference
CircRNAseq	Bowtie	[15]
Segemehl	Segemehl	[102]
Guo 2014	Bowtie	[103]
KNIFE	Bowtie2	[104]
DCC	STAR	[105]

5. Conclusions

In the past few years, the advancement of such high-throughput technologies as sequencing and omics techniques led to the identification of circRNAs. These conserved endogenous RNAs have an extensive distribution, even though they present tissue-specific expression, cell type specificity, and multiple functions. Intriguingly, dysregulated expression of circRNAs has been identified in various cancers. Even though little is known about their function and molecular mechanisms during cancer initiation, progression, and metastasis, circRNAs have a great potential as cancer biomarkers. Indeed, the emerging involvement of circRNA deregulation in cancer pathogenesis has opened promising opportunities for their clinical application in tumor diagnosis, outcome prediction, and therapy. Even though circRNAs arouse considerable interest and are under the magnifying glass, their characterization and study are still in their infancy, and many questions still remain unanswered. It is highly unlikely that any single circRNA, despite being highly sensitive and specific, exists in all cancer types, and one specific circRNA is usually insufficient to predict or monitor cancer; therefore, integrating a panel of cancer-associated circRNAs as biomarkers or signatures might be a reasonable method in future clinical practice, in particular in pediatric cancer. Indeed, the current challenges in the field of childhood cancer diseases include identification of novel biomarkers that may allow a more accurate risk stratification. These circular molecules, specific and extremely stable, implicated in the regulation of physiological processes and deregulated in several pathologies, such as cancer, serve as stable clinical biomarkers of disease and also provide new potential therapeutic targets in the pediatric setting.

Author Contributions: Writing—original draft preparation, A.G. and M.C.; writing—review and editing, A.G., M.C. and A.P.; supervision, A.D.G.; All authors have read and agreed to the published version of the manuscript.

Funding: This research received no external funding.

Institutional Review Board Statement: Not applicable.

Informed Consent Statement: Not applicable.

Data Availability Statement: Not applicable.

Acknowledgments: Marta Colletti was supported by Fondazione Umberto Veronesi. The authors are thankful to Fondazione AIRC per la Ricerca sul Cancro for funding (IG-25964 to Angela Di Giannatale). This work was also supported by the Italian Ministry of Health with the Current Research funds.

Conflicts of Interest: The authors declare no conflict of interest.

References

1. Kos, A.; Dijkema, R.; Arnberg, A.C.; van der Meide, P.H.; Schellekens, H. The hepatitis delta (delta) virus possesses a circular RNA. *Comp. Study Nat.* **1986**, *323*, 558–560. [CrossRef] [PubMed]
2. Ford, E.; Ares, M., Jr. Synthesis of circular RNA in bacteria and yeast using RNA cyclase ribozymes derived from a group I intron of phage T4. *Proc. Natl. Acad. Sci. USA* **1994**, *91*, 3117–3121. [CrossRef] [PubMed]

3. Grabowski, P.J.; Zaug, A.J.; Cech, T.R. The intervening sequence of the ribosomal RNA precursor is converted to a circular RNA in isolated nuclei of Tetrahymena. *Cell* **1981**, *23*, 467–476. [CrossRef] [PubMed]
4. Capel, B.; Swain, A.; Nicolis, S.; Hacker, A.; Walter, M.; Koopman, P.; Goodfellow, P.; Lovell-Badge, R. Circular transcripts of the testis-determining gene Sry in adult mouse testis. *Cell* **1993**, *73*, 1019–1030. [CrossRef] [PubMed]
5. Yu, C.Y.; Kuo, H.C. The emerging roles and functions of circular RNAs and their generation. *J. Biomed. Sci.* **2019**, *26*, 29. [CrossRef]
6. Memczak, S.; Jens, M.; Elefsinioti, A.; Torti, F.; Krueger, J.; Rybak, A.; Maier, L.; Mackowiak, S.D.; Gregersen, L.H.; Munschauer, M.; et al. Circular RNAs are a large class of animal RNAs with regulatory potency. *Nature* **2013**, *495*, 333–338. [CrossRef]
7. Zhang, Y.; Zhao, Y.; Liu, Y.; Wang, M.; Yu, W.; Zhang, L. Exploring the regulatory roles of circular RNAs in Alzheimer's disease. *Transl. Neurodegener.* **2020**, *9*, 35. [CrossRef]
8. Li, J.; Sun, D.; Pu, W.; Wang, J.; Peng, Y. Circular RNAs in Cancer: Biogenesis, Function, and Clinical Significance. *Trends Cancer* **2020**, *6*, 319–336. [CrossRef]
9. Yin, Y.; Long, J.; He, Q.; Li, Y.; Liao, Y.; He, P.; Zhu, W.J. Emerging roles of circRNA in formation and progression of cancer. *Cancer* **2019**, *10*, 5015–5021. [CrossRef]
10. Wang, Y.; Mo, Y.; Gong, Z.; Yang, X.; Yang, M.; Zhang, S.; Xiong, F.; Xiang, B.; Zhou, M.; Liao, Q.; et al. Circular RNAs in human cancer. *Mol. Cancer* **2017**, *16*, 25. [CrossRef]
11. Tang, X.; Ren, H.; Guo, M.; Qian, J.; Yang, Y.; Gu, C. Review on circular RNAs and new insights into their roles in cancer. *Comput. Struct. Biotechnol. J.* **2021**, *19*, 910–928. [CrossRef]
12. Ivanov, A.; Memczak, S.; Wyler, E.; Torti, F.; Porath, H.T.; Orejuela, M.R.; Piechotta, M.; Levanon, E.Y.; Landthaler, M.; Dieterich, C.; et al. Analysis of intron sequences reveals hallmarks of circular RNA biogenesis in animals. *Cell Rep.* **2015**, *10*, 170–177. [CrossRef]
13. Lander, E.S.; Linton, L.M.; Birren, B.; Nusbaum, C.; Zody, M.C.; Baldwin, J.; Devon, K.; Dewar, K.; Doyle, M.; Fitzhugh, W.; et al.; International Human Genome Sequencing Consortium. Initial sequencing and analysis of the human genome. *Nature* **2001**, *409*, 860–921.
14. Dong, R.; Ma, X.K.; Chen, L.L.; Yang, L. Increased complexity of circRNA expression during species evolution. *RNA Biol.* **2017**, *14*, 1064–1074. [CrossRef]
15. Jeck, W.R.; Sorrentino, J.A.; Wang, K.; Slevin, M.K.; Burd, C.E.; Liu, J.; Marzluff, W.F.; Sharpless, N.E. Circular RNAs are abundant, conserved, and associated with ALU repeats. *RNA* **2013**, *19*, 141–157. [CrossRef]
16. Zhang, X.O.; Wang, H.B.; Zhang, Y.; Lu, X.; Chen, L.L.; Yang, L. Complementary sequence-mediated exon circularization. *Cell* **2014**, *159*, 134–147. [CrossRef]
17. Wilusz, J.E. Repetitive elements regulate circular RNA biogenesis. *Mob. Genet. Elem.* **2015**, *5*, 1–7. [CrossRef]
18. Zhang, Y.; Xue, W.; Xiang Li, X.; Zhang, J.; Chen, S.; Zhang, J.L.; Yang, L.; Chen, L.L. The Biogenesis of Nascent Circular RNAs. *Cell Rep.* **2016**, *15*, 611–624. [CrossRef]
19. Athanasiadis, A.; Rich, A.; Maas, S. Widespread A-to-I RNA editing of Alu-containing mRNAs in the human transcriptome. *PLoS Biol.* **2004**, *2*, e391. [CrossRef]
20. Tang, M.; Lv, Y. The Role of N6 -Methyladenosine Modified Circular RNA in Pathophysiological Processes. *Int. J. Biol. Sci.* **2021**, *17*, 2262–2277. [CrossRef]
21. Kristensen, L.S.; Andersen, M.S.; Stagsted, L.V.W.; Ebbesen, K.K.; Hansen, T.B.; Kjems, J. The biogenesis, biology and characterization of circular RNAs. *Nat. Rev. Genet.* **2019**, *20*, 675–691. [CrossRef] [PubMed]
22. Li, X.; Yang, L.; Chen, L.L. The Biogenesis, Functions, and Challenges of Circular RNAs. *Mol. Cell* **2018**, *71*, 428–442. [CrossRef] [PubMed]
23. Zhou, Q.; Xie, D.; Wang, R.; Liu, L.; Yu, Y.; Tang, X.; Hu, Y.; Cui, D. The emerging landascape of exosomal CircRNAs in solid cancers and hematological malignancies. *Biomark. Res.* **2022**, *10*, 28. [CrossRef] [PubMed]
24. Wan, L.; Zhang, L.; Fan, K.; Cheng, Z.X.; Sun, Q.C.; Wang, J.J. Circular RNA-ITCH Suppresses Lung Cancer Proliferation via Inhibiting the Wnt/β-Catenin Pathway. *Biomed. Res. Int.* **2016**, *2016*, 1579490. [CrossRef] [PubMed]
25. Wang, X.; Zhang, Y.; Huang, L.; Zhang, J.; Pan, F.; Li, B.; Yan, Y.; Jia, B.; Liu, H.; Li, S.; et al. Decreased expression of hsa_circ_001988 in colorectal cancer and its clinical significances. *Int. J. Clin. Exp. Pathol.* **2015**, *8*, 16020–16025. [PubMed]
26. Ahmed, I.; Karedath, T.; Andrews, S.S.; Al-Azwani, I.K.; Mohamoud, Y.A.; Querleu, D.; Rafii, A.; Malek, J.A. Altered expression pattern of circular RNAs in primary and metastatic sites of epithelial ovarian carcinoma. *Oncotarget* **2016**, *7*, 36366–36381. [CrossRef]
27. Hansen, T.B.; Kjems, J.; Damgaard, C.K. Circular RNA and miR-7 in cancer. *Cancer Res.* **2013**, *73*, 5609–5612. [CrossRef]
28. Li, J.; Huang, C.; Zou, Y.; Ye, J.; Yu, J.; Gui, Y. CircTLK1 promotes the proliferation and metastasis of renal cell carcinoma by sponging miR-136-5p. *Mol. Cancer* **2020**, *19*, 103. [CrossRef]
29. Du, W.W.; Zhang, C.; Yang, W.; Yong, T.; Awan, F.M.; Yang, B.B. Identifying and Characterizing circRNA-Protein Interaction. *Theranostics* **2017**, *7*, 4183–4191. [CrossRef]
30. Lv, T.; Miao, Y.F.; Jin, K.; Han, S.; Xu, T.Q.; Qiu, Z.L.; Zhang, X.H. Dysregulated circular RNAs in medulloblastoma regulate proliferation and growth of tumor cells via host genes. *Cancer Med.* **2018**, *7*, 6147–6157. [CrossRef]
31. Wang, X.; Xu, D.; Pei, X.; Zhang, Y.; Zhang, Y.; Gu, Y.; Li, Y. CircSKA3 Modulates FOXM1 to Facilitate Cell Proliferation, Migration, and Invasion While Confine Apoptosis in Medulloblastoma via miR-383-5p. *Cancer Manag. Res.* **2020**, *12*, 13415–13426. [CrossRef]

32. Zhao, X.; Guan, J.; Luo, M. Circ-SKA3 upregulates ID3 expression by decoying miR-326 to accelerate the development of medulloblastoma. *J. Clin. Neurosci.* **2021**, *86*, 87–96. [CrossRef]
33. Liu, X.C.; Wang, F.C.; Wang, J.H.; Zhao, J.Y.; Ye, S.Y. The Circular RNA circSKA3 Facilitates the Malignant Biological Behaviors of Medulloblastoma via miR-520 h/CDK6 Pathway. *Mol. Biotechnol.* **2022**, *64*, 1022–1033. [CrossRef]
34. Ahmadov, U.; Bendikas, M.M.; Ebbesen, K.K.; Sehested, A.M.; Kjems, J.; Broholm, H.; Kristensen, L.S. Distinct circular RNA expression profiles in pediatric ependymomas. *Brain Pathol.* **2021**, *31*, 387–392. [CrossRef]
35. Alexander, F. Neuroblastoma. *Urol. Clin. N. Am.* **2000**, *27*, 383–392. [CrossRef]
36. Chen, Y.; Yang, F.; Fang, E.; Xiao, W.; Mei, H.; Li, H.; Li, D.; Song, H.; Wang, J.; Hong, M.; et al. Circular RNA circAGO2 drives cancer progression through facilitating HuR-repressed functions of AGO2-miRNA complexes. *Cell Death Differ.* **2019**, *26*, 1346–1364. [CrossRef]
37. Li, H.; Yang, F.; Hu, A.; Wang, X.; Fang, E.; Chen, Y.; Li, D.; Song, H.; Wang, J.; Guo, Y.; et al. Therapeutic targeting of circ-CUX1/EWSR1/MAZ axis inhibits glycolysis and neuroblastoma progression. *EMBO Mol. Med.* **2019**, *11*, e10835. [CrossRef]
38. Liu, K.C.; Lin, B.S.; Zhao, M.; Wang, K.Y.; Lan, X.P. Cutl1: A potential target for cancer therapy. *Cell Signal.* **2013**, *25*, 349–354. [CrossRef]
39. Zhang, X.; Zhang, J.; Liu, Q.; Zhao, Y.; Zhang, W.; Haiyan Yang, H. Circ-CUX1 Accelerates the Progression of Neuroblastoma via miR-16-5p/DMRT2 Axis. *Neurochem. Res.* **2020**, *45*, 2840–2855. [CrossRef]
40. Wang, Y.; Niu, Q.; Dai, J.; Shi, H.; Zhang, J. circCUX1 promotes neuroblastoma progression and glycolysis by regulating the miR-338-3p/PHF20 axis. *Gen. Physiol. Biophys.* **2021**, *40*, 17–29.
41. Yang, J.; Yu, L.; Yan, J.; Xiao, Y.; Li, W.; Xiao, J.; Lei, J.; Xiang, D.; Zhang, S.; Yu, X. Circular RNA DGKB Promotes the Progression of Neuroblastoma by Targeting miR-873/GLI1 Axis. *Front. Oncol.* **2020**, *10*, 1104. [CrossRef] [PubMed]
42. Harenza, J.L.; Diamond, M.A.; Adams, R.N.; Song, M.M.; Davidson, H.L.; Hart, L.S.; Dent, M.H.; Fortina, P.; Reynolds, C.P.; Maris, J.M. Transcriptomic profiling of 39 commonly-used neuroblastoma cell lines. *Sci. Data* **2017**, *4*, 170033. [CrossRef] [PubMed]
43. Zhang, L.; Zhou, H.; Li, J.; Wang, X.; Zhang, X.; Shi, T.; Feng, G. Comprehensive Characterization of Circular RNAs in Neuroblastoma Cell Lines. *Technol. Cancer Res. Treat.* **2020**, *19*, 1533033820957622. [CrossRef] [PubMed]
44. Li, Y.; Shang, Y.M.; Wang, Q.W. MicroRNA-21 promotes the proliferation and invasion of neuroblastoma cells through targeting CHL1. *Minerva Med.* **2016**, *107*, 287–293. [PubMed]
45. Lin, W.; Wang, Z.; Wang, J.; Yan, H.; Han, Q.; Yao, W.; Li, K. circRNA-TBC1D4, circRNA-NAALAD2 and circRNA-TGFBR3: Selected Key circRNAs in Neuroblastoma and Their Associations with Clinical Features. *Cancer Manag. Res.* **2021**, *13*, 4271–4281. [CrossRef]
46. Chen, Y.; Lin, L.; Hu, X.; Li, Q.; Wu, M. Silencing of circular RNA circPDE5A suppresses neuroblastoma progression by targeting the miR-362-5p/NOL4L axis. *Int. J. Neurosci.* **2021**, 1–11. [CrossRef]
47. Fang, Y.; Yao, Y.; Mao, K.; Yanyan Zhong, Y.; Xu, Y. Circ_0132817 facilitates cell proliferation, migration, invasion and glycolysis by regulating the miR-432-5p/NOL4L axis in neuroblastoma. *Exp. Brain Res.* **2021**, *239*, 1841–1852. [CrossRef]
48. Tang, J.; Liu, F.; Huang, D.; Zhao, C.; Liang, J.; Wang, F.; Zeng, J.; Zhang, M.; Zhai, X.; Li, L. circ0125803 facilitates tumor progression by sponging miR-197-5p and upregulating E2F1 in neuroblastoma. *Pathol. Res. Pract.* **2022**, *233*, 153857. [CrossRef]
49. Yang, J.; Liu, B.; Xu, Z.; Feng, M. Silencing of Circ_0135889 Restrains Proliferation and Tumorigenicity of Human Neuroblastoma Cells. *J. Surg. Res.* **2022**, *279*, 135–147. [CrossRef]
50. Jin, Q.; Li, J.; Yang, F.; Feng, L.; Du, X. Circular RNA circKIF2A Contributes to the Progression of Neuroblastoma Through Regulating PRPS1 Expression by Sponging miR-377-3p. *Biochem. Genet.* **2022**, *60*, 1380–1401. [CrossRef]
51. Karami Fath, M.; Pourbagher Benam, S.; Salmani, K.; Naderi, S.; Fahham, Z.; Ghiabi, S.; Houshmand Kia, S.A.; Naderi, M.; Darvish, M.; Barati, G. Circular RNAs in neuroblastoma: Pathogenesis, potential biomarker, and therapeutic target. *Pathol. Res. Pract.* **2022**, *238*, 154094. [CrossRef]
52. Ries, L.A.G.; Harkins, D.; Krapcho, M.; Mariotto, A.; Miller, B.A.; Feuer, E.J.; Clegg, L.; Eisner, M.P.; Horner, M.J.; Howlader, N.; et al. SEER Cancer Statistics Review, 1975–2003; National Cancer Institute: Bethesda, MD, USA. Available online: http://seer.cancer.gov/csr/1975_2003 (accessed on 10 June 2011).
53. Pastore, G.; Peris-Bonet, R.; Carli, M.; Martínez-García, C.; Sánchez de Toledo, J.; Steliarova-Foucher, E. Childhood soft tissue sarcomas incidence and survival in European children (1978–1997): Report from the Automated Childhood Cancer Information System project. *Eur. J. Cancer* **2006**, *42*, 2136. [CrossRef]
54. Sun, X.; Guo, W.; Shen, J.K.; Mankin, H.J.; Hornicek, F.J.; Duan, Z. Rhabdomyosarcoma: Advances in molecular and cellular biology. *Sarcoma* **2015**, *2015*, 232010. [CrossRef]
55. Rossi, F.; Legnini, I.; Megiorni, F.; Colantoni, A.; Santini, T.; Morlando, M.; Di Timoteo, G.; Dattilo, D.; Dominici, C.; Bozzoni, I. Circ-ZNF609 regulates G1-S progression in rhabdomyosarcoma. *Oncogene* **2019**, *38*, 3843–3854. [CrossRef]
56. Rossi, F.; Centrón-Broco, A.; Dattilo, D.; Di Timoteo, G.; Guarnacci, M.; Colantoni, A.; Beltran Nebot, M.; Bozzoni, I. CircVAMP3: A circRNA with a Role in Alveolar Rhabdomyosarcoma Cell Cycle Progression. *Genes* **2021**, *12*, 985. [CrossRef]
57. Marko, T.A.; Diessner, B.J.; Spector, L.G. Prevalence of Metastasis at Diagnosis of Osteosarcoma: An International Comparison. *Pediatr. Blood Cancer.* **2016**, *63*, 1006–1011. [CrossRef]
58. Wang, J.; Ni, J.; Song, D.; Ding, M.; Huang, J.; Li, W.; He, G. The regulatory effect of has-circ-0001146/miR-26a-5p/MNAT1 network on the proliferation and invasion of osteosarcoma. *Biosci. Rep.* **2020**, *40*, BSR20201232. [CrossRef]

59. Zhang, C.; Na, N.; Liu, L.; Qiu, Y. CircRNA hsa_circ_0005909 Promotes Cell Proliferation of Osteosarcoma Cells by Targeting miR-338-3p/HMGA1 Axis. *Cancer Manag. Res.* **2021**, *13*, 795–803. [CrossRef]
60. Cao, J.; Huang, Z.; Ou, S.; Wen, F.; Yang, G.; Miao, Q.; Zhang, H.; Wang, Y.; He, X.; Shan, Y.; et al. circ0093740 Promotes Tumor Growth and Metastasis by Sponging miR-136/145 and Upregulating DNMT3A in Wilms Tumor. *Front. Oncol.* **2021**, *11*, 647352 [CrossRef]
61. Zhou, R.; Jia, W.; Gao, X.; Deng, F.; Fu, K.; Zhao, T.; Li, Z.; Fu, W.; Liu, G. CircCDYL Acts as a Tumor Suppressor in Wilms' Tumor by Targeting miR-145-5p. *Front. Cell Dev. Biol.* **2021**, *9*, 668947. [CrossRef]
62. Xu, J.; Hao, Y.; Gao, X.; Wu, Y.; Ding, Y.; Wang, B. CircSLC7A6 promotes the progression of Wilms' tumor via microRNA-107/ABL proto-oncogene 2 axis. *Bioengineered* **2022**, *13*, 308–318. [CrossRef] [PubMed]
63. Hiyama, E. Pediatric hepatoblastoma: Diagnosis and treatment. *Transl. Pediatr.* **2014**, *3*, 293–299. [PubMed]
64. Kremer, N.; Walther, A.E.; Tiao, G.M. Management of hepatoblastoma: An update. *Curr. Opin. Pediatr.* **2014**, *26*, 362–369. [CrossRef] [PubMed]
65. Liu, B.H.; Zhang, B.B.; Liu, X.Q.; Zheng, S.; Dong, K.R.; Dong, R. Expression Profiling Identifies Circular RNA Signature in Hepatoblastoma. *Cell Physiol. Biochem.* **2018**, *45*, 706–719. [CrossRef] [PubMed]
66. Song, H.; Bian, Z.X.; Li, H.Y.; Zhang, Y.; Ma, J.; Chen, S.H.; Zhu, J.B.; Zhang, X.; Wang, J.; Gu, S.; et al. Characterization of hsa_circ_0000594 as a new biomarker and therapeutic target for hepatoblastoma. *Eur. Rev. Med. Pharm. Sci.* **2019**, *23*, 8274–8286.
67. Zhen, N.; Gu, S.; Ma, J.; Zhu, J.; Yin, M.; Xu, M.; Wang, J.; Huang, N.; Cui, Z.; Bian, Z.; et al. CircHMGCS1 Promotes Hepatoblastoma Cell Proliferation by Regulating the IGF Signaling Pathway and Glutaminolysis. *Theranostics* **2019**, *9*, 900–919. [CrossRef]
68. Glazar, P.; Papavasileiou, P.; Rajewsky, N. Circbase: A database for circular RNAs. *RNA* **2014**, *20*, 1666–1670. [CrossRef]
69. Liu, Y.; Song, J.; Liu, Y.; Zhou, Z.; Wang, X. Transcription activation of circ-STAT3 induced by Gli2 promotes the progression of hepatoblastoma via acting as a sponge for miR-29a/b/c-3p to upregulate STAT3/Gli2. *J. Exp. Clin. Cancer Res.* **2020**, *39*, 101. [CrossRef]
70. Luping, C.; Juanyi, S.; Yaohao, W.; Ronglin, Q.; Lexiang, Z.; Lei, L.; Jianhang, S.; Minyi, L.; Xiaogeng, D. CircRNA CDR1as promotes hepatoblastoma proliferation and stemness by acting as a miR-7-5p sponge to upregulate KLF4 expression. *Aging (Albany NY)* **2020**, *12*, 19233–19253.
71. Li, X.; Wang, H.; Liu, Z.; Abudureyimu, A. CircSETD3 (Hsa_circ_0000567) Suppresses Hepatoblastoma Pathogenesis viaTargeting the miR-423-3p/Bcl-2-Interacting Mediator of Cell Death Axis. *Front. Genet.* **2021**, *12*, 724197.
72. Xing, L.; Zhang, L.; Feng, Y.; Cui, Z.; Ding, L. Downregulation of circular RNA hsa_circ_0001649 indicates poor prognosis for retinoblastoma and regulates cell proliferation and apoptosis via AKT/mTOR signaling pathway. *Biomed. Pharmacother.* **2018**, *105*, 326–333. [CrossRef]
73. Lyu, J.; Wang, Y.; Zheng, Q.; Hua, P.; Zhu, X.; Li, J.; Li, J.; Ji, X.; Zhao, P. Reduction of circular RNA expression associated with human retinoblastoma. *Exp. Eye Res.* **2019**, *184*, 278–285. [CrossRef]
74. Fu, C.; Wang, S.; Jin, L.; Zhang, M.; Li, M. CircTET1 Inhibits Retinoblastoma Progression via Targeting miR-492 and miR-494-3p through Wnt/beta-catenin Signaling Pathway. *Curr. Eye Res.* **2021**, *46*, 978–987. [CrossRef]
75. Wenbo, Z.; Shuai, W.; Tingyu, Q.; Wenzhan, W. Circular RNA (circ-0075804) promotes the proliferation of retinoblastoma via combining heterogeneous nuclear ribonucleoprotein K (HNRNPK) to improve the stability of E2F transcription factor 3 E2F3. *J. Cell. Biochem.* **2020**, *121*, 3516–3525.
76. Du, S.; Wang, S.; Zhang, F.; Lv, Y. SKP2, positively regulated by circ_ODC1/miR-422a axis, promotes the proliferation of retinoblastoma.]. *J. Cell. Biochem.* **2020**, *121*, 322–331. [CrossRef]
77. Chen, N.N.; Chao, D.L.; Li, X.G. Circular RNA has_circ_0000527 participates in proliferation, invasion and migration of retinoblastoma cells via miR-646/BCL-2 axis. *Cell Biochem. Funct.* **2020**, *38*, 1036–1046. [CrossRef]
78. Zhang, L.; Wu, J.; Li, Y.; Jiang, Y.; Wang, L.; Chen, Y.; Lv, Y.; Zou, Y.; Ding, X. Circ_0000527 promotes the progression of retinoblastoma by regulating miR-646/LRP6 axis. *Cancer Cell Int.* **2020**, *20*, 301. [CrossRef]
79. Yu, B.; Zhao, J.; Dong, Y. Circ_0000527 Promotes Retinoblastoma Progression through Modulating miR-98-5p/XIAP Pathway. *Curr. Eye Res.* **2021**, *46*, 1414–1423. [CrossRef]
80. Liu, H.; Yuan, H.; Xu, D.; Chen, K.; Tan, N.; Zheng, Q.-J. Circular RNA circ_0000034 upregulates STX17 level to promote human retinoblastoma development via inhibiting miR-361-3p. *Eur. Rev. Med. Pharm. Sci.* **2020**, *23*, 12080–12092.
81. Zhe, S.; Ai, Z.; Mingyu, H.; Tao, J. Circular RNA hsa_circ_0000034 promotes the progression of retinoblastoma via sponging microRNA-361-3p.]. *Bioengineered* **2020**, *1*, 949–957.
82. Xu, L.; Long, H.; Zhou, B.; Jiang, H.; Cai, M. CircMKLN1 Suppresses the Progression of Human Retinoblastoma by Modulation of miR-425-5p/PDCD4 Axis. *Curr. Eye Res.* **2021**, *46*, 1751–1761. [CrossRef] [PubMed]
83. Zheng, T.; Chen, W.; Wang, X.; Cai, W.; Wu, F.; Lin, C. Circular RNA circ-FAM158A promotes retinoblastoma progression by regulating miR-138-5p/SLC7A5 axis. *Exp. Eye Res.* **2021**, *11*, 108650. [CrossRef] [PubMed]
84. Jiang, G.; Qu, M.; Kong, L.; Song, X.; Jiang, S. hsa_circ_0084811 Regulates Cell Proliferation and Apoptosis in Retinoblastoma through miR-18a-5p/miR-18b-5p/E2F5 Axis. *Biomed. Res. Int.* **2022**, *2022*, 6918396. [CrossRef]
85. Zhou, X.; Zhan, L.; Huang, K.; Wang, X. The functions and clinical significance of circRNAs in hematological malignancies. *J. Hematol. Oncol* **2020**, *13*, 138. [CrossRef] [PubMed]

6. Zhu, C.; Guo, A.; Sun, B.; Zhou, Z. Comprehensive elaboration of circular RNA in multiple myeloma. *Front. Pharm.* **2022**, *13*, 971070. [CrossRef]

7. Allegra, A.; Cicero, N.; Tonacci, A.; Musolino, C.; Gangemi, S. Circular RNA as a Novel Biomarker for Diagnosis and Prognosis and Potential Therapeutic Targets in Multiple Myeloma. *Cancers* **2022**, *14*, 1700. [CrossRef]

8. Hu, J.; Han, Q.; Gu, Y.; Ma, J.; McGrath, M.; Fengchang Qiao, F.; Chen, B.; Song, C.; Ge, Z. Circular RNA PVT1 expression and its roles in acute lympho-blastic leukemia. *Epigenomics* **2018**, *10*, 723–732. [CrossRef]

9. Gaffo, E.; Boldrin, E.; Dal Molin, A.; Bresolin, S.; Bonizzato, A.; Trentin, L.; Frasson, C.; Debatin, K.M.; Meyer, L.H.; Kronnie, G.T.; et al. Circular RNA differential expression in blood cell populations and exploration of circRNA deregulation in pediatric acute lymphoblastic leukemia. *Sci. Rep.* **2019**, *9*, 14670. [CrossRef]

0. Liu, X.; Liu, X.; Cai, M.; Luo, A.; He, Y.; Liu, S.; Zhang, X.; Yang, X.; Xu, L.; Jiang, H. CircRNF220, not its linear cognate gene RNF220, regulates cell growth and is associated with relapse in pediatric acute myeloid leukemia. *Mol. Cancer* **2021**, *20*, 139. [CrossRef]

1. Dal Molin, A.; Hofmans, M.; Gaffo, E.; Bu-ratin, A.; Cavé, H.; Flotho, C.; de Haas, V.; Niemeyer, C.M.; Stary, J.; Van Vlierberghe, P.; et al. CircR-NAs Dysregulated in Juvenile Myelomonocytic Leukemia: CircMCTP1 Stands Out. *Front. Cell Dev. Biol.* **2021**, *8*, 613540. [CrossRef]

2. Cocquet, J.; Chong, A.; Zhang, G.; Veitia, R.A. Reverse transcriptase template switching and false alternative transcripts. *Genomics* **2006**, *88*, 127–131. [CrossRef]

3. Szabo, L.; Salzman, J. Detecting circular RNAs: Bioinformatic and experimental challenges. *Nat. Rev. Genet.* **2016**, *17*, 679–692. [CrossRef]

4. Hansen, T.B.; Venø, M.T.; Damgaard, C.K.; Kjems, J. Comparison of circular RNA prediction tools. *Nucleic Acids Res.* **2016**, *44*, e58. [CrossRef]

5. Chen, I.; Chia-Ying Chen, C.Y.; Chuang, T.J. Biogenesis, identification, and function of exonic circular RNAs. *Wiley Interdiscip. Rev. RNA* **2015**, *6*, 563–679. [CrossRef]

6. Chen, L.; Wang, C.; Sun, H.; Wang, J.; Liang, Y.; Wang, Y.; Wong, G. The bioinformatics toolbox for circRNA discovery and analysis. *Brief. Bioinform.* **2021**, *22*, 1706–1728. [CrossRef]

7. Wang, K.; Singh, D.; Zeng, Z.; Coleman, S.J.; Huang, Y.; Savich, G.L.; He, X.; Mieczkowski, P.; Grimm, S.A.; Perou, C.M.; et al. MapSplice: Accurate mapping of RNA-seq reads for splice junction discovery. *Nucleic Acids Res.* **2010**, *38*, e178. [CrossRef]

8. Westholm, J.O.; Miura, P.; Sara Olson, S.; Shenker, S.; Joseph, B.; Sanfilippo, P.; Celniker, S.E.; Graveley, B.R.; Lai, E.C. Genome-wide analysis of drosophila circular RNAs reveals their structural and sequence properties and age-dependent neural accumulation. *Cell Rep.* **2014**, *9*, 1966–1980. [CrossRef]

9. Gao, Y.; Wang, J.; Zhao, F. CIRI: An efficient and unbiased algorithm for de novo circular RNA identification. *Genome Biol.* **2015**, *16*, 4. [CrossRef]

00. Salzman, J.; Gawad, C.; Wang, P.L.; Lacayo, N.; Brown, P.O. Circular RNAs are the predominant transcript isoform from hundreds of human genes in diverse cell types. *PLoS ONE* **2012**, *7*, e30733. [CrossRef]

01. Salzman, J.; Chen, R.E.; Olsen, M.N.; Wang, P.L.; Brown, P.O. Cell-type specific features of circular RNA expression. *PLoS Genet.* **2013**, *9*, e1003777. [CrossRef]

02. Hoffmann, S.; Otto, C.; Doose, G.; Tanzer, A.; Langenberger, D.; Christ, S.; Kunz, M.; Holdt, L.M.; Teupser, D.; Hackermüller, J.; et al. A multi-split mapping algorithm for circular RNA, splicing, trans-splicing and fusion detection. *Genome Biol.* **2014**, *15*, R34. [CrossRef] [PubMed]

03. Guo, J.O.; Agarwal, V.; Guo, H.; Bartel, D.P. Expanded identification and characterization of mammalian circular RNAs. *Genome Biol.* **2014**, *15*, 409. [CrossRef] [PubMed]

04. Szabo, L.; Morey, R.; Palpant, N.J.; Wang, P.L.; Afari, N.; Jiang, C.; Parast, M.M.; Murry, C.E.; Laurent, L.C.; Salzman, J. Statistically based splicing detection reveals neural enrichment and tissue-specific induction of circular RNA during human fetal development. *Genome Biol.* **2015**, *16*, 126. [CrossRef] [PubMed]

05. Cheng, J.; Metge, F.; Dieterich, C. Specific identification and quantification of circular RNAs from sequencing data. *Bioinformatics* **2016**, *32*, 1094–1096. [CrossRef] [PubMed]

biomedicines

Article

MicroRNA Profiling Shows a Time-Dependent Regulation within the First 2 Months Post-Birth and after Mild Neonatal Hypoxia in the Hippocampus from Mice

Aisling Leavy [1], Gary P. Brennan [2] and Eva M. Jimenez-Mateos [1,*

[1] Discipline of Physiology, School of Medicine, Trinity College Dublin, The University of Dublin, D02 R590 Dublin, Ireland
[2] Conway Institute, School of Biomolecular and Biomedical Science, University College Dublin, Belfield, D04 C7X2 Dublin, Ireland
* Correspondence: jimeneze@tcd.ie

Abstract: Brain development occurs until adulthood, with time-sensitive processes happening during embryo development, childhood, and puberty. During early life and childhood, dynamic changes in the brain are critical for physiological brain maturation, and these changes are tightly regulated by the expression of specific regulatory genetic elements. Early life insults, such as hypoxia, can alter the course of brain maturation, resulting in lifelong neurodevelopmental conditions. MicroRNAs are small non-coding RNAs, which regulate and coordinate gene expression. It is estimated that one single microRNA can regulate the expression of hundreds of protein-coding genes.. Uncovering the miRNome and microRNA-regulated transcriptomes may help to understand the patterns of genes regulating brain maturation, and their contribution to neurodevelopmental pathologies following hypoxia at Postnatal day 7. Here, using a PCR-based platform, we analyzed the microRNA profile postnatally in the hippocampus of control mice at postnatal day 8, 14, and 42 and after hypoxia at postnatal day 7, to elucidate the set of microRNAs which may be key for postnatal hippocampus maturation. We observed that microRNAs can be divided in four groups based on their temporal expression. Further after an early life insult, hypoxia at P7, 15 microRNAs showed a misregulation over time, including Let7a. We speculated that the transcriptional regulator c-myc is a contributor to this process. In conclusion, here, we observed that microRNAs are regulated postnatally in the hippocampus and alteration of their expression after hypoxia at birth may be regulated by the transcriptional regulator c-myc.

Keywords: microRNAs1; hippocampus; hypoxia

Citation: Leavy, A.; Brennan, G.P.; Jimenez-Mateos, E.M. MicroRNA Profiling Shows a Time-Dependent Regulation within the First 2 Months Post-Birth and after Mild Neonatal Hypoxia in the Hippocampus from Mice. *Biomedicines* **2022**, *10*, 2740. https://doi.org/10.3390/ biomedicines10112740

Academic Editors: Milena Rizzo and Elena Levantini

Received: 28 September 2022
Accepted: 26 October 2022
Published: 28 October 2022

Publisher's Note: MDPI stays neutral with regard to jurisdictional claims in published maps and institutional affiliations.

1. Introduction

Human brain development begins in the 3rd week of gestation and continuing until adulthood [1], with dynamic changes in the brain happening through childhood and adolescence. The complex process that is postnatal brain maturation involves the formation and maturation of tracts, combined with regressive processes including apoptosis of brain cells [2,3]. Whilst the generation and migration of neurons is mainly a prenatal event, the neuron-glia interactions and the functional organization of neural circuits happens during postnatal life [4]. In particular, in humans during school-age years (4–11 years old), maturation of the brain's connectivity tract is prominent [5]. In the postnatal period, neurogenesis is limited and focused within specific areas, such as the olfactory bulb and the dentate gyrus of the hippocampus [6]. Gliogenesis, on the other hand, is very prominent during the postnatal period. Glial progenitors proliferate in the forebrain subventricular zone (SVZ) and migrate to the cortex, striatum, and hippocampus. Here, they differentiate into oligodendrocytes and astrocytes, forming the myelin sheaths and providing homeostatic support to neurons and enabling the maturation of neuronal circuits [7,8]. Furthermore, synaptic pruning and production of neuronal connections happen prominently during the first years of life. In fact, across the whole brain, the number of synapses is double in childhood than in adulthood, demonstrating that the processes of synaptic remodelling are very active during childhood and adolescence, and these processes involve the generation and, removal of synapses [4]. Despite the current understanding of the overall process, the molecular mechanisms dictating postnatal brain are not well understood.

MicroRNAs are small non-coding RNAs (22–24 nucleotides) that regulate gene expression post-transcriptionally, by inhibiting translation or promoting mRNA degradation through 3′UTR binding of the target mRNA [9]. One single microRNA can regulate the translation of hundreds of protein-coding mRNAs. Although, bioinformatics predictor models show that most of the protein-coding genes are regulated by combinations of microRNAs, rather than one single microRNA, therefore, unravelling microRNA transcriptomes are necessary to understand gene regulation of complex processes [9]. Importantly, microRNAs can be grouped in families, these microRNAs share target genes, and have reductant functions [10].

MicroRNAs are essential for normal development, using genetic techniques it was observed that disruption of microRNA production results in embryonic lethality in zebrafish, worms, flies, and mice, mainly due to neuronal deficits [11–14]. MicroRNAs are widely implicated in the regulation and homeostasis of early brain development, including embryonic neurogenesis and neuronal differentiation. The complexity of brain maturation, with several processes happening over years in humans, suggests that families of microRNAs may be involved in directing brain development and maturation [15,16], however only few studies have analysed the role of microRNAs in specific areas of the brain postnatally, e.g., hypothalamus [17]. Nevertheless, the specific function of a single microRNA may not be critical, as these microRNAs may have very high temporal specificity and cellular expression, and their function may be compensated by other cells in complex multicellular organisms [18]. To fully understand the normal development of complex organs, such the brain, it is important to analyse the expression of microRNAs at several time points.

C-myc is a highly conserved transcription factor belonging to the MYC family of basic helix-loop-helix transcription factors and controls the transcriptional networks that govern cell growth, division, differentiation, and death [19]. MYC regulates extensive transcriptional programs and drives cells to pre-determined cellular states by amplifying cell-specific transcriptional programs [19]. MYC mRNA has been identified in all multiple neural cell types including, neurons, radial glia cells, and oligodendrocytes, suggesting it has a key function in the formation of brain circuits. Transcriptional control by MYC is complex, it can promote and repress transcription depending on the cellular context [19]. In this aspect, MYC protein inhibits the transcription of several microRNA clusters [19], demonstrating that activation of MYC will result in the amplification of signaling pathways. In fact, in human hepatocellular carcinoma cells, MYC represses the expression of the let-7a cluster by binding to the non-canonical E-box [20] and, in the brain, MYC represses the expression of the miR-23 cluster, which may result in a reduction in myelination [21].

In this study, we aim to evaluate how microRNA expression changes over time in the hippocampus and how an early life insult, hypoxia at Postnatal day 7 may result in the misregulation of microRNAs and contribute to the neurological outcomes. To this aim we analyzed the microRNA profile in the hippocampus from Postnatal day 8 (P8) to Postnatal day 42 (P42) in mice, and how microRNA expression changed after perinatal stress using a mouse model of hypoxia at Postnatal day 7 (P7). In control, we observed that microRNAs can be divided into 4 different groups, microRNAs with stable expression during aging, microRNAs with either increased or decreased expression between ages, and microRNAs with a peak in the middle time point (P14) compared to P8 and P42. Additionally, we observed that after hypoxia at Postnatal day 7, a sub-set of microRNAs changed the expression over time, including the let-7a cluster. Here, we speculated that while these happen at specific time points, the overall summation and accumulation may contribute to the pathology after hypoxia in the neonatal period. Finally, bioinformatics tools were validated using chromatin immunoprecipitation (ChiP) assay, and we identified that the transcriptional controller c-myc may regulate the expression of two of these microRNA clusters, let-7a and miR-23.

2. Materials and Methods

2.1. Mouse Model of Neonatal Hypoxia

A total of 70 mice were used in this study. All animal procedures were performed in accordance with the principles of European Communities Council Directive (86/609/EEC, 2010/63/EU), under license (REC#1203b or REC#P140) from the Department of Health and Health Products Regulatory Authority (Ireland) and procedures were approved by the Research Ethics Committee of the Royal College of Surgeons in Ireland and Trinity College Dublin. Neonatal litters of C57BL/6J mice [weight, 4–6 g; age, postnatal day 6.5–7.5 (P7)], were obtained from the Biomedical Research Facility, RCSI or CMU at TCD. Pups were kept with their dams in a barrier-controlled facility on a 12 h light–dark (7 a.m.–7 p.m.) standard cycle with access to food and water ad libitum. All experiments were performed during the light cycle.

Hypoxia in mouse pups was carried out as previously described [22]. Male and female pups from the same litter were randomly placed in a clear hypoxic chamber and exposed to a premixed gas containing 5% O2/95% N2 for 15 min at 34 °C and 80% humidity. Under these conditions, 97% of the pups develop behavioural seizures during and after hypoxia. All animals were observed during the 15 min of hypoxia and 15 min post-hypoxia before returning to the dam. Both sexes were used for consecutive studies.

2.2. RNA Isolation and Open Array

For each time point and experimental condition, mice were perfused with cold PBS and both hippocampi were isolated and pooled together for RNA extraction. Total RNA was extracted using the Trizol method [22–24]. The quality and quantity of RNA were measured using a Nanodrop Spectrophotometer (Thermo Scientific, Waltham, MA, USA) and samples with an absorbance ratio at 260/280 between 1.8–2.2 were considered acceptable.

500 ng of RNA was processed by reverse transcriptase and pre-amplification steps following the manufacturer's protocol (Applied Biosystems, Waltham, MA, USA). The pre-amplification reaction was mixed with TaqMan OpenArray Real-Time PCR Master mix (1:1). The mix was loaded onto the OpenArray custom-designed panel (215 mature highly expressed in brain miRNAs) and ran using a QuantStudio 12 K Flex PCR (Life Technologies, Carlsbad, CA, USA). The number of miRNA identified in each sample showed good consistency: 136, 132, and 146 in control samples (P8, P14, and P42 respectively), and 153, 131, and 147 in hypoxia samples (P8, P14, and P42 respectively). Any microRNA with a Ct value above 28 was considered "not detected" and it must be expressed in 3 out of 4 samples to be considered detected within a group. After this selection criteria, 157 microRNAs were detected from the original 215 miRNA. The mean of the Ct values was (i) in control samples, P8: 20.87, P14: 20.15, and P42: 20.92 and (ii) in hypoxic samples, P8: 20.64, P14: 20.35, and P42: 19.29 (Figure 1A).

The Geometric Mean Normalisation Method (GMN) and $2^{-\Delta\Delta CT}$ was used to determine the relative quantification of each target miRNA expression, using for each microRNA the average of control samples at P8 as a reference to calculate the $\Delta\Delta CT$.

miRNAs with a fold change value of under 0.6 or over 1.5 were deemed significant. Heatmaps were compiled to represent the fold change values for each miRNA sample group over the three time points, allowing miRNAs to be grouped according to their trends in expression over time (Figure 2).

2.3. Ambulation Score

Pups were placed in a flat, enclosed area and recorded for three minutes. Immobile pups were encouraged to move by gently prodding the back of the pup. The ambulation scoring scale was adapted from Feather-Schussler et al. [25]: 0 = no movement, 0.5 = lots of turning, encouragement needed, little or no straight crawling, 1 = crawling with asymmetric limb movement, 1.5 = crawling, mostly asymmetric with sporadically symmetric limb movement, 2 = slow crawling with symmetric limb movement, 2.5 = fast crawling with symmetric limb movement, and 3 = walking with symmetric limb movement. Symmetric limb movement was noted when the pups' hind paws met the front paws when moving in a fluid and continuous motion. Asymmetric limb movement was described as erratic placement of front and hind paws when moving, in addition to a lack of fluidity. Crawling was differentiated from walking based on what portion of the hindlimb was in contact with the ground when ambulating. During crawling, the entire hind paw and heel make contact with the ground. While only the toes come into contact with the ground. The maximum score during the 3 min was given [25].

2.4. RT-PCR of Primary Let-7a Cluster Sequence

250 ng of total RNA was retro transcribed using the High-Capacity cDNA Reverse Transcription kit (Thermo Fisher, Cat n: 4368814) following the manufacturer's protocol. Two microlitres of RT product were amplified using TaqMan Fast Universal Master Mix 2x (Thermo Fisher, Cat n: 4352042) following manufacturer instructions. Pre-designed primers for the primary sequence and the housekeeping gene (actin) were purchased from Thermo Fisher (Let7a primers: Mm03306744pri and Actin: Mm00306184pri). Let7a levels were normalized to actin (housekeeping gene) and data was represented as a relative expression to the control group using the $2^{-\Delta\Delta Ct}$ method [23].

2.5. Identification of c-Myc Binding Sites in the Promoter Regions

The promoter region from the 15 misregulated microRNAs over time in the hypoxia group compared to the control was examined for potential Myc binding sites. Potential C-Myc binding sites were identified by searching for Myc-binding motifs within 5000 bases upstream of the transcriptional start sites of genes of interest, using the UCSC genome browser (version: GRCm39/mm39). Primers were then designed to flank the predicted Myc binding sites.

2.6. Chromatin Immunoprecipitation (ChIP)

ChIP was performed as previously described [23]. Briefly, both hippocampi were extracted 72 h after hypoxia or normoxia conditions. Both hippocampi were pooled and homogenized in 1% formaldehyde (Thermo Fisher Scientific, Rockford, IL, USA) and incubated at room temperature for 10 min. Formaldehyde was quenched with 0.125 M glycine (Sigma-Aldrich Ltd., Wicklow, Ireland). Samples were then incubated in a hypotonic buffer (20 mM Tris-HCl, 10 mM NaCl, 3 mM MgCl2), lysed in a dounce homogenizer and rotated at 4 °C for 20 min. 1% NP-40 (MP) was added to each sample, then samples were centrifuged at 15,000 rpm for 5 min at 4 °C and chromatin was sheared in a Diagenode Pico sonicator. To ensure appropriate shearing profiles, DNA was analyzed on a 2% agarose gel and visualized on a UV box (VWR International Ltd., Dublin, Ireland). Then, samples were incubated with magnetic Dynabeads® (Thermo Fisher Scientific, Rockford, IL, USA) which had been pre-incubated overnight with 5 μg of c-myc antibody (Merck Millipore, Billerica, MA, USA) or IgG control (Cell Signaling, Danvers, MA, USA). Beads and DNA were incubated at 4 °C for 1 h. Magnetic beads and bound material was washed with RIPA buffer to remove loosely bound DNA/proteins. Beads were then separated from complexed DNA using Chelex reagent (Sigma-Aldrich Ltd., Wicklow, Ireland) and heated to 100 °C for 10 min. Samples were then incubated at room temperature for a further 10 min before being centrifuged at 15,000 rpm for 5 min. The supernatant was transferred to a new tube, stored at 4 °C and quantified by qPCR with gene-specific primers. Transcription factor occupancy in control and post-hypoxia were normalized to IgG binding to the DNA and calculated as a percentage of total input.

Primer sequence: miR23b-27 site 1 (F: 5′GCAGTGGGTGCCTGTAAG3′, R: 5′CTGTTT GGCTCCGTTTCG3′); miR23b-27 site 2 (F: 5′GGGACTTGATGAAATGAGC3′, R: 5′TGAAC TTAACTGTGGGTTG3′); Let-7a-1 site 1 (F: 5′TCTCAGGGCCAGAACACTTG3′, R: 5′TGG TGCTAAAAGGCAACCCA3′); Let-7a-1 site 2 (F: 5′TGTCTCATGAACATCTTTTCCTACT3′ R: 5′GCTGGTGTGTGATATCGAGC3′).

2.7. Statistical Analysis

All data are presented as box and whiskers. Two group comparisons were made using an unpaired Student's two-tailed *t*-test (GraphPad Prism). Multiple group comparison was made using One-way Anova, and Bonferroni post hoc test. Significance was accepted a $p < 0.05$.

3. Results

3.1. MicroRNA Expression Profiles in the Postnatal Hippocampus

Postnatal brain development is critical for the formation and maturation of neuronal circuits. Whilst the overall process of postnatal brain maturation is understood, little is known about the molecular mechanisms guiding postnatal brain maturation. Here, we examined the regulation of 215 microRNAs on postnatal days 8, 14, and 42 to evaluate their expression over the first weeks of life. 155 microRNAs were expressed within at least one condition (Figure 1A), and from those 130 microRNAs were detected in all samples (Figure 1B). One microRNA was only detected at P8 (miR-10b-5p (Figure 1B)), 5 microRNAs were only detected at P14 (e.g., miR-191-3p, miR-216a-5p (Figure 1B)), and one microRNA was only detected at P42 (miR-770-5p (Figure 1B)). Furthermore, 5 microRNAs were detected at the earlier time points (P8 and P14, (Figure 1B)) and, 13 microRNAs were detected at the later timepoints, P14 and P42 (Figure 1B). No microRNAs were detected in both P8 and P42 exclusively.

3.2. MicroRNAs Regulation between Ages in the Postnatal Hippocampus in Mice

Then, we analyzed if we could observe trends in the expression of microRNAs between ages (Figure 2). Detected microRNAs were divided into four distinct groups: Group I microRNAs with stable expression across all ages (Figure 2A); Group II, microRNAs with increased expression at P14 and P42 compared to P8 (Figure 2B); Group III, microRNAs with a peak in expression at P14, with higher or lower expression at P14 compared to P8 and P42 (Figure 2C); and Group IV, microRNAs with peak expression at P8, i.e., lower expression in the latest time point compared to P8 (Figure 2D). Among the 155 microRNAs expressed in all samples, 43 microRNAs were in Group I, and had stable microRNA expression between ages (Figure 2A) including miR-142-3p and miR-148a-3p (Figure 2E). 43 microRNAs were included in Group II with a higher expression in the later time points compared with the earlier ones (Figure 2B), for example, miR-150-5p and miR-219a-5p (Figure 2F); 25 microRNAs had a peak at P14 (higher or lower expression than in P8 and P42) (Group III, Figure 2D), including miR-323a-3p and miR-369-3p (Figure 2G). Finally, in Group IV, 27 microRNAs had a decreased expression over time (Figure 2D), such as miR-106b-5p and miR-19a-3p (Figure 2H).

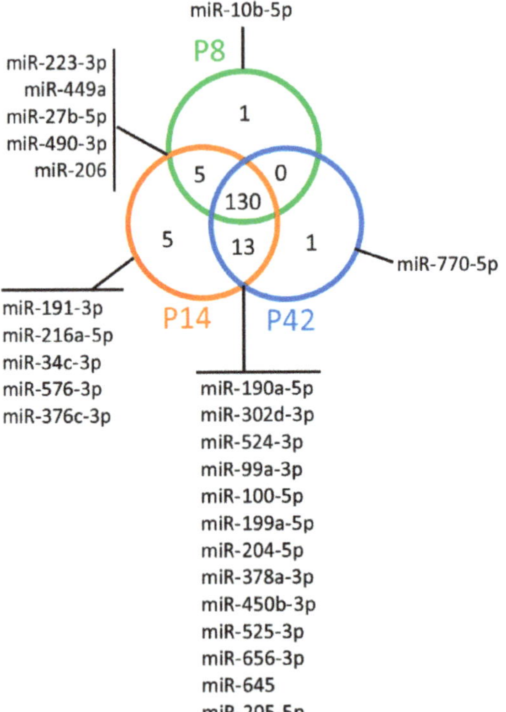

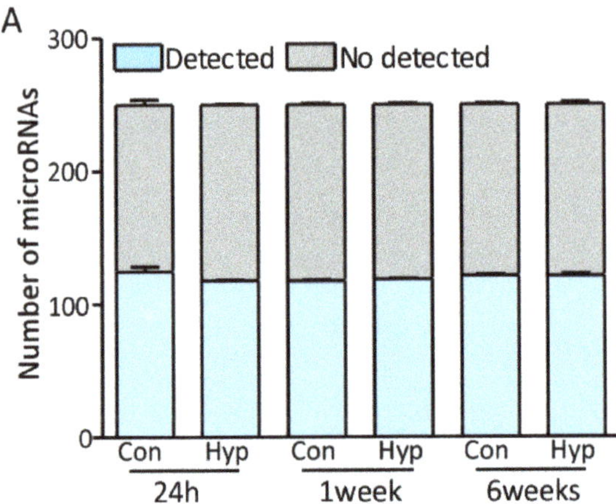

Figure 1. MicroRNA profile in the hippocampus and comparison between ages. (**A**) Graphs show the average number of detected microRNAs (blue) and non-detected (grey), in each experimental group. Con: Control; Hyp: Hypoxia; 24 h: 24 h post-hypoxia (Postnatal day 8, *n* = 4); 1 week: 1 week post-hypoxia (Postnatal day 14, *n* = 4); 6 weeks: 6 weeks post-hypoxia (Postnatal day 42, *n* = 4). (**B**) Ven diagram shows the average of number of detected microRNAs at P8 (green), P14 (orange) and P42 (blue) in the control group.

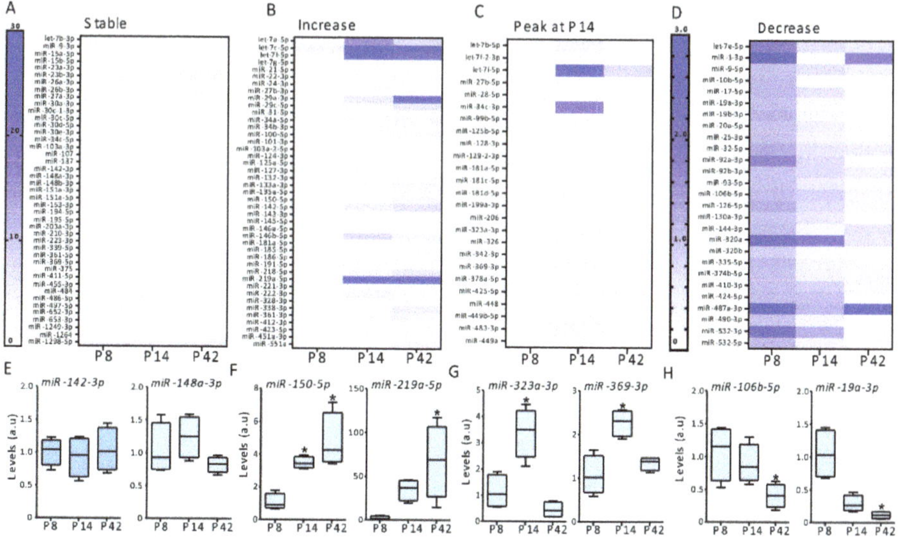

Figure 2. Expression of microRNAs between ages. (**A**) Heat map shows relative expression levels of microRNAs with stable expression at postnatal age 8, 14 and 42 ($n = 4$/group). (**B**) Heat map shows relative expression of microRNAs with increase expression between ages ($n = 4$/group). (**C**) Heat map shows relative expression of microRNAs with a peak in expression at P14 compared to P8 and P42 ($n = 4$/group). (**D**) Heat map shows relative expression of microRNAs with decrease expression between ages ($n = 4$/group). (**E**) Graphs show relative expression of two microRNAs with stable expression between ages miR-142-3p (left) and miR-148a-3p (right) ($n = 4$/group). (**F**) Graphs show the relative expression miR-150-5p (left) and miR-219a-5p (right) which shows an increase expression between ages ($n = 4$/group). (**G**) Graphs show the relative expression of miR-323a-3p (left) and miR-369-3p (right) at P8, P14 and P42 ($n = 4$/group). (**H**) Graphs show the relative expression of two microRNA with decrease expression between ages, miR-106b-5p (left) and miR-19a-3p (right) ($n = 4$/group). One-way ANOVA, * $p < 0.05$.

3.3. Differentially Regulated microRNAs after Hypoxia at P7 in Mice

Subsequently, we evaluated if early life stress may disrupt normal developmental miRNA expression profiles. Previously we have shown that pups under hypoxia develop long-term neurological outcomes, including, anxiety-like behavior, hippocampal dysfunction, and a lower threshold to develop seizures [22]. Corroborating these results, P7 pups were subjected to hypoxia (Figure 3A),ambulation test was performed 72 h post-procedure to evaluate neurodevelopment, pups post hypoxia received lower ambulation scores than the control pups (Figure 3B). This demonstrates that hypoxia induces a delay on neurodevelopment. In a separate set of mice, P7 pups were subjected to hypoxia, and hippocampi were dissected 24 h, 7 days, and 5 weeks post-hypoxia corresponding to P8, P14 and P42, respectively, in the control group (Figure 3A). 32 microRNAs were differentially regulated at any given time point. Not surprisingly, the greatest changes in expression were observed 24 h after hypoxia; 22 microRNAs were up-regulated, and no microRNAs were observed to be down-regulated (Figure 3C). Among these microRNAs, we observed an increase in the levels of miR-146b-3p, an inflammation regulated microRNA (Figure 3D), and let-7a-5p, a developmentally regulated microRNA (Figure 3E). 7 microRNAs were differentially expressed 1-week post-hypoxia (Figure 3F); 3 microRNAs were up-regulated, and 4 microRNAs were down-regulated, including miR-29c-5p and miR-532-5p (Figure 3G,H). Finally, 3 microRNAs were dysregulated 5 weeks post-hypoxia (Figure 3I); 2 microRNAs were up-regulated, let-7f-5p (Figure 3K) and miR-124-3p (Figure 3K), and one microRNA was down-regulated. Further analysis showed that the majority of these microRNAs showed a similar trend in expression over time in the control group, such as miR-532-5p (Figure 3H). However, a subset of microRNAs showed a different trend in expression overtime post-hypoxia compared to the control group, even if they were dysregulated in one single time point.

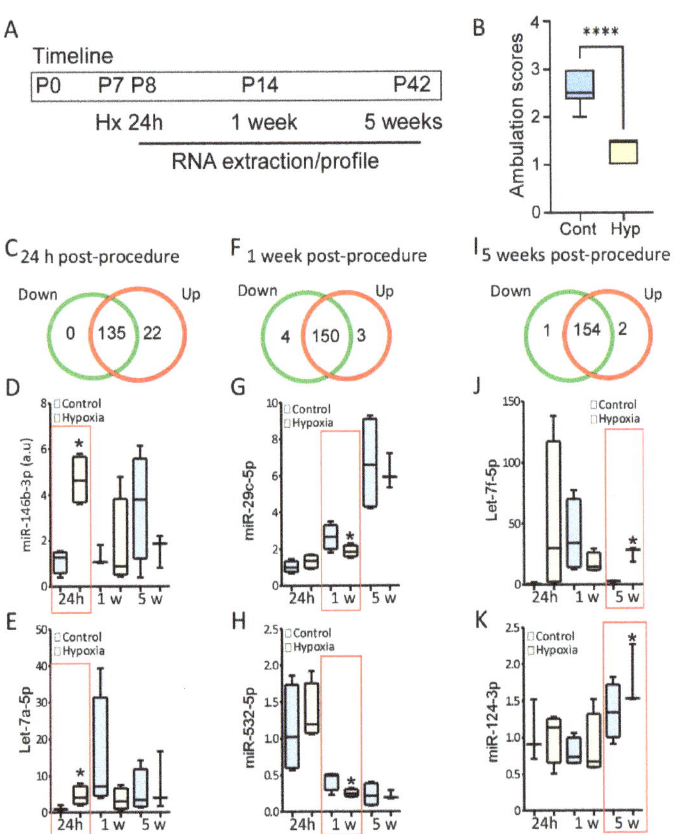

Figure 3. MicroRNA expression post-hypoxia compared to control at different age. (**A**) Schematic showing the timeline of the experimental procedures. (**B**) Pups subjected to hypoxia have a lower ambulation score 72 h post-hypoxia compared to the control pups, demonstrating that hypoxia results in a neurodevelopmental delay. (**C**) Venn-Diagram shows the number of down-regulated (left green), no-regulated (middle, green and red area) and up-regulated (right red) 24 h post-procedure. (**D,E**) Graphs show relative expression of miR-146b-3p (**D**) and let-7a-5p (**E**), two of the microRNAs up-regulated 24 h post procedure ($n = 4$/control group; $n = 4$, 24 h and 1 week post-hypoxia; $n = 3$, 5 weeks post-hypoxia. (**F**) Venn-Diagram shows the number of down-regulated (left green), no-regulated (middle, green and red area) and up-regulated (right red) 1 week post-procedure. (**G,H**) Graphs show relative expression of miR-29c-5p ($n = 3$–4/group) (**G**) and miR-532-5p ($n = 4$/control group; $n = 4$, 24 h and 1 week post-hypoxia; $n = 3$, 5 weeks post-hypoxia) (**H**), two microRNAs with lower expression 1 week post procedure. (**I**) Venn-Diagram shows the number of down-regulated (left green), no-regulated (middle, green and red area) and up-regulated (right red) 6 week post-procedure. (**J,K**) Graphs show relative expression of let-7f-5p ($n = 4$/control group; $n = 4$, 24 h and 1 week post-hypoxia; $n = 3$, 5 weeks post-hypoxia) (**J**) and miR-124-3p ($n = 4$/control group; $n = 4$, 24 h and 1 week post-hypoxia; $n = 3$, 5 weeks post-hypoxia) (**K**), the two microRNAs up-regulated 5 week post-procedure. ($n = 4$/control group; $n = 4$, 24 h and 1 week post-hypoxia; $n = 3$, 5 weeks post-hypoxia). One-way Anova, * $p < 0.05$. **** $p < 0.001$.

Following this, we decided to analyze if microRNA showed an overall different trend in regulation through the ages post-hypoxia compared to the control group (Figure 4A). We observed that 15 microRNAs, including let-7a (Figure 3E), let-7f (Figure 3J), let-7g, and miR-222 have increased levels over time or showed a peak at P14 in the control group; however, after hypoxia, these microRNAs had a stable expression or decreased expression over time (Figure 4B,C).

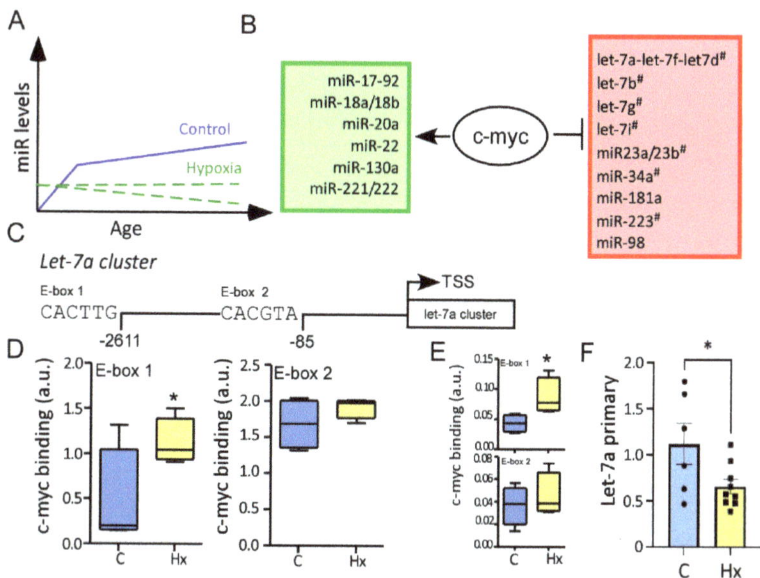

Figure 4. C-myc repress the expression of let-7a and miR-23a cluster. (**A**) Schematic diagram representing the main response of microRNA regulation post-hypoxia, even if the microRNA is up-regulated at one single time point, e.g., 24 h, overall the microRNA expression is stable or down regulated, in contrast these microRNAs have an increase expression between ages in the control group. (**B**) List of published microRNAs regulated by c-myc, either enhance (green) or repress (red) their expression. (**C**) Schematic showing c-myc E-Box consensus sites in the Let7a cluster promoter region in the mouse genome. (**D**) Increased c-myc occupancy of c-myc consensus sites in the let-7a promoter in the hippocampus from control and hypoxia mice 72 h post-procedure ($n = 4$/group). (**E**) Increased c-myc occupancy of c-myc consensus sites in the miR-23a promoter in the hippocampus from control and hypoxia mice 72 h post-procedure ($n = 4$/group). Note: In both promoter c-myc has a higher occupancy on the E-box 1 region and not in the E-box 2 region. (**F**). Relative expression of the primary sequence of Let-7a in hippocampus in control and hypoxia group 72 h post procedure $n = 6$, control $n = 9$, hypoxia. Un-paired *t*-test, * $p < 0.05$.

3.4. MYC Has a High Affinity for Let-7a Promoter in the Hippocampus after Hypoxia

To evaluate if these 15 microRNAs have a common regulator, bioinformatics evaluated common transcriptional sequences at the promoter sites. Bioinformatics analysis showed that these 15 microRNAs had canonical and non-canonical binding sites for c-myc within the promoter areas (Figure 4B). ChIP assays against c-myc were performed in hippocampus 72 h post-procedure (hypoxia or control) and the affinity to let-7a and miR-23 was analyzed. As previously seen in other cell types, c-myc binding was increased post-hypoxia on the first non-canonical Ebox within the Let7a cluster (Figure 4C), and similar results were found when the miR-23a cluster was evaluated (Figure 4D). To further demonstrate that c-myc acts as a transcriptional repressor, we evaluated the primary sequence of let7a, corroborating c-myc affinity results, let7a primary sequence was downregulated in the hypoxia group compared to the control group 72 h post-hypoxia (Figure 4E).

4. Discussion

In the current study, we created a neurodevelopmental microRNA profile of 215 brain-expressed microRNAs in hippocampus at 3 different ages, postnatal days 8, 14 and 42. We showed that microRNAs cluster according to expression over time into 4 different groups, stable, up-regulated, down-regulated and showing a peak at the middle time point. Further, when pups were subjected to hypoxia at Postnatal day 7, the regulation of microRNAs was altered, with 15 of these microRNAs regulated by the transcription factors c-myc.

Corroborating these results, previous studies analyzed the role of microRNAs on the development of the layers of the medial entorhinal cortex postnatally, and age-specific microRNA regulation was identified in rats [26]. MiR-20a, miR-29c, miR-132 and miR-219-5p in particular, were found to be up-regulated in the medial entorhinal cortex, suggesting that these microRNAs may be crucial for neuronal and glial maturation postnatally, including neuronal maturation, dendritic spines morphology and inhibit apoptosis in neurons [27,28], interestengly miR-219 has a key role on myelin formation by regulating oligodendrocytes maturation [29]. This suggests that these microRNAs may be key to the maturation and establishment of neuronal circuits independent of the brain region and species.

Whilst, neurogenesis is more prominent during prenatal development, glia-genesis mainly occurs during postnatal development. In our study and previously, miR-20a was observed to be downregulated over time [26]. MiR-20a is critical for brain maturation during embryogenesis by regulating axonal growth and dendritogenesis, and its expression is time-dependent [30]. Current results support the idea that miR-20a is controlled during early brain development and its expression is downregulated postnatally [30]. Let-7a and miR-125, are microRNAs involved in astrocyte and microglia homeostasis and were observed to be up-regulated between ages in our study, strengthening the hypothesis that glia maturation is more prominent postnatally.

Postnatal brain development is a time-dependent process, genes are expressed in very specific patterns, and the accumulation of malfunctioning genes may cause neurological conditions. After neonatal hypoxia, we observed up-regulation or downregulation of microRNAs at one specific time, but more importantly, we determined patterns of expression with microRNAs being up- or down-regulated over time. We suggest that these cumulative effects may be more detrimental over time than at a single time point. Further, c-myc is a master regulator of gene expression, and through c-myc, the detrimental signal may be amplified and alter several pathways in parallel, thus reducing the possibility of compensatory signals to maintain homeostasis. Finally, c-myc is a transcription factor that previously has been involved in hypoxic responses [31]. Whilst this work has been mainly done in cancer, we speculate that will also be implicated in the brain.

The current study presents some limitations, first, we analysed a small number of microRNAs, selected by their expression on the postanal brain, we cannot discard that important microRNAs involved mainly in the maintenance of brain function, may have a higher expression at P42, and were not evaluated in the current study. Future studies should evaluate microRNA expression in elder mice to consider that microRNA expression may also be regulated from adulthood to the elderly. In the current study we analyzed male and female mice together, we observed no differences in expression between mice, however we cannot discard sex differences. Further studies should evaluate sex differences during postnatal brain development to evaluate the effect of sex in microRNA expression. Finally, we focus on the hippocampus, a vulnerable region to hypoxia, similar changes in specific microRNAs have been seen in other brain regions, e.g., entorhinal cortex [26], we can speculate that common mechanisms are activated independent of the region, however, it will be necessary to evaluate their regulation in parallel to determine if the time scale of brain maturation is similar.

5. Conclusions

Postnatal brain development is a sensitive time regulated process, and while acute dysregulation of a single microRNA may be compensated by other pathways, the accumulation effects of increasing dysregulation between ages could have deleterious consequences, particularly, when microRNAs can regulate thousands of pathways and may amplify the detrimental signals. In this study, we showed how microRNAs are expressed across different ages postnatally in the hippocampus, which may be critical for understanding the underlying mechanisms behind the correct brain maturation and neuronal circuits. We also observed that 15 microRNAs have an altered expression over time following hypoxia possibly due to the activity of c-myc, and these changes may result in long-lasting changes during brain maturation.

Author Contributions: Data collection and analysis of microRNAs, A.L. Chromatin Immunoprecipitation and data analysis, G.P.B. Conceptualization and design, E.M.J.-M. All authors have read and agreed to the published version of the manuscript.

Funding: This work was supported by Provost Predoctoral Fellow to A.L., by 13/SIRG/2114 (T), Wellcome Trust-Ph2IISF Award to E.M.J.-M. and by 18/SIRG/5646 to G.P.B.

Institutional Review Board Statement: The animal study protocol was approved by the Institutional Review Board (or Ethics Committee) of Trinity College Dublin (REC#P140).

Informed Consent Statement: Not applicable.

Data Availability Statement: The data presented in this study are available on request from the corresponding author.

Conflicts of Interest: The authors declare no conflict of interest.

References

1. Stiles, J. Brain development and the nature versus nurture debate. *Prog. Brain. Res.* **2011**, *189*, 3–22. [PubMed]
2. Paus, T.; Keshavan, M.; Giedd, J.N. Why do many psychiatric disorders emerge during adolescence? *Nat. Rev. Neurosci.* **2008**, *9*, 947–957. [CrossRef] [PubMed]
3. Toga, A.W.; Thompson, P.M.; Sowell, E.R. Mapping brain maturation. *Trends Neurosci.* **2006**, *29*, 148–159. [CrossRef] [PubMed]
4. Stiles, J.; Jernigan, T.L. The basics of brain development. *Neuropsychol. Rev.* **2010**, *20*, 327–348. [CrossRef]
5. Lebel, C.; Walker, L.; Leemans, A.; Phillips, L.; Beaulieu, C. Microstructural maturation of the human brain from childhood to adulthood. *Neuroimage* **2008**, *40*, 1044–1055. [CrossRef] [PubMed]
6. Jurkowski, M.P.; Bettio, L.; KWoo, E.; Patten, A.; Yau, S.Y.; Gil-Mohapel, J. Beyond the Hippocampus and the SVZ: Adult Neurogenesis Throughout the Brain. *Front. Cell Neurosci.* **2020**, *14*, 576444. [CrossRef]
7. Cayre, M.; Canoll, P.; Goldman, J.E. Cell migration in the normal and pathological postnatal mammalian brain. *Prog. Neurobiol.* **2009**, *88*, 41–63. [CrossRef]
8. McTigue, D.M.; Tripathi, R.B. The life, death, and replacement of oligodendrocytes in the adult CNS. *J. Neurochem.* **2008**, *107*, 1–19. [CrossRef]
9. Dexheimer, P.J.; Cochella, L. MicroRNAs: From Mechanism to Organism. *Front. Cell Dev. Biol.* **2020**, *8*, 409. [CrossRef]
10. Alvarez-Saavedra, E.; Horvitz, H.R. Many families of C. *elegans* microRNAs are not essential for development or viability. *Curr. Biol.* **2010**, *20*, 367–373. [CrossRef]
11. Alisch, R.S.; Jin, P.; Epstein, M.; Caspary, T.; Warren, S.T. Argonaute2 is essential for mammalian gastrulation and proper mesoderm formation. *PLoS Genet.* **2007**, *3*, e227. [CrossRef] [PubMed]
12. Bernstein, E.; Kim, S.Y.; Carmell, M.A.; Murchison, E.P.; Alcorn, H.; Li, M.Z.; Mills, A.A.; Elledge, S.J.; Anderson, K.V.; Hannon, G.J. Dicer is essential for mouse development. *Nat. Genet.* **2003**, *35*, 215–217. [CrossRef] [PubMed]
13. Kataoka, Y.; Takeichi, M.; Uemura, T. Developmental roles and molecular characterization of a Drosophila homologue of Arabidopsis Argonaute1, the founder of a novel gene superfamily. *Genes Cells* **2001**, *6*, 313–325. [CrossRef] [PubMed]
14. Vasquez-Rifo, A.; Jannot, G.; Armisen, J.; Labouesse, M.; Bukhari, S.I.; Rondeau, E.L.; Miska, E.A.; Simard, M.J. Developmental characterization of the microRNA-specific C. *elegans* Argonautes alg-1 and alg-2. *PLoS ONE* **2012**, *7*, e33750. [CrossRef]
15. Hornstein, E.; Shomron, N. Canalization of development by microRNAs. *Nat. Genet.* **2006**, *38*, S20–S24. [CrossRef]
16. Ilbay, O.; Ambros, V. Pheromones and Nutritional Signals Regulate the Developmental Reliance on let-7 Family MicroRNAs in C. *elegans. Curr. Biol.* **2019**, *29*, 1735–1745.e4. [CrossRef]

27. Doubi-Kadmiri, S.; Benoit, C.; Benigni, X.; Beaumont, G.; Vacher, C.M.; Taouis, M.; Baroin-Tourancheau, A.; Amar, L. Substantial and robust changes in microRNA transcriptome support postnatal development of the hypothalamus in rat. *Sci. Rep.* **2016**, *6*, 24896. [CrossRef]
28. Alberti, C.; Cochella, L. A framework for understanding the roles of miRNAs in animal development. *Development* **2017**, *144*, 2548–2559. [CrossRef]
29. Zaytseva, O.; Kim, N.H.; Quinn, L.M. MYC in Brain Development and Cancer. *Int. J. Mol. Sci.* **2020**, *21*, 7742. [CrossRef]
30. Wang, Z.; Lin, S.; Li, J.J.; Xu, Z.; Yao, H.; Zhu, X.; Xie, D.; Shen, Z.; Sze, J.; Li, K.; et al. MYC protein inhibits transcription of the microRNA cluster MC-let-7a-1~let-7d via noncanonical E-box. *J. Biol. Chem.* **2011**, *286*, 39703–39714. [CrossRef]
31. Lin, S.T.; Huang, Y.; Zhang, L.; Heng, M.Y.; Ptacek, L.J.; Fu, Y.H. MicroRNA-23a promotes myelination in the central nervous system. *Proc. Natl. Acad. Sci. USA* **2013**, *110*, 17468–17473. [CrossRef] [PubMed]
32. Quinlan, S.; Merino-Serrais, P.; Di Grande, A.; Dussmann, H.; Prehn, J.H.M.; Ni Chonghaile, T.; Henshall, D.C.; Jimenez-Mateos, E.M. The Anti-inflammatory Compound Candesartan Cilexetil Improves Neurological Outcomes in a Mouse Model of Neonatal Hypoxia. *Front. Immunol.* **2019**, *10*, 1752. [CrossRef] [PubMed]
33. Engel, T.; Brennan, G.P.; Sanz-Rodriguez, A.; Alves, M.; Beamer, E.; Watters, O.; Henshall, D.C.; Jimenez-Mateos, E.M. A calcium-sensitive feed-forward loop regulating the expression of the ATP-gated purinergic P2X7 receptor via specificity protein 1 and microRNA-22. *Biochim. Biophys. Acta Mol. Cell Res.* **2017**, *1864*, 255–266. [CrossRef]
34. Quinlan, S.M.M.; Rodriguez-Alvarez, N.; Molloy, E.J.; Madden, S.F.; Boylan, G.B.; Henshall, D.C.; Jimenez-Mateos, E.M. Complex spectrum of phenobarbital effects in a mouse model of neonatal hypoxia-induced seizures. *Sci. Rep.* **2018**, *8*, 9986. [CrossRef] [PubMed]
35. Feather-Schussler, D.N.; Ferguson, T.S. A Battery of Motor Tests in a Neonatal Mouse Model of Cerebral Palsy. *J. Vis. Exp.* **2016**, e53569. [CrossRef]
36. Olsen, L.C.; O'Reilly, K.C.; Liabakk, N.B.; Witter, M.P.; Saetrom, P. MicroRNAs contribute to postnatal development of laminar differences and neuronal subtypes in the rat medial entorhinal cortex. *Brain. Struct. Funct.* **2017**, *222*, 3107–3126. [CrossRef]
37. Kole, A.J.; Swahari, V.; Hammond, S.M.; Deshmukh, M. miR-29b is activated during neuronal maturation and targets BH3-only genes to restrict apoptosis. *Genes Dev.* **2011**, *25*, 125–130. [CrossRef]
38. Lippi, G.; Steinert, J.R.; Marczylo, E.L.; D'Oro, S.; Fiore, R.; Forsythe, I.; Schratt, G.; Zoli, M.; Nicotera, P.; Young, K.W. Targeting of the Arpc3 actin nucleation factor by miR-29a/b regulates dendritic spine morphology. *J. Cell Biol.* **2011**, *194*, 889–904. [CrossRef]
39. Bruinsma, I.B.; Van Dijk, M.; Bridel, C.; Van De Lisdonk, T.; Haverkort, S.Q.; Runia, T.F.; Steinman, L.; Hintzen, R.Q.; Killestein, J.; Verbeek, M.M.; et al. Regulator of oligodendrocyte maturation, miR-219, a potential biomarker for MS. *J. Neuroinflamm.* **2017**, *14*, 235. [CrossRef]
40. Arzhanov, I.; Sintakova, K.; Romanyuk, N. The Role of miR-20 in Health and Disease of the Central Nervous System. *Cells* **2022**, *11*, 1525. [CrossRef]
41. Kim, J.W.; Gao, P.; Liu, Y.C.; Semenza, G.L.; Dang, C.V. Hypoxia-inducible factor 1 and dysregulated c-Myc cooperatively induce vascular endothelial growth factor and metabolic switches hexokinase 2 and pyruvate dehydrogenase kinase 1. *Mol. Cell. Biol.* **2007**, *27*, 7381–7393. [CrossRef] [PubMed]

Review

Non Coding RNAs as Regulators of Wnt/β-Catenin and Hippo Pathways in Arrhythmogenic Cardiomyopathy

Marina Piquer-Gil [1], Sofía Domenech-Dauder [1], Marta Sepúlveda-Gómez [1], Carla Machí-Camacho [1], Aitana Braza-Boïls [1,2,*] and Esther Zorio [1,2,3,*]

1   Unit of Inherited Cardiomyopathies and Sudden Death (CaFaMuSMe), Health Research Institute La Fe, 46026 Valencia, Spain
2   Center for Biomedical Network Research on Cardiovascular Diseases (CIBERCV), 28015 Madrid, Spain
3   Cardiology Department, Hospital Universitario y Politécnico La Fe, 46026 Valencia, Spain
*   Correspondence: aitana_braza@iislafe.es (A.B.-B.); zorio_est@gva.es (E.Z.);
    Tel.: +34-9612-46641 (A.B.-B.); +34-9612-44790 (E.Z.)

Abstract: Arrhythmogenic cardiomyopathy (ACM) is an inherited cardiomyopathy histologically characterized by the replacement of myocardium by fibrofatty infiltration, cardiomyocyte loss, and inflammation. ACM has been defined as a desmosomal disease because most of the mutations causing the disease are located in genes encoding desmosomal proteins. Interestingly, the instable structure of these intercellular junctions in this disease are closely related to a perturbed Wnt/β-catenin pathway. Imbalance in the Wnt/β-catenin signaling and also in the crosslinked Hippo pathway leads to the transcription of proadipogenic and profibrotic genes. Aiming to shed light on the mechanisms by which Wnt/β-catenin and Hippo pathways modulate the progression of the pathological ACM phenotype, the study of non-coding RNAs (ncRNAs) has emerged as a potential source of actionable targets. ncRNAs comprise a wide range of RNA species (short, large, linear, circular) which are able to finely tune gene expression and determine the final phenotype. Some share recognition sites, thus referred to as competing endogenous RNAs (ceRNAs), and ensure a coordinating action. Recent cancer research studies regarding the key role of ceRNAs in Wnt/β-catenin and Hippo pathways modulation pave the way to better understanding the molecular mechanisms underlying ACM.

Keywords: arrhythmogenic cardiomyopathy; ncRNA; lncRNA; miRNA; circRNA; pseudogene; ceRNA; Wnt/β-catenin pathway; Hippo pathway

Citation: Piquer-Gil, M.; Domenech-Dauder, S.; Sepúlveda-Gómez, M.; Machí-Camacho, C.; Braza-Boïls, A.; Zorio, E. Non Coding RNAs as Regulators of Wnt/β-Catenin and Hippo Pathways in Arrhythmogenic Cardiomyopathy. *Biomedicines* **2022**, *10*, 2619. https://doi.org/10.3390/biomedicines10102619

Academic Editors: Milena Rizzo and Elena Levantini

Received: 31 August 2022
Accepted: 14 October 2022
Published: 18 October 2022

Publisher's Note: MDPI stays neutral with regard to jurisdictional claims in published maps and institutional affiliations.

1. Arrhythmogenic Cardiomyopathy

Arrhythmogenic cardiomyopathy (ACM) is a rare disease with an estimated prevalence between 1:1000 and 1:5000 (OMIM #107970; ORPHA247) characterized by fibrofatty replacement of the ventricular myocardium and a high ventricular arrhythmia burden with increased risk of sudden cardiac death [1–6], sometimes as the first manifestation of the disease. ACM has been defined as a desmosomal disease caused by mutations in any of the five genes encoding desmosomal proteins: desmocollin-2 (*DSC2*), desmoglein-2 (*DSG2*), desmoplakin (*DSP*), plakoglobin (*JUP*), and plakophilin-2 (*PKP2*). Mutations in these genes are responsible of more than 60% of genotyped ACM cases. However, mutations in 14 non-desmosomal genes have also been identified in 5% of patients, namely in α-actinin-2 (*ACTN2*), cadherin-2 (*CDH2*), αT-catenin (*CTNNA3*), desmin (*DES*), filamin C (*FLNC*), LIM domain binding protein 20 (*LDB3*), lamin A/C (*LMNA*), transmembrane protein 43 (*TMEM43*), transforming growth factor beta 3 (*TGFB3*), tight junction protein 1(*TJP1*), titin (*TTN*), RNA-binding motif protein 20 (*RBM20*), sodium voltage-gated channel, alpha subunit 5 (*SCN5A*), and phospholamban (*PLN*) (Table 1) [1,3–7]. Historically ACM caused by mutations in desmosomal genes was associated with isolated or predominantly right ventricular disease (RV-ACM). However, recent studies have highlighted a

high prevalence of left ventricular involvement along with isolated and left-dominant ACM forms (LV-ACM) in patients carrying *FLNC*, *DSP*, or *DSG2* mutations, for instance [3,8–10].

Table 1. List of genes associated with arrhythmogenic cardiomyopathy.

Cell Structure	Gene	Protein	Chromosomal Location	Anatomic Affection	Related to Wnt/β-Catenin
Desmosome	*DSC2*	Desmocollin-2	18q12.1	RV/BiV	
	DSG2	Desmoglein-2	18q12.1	RV/LV/BiV	
	DSP	Desmoplakin	6p24.3	LV/BiV	[11,12]
	JUP	Plakoglobin	17q21.2	RV/BiV	[13]
	PKP2	Plakophilin-2	12p11.21	RV/BiV	[14–16]
Cytoskeleton	*ACTN2*	α-actinin-2	1q43	LV	[17]
	CDH2	Cadherin-2	18q12.1	RV/BiV	
	CTNNA3	Alpha-3 catenin	10q21.3	RV/BiV	
	DES	Desmin	2q35	LV/BiV	
	FLNC	Filamin C	7q32.1	LV	
	LDB3	LIM domain binding 3	10q23.2	RV	
	LMNA	Lamin A/C	1q22	LV/BiV	
	TEMEM43	Transmembrane protein 43	3p25.1	RV/BiV	
	TGFB3	Transforming growth factor beta-3	14q24.3	RV	
	TJP1	Tight junction protein 1	15q13.1	RV/BiV	
	TTN	Titin	2q31.2	RV/LV/BiV	
	RBM20	RNA-binding motif protein 20	10q25.2	LV	
Sodium transport	*SCN5A*	Sodium voltage-gated channel alpha subunit 5	3p22.2	RV/LV/BiV	
Calcium homeostasis	*PLN*	Phospholamban	6q22.31	LV/BiV	

RV: right ventricular involvement; LV: left ventricular involvement; BiV: biventricular involvement.

ACM is characterized by the fibrofatty replacement of the myocardium, typically progressing from the epicardium to the endocardium, and transmural fibrofatty infiltration can also be observed in advanced stages of the disease [1,4–6,10,18].

The electrical heterogeneity created by these pathological changes generates circuits for re-entrant ventricular tachyarrhythmias. Thus, the hallmarks of ACM pathogenesis include cardiomyocyte loss, fibrosis, adipogenesis, inflammation, and arrhythmogenesis [4]. All these cellular mechanisms are interrelated and are orchestrated by the same signaling pathways, which in turn are also connected.

Desmosomes are crucial for myocardium integrity and function, because they ensure cell-to-cell communication as well as intracellular signaling through the Wnt/β-catenin and Hippo pathways [18].

2. Dysregulated Signaling Pathways in ACM

The inhibition of the canonical Wnt/β-catenin pathway and the activation of the Hippo and TFGβ pathways appear to play key roles in ACM physiopathology by regulating the adipogenic and fibrotic cascades [4,18–22].

2.1. Wnt/β-Catenin Pathway in ACM

The Wnt/β-catenin pathway orchestrates heart development and regulates cardiac tissue homeostasis in adults.

Briefly, the activation of the membrane Frizzled receptor (Fz) by WNT ligands triggers the recruitment of low-density lipoprotein receptor-related proteins 5 and 6 (LRP5/6) and Dishevelled (DVL). This macromolecular complex also assembles cytoplasmic proteins, such as Axin and glycogen synthase kinase 3β (GSK3β), and releases free β-catenin in the cytoplasm. In its unphosphorylated form, free β-catenin escapes destruction and translocates into the nucleus. Once there, it binds the transcription factors TCF/LEF to activate the transcription of genes responsible for adipogenesis, fibrosis, cell cycle and cell differentiation (Figure 1a). In contrast, in the absence of WNT ligands, Fz and LRP5/6 receptors remain uncoupled at the membrane, whereas Axin and GSK3β build the degradation complex in the cytoplasm, together with adenomatous polyposis coli (APC) and casein kinase 1 (CK1). This protein complex allows GSK3β to phosphorylate β-catenin. Phosphorylation marks on β-catenin are detected by the ubiquitin ligase beta-transducin repeat containing E3 (β-TrCP), and trigger its ubiquitination and subsequent degradation (Figure 1b), so that no profibrotic or proadipogenic genes are expressed [4,18,19,21,23].

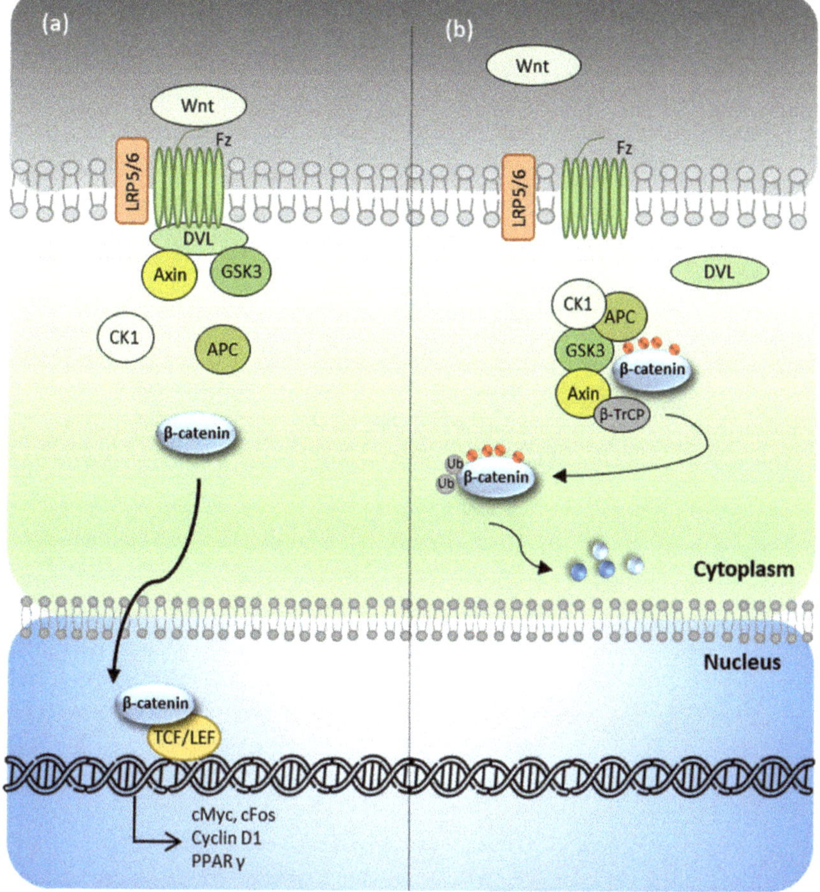

Figure 1. Wnt/β-catenin pathway. (**a**) Wnt/β-catenin pathway in the presence of Wnt ligand. (**b**) Wnt/β-catenin pathway in the absence of Wnt ligand. APC: adenomatous polyposis coli; β-TrCP: beta-transducin repeat containing E3 ubiquitin protein ligase; CK1: casein kinase 1; DVL: Dishevelled; Fz: Frizzled receptor; GSK3β: glycogen synthase kinase 3β; LRP5/6: low-density lipoprotein receptor-related proteins 5 and 6; PPARγ: peroxisome proliferator-activated receptor gamma; TCF/LEF: T cell factor/lymphoid enhancer factor family.

The canonical Wnt signaling pathway has been fully accepted as partially responsible for ACM pathogenesis [4,18,19,21,23–25] according to the "dysregulated signalling hypothesis" postulated by Prof Marian et al. [26], and further supported by the structural resemblance between β-catenin and plakoglobin (also known as γ-catenin). They demonstrated that the desmosomal plakoglobin translocated into the nucleus in an in vitro ACM model, finally leading a shift in the expression of Wnt-dependent genes and an increase in intracellular fat droplets [19]. The altered integrity of the desmosomes in ACM cases may increase the pool of cytoplasmatic plakoglobin, which favors its nuclear translocation and slow β-catenin degradation at the proteosome. Nuclear plakoglobin interacts with transcription factors, with the final effect of an unbalanced Wnt/β-catenin pathway yielding an increased transcription of proadipogenic and profibrotic genes [27]. In recent years, direct and indirect evidence for suppressed Wnt signaling has been gathered from several cellular and/or animal models of ACM-causing mutations in *JUP*, *PKP2*, *DSG*, and *DSP* genes (Table 2) [13,16,19,24,25,28,29].

Table 2. Main findings supporting the role of the Wnt/β-catenin pathway in different genetic scenarios.

Gene	Model Used	Main Finding	Wnt Alteration
PKP2	hiPSC-CM [14] hiPSCs-CM from ACM patients with *PKP2* mutations [16] *PKP2* knock-down HL-1 [15]	*PKP2* mutated is associated with increased lipogenesis PG translocates into the nucleus and competes with β-catenin Activation of the Hippo pathway	↓ β-catenin in nucleus ↓ β-catenin activity ↓ Transcription of canonical Wnt target genes
DSP	*DSP*$^\pm$ mice [11] FAPs from *DSP*$^\pm$ mice [13]	Adipocytes in ACM originate from second heart field cardiac progenitors FAPs express desmosome proteins and differentiate into adipocytes through Wnt pathway-dependent mechanisms.	Suppressed expression of c-Myc ↓ expression of Wnt target genes
DSP	dsp knock-down *Danio rerio* (zebrafish) [24]	Wnt, TGFβ, and Hippo pathways were deregulated	Wnt pathway was the most affected pathway
JUP	Mouse model overexpressing truncated *JUP* [13]	Nuclear PG is essential for differentiation into adipocytes	↓ Wnt target genes
DSG	*DSG* mut mice, *Danio rerio* (zebrafish) and ACM patients [25]	Reduced expression of *Lgals3/GAL* genes	*Lgals3/GAL* regulate Wnt, Hippo, and TGFβ pathways

CM: cardiomyocyte; FAP: fibroadipose progenitor; hiPSC: human induced pluripotent stem cell; PG: plakoglobin; PKP2: plakophilin-2; DSP: desmoplakin; GAL: galectin-3; ACM: arrhythmogenic cardiomyopathy. Arrows denote trend.

2.2. Hippo Pathway in ACM

Closely linked to the Wnt/β-catenin pathway, the Hippo pathway has also been regarded [24] as an important regulator of adipogenesis, cell proliferation and differentiation. Yes-associated protein (YAP) and transcriptional coactivator with PDZ-binding motif (TAZ) represent the core of this pathway, which is controlled by two Ser/Thr kinases, namely Ste20-like kinase1/2 (MST1/2) and large tumor suppressor kinase 1/2 (LATS1/2), and their cofactors Salvador1 (SAV1) and monopolar spindle-one-binder protein (MOB1) [30,31]. MST1/2 and its cofactor SAV1 phosphorylate LATS1/2 and MOB1, which in turn phosphorylate YAP and TAZ ensuring their degradation in the cytoplasm. Moreover, the activity of MST1/2 and LATS1/2 is regulated by neurofibromin 2 (NF2) which, in turn, could be inactivated by the phosphorylated form of PKCα [32]. Briefly, upon YAP Ser127 and Ser381 phosphorylation, the ubiquitination and degradation of YAP is promoted through its interaction with CK1 and the ubiquitin ligase β-TrCP [30,31] (Figure 2a). In contrast, if the Hippo pathway is not activated, YAP/TAZ complex accumulates in the cytoplasm instead of being degraded, allowing YAP/TAZ to translocate into the nucleus where it binds to the transcription factors TEAD/TEF or SMAD to promote the expression of proliferation and differentiation genes (Figure 2b).

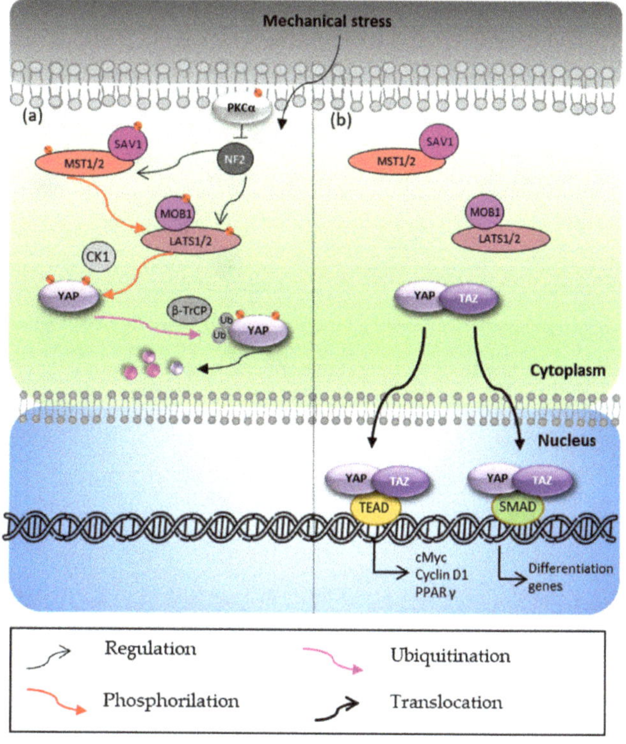

Figure 2. Hippo pathway. (**a**) Hippo pathway in the presence of stimulus. (**b**) Hippo pathway in the absence of stimulus. CK1: casein kinase; β-TrCP: beta-transducin repeat containing E3 ubiquitin protein ligase; SAV1: Salvador1; MST1/2: Ste20-like kinase 1/2; MOB1: monopolar spindle-one binder protein 1; LATS1/2: large tumor suppressor kinase 1/2; SMAD: suppressor of mothers against decapentaplegic; TAZ: transcriptional coactivator with PDZ-binding motif; TEAD: transcriptional enhanced associate domain Ub: ubiquitination marks; YAP: Yes-associated protein. Red points indicate phosphorylation sites.

In ACM pathogenesis, the abnormal cell–cell interaction unbalances the Hippo pathway, increasing the activity of the previously mentioned kinases [33]. Cell remodelling and mechanotransduction signals further promote the phosphorylation of YAP/TAZ by the ACM-related reduction PKCα in the membrane, and the subsequent higher level of NF2 activity [15]. All these steps result in phosphorylated YAP/TAZ cytoplasmatic sequestration and inhibition of the expression of the Hippo target genes. Several authors have studied the crosstalk between the Wnt/β-catenin and Hippo pathways. Although not all the molecular mechanisms are fully understood, the recognized effect in ACM can be summarized as reduced Wnt/β-catenin and Hippo/YAP signaling with a subsequent increase in proadipogenic gene expression [4,15,18,21,34]. Firstly, as previously mentioned, cytoplasmic PG competes with β-catenin to bind TCF/LEF, although in a weaker manner, with a net effect of suppressing of the Wnt/β-catenin pathway and inducing a shift from a myocardiogenic to an adipogenic phenotype, by reducing the expression of cyclin-D1 and c-Myc and enhancing the transcription of proadipogenic and apoptotic genes [13,19,29] (Figure 3a) Secondly, experiments with coimmunoprecipitation demonstrated that phospho-YAP binds phospho-β-catenin (Figure 3b) or PG (Figure 3c) in ACM, although the functional significance of these interactions remains unclear [15]. Finally, both pathways can also be interconnected by TAZ activity. Varelas et al. demonstrated that phospho-TAZ binds and blocks Dvl, negatively regulating the Wnt/β-catenin pathway [35] (Figure 3d).

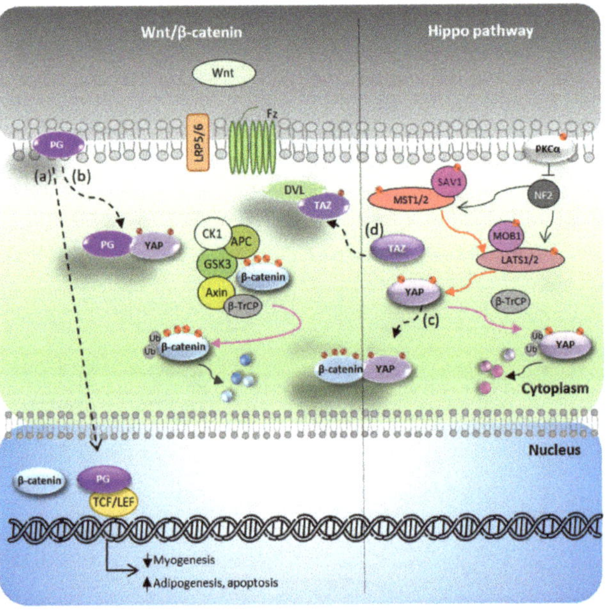

Figure 3. Interconnection between Wnt/β-catenin and Hippo pathways in the pathologic ACM scenario. (**a**) Free plakoglobin translocates into the nucleus and competes with β-catenin to bind the transcription factors TFC/LEF, although in a weaker manner, thus producing a shift from myogenic to adipogenic cell phenotype; (**b**) Free plakoglobin binds and blocks phospho-YAP, impairing its degradation; (**c**) phospho-YAP binds phospho-β-catenin, avoiding its ubiquitination; (**d**) phospho-TAZ binds Dvl, blocks β-catenin translocation to the nucleus, and impairs gene transcription. APC: adenomatous polyposis coli; β-TrCP: beta-transducin repeat containing E3 ubiquitin protein ligase; CK1: casein kinase 1; GSK3β: glycogen synthase kinase-3 beta; SAV1: Salvador1; MST1/2: Ste20-like kinase 1/2; MOB1: monopolar spindle-one-binder protein 1; PG: Plakoglobin; LATS1/2: large tumor suppressor kinase 1/2; NF2: neurofibromatosis type 2; TAZ: transcriptional coactivator with PDZ-binding motif; Ub: ubiquitination marks; YAP: Yes-associated protein. Discontinuous arrows indicate pathological ACM situations.

3. Non-Coding RNAs

Notwithstanding that 80% of DNA is transcribed, less than 2% encodes for proteins. There is an increasing interest in unravelling the roles of the remaining 98% of the genome, previously named junk DNA (including introns and spacer DNA). The potential role as regulator of a wide range of biological functions is well established for at least a fraction of this junk DNA, i.e., the well-named non-coding RNAs (ncRNAs). Indeed, ncRNAs are known to control cellular behavior through modulation of gene expression at a transcriptional or post-transcriptional level. Moreover, ncRNAs orchestrate intracellular pathways in a fine-tuned manner due to their ability to bind proteins and/or other RNA species [36–39].

Although ncRNAs do not encode for proteins, they can be transcribed, either constitutively or under regulation (Figure 4). The housekeeping ncRNAs comprise several species, including small nucleolar RNAs (snoRNA), ribosomal RNA (rRNA), or transfer RNA (tRNA). snoRNAs (60–300 bp), usually encoded in intronic regions, are among the small nucleolar riboproteins responsible for rRNA modifications. rRNA represents 80% of total cellular RNA and together with tRNAs makes possible protein translation [36,38,40,41].

Meanwhile, the regulatory ncRNAs are heterogeneous in size and form, including small microRNAs (miRNAs) or piwi-interacting RNAs (piRNAs), and larger long non-coding RNAs (lncRNAs), circular RNAs (circRNAs), or pseudogenes. Their characteristics and functions are detailed in the following.

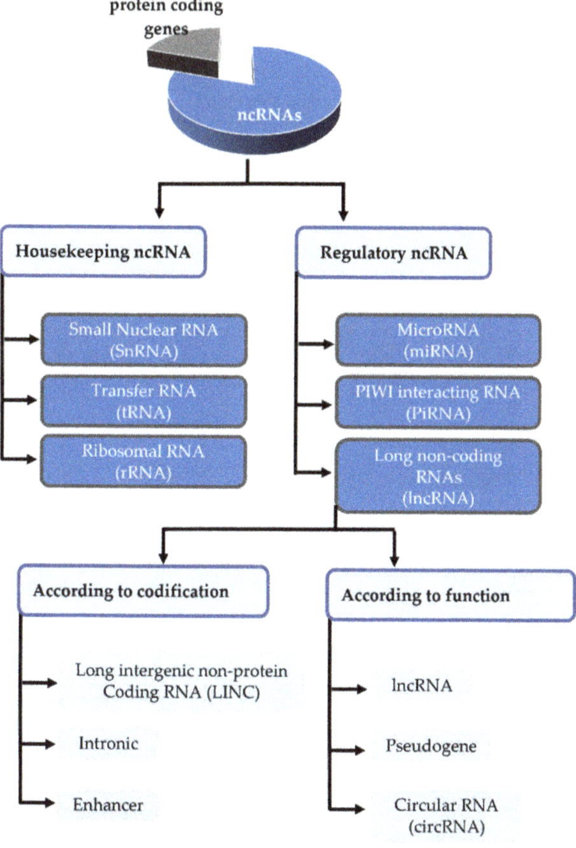

Figure 4. Classification of the non-coding transcriptome.

In recent years, a huge effort has been made by the scientific community to study the effects of individual miRNAs on their target mRNAs, aiming to improve knowledge of the complex mechanisms governing these interactions. However, the paradigm has changed, and recent findings supporting the multiple and interconnected functions of different ncRNAs species within the networks called competing endogenous RNAs (ceRNAs) have shed light on the molecular mechanisms underlying certain complex diseases. ACM is a far from simple disease and thus, ncRNAs may present a promising opportunity to harmonize the complex interactions underlying the Wnt/β-catenin and Hippo pathways.

Salmena et al. postulated the hypothesis by which ncRNAs interact among themselves, forming a regulatory network focused on communication mediated by MicroRNA response elements (MREs) called ceRNAs (Figure 5) [42]. The canonical role of these MicroRNA response elements (MREs) is to bind miRNAs and to interfere with the translation of their target mRNAs (Figure 5a,b). However, the complexity of their mechanisms of action required the definition of the new concept of ceRNAs, to better understand the interaction among different types of ncRNAs. For example, lncRNAs (including circRNAs) may behave like sponges for miRNAs, which then become unable to bind their target mRNAs so that the protein translation remains preserved (Figure 5c). The second model of ceRNAs would apply in situations where miRNAs lose their inhibitory action on their target mRNAs because the MREs present in the 3′UTR are occupied and blocked by lncRNAs (Figure 5d). Finally, ceRNAs can include pseudogenes that present similar MREs to their original genes and consequently are able to bind their target miRNAs, hampering their inhibitory effect on the translation of the original gene (Figure 5e).

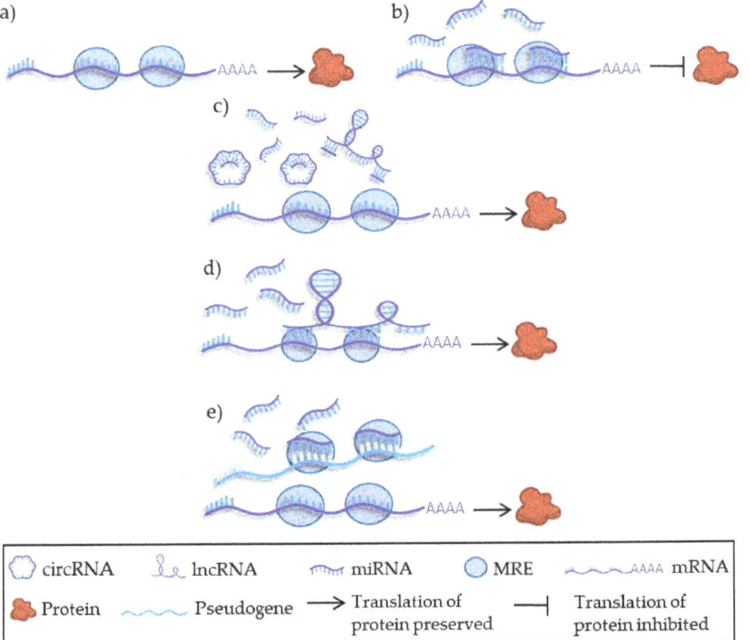

Figure 5. Main models of competing endogenous RNA (ceRNAs). (**a**) mRNAs are rich in MREs; (**b**) miRNAs bind MREs in the 3′UTR of their target mRNAs and inhibit protein translation; (**c**) ncRNAs such as lncRNAs or circRNAs are rich in MREs and can gather miRNAs, reducing the MiRNAs' inhibitory effect on the translation. (**d**) lncRNAs can block the MREs in mRNAs, impairing the binding of miRNAs and reducing the negative influence on protein translation. (**e**) Pseudogenes can also bind miRNAs, avoiding their effect on the original gene and allowing their translation. circRNA: circular RNA; lncRNA: long non-coding RNA; miRNA: microRNA; MRE: MicroRNA response element; mRNA: messenger RNA.

This plethora of functions has placed ncRNAs in the research spotlight, as promising key molecules to increase knowledge and understanding of biological and pathophysiological processes and to pave the way for design of new therapeutic strategies.

3.1. Biogenesis and Function of microRNAs (miRNAs)

The best-studied kind of ncRNA are the small single-strand sequences (20–22 nt) named miRNAs [43].

They can be encoded in intronic regions of their target mRNAs or in intergenic positions. miRNAs are transcribed into the nucleus as pri-miRNAs (more than 1000 nt) by the enzymes RNA polymerase II and III. Still in the nucleus, Drosha processes them into 60–120 nt pre-miRNAs. This stem-loop RNA is exported to the cytoplasm by exportin 5 and Ran-GTPase. Then, the pre-miRNA is processed by Dicer into a short double-strand molecule, which is recruited by the RISC complex in order to select one of the strands to interact with Argonaute proteins (Ago) and finally exert its function by targeting the 3′UTR of its target mRNA [36–38,44,45] (Figure 6).

miRNAs can bind their complementary MRE in the 3′UTR of their target mRNA to inhibit gene translation. However, they can also bind MREs in the 5′UTR or even in coding regions, although less frequently. If the seed sequence and the MRE bind with perfect complementarity, then the mRNA molecule's fate is degradation. However, if the complementarity is imperfect, which is more common, then the translation of the target mRNA is inhibited [37,38,44,45].

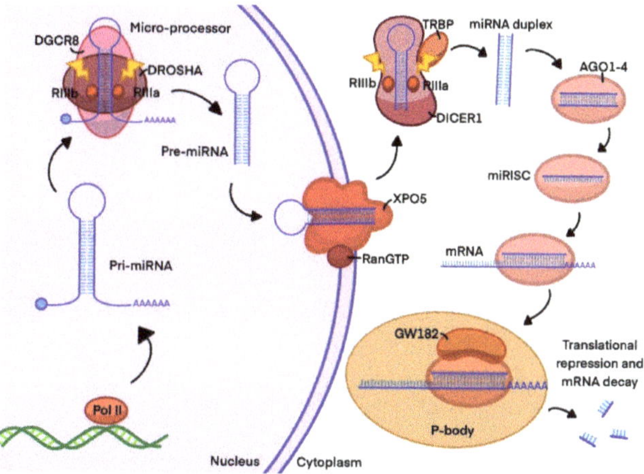

Figure 6. Biogenesis of miRNAs. miRNAs are transcribed as hairpin pri-miRNAs (1000 nt) by the RNA polymerases II and III. Drosha processes pri-miRNAs into shorter pre-miRNAs (60–120 nt). This stem-loop RNA is exported to the cytoplasm by exportin 5 and Ran-GTPase. Then, the pre-miRNA is processed by Dicer into a short double-strand RNA molecule, which is recruited by the RISC complex to select one of the strands to interact with the protein Argonaute (Ago), to exert its function by targeting the 3′UTR target mRNA.

Interestingly, their expression is cell-specific and their final effect is complex because they can regulate different pathways in a coordinated manner by blocking the translation of certain mRNAs at different levels of the same signaling pathway [45,46]. Furthermore, several miRNAs can cooperate to target a given mRNA, and a single miRNA can target several lncRNAs and different mRNAs from up- or downstream in the same pathway, thus acting as master regulators of gene transcription [36,38,47].

Many studies have addressed the role of miRNAs in pathological cardiac processes such as fibrosis after myocardium infarction (MI), hypertrophy, atrial fibrillation, or heart failure [48–58]. Regarding ACM, very few studies have been published. Those that have were focused on different samples and employed different technologies, making it difficult to draw conclusions about the effect of miRNAs on ACM pathogenesis. Remarkably, none of the previous studies linked miRNAs with the Wnt/β-catenin or Hippo pathways. Firstly, Rainer et al. explored miRNA differences by using Taqman low-density arrays (768 miRNAs included) on cardiac stem cells isolated from three biopsies from ACM patients and three from explanted hearts, yielding three miRNAs with significant differences in their expression (miR-520c-3p, miR-29b-3p, and miR-1183) [59]. Secondly, Bueno Marinas studied nine ACM cases and four controls (RV tissue and plasma) by means of an 84-miRNA cardiac-related array. After validation in a larger cohort of samples, they concluded that a group of six miRNAs (miR-122-5p, miR-133a-3p, miR-133b, miR-142-3p, miR-182-5p, and miR-183-5p) presented high discriminatory diagnostic power in ACM patients [60]. Thirdly, Sacchetto et al. screened 754 miRNAs in 21 pooled plasmas from ACM patients and 20 controls. They found five differently expressed miRNAs (miR-20b miR-505, miR-250c, miR-590-5p, and miR-185-5p) [61]. Finally, Khudiakov et al. focused their study on the miRNA composition of pericardial fluid from six ACM patients and three controls by using small RNA sequencing. They identified five miRNAs (miR-1-3p, has-miR-21-5p, miR-122-5p, miR-206, and miR-3679-5p) with different levels between the two clinical groups [56].

3.2. Biogenesis and Functions of PIWI-Interacting RNAs (piRNA)

piRNAs are small ncRNAs (24–30 nt) that also regulate gene expression but without the participation of DICER in their biogenesis. They can be encoded in transposable elements in either non-coding or coding gene regions. RNAPol II is responsible for producing long single-stranded RNA precursors which are matured into fragments of 25–40 nt by the endonuclease MitoPLD/Zucchini (Zuc). The final length of the mature piRNA depends on the target PIWI proteins to which the piRNA binds [38,62].

The wide range of ways by which piRNAs can regulate gene expression suggests that these piRNAs are more powerful regulators than the better-known miRNAs. The mechanisms of action of these ncRNAs require the association of ncRNA to the PIWI proteins to suppress expression of transposable elements, and their transcription could be regulated by DNA methylation or histone modification [62].

3.3. Biogenesis and Functions of Long Non-Coding RNAs (lncRNA)

lncRNAs are the longest ncRNAs, representing a heterogeneous group characterized just by their length (more than 200 nt). There are many differences among them, according to the location in which they are encoded, post-transcriptional modifications, binding properties, final function, etc. (Figure 4). Similarly to protein-coding DNA, the DNA fragments that encode these lncRNAs contain multiple exons (although less than protein-coding transcripts). When the RNA Pol II transcribes the sequence, the resulting RNA undergoes splicing and usually includes a 5′ cap and 3′ poly(A) tail [36,40,63,64]. The expression of lncRNAs can be activated or inhibited by epigenetic mechanisms. Attending to their location within the genome they can be identified with their own nomenclature as enhancers, antisense, intronic, or intergenic, also known as long intergenic non-protein coding RNA (LINCRNA) [36,38,65].

lncRNAs can exert local effects on sites close to their transcription sites, known as cis effects. Thus, they are able to regulate the expression of neighboring genes at an epigenetic, transcriptional, or post-transcriptional level. Due to their ability to bind proteins (e.g., transcription factors, chromatin/DNA modifying enzymes, or RNA-binding protein (RBPs)) or other nucleic acids (DNA, mRNA, or ncRNAs), they act as scaffold ensuring the proximity among all payers in this complex regulation. Additionally, they can also act at distant genomic or cellular locations (trans effect) as cell-to-cell communicators [40].

Many different functions have been attributed to lncRNAs (Figure 7) [40,47,65]. Firstly, lncRNAs can regulate gene expression by controlling the recruitment of essential proteins for transcription regulation (e.g., methyltransferases, demethylases, acetyltransferases, or deacetylases) (Figure 7a,b). Secondly, lncRNAs can interact with other RNA species such as mRNAs, regulating their splicing or stability and, as a consequence, their translation into proteins (Figure 7c,d). Finally, the presence of repeated MREs in lncRNAs allows them to act as sponges for miRNAs by hijacking them (and subsequently inactivating them) or by blocking the binding sites in the 3′UTR of their target mRNAs (Figure 7e) [36,38,40,65].

An example of a well-known ceRNA orchestrated by a lncRNA is the overexpression of MIAT (myocardial infarction-associated transcript) in myocardial infarction. This lncRNA can sponge the negative regulator miR-24, and subsequently induce an overexpression of miR-24 target mRNAs, Furin and TGF-β1, in cardiac fibroblasts [66]. Another example is the ceRNA coordinated by the lncRNA HOX transcript antisense RNA (HOTAIR), which acts as a sponge for miR-613, negatively regulating the expression of connexin-43 in atrial fibrillation [67].

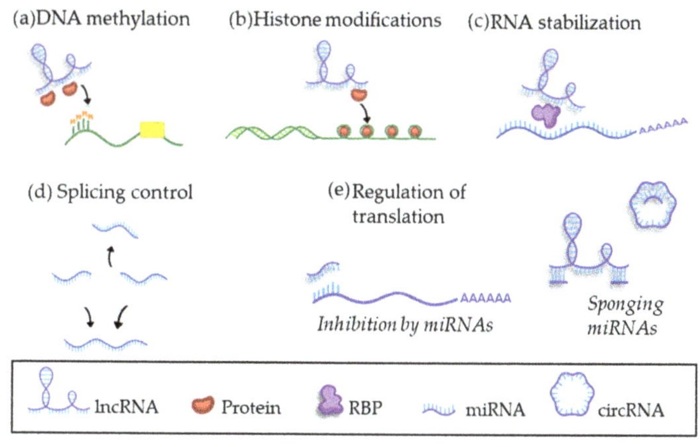

Figure 7. Main mechanisms of action of lncRNAs. (**a–c**) lncRNAs bind key proteins to promote epigenetic mechanisms such as (**a**) DNA methylatyion if lncRNA binds DNA transferases; (**b**) Histone acetylation if it binds histone acetylases/deacetylases, or (**c**) stabilizing mRNA by binding riboproteins. (**d,e**) lncRNA can bind other types of ncRNAs to regulate (**d**) alternative splicing, or (**e**) by blocking MREs in mRNAs or sponging miRNAs. lncRNA: long non-coding RNA; miRNA: microRNA; RBP: RNA binding protein.

3.3.1. Biogenesis and Functions of Circular RNAs (circRNAs)

circRNAs are a kind of covalently closed single-stranded lncRNAs generated by reverse splicing of mRNA precursors (pre-mRNAs), which means that the 5′ splice site is directly joined with a downstream 3′ splice site to be covalently bonded [44,68–72]. Notably, circRNAs can be hosted in coding genes whose levels are in turn regulated by those circRNAs. This is the case for the most abundant circRNA in the heart, the circSLC8A1, a single-exon circRNA codified in exon 2 of the SLC8A1 gene which encodes the Na^+/Ca^{++} exchanger [63,73]. It is also worth noting the characterization of 402 circRNAs contained within the huge TTN gene and containing from 1 to 153 exons [63].

Interestingly, circRNA expression is mostly tissue-specific due to the highly cell-type-specific manner of interaction with RBP [68]. Due to their special circular conformation, lacking 3′ poly(A) tail and 5′ cap structures, circRNAs are more resistant to degradation by exonucleases than linear sequences. Consequently, they have a longer half-life suitable for measurement in biofluids [44,68,74].

The most studied mechanism of action of circRNAs is their role as miRNA sponges. This feature stems from their ability to present repeated MREs, acting as effective sponges of miRNAs, or even blocking their complementary sequences in the 3′UTR of the target mRNAs (Figure 7) [40,47,63,68,74]. In addition to behaving as a miRNA sponge, circRNAs can bind proteins and so act as a scaffold to promote protein–protein interactions, or can sponge target proteins to modify their fate [40,47,65,68,74]. Two circRNAs widely expressed in the heart are heart-related circRNA (HRCR) and CDR1. Both act as miRNA sponges; HRCR was observed to bind miR-223 in a murine model of cardiac hypertrophy and heart failure [75], while CDR1 mediates post-myocardial infarction damage in mice by sponging miR-7 [76].

Finally, as an exception to the pivotal definition of ncRNAs, cases have been described in which circRNAs can be translated into proteins. This is the case with CircAXIN1, which encodes for an axin-like protein named AXIN1-2955aa that promotes gastric cancer development through Wnt/β-catenin pathway destabilization [77].

3.3.2. Pseudogenes, Biogenesis and Functions

The pseudogenes represent another kind of lncRNA, including similar sequences to those of coding genes but lacking the possibility of being translated. GENECODE v.31 identified 13,000 pseudogenes [78], although they could be considered relics of evolution or unprofitable DNA. However, although these sequences resemble the original gene, they usually harbor premature stop codons, deletions, insertions or even frameshift mutations that impair their translation into proteins [78–80]. These are the non-processed pseudogenes. In contrast, pseudogenes can also result from processed genetic elements such as the insertion of the reverse transcription of the original mRNA (retrotransposition) [80,81].

The transcription of pseudogenes is regulated by proximal regulatory elements due to the lack of canonical promoters, and their expression seems to be tissue-dependent and associated with pathologic conditions (e.g., cancer) [40,81].

Their similar sequence to the original genes allows them to act as ceRNAs by modulating the effect of other ncRNAs on the expression of their target genes [40,47,65]. PTEN and its pseudogene PTENP1 have been widely studied because of their implication in cancer pathogenesis. The main difference between PTENP1 and PTEN results from the lack of an initiation codon due to a missense mutation impairing its translation, and the 3′UTR of the pseudogene is 1 kb shorter than in the original sequence. Despite the shorter length of the 3′UTR, PTENP1 conserves MREs to bind miR-17, miR-21, miR-214, miR-19, and miR-26. Thus, PTENP1 3′UTR functions as a sponge for PTEN-targeting miRNAs [81]. For this reason, copy number variations in pseudogenes can strongly interfere with the pathogenesis of many diseases. Indeed, it has been reported that an increase in the copy number of the pseudogene NOTCH2NL is associated with autism, whereas a reduction in the copy number of this pseudogene is related to schizophrenia [82].

Study of the role of ncRNAs in pathophysiology requires a multifactorial approach to assess the degree of implication of each player in these complex scenarios orchestrated by ceRNAs.

4. Non-Coding RNAs Databases

The scientific community has devoted huge efforts to better understanding and classifying ncRNA species and their potential interconnections. Recent developments have rendered previous approaches too simple and obsolete. It is no longer valid to explain the molecular pathogenesis of a complex disease only based upon the effect of a single miRNA on its target mRNAs, as the first studies on the topic did. Nowadays, the study of ncRNAs requires an open-minded approach in which proteins, RBP, nucleic acids, and ncRNAs act within interconnected ceRNA networks.

In order to update the published information regarding the validated connections among the aforementioned ncRNAs, many online databases have been created (Table 3).

Table 3. Online databases containing information about ncRNAs.

	Database	ncRNA Type	Year	N° ncRNAs	Species
General	NCBI	All	2021		From virus to Hsa
	RNACentral [83]		2021	±18,000,000	From virus to Hsa
	RFam [84]		2021	±14,800	From virus to Hsa
Small ncRNA	miRbase [85]	miRNA	2019	±38,600	271 species
	Diana-TarBase [86]	miRNA	2018	±1,000,000	18 species
	CircBase [87]	cicRNA	2014	±50,000	6 species
	CircInteractome [88]	cirRNA	2016	±117,000	Hsa
	Snopy [89]	snoRNA	2013	±13,800	41 species
lncRNA	Noncode [90]	lncRNA	2021	±645,000	Hsa, Mmu
	Diana-LNBase [91]		2020	±500,000	19 animals, 23 plants

Hsa: Homo sapiens; Mmu: Mus musculus.

5. Non-Coding RNAs in ACM

Only a few studies have focused on the role of ncRNAs in ACM, and all have restricted their results to one species, namely miRNAs, concluding that they emerge as potentially useful biomarkers [51–57,59,61,92,93]. To the best of our knowledge, no studies have addressed the regulation of the Wnt/Hippo pathway by ncRNAs in the scenario of ACM. However, due to the importance of both pathways in the pathogenesis of many cancers, other studies have shed light on the issue (Table 4).

Table 4. Main competing endogenous RNAs (ceRNAs) studied in cancer research that regulate the most affected pathways involved in arrhythmogenic cardiomyopathy.

	Direct Target	Indirect Target	Mechanism of Action	Potential Effect on ACM Pathogenesis
MALAT1	miR-145 [94] miR-382-3p [95]	TGFB Resistin	miRNA sponge miRNA sponge	Fibrosis Adipogenesis
HOTAIR	miR-34a [96] miR-613 [67]	Sirtuin 1 Connexin-43	miRNA sponge miRNA sponge	Adipogenesis Conexome structure
MIAT	miR-24 [66]	TGFB1, Furin	miRNA sponge	Fibrosis
KCNQ1OT1	miR-214-3p [97]	TGFB1	miRNA sponge	Fibrosis
MRLP23-AS1	miR-30b [98]	β-catenin	miRNA sponge	Wnt pathway
lincRNA-p21	β-catenin (mRNA) [99]		Inhibition of translation	Wnt pathway
LSINCT5	miR-30a [100]	β-catenin	miRNA sponge	Wnt pathway
LINC00467	Binding EZH2 [101]	DKK1	Regulation of transcription by epigenetic mechanisms	Wnt pathway
DANCR	β-catenin (mRNA) [102,103] miR-216a [104] miR-214, -320a [102,103]	 β-catenin β-catenin	Inhibition of translation Inhibition of translation Inhibition of translation	Wnt pathway Wnt pathway Wnt pathway
UCA1	miR-18a [105]	YAP1	miRNA sponge	Hippo pathway
FLVCR1-AS1	miR-513 [106]	YAP1	miRNA sponge	Hippo pathway
RP11-323N12.5	binding C-Myc to YAP1 promoter [107]		Transcription of YAP1 mRNA	Hippo pathway
circYap	YAP1 mRNA and translation protein complex [108]		Inhibition of translation	Hippo pathway
circRNA_010567	miR141 [109]	TGFB1	miRNA sponge	Fibrosis
circAXIN1	[77]		Translate into Axin protein named AXIN1-29545aa	Wnt pathway

DANCR: Differentiation antagonizing non-protein coding RNA; HOTAIR: lncRNA-HOX transcript antisense RNA; MALAT1: lncRNA metastasis-associated lung adenocarcinoma transcript 1; MIAT: myocardial infarction associated transcript; UCA1: urothelial cancer associated 1.

5.1. Non-Coding RNAs in the Wnt/β-Catenin Pathway

Recent research on osteosarcoma has suggested that lncRNA MRLP23-AS1 might be overexpressed and act as a sponge for miR-30b, so that the miR-30b inhibitory effect on its target mRNA (myosin heavy chain 9, MYH9) is reduced to provoke an increase of the final level of this protein. Meanwhile, MYH9 is a trans-activator of the expression of β-catenin, thus exerting a final activating effect on the Wnt/β-catenin pathway [98]. Moreover, β-catenin is also indirectly regulated by lncRNA LSINCT5. This lncRNA has been shown to be upregulated in breast cancer and is capable of sponging miR-30a, which inhibits β-catenin translation [100]. DANCR (differentiation antagonizing non-protein coding RNA) is one of the better studied lncRNAs, because of its implication in the regulation of the

Wnt/β-catenin signalling pathway [102,110]. The lncRNA DANCR regulates β-catenin translation not only indirectly by sponging miR-216a [104], miR-214, or miR-320a [102,103] and modulating the inhibitory effect of these miRNAs on the mRNA of β-catenin, but also directly by binding β-catenin mRNA so that MREs are blocked and there is no inhibitory effect on translation [102,103].

Yang et al. went further in their study in which they corroborated that lncRNA long intergenic non-protein coding RNA 467 (LINC00467) was upregulated in lung adenocarcinoma mediated by STAT1. Moreover, LINC00467 was responsible for the downregulation of the Wnt/β-catenin pathway inhibitor, DKK1, by recruiting the histone methyltransferase EZH2 and epigenetically silencing its expression downstream. In this way, by reduction of the inhibitory effect of DKK, the Wnt/β-catenin pathway was overactivated in lung adenocarcinoma [101].

5.2. Non-Coding RNAs in the Hippo Pathway

The Hippo pathway has also been widely studied in the pathogenesis of several cancers, because of its crosslink with the Wnt-β-catenin pathway. Several studies have demonstrated that certain miRNAs are responsible for the regulation of YAP, such as miR-550a-3-5p [111], -27b-3p [112], -195-5p [113], or TAZ by miR-9-3p [114].

Additionally, the Hippo pathway might be also regulated indirectly. The translation of the kinase responsible for YAP phosphorylation, LATS2, is regulated by miR-103a-3p and -429 in colorectal cancer, so YAP phosphorylation is reduced and subsequently its nuclear localization increases [115]. Probably, as already mentioned, the regulation of the final levels of these miRNAs are further regulated at a much higher level of complexity by ceRNAs that accurately determine the finely tuned regulation of this pathway. In this respect, the function of lncRNA FLVCR1-AS1 seems to regulate the Hippo pathway in ovarian serous cancer by sponging a negative regulator of the translation of YAP, the miR-513 [106].

Interestingly, YAP/TAZ can also modulate the transcription of other genes by binding proteins or even their mRNAs. In gastric cancer, regulation of the expression of YAP1 and lncRNA RP11-323N12.5 seems to be bidirectional. The expression of RP11-323N12.5 seems to be regulated by YAP/TAZ/TEAD, while RP11-323N12.5 activates YAP1 transcription by acting as scaffold and promoting the binding of the transcription factor c-MYC to the YAP1 promoter [107]. However, its circular RNA (circYap) negatively regulates YAP translation due to its ability to bind YAP mRNA and the translation initiation proteins (eIF4G and PABP), as has been described in breast cancer where overexpression of circYap reduces YAP translation by suppressing the formation of transcriptional initiation machinery without modifying YAP mRNA levels [108].

6. Concluding Remarks

From a clinical standpoint, the extraordinary complexity of ACM is supported by the increasing list of causal genes, lists of major and minor diagnostic criteria, the discussion about the criteria to be used (Task Force of Padua), and the wide range of phenotypic features involving the right ventricle, the left ventricle, or both, with or without relevant inflammation or even with associated non-compaction. From a molecular perspective, the complexity grows exponentially in terms of interwoven pathways underlying the well-known fibrofatty replacement of the ventricular myocardium, of which the most widely accepted are the Wnt/β-catenin and Hippo pathways. The ultimate mechanisms orchestrating all the necessary elements to produce ACM remain broadly unknown. A few studies have reported scarce results regarding measurement of miRNAs in ACM, and none of them linked the singular differences with molecular pathways. However, the other ncRNAs have not yet been addressed in ACM. Taking advantage of great efforts made in cancer research, we have presented in this paper evidence of the role of ncRNAs in modulating the Wnt/β-catenin and Hippo pathways. A challenge is thus presented. Can the presented interactions or others in the same scenario help to improve understanding of ACM pathophysiology? Will revealing ceRNA networks involved in ACM be able to open avenues leading to the development of new therapeutic options for patients? A new era of research has just begun.

Author Contributions: Conceptualization, A.B.-B. and E.Z.; writing—original draft preparation M.P.-G., S.D.-D., M.S.-G., C.M.-C.; writing—review and editing, M.P.-G., A.B.-B. and E.Z.; graphical design, M.P.-G., S.D.-D. and A.B.-B.; funding acquisition, A.B.-B. and E.Z. All authors have read and agreed to the published version of the manuscript.

Funding: This research was funded by Instituto de Salud Carlos III and FEDER Union Europea, "Una forma de hacer Europa" (PI18/01582, PI21/01282). M.P.-G. is supported by Agencia Valenciana de Innovación (OGMIOS. INNEST/2021/55). A.B.-B. is supported by Marató TV3 (736/C/2020). Health Research Institute La Fe. Memorial Nacho Barberá.

Conflicts of Interest: The authors declare no conflict of interest.

References

1. Miles, C.; Finocchiaro, G.; Papadakis, M.; Gray, B.; Westaby, J.; Ensam, B.; Basu, J.; Parry-Williams, G.; Papatheodorou, E.; Paterson, C.; et al. Sudden Death and Left Ventricular Involvement in Arrhythmogenic Cardiomyopathy. *Circulation* **2019**, *139*, 1786–1797. [CrossRef] [PubMed]
2. Protonotarios, A.; Elliott, P.M. Arrhythmogenic Cardiomyopathies (ACs): Diagnosis, Risk Stratification and Management. *Heart* **2019**, *105*, 1117–1128. [CrossRef] [PubMed]
3. Protonotarios, A.; Bariani, R.; Cappelletto, C.; Pavlou, M.; García-García, A.; Cipriani, A.; Protonotarios, I.; Rivas, A.; Wittenberg, R.; Graziosi, M.; et al. Importance of Genotype for Risk Stratification in Arrhythmogenic Right Ventricular Cardiomyopathy Using the 2019 ARVC Risk Calculator. *Eur. Heart J.* **2022**, *43*, 3053–3067. [CrossRef] [PubMed]
4. Austin, K.M.; Trembley, M.A.; Chandler, S.F.; Sanders, S.P.; Saffitz, J.E.; Abrams, D.J.; Pu, W.T. Molecular Mechanisms of Arrhythmogenic Cardiomyopathy. *Nat. Rev. Cardiol.* **2019**, *16*, 519–537. [CrossRef]
5. Corrado, D.; Basso, C. Arrhythmogenic Left Ventricular Cardiomyopathy. *Heart* **2022**, *108*, 733–743. [CrossRef]
6. Patel, V.; Asatryan, B.; Siripanthong, B.; Munroe, P.B.; Tiku-Owens, A.; Lopes, L.R.; Khanji, M.Y.; Protonotarios, A.; Santangeli, P.; Muser, D.; et al. State of the Art Review on Genetics and Precision Medicine in Arrhythmogenic Cardiomyopathy. *Int. J. Mol. Sci.* **2020**, *21*, 6615. [CrossRef]
7. Stevens, T.L.; Wallace, M.J.; el Refaey, M.; Roberts, J.D.; Koenig, S.N.; Mohler, P.J. Arrhythmogenic Cardiomyopathy: Molecular Insights for Improved Therapeutic Design. *J. Cardiovasc. Dev. Dis.* **2020**, *7*, 21. [CrossRef]
8. Akhtar, M.M.; Lorenzini, M.; Pavlou, M.; Ochoa, J.P.; O'Mahony, C.; Restrepo-Cordoba, M.A.; Segura-Rodriguez, D.; Bermúdez-Jiménez, F.; Molina, P.; Cuenca, S.; et al. Association of Left Ventricular Systolic Dysfunction among Carriers of Truncating Variants in Filamin C With Frequent Ventricular Arrhythmia and End-Stage Heart Failure. *JAMA Cardiol.* **2021**, *6*, 891–901. [CrossRef]
9. Ortiz-Genga, M.F.; Cuenca, S.; Dal Ferro, M.; Zorio, E.; Salgado-Aranda, R.; Climent, V.; Padrón-Barthe, L.; Duro-Aguado, I.; Jiménez-Jáimez, J.; Hidalgo-Olivares, V.M.; et al. Truncating FLNC Mutations Are Associated With High-Risk Dilated and Arrhythmogenic Cardiomyopathies. *J. Am. Coll. Cardiol.* **2016**, *68*, 2440–2451. [CrossRef]
10. Bueno-Beti, C.; Asimaki, A. Histopathological Features and Protein Markers of Arrhythmogenic Cardiomyopathy. *Front. Cardiovasc. Med.* **2021**, *8*, 746321. [CrossRef]
11. Lombardi, R.; Dong, J.; Rodriguez, G.; Bell, A.; Leung, T.K.; Schwartz, R.J.; Willerson, J.T.; Brugada, R.; Marian, A.J. Genetic Fate Mapping Identifies Second Heart Field Progenitor Cells as a Source of Adipocytes in Arrhythmogenic Right Ventricular Cardiomyopathy. *Circ. Res.* **2009**, *104*, 1076–1084. [CrossRef]
12. Lombardi, R.; Chen, S.N.; Ruggiero, A.; Gurha, P.; Czernuszewicz, G.Z.; Willerson, J.T.; Marian, A.J. Cardiac Fibro-Adipocyte Progenitors Express Desmosome Proteins and Preferentially Differentiate to Adipocytes upon Deletion of the Desmoplakin Gene. *Circ. Res.* **2016**, *119*, 41–54. [CrossRef]
13. Lombardi, R.; da Graca Cabreira-Hansen, M.; Bell, A.; Fromm, R.R.; Willerson, J.T.; Marian, A.J. Nuclear Plakoglobin Is Essential for Differentiation of Cardiac Progenitor Cells to Adipocytes in Arrhythmogenic Right Ventricular Cardiomyopathy. *Circ. Res.* **2011**, *109*, 1342–1353. [CrossRef]
14. Caspi, O.; Huber, I.; Gepstein, A.; Arbel, G.; Maizels, L.; Boulos, M.; Gepstein, L. Modeling of Arrhythmogenic Right Ventricular Cardiomyopathy with Human Induced Pluripotent Stem Cells. *Circ. Cardiovasc. Genet.* **2013**, *6*, 557–568. [CrossRef]
15. Chen, S.N.; Gurha, P.; Lombardi, R.; Ruggiero, A.; Willerson, J.T.; Marian, A.J. The Hippo Pathway Is Activated and Is a Casual Mechanism for Adipogenesis in Arrhythmogenic Cardiomyopathy. *Circ. Res.* **2014**, *114*, 454–468. [CrossRef]
16. Kim, C.; Wong, J.; Wen, J.; Wang, S.; Wang, C.; Spiering, S.; Kan, N.G.; Forcales, S.; Puri, P.L.; Leone, T.C.; et al. Studying Arrhythmogenic Right Ventricular Dysplasia with Patient-Specific IPSCs. *Nature* **2013**, *494*, 105–110. [CrossRef]
17. Good, J.-M.; Fellmann, F.; Bhuiyan, Z.A.; Rotman, S.; Pruvot, E.; Schläpfer, J. ACTN2 Variant Associated with a Cardiac Phenotype Suggestive of Left-Dominant Arrhythmogenic Cardiomyopathy. *Hearth Case Rep.* **2020**, *6*, 15–19. [CrossRef]

18. Kohela, A.; van Rooij, E. Fibro-Fatty Remodelling in Arrhythmogenic Cardiomyopathy. *Basic Res. Cardiol.* **2022**, *117*, 22. [CrossRef]
19. Garcia-Gras, E.; Lombardi, R.; Giocondo, M.J.; Willerson, J.T.; Schneider, M.D.; Khoury, D.S.; Marian, A.J. Suppression of Canonical Wnt/β-Catenin Signaling by Nuclear Plakoglobin Recapitulates Phenotype of Arrhythmogenic Right Ventricular Cardiomyopathy. *J. Clin. Investig.* **2006**, *116*, 2012–2021. [CrossRef]
20. Rouhi, L.; Fan, S.; Cheedipudi, S.M.; Braza-Boïls, A.; Molina, M.S.; Yao, Y.; Robertson, M.J.; Coarfa, C.; Gimeno, J.R.; Molina, P.; et al. The EP300/TP53 Pathway, a Suppressor of the Hippo and Canonical WNT Pathways, Is Activated in Human Hearts with Arrhythmogenic Cardiomyopathy in the Absence of Overt Heart Failure. *Cardiovasc. Res* **2022**, *118*, 1466–1478. [CrossRef]
21. Mazurek, S.; Kim, G.H. Genetic and Epigenetic Regulation of Arrhythmogenic Cardiomyopathy. *Biochim. Biophys. Acta (BBA) Mol. Basis Dis.* **2017**, *1863*, 2064–2069. [CrossRef]
22. Gao, S.; Puthenvedu, D.; Lombardi, R.; Chen, S.N. Established and Emerging Mechanisms in the Pathogenesis of Arrhythmogenic Cardiomyopathy: A Multifaceted Disease. *Int. J. Mol. Sci.* **2020**, *21*, 6320. [CrossRef]
23. Lorenzon, A.; Calore, M.; Poloni, G.; de Windt, L.J.; Braghetta, P.; Rampazzo, A. Wnt/β-Catenin Pathway in Arrhythmogenic Cardiomyopathy. *Oncotarget* **2017**, *8*, 60640–60655. [CrossRef]
24. Giuliodori, A.; Beffagna, G.; Marchetto, G.; Fornetto, C.; Vanzi, F.; Toppo, S.; Facchinello, N.; Santimaria, M.; Vettori, A.; Rizzo, S.; et al. Loss of Cardiac Wnt/β-Catenin Signalling in Desmoplakin-Deficient AC8 Zebrafish Models Is Rescuable by Genetic and Pharmacological Intervention. *Cardiovasc. Res.* **2018**, *114*, 1082–1097. [CrossRef]
25. Cason, M.; Celeghin, R.; Marinas, M.B.; Beffagna, G.; della Barbera, M.; Rizzo, S.; Remme, C.A.; Bezzina, C.R.; Tiso, N.; Bauce, B.; et al. Novel Pathogenic Role for Galectin-3 in Early Disease Stages of Arrhythmogenic Cardiomyopathy. *Heart Rhythm.* **2021**, *18*, 1394–1403. [CrossRef]
26. Peifer, M.; Berg, S.; Reynolds, A.B. A Repeating Amino Acid Motif Shared by Proteins with Diverse Cellular Roles. *Cell* **1994**, *76*, 789–791. [CrossRef]
27. Asimaki, A.; Protonotarios, A.; James, C.A.; Chelko, S.P.; Tichnell, C.; Murray, B.; Tsatsopoulou, A.; Anastasakis, A.; te Riele, A.; Kléber, A.G.; et al. Characterizing the Molecular Pathology of Arrhythmogenic Cardiomyopathy in Patient Buccal Mucosa Cells. *Circ. Arrhythm. Electrophysiol.* **2016**, *9*, e003688. [CrossRef]
28. Asimaki, A.; Kleber, A.G.; Saffitz, J.E. Pathogenesis of Arrhythmogenic Cardiomyopathy. *Can. J. Cardiol.* **2015**, *31*, 1313–1324. [CrossRef]
29. Hoorntje, E.T.; te Rijdt, W.P.; James, C.A.; Pilichou, K.; Basso, C.; Judge, D.P.; Bezzina, C.R.; van Tintelen, J.P. Arrhythmogenic Cardiomyopathy: Pathology, Genetics, and Concepts in Pathogenesis. *Cardiovasc. Res.* **2017**, *113*, 1521–1531. [CrossRef]
30. Kulaberoglu, Y.; Lin, K.; Holder, M.; Gai, Z.; Gomez, M.; Assefa Shifa, B.; Mavis, M.; Hoa, L.; Sharif, A.A.D.; Lujan, C.; et al. Stable MOB1 Interaction with Hippo/MST Is Not Essential for Development and Tissue Growth Control. *Nat. Commun.* **2017**, *8*, 695. [CrossRef]
31. Zhang, Y.; Wang, Y.; Ji, H.; Ding, J.; Wang, K. The Interplay between Noncoding RNA and YAP/TAZ Signaling in Cancers: Molecular Functions and Mechanisms. *J. Exp. Clin. Cancer Res.* **2022**, *41*, 202. [CrossRef] [PubMed]
32. Zhang, N.; Bai, H.; David, K.K.; Dong, J.; Zheng, Y.; Cai, J.; Giovannini, M.; Liu, P.; Anders, R.A.; Pan, D. The Merlin/NF2 Tumor Suppressor Functions through the YAP Oncoprotein to Regulate Tissue Homeostasis in Mammals. *Dev. Cell* **2010**, *19*, 27–38. [CrossRef] [PubMed]
33. Hu, Y.; Pu, W.T. Hippo Activation in Arrhythmogenic Cardiomyopathy. *Circ. Res.* **2014**, *114*, 402–405. [CrossRef] [PubMed]
34. Li, J.; Swope, D.; Raess, N.; Cheng, L.; Muller, E.J.; Radice, G.L. Cardiac Tissue-Restricted Deletion of Plakoglobin Results in Progressive Cardiomyopathy and Activation of β-Catenin Signaling. *Mol. Cell Biol.* **2011**, *31*, 1134–1144. [CrossRef]
35. Varelas, X.; Miller, B.W.; Sopko, R.; Song, S.; Gregorieff, A.; Fellouse, F.A.; Sakuma, R.; Pawson, T.; Hunziker, W.; McNeill, H.; et al. The Hippo Pathway Regulates Wnt/β-Catenin Signaling. *Dev. Cell* **2010**, *18*, 579–591. [CrossRef]
36. Goodall, G.J.; Wickramasinghe, V.O. RNA in Cancer. *Nat. Rev. Cancer* **2021**, *21*, 22–36. [CrossRef]
37. Winkle, M.; El-Daly, S.M.; Fabbri, M.; Calin, G.A. Noncoding RNA Therapeutics—Challenges and Potential Solutions. *Nat. Rev. Drug Discov.* **2021**, *20*, 629–651. [CrossRef]
38. Esteller, M. Non-Coding RNAs in Human Disease. *Nat. Rev. Genet.* **2011**, *12*, 861–874. [CrossRef]
39. Ramón y Cajal, S.; Segura, M.F.; Hümmer, S. Interplay Between NcRNAs and Cellular Communication: A Proposal for Understanding Cell-Specific Signaling Pathways. *Front. Genet.* **2019**, *10*, 281. [CrossRef]
40. Slack, F.J.; Chinnaiyan, A.M. The Role of Non-Coding RNAs in Oncology. *Cell* **2019**, *179*, 1033–1055. [CrossRef]
41. Fiedler, J.; Baker, A.H.; Dimmeler, S.; Heymans, S.; Mayr, M.; Thum, T. Non-Coding RNAs in Vascular Disease—From Basic Science to Clinical Applications: Scientific Update from the Working Group of Myocardial Function of the European Society of Cardiology. *Cardiovasc. Res.* **2018**, *114*, 1281–1286. [CrossRef]
42. Salmena, L.; Poliseno, L.; Tay, Y.; Kats, L.; Pandolfi, P.P. A CeRNA Hypothesis: The Rosetta Stone of a Hidden RNA Language? *Cell* **2011**, *146*, 353–358. [CrossRef]
43. Bartel, D.P. MicroRNAs: Target Recognition and Regulatory Functions. *Cell* **2009**, *136*, 215–233. [CrossRef]
44. Garbo, S.; Maione, R.; Tripodi, M.; Battistelli, C. Next RNA Therapeutics: The Mine of Non-Coding. *Int. J. Mol. Sci.* **2022**, *23*, 7471. [CrossRef]
45. Ambros, V. The Functions of Animal MicroRNAs. *Nature* **2004**, *431*, 350–355. [CrossRef]
46. Barwari, T.; Joshi, A.; Mayr, M. MicroRNAs in Cardiovascular Disease. *J. Am. Coll. Cardiol.* **2016**, *68*, 2577–2584. [CrossRef]

47. Xiong, W.; Yao, M.; Yang, Y.; Qu, Y.; Qian, J. Implication of Regulatory Networks of Long Noncoding RNA/Circular RNA MiRNA-MRNA in Diabetic Cardiovascular Diseases. *Epigenomics* **2020**, *12*, 1929–1947. [CrossRef]
48. Luo, F.; Liu, W.; Bu, H. MicroRNAs in Hypertrophic Cardiomyopathy: Pathogenesis, Diagnosis, Treatment Potential and Roles as Clinical Biomarkers. *Heart Fail. Rev.* **2022**, *27*, 2211–2221. [CrossRef]
49. Varzideh, F.; Kansakar, U.; Donkor, K.; Wilson, S.; Jankauskas, S.S.; Mone, P.; Wang, X.; Lombardi, A.; Santulli, G. Cardiac Remodeling After Myocardial Infarction: Functional Contribution of MicroRNAs to Inflammation and Fibrosis. *Front. Cardiovasc. Med.* **2022**, *9*, 863238. [CrossRef]
50. Tanase, D.M.; Gosav, E.M.; Petrov, D.; Teodorescu, D.-S.; Buliga-Finis, O.N.; Ouatu, A.; Tudorancea, I.; Rezus, E.; Rezus, C. MicroRNAs (MiRNAs) in Cardiovascular Complications of Rheumatoid Arthritis (RA): What Is New? *Int. J. Mol. Sci.* **2022**, *23*, 5254. [CrossRef]
51. Zhang, H.; Liu, S.; Dong, T.; Yang, J.; Xie, Y.; Wu, Y.; Kang, K.; Hu, S.; Gou, D.; Wei, Y. Profiling of Differentially Expressed MicroRNAs in Arrhythmogenic Right Ventricular Cardiomyopathy. *Sci. Rep.* **2016**, *6*, 28101. [CrossRef]
52. Mazurek, S.R.; Calway, T.; Harmon, C.; Farrell, P.; Kim, G.H. MicroRNA-130a Regulation of Desmocollin 2 in a Novel Model of Arrhythmogenic Cardiomyopathy. *Microrna* **2017**, *6*, 143–150. [CrossRef]
53. Sommariva, E.; D'Alessandra, Y.; Farina, F.M.; Casella, M.; Cattaneo, F.; Catto, V.; Chiesa, M.; Stadiotti, I.; Brambilla, S.; dello Russo, A.; et al. MiR-320a as a Potential Novel Circulating Biomarker of Arrhythmogenic CardioMyopathy. *Sci. Rep.* **2017**, *7*, 4802. [CrossRef]
54. Yamada, S.; Hsiao, Y.-W.; Chang, S.-L.; Lin, Y.-J.; Lo, L.-W.; Chung, F.-P.; Chiang, S.-J.; Hu, Y.-F.; Tuan, T.-C.; Chao, T.-F.; et al. Circulating MicroRNAs in Arrhythmogenic Right Ventricular Cardiomyopathy with Ventricular Arrhythmia. *Europace* **2018**, *20*, f37–f45. [CrossRef]
55. Patanè, S.; Patanè, F. Arrhythmogenic Right Ventricular Cardiomyopathy: Implications of next-Generation Sequencing and of MiRNAs Regulation in Appropriate Understanding and Treatment. *Europace* **2018**, *20*, 729. [CrossRef]
56. Khudiakov, A.A.; Smolina, N.A.; Perepelina, K.I.; Malashicheva, A.B.; Kostareva, A.A. Extracellular MicroRNAs and Mitochondrial DNA as Potential Biomarkers of Arrhythmogenic Cardiomyopathy. *Biochemistry* **2019**, *84*, 272–282. [CrossRef]
57. Chiti, E.; di Paolo, M.; Turillazzi, E.; Rocchi, A. MicroRNAs in Hypertrophic, Arrhythmogenic and Dilated Cardiomyopathy. *Diagnostics* **2021**, *11*, 1720. [CrossRef]
58. Yousefi, F.; Shabaninejad, Z.; Vakili, S.; Derakhshan, M.; Movahedpour, A.; Dabiri, H.; Ghasemi, Y.; Mahjoubin-Tehran, M.; Nikoozadeh, A.; Savardashtaki, A.; et al. TGF-β and WNT Signaling Pathways in Cardiac Fibrosis: Non-Coding RNAs Come into Focus. *Cell Commun. Signal.* **2020**, *18*, 87. [CrossRef]
59. Rainer, J.; Meraviglia, V.; Blankenburg, H.; Piubelli, C.; Pramstaller, P.P.; Paolin, A.; Cogliati, E.; Pompilio, G.; Sommariva, E.; Domingues, F.S.; et al. The Arrhythmogenic Cardiomyopathy-Specific Coding and Non-Coding Transcriptome in Human Cardiac Stromal Cells. *BMC Genom.* **2018**, *19*, 491. [CrossRef]
60. Bueno Marinas, M.; Celeghin, R.; Cason, M.; Thiene, G.; Basso, C.; Pilichou, K. The Role of MicroRNAs in Arrhythmogenic Cardiomyopathy: Biomarkers or Innocent Bystanders of Disease Progression? *Int. J. Mol. Sci.* **2020**, *21*, 6434. [CrossRef]
61. Sacchetto, C.; Mohseni, Z.; Colpaert, R.M.W.; Vitiello, L.; de Bortoli, M.; Vonhögen, I.G.C.; Xiao, K.; Poloni, G.; Lorenzon, A.; Romualdi, C.; et al. Circulating MiR-185-5p as a Potential Biomarker for Arrhythmogenic Right Ventricular Cardiomyopathy. *Cells* **2021**, *10*, 2578. [CrossRef] [PubMed]
62. Hanusek, K.; Poletajew, S.; Kryst, P.; Piekiełko-Witkowska, A.; Bogusławska, J. PiRNAs and PIWI Proteins as Diagnostic and Prognostic Markers of Genitourinary Cancers. *Biomolecules* **2022**, *12*, 186. [CrossRef] [PubMed]
63. Iyer, M.K.; Niknafs, Y.S.; Malik, R.; Singhal, U.; Sahu, A.; Hosono, Y.; Barrette, T.R.; Prensner, J.R.; Evans, J.R.; Zhao, S.; et al. The Landscape of Long Noncoding RNAs in the Human Transcriptome. *Nat. Genet.* **2015**, *47*, 199–208. [CrossRef] [PubMed]
64. Qian, X.; Zhao, J.; Yeung, P.Y.; Zhang, Q.C.; Kwok, C.K. Revealing LncRNA Structures and Interactions by Sequencing-Based Approaches. *Trends Biochem. Sci.* **2019**, *44*, 33–52. [CrossRef]
65. Kondo, Y.; Shinjo, K.; Katsushima, K. Long Non-coding RNAs as an Epigenetic Regulator in Human Cancers. *Cancer Sci.* **2017**, *108*, 1927–1933. [CrossRef]
66. Qu, X.; Du, Y.; Shu, Y.; Gao, M.; Sun, F.; Luo, S.; Yang, T.; Zhan, L.; Yuan, Y.; Chu, W.; et al. MIAT Is a Pro-Fibrotic Long Non-Coding RNA Governing Cardiac Fibrosis in Post-Infarct Myocardium. *Sci. Rep.* **2017**, *7*, 42657. [CrossRef]
67. Dai, W.; Chao, X.; Li, S.; Zhou, S.; Zhong, G.; Jiang, Z. Long Noncoding RNA HOTAIR Functions as a Competitive Endogenous RNA to Regulate Connexin43 Remodeling in Atrial Fibrillation by Sponging MicroRNA-613. *Cardiovasc. Ther.* **2020**, *2020*, 5925342. [CrossRef]
68. Zhao, X.; Zhong, Y.; Wang, X.; Shen, J.; An, W. Advances in Circular RNA and Its Applications. *Int. J. Med. Sci.* **2022**, *19*, 975–985. [CrossRef]
69. Li, Z.; Huang, C.; Bao, C.; Chen, L.; Lin, M.; Wang, X.; Zhong, G.; Yu, B.; Hu, W.; Dai, L.; et al. Exon-Intron Circular RNAs Regulate Transcription in the Nucleus. *Nat. Struct. Mol. Biol.* **2015**, *22*, 256–264. [CrossRef]
70. Zhang, Y.; Zhang, X.-O.; Chen, T.; Xiang, J.-F.; Yin, Q.-F.; Xing, Y.-H.; Zhu, S.; Yang, L.; Chen, L.-L. Circular Intronic Long Noncoding RNAs. *Mol. Cell* **2013**, *51*, 792–806. [CrossRef]
71. Jeck, W.R.; Sorrentino, J.A.; Wang, K.; Slevin, M.K.; Burd, C.E.; Liu, J.; Marzluff, W.F.; Sharpless, N.E. Circular RNAs Are Abundant, Conserved, and Associated with ALU Repeats. *RNA* **2013**, *19*, 141–157. [CrossRef]

Biomedicines **2022**, *10*, 2619

72. Chen, L.-L. The Expanding Regulatory Mechanisms and Cellular Functions of Circular RNAs. *Nat. Rev. Mol. Cell Biol* **2020**, *21*, 475–490. [CrossRef]
73. Tan, W.L.W.; Lim, B.T.S.; Anene-Nzelu, C.G.O.; Ackers-Johnson, M.; Dashi, A.; See, K.; Tiang, Z.; Lee, D.P.; Chua, W.W.; Luu, T.D.A.; et al. A Landscape of Circular RNA Expression in the Human Heart. *Cardiovasc. Res* **2017**, *113*, 298–309. [CrossRef]
74. Mester-Tonczar, J.; Hašimbegović, E.; Spannbauer, A.; Traxler, D.; Kastner, N.; Zlabinger, K.; Einzinger, P.; Pavo, N.; Goliasch, G.; Gyöngyösi, M. Circular RNAs in Cardiac Regeneration: Cardiac Cell Proliferation, Differentiation, Survival, and Reprogramming. *Front. Physiol.* **2020**, *11*, 580465. [CrossRef]
75. Wang, K.; Long, B.; Liu, F.; Wang, J.-X.; Liu, C.-Y.; Zhao, B.; Zhou, L.-Y.; Sun, T.; Wang, M.; Yu, T.; et al. A Circular RNA Protects the Heart from Pathological Hypertrophy and Heart Failure by Targeting MiR-223. *Eur. Heart J.* **2016**, *37*, 2602–2611. [CrossRef]
76. Geng, H.-H.; Li, R.; Su, Y.-M.; Xiao, J.; Pan, M.; Cai, X.-X.; Ji, X.-P. The Circular RNA Cdr1as Promotes Myocardial Infarction by Mediating the Regulation of MiR-7a on Its Target Genes Expression. *PLoS ONE* **2016**, *11*, e0151753. [CrossRef]
77. Peng, Y.; Xu, Y.; Zhang, X.; Deng, S.; Yuan, Y.; Luo, X.; Hossain, M.T.; Zhu, X.; Du, K.; Hu, F.; et al. A Novel Protein AXIN1-295aa Encoded by CircAXIN1 Activates the Wnt/β-Catenin Signaling Pathway to Promote Gastric Cancer Progression. *Mol. Cancer* **2021**, *20*, 158. [CrossRef]
78. Pei, B.; Sisu, C.; Frankish, A.; Howald, C.; Habegger, L.; Mu, X.J.; Harte, R.; Balasubramanian, S.; Tanzer, A.; Diekhans, M.; et al. The GENCODE Pseudogene Resource. *Genome Biol.* **2012**, *13*, R51. [CrossRef]
79. Jacq, C.; Miller, J.R.; Brownlee, G.G. A Pseudogene Structure in 5S DNA of Xenopus Laevis. *Cell* **1977**, *12*, 109–120. [CrossRef]
80. Cheetham, S.W.; Faulkner, G.J.; Dinger, M.E. Overcoming Challenges and Dogmas to Understand the Functions of Pseudogenes. *Nat. Rev. Genet.* **2020**, *21*, 191–201. [CrossRef]
81. Poliseno, L.; Salmena, L.; Zhang, J.; Carver, B.; Haveman, W.J.; Pandolfi, P.P. A Coding-Independent Function of Gene and Pseudogene MRNAs Regulates Tumour Biology. *Nature* **2010**, *465*, 1033–1038. [CrossRef]
82. Fiddes, I.T.; Lodewijk, G.A.; Mooring, M.; Bosworth, C.M.; Ewing, A.D.; Mantalas, G.L.; Novak, A.M.; van den Bout, A.; Bishara, A.; Rosenkrantz, J.L.; et al. Human-Specific NOTCH2NL Genes Affect Notch Signaling and Cortical Neurogenesis. *Cell* **2018**, *173*, 1356–1369.e22. [CrossRef]
83. Sweeney, B.A.; Petrov, A.I.; Ribas, C.E.; Finn, R.D.; Bateman, A.; Szymanski, M.; Karlowski, W.M.; Seemann, S.E.; Gorodkin, J.; Cannone, J.J.; et al. RNAcentral 2021: Secondary Structure Integration, Improved Sequence Search and New Member Databases. *Nucleic Acids Res.* **2021**, *49*, D212–D220. [CrossRef]
84. Kalvari, I.; Nawrocki, E.P.; Ontiveros-Palacios, N.; Argasinska, J.; Lamkiewicz, K.; Marz, M.; Griffiths-Jones, S.; Toffano-Nioche, C.; Gautheret, D.; Weinberg, Z.; et al. Rfam 14: Expanded Coverage of Metagenomic, Viral and MicroRNA Families. *Nucleic Acids Res.* **2021**, *49*, D192–D200. [CrossRef] [PubMed]
85. Kozomara, A.; Birgaoanu, M.; Griffiths-Jones, S. MiRBase: From MicroRNA Sequences to Function. *Nucleic Acids Res.* **2019**, *47*, D155–D162. [CrossRef] [PubMed]
86. Karagkouni, D.; Paraskevopoulou, M.D.; Chatzopoulos, S.; Vlachos, S.; Tastsoglou, S.; Kanellos, I.; Papadimitriou, D.; Kavakiotis, I.; Maniou, S.; Skoufos, G.; et al. DIANA-TarBase v8: A Decade-Long Collection of Experimentally Supported MiRNA. *Gene Interact.* **2018**, *46*, 239–245. [CrossRef]
87. Glažar, P.; Papavasileiou, P.; Rajewsky, N. CircBase: A Database for Circular RNAs. *RNA* **2014**, *20*, 1666–1670. [CrossRef] [PubMed]
88. Dudekula, D.B.; Panda, A.C.; Grammatikakis, I.; De, S.; Abdelmohsen, K.; Gorospe, M. CircInteractome: A Web Tool for Exploring Circular RNAs and Their Interacting Proteins and MicroRNAs. *RNA Biol.* **2016**, *13*, 34–42. [CrossRef]
89. Yoshihama, M.; Nakao, A.; Kenmochi, N. SnOPY: A Small Nucleolar RNA Orthological Gene Database. *BMC Res. Notes* **2013**, *6*, 1. [CrossRef]
90. Zhao, L.; Wang, J.; Li, Y.; Song, T.; Wu, Y.; Fang, S.; Bu, D.; Li, H.; Sun, L.; Pei, D.; et al. NONCODEV6: An Updated Database Dedicated to Long Non-Coding RNA Annotation in Both Animals and Plants. *Nucleic Acids Res.* **2021**, *49*, D165–D171. [CrossRef]
91. Karagkouni, D.; Paraskevopoulou, M.D.; Tastsoglou, S.; Skoufos, G.; Karavangeli, A.; Pierros, V.; Zacharopoulou, E.; Hatzigeorgiou, A.G. DIANA-LncBase v3: Indexing Experimentally Supported MiRNA Targets on Non-Coding Transcripts. *Nucleic Acids Res.* **2020**, *48*, D101–D110. [CrossRef]
92. Bueno Marinas, M.; Celeghin, R.; Cason, M.; Bariani, R.; Frigo, A.C.; Jager, J.; Syrris, P.; Elliott, P.M.; Bauce, B.; Thiene, G.; et al. A MicroRNA Expression Profile as Non-Invasive Biomarker in a Large Arrhythmogenic Cardiomyopathy Cohort. *Int. J. Mol. Sci.* **2020**, *21*, 1536. [CrossRef]
93. Khudiakov, A.A.; Panshin, D.D.; Fomicheva, Y.; Knyazeva, A.A.; Simonova, K.A.; Lebedev, D.S.; Mikhaylov, E.N.; Kostareva, A.A. Different Expressions of Pericardial Fluid MicroRNAs in Patients with Arrhythmogenic Right Ventricular Cardiomyopathy and Ischemic Heart Disease Undergoing Ventricular Tachycardia Ablation. *Front. Cardiovasc. Med.* **2021**, *8*, 647812. [CrossRef]
94. Che, H.; Wang, Y.; Li, H.; Li, Y.; Sahil, A.; Lv, J.; Liu, Y.; Yang, Z.; Dong, R.; Xue, H.; et al. Melatonin Alleviates Cardiac Fibrosis via Inhibiting LncRNA MALAT1/MiR-141-mediated NLRP3 Inflammasome and TGF-β1/Smads Signaling in Diabetic Cardiomyopathy. *FASEB J.* **2020**, *34*, 5282–5298. [CrossRef]
95. Liu, S.-X.; Zheng, F.; Xie, K.-L.; Xie, M.-R.; Jiang, L.-J.; Cai, Y. Exercise Reduces Insulin Resistance in Type 2 Diabetes Mellitus via Mediating the LncRNA MALAT1/MicroRNA-382-3p/Resistin Axis. *Mol. Ther. Nucleic Acids* **2019**, *18*, 34–44. [CrossRef]

96. Gao, L.; Wang, X.; Guo, S.; Xiao, L.; Liang, C.; Wang, Z.; Li, Y.; Liu, Y.; Yao, R.; Liu, Y.; et al. LncRNA HOTAIR Functions as a Competing Endogenous RNA to Upregulate SIRT1 by Sponging MiR-34a in Diabetic Cardiomyopathy. *J. Cell Physiol.* **2019**, *234*, 4944–4958. [CrossRef]

97. Yang, F.; Qin, Y.; Wang, Y.; Li, A.; Lv, J.; Sun, X.; Che, H.; Han, T.; Meng, S.; Bai, Y.; et al. LncRNA KCNQ1OT1 Mediates Pyroptosis in Diabetic Cardiomyopathy. *Cell Physiol. Biochem.* **2018**, *50*, 1230–1244. [CrossRef]

98. Zhang, H.; Liu, S.; Tang, L.; Ge, J.; Lu, X. Long Non-Coding RNA (LncRNA) MRPL23-AS1 Promotes Tumor Progression and Carcinogenesis in Osteosarcoma by Activating Wnt/β-Catenin Signaling via Inhibiting MicroRNA MiR-30b and Upregulating Myosin Heavy Chain 9 (MYH9). *Bioengineered* **2021**, *12*, 162–171. [CrossRef]

99. Yoon, J.-H.; Abdelmohsen, K.; Srikantan, S.; Yang, X.; Martindale, J.L.; De, S.; Huarte, M.; Zhan, M.; Becker, K.G.; Gorospe, M LincRNA-P21 Suppresses Target MRNA Translation. *Mol. Cell* **2012**, *47*, 648–655. [CrossRef]

100. Zhang, G.; Song, W. Long Non-Coding RNA LSINCT5 Inactivates Wnt/β-Catenin Pathway to Regulate MCF-7 Cell Proliferation and Motility through Targeting the MiR-30a. *Ann. Transl. Med.* **2020**, *8*, 1635. [CrossRef]

101. Yang, J.; Liu, Y.; Mai, X.; Lu, S.; Jin, L.; Tai, X. STAT1-Induced Upregulation of LINC00467 Promotes the Proliferation Migration of Lung Adenocarcinoma Cells by Epigenetically Silencing DKK1 to Activate Wnt/β-Catenin Signaling Pathway. *Biochem. Biophys. Res. Commun.* **2019**, *514*, 118–126. [CrossRef]

102. Farooqi, A.A.; Mukhanbetzhanovna, A.A.; Yilmaz, S.; Karasholakova, L.; Yulaevna, I.M. Mechanistic Role of DANCR in the Choreography of Signaling Pathways in Different Cancers: Spotlight on Regulation of Wnt/β-Catenin and JAK/STAT Pathways by Oncogenic Long Non-Coding RNA. *Noncoding RNA Res.* **2021**, *6*, 29–34. [CrossRef]

103. Pan, L.; Liang, W.; Gu, J.; Zang, X.; Huang, Z.; Shi, H.; Chen, J.; Fu, M.; Zhang, P.; Xiao, X.; et al. Long Noncoding RNA DANCR Is Activated by SALL4 and Promotes the Proliferation and Invasion of Gastric Cancer Cells. *Oncotarget* **2018**, *9*, 1915–1930. [CrossRef]

104. Yu, J.E.; Ju, J.A.; Musacchio, N.; Mathias, T.J.; Vitolo, M.I. Long Noncoding RNA DANCR Activates Wnt/β-Catenin Signaling through MiR-216a Inhibition in Non-Small Cell Lung Cancer. *Biomolecules* **2020**, *10*, 1646. [CrossRef]

105. Zhu, H.; Bai, W.; Ye, X.; Yang, A.; Jia, L. Long Non-Coding RNA UCA1 Desensitizes Breast Cancer Cells to Trastuzumab by Impeding MiR-18a Repression of Yes-Associated Protein 1. *Biochem. Biophys. Res. Commun.* **2018**, *496*, 1308–1313. [CrossRef]

106. Yan, H.; Li, H.; Silva, M.A.; Guan, Y.; Yang, L.; Zhu, L.; Zhang, Z.; Li, G.; Ren, C. LncRNA FLVCR1-AS1 Mediates MiR-513/YAP1 Signaling to Promote Cell Progression, Migration, Invasion and EMT Process in Ovarian Cancer. *J. Exp. Clin. Cancer Res.* **2019**, *38*, 356. [CrossRef]

107. Wang, J.; Huang, F.; Shi, Y.; Zhang, Q.; Xu, S.; Yao, Y.; Jiang, R. RP11-323N12.5 Promotes the Malignancy and Immunosuppression of Human Gastric Cancer by Increasing YAP1 Transcription. *Gastric. Cancer* **2021**, *24*, 85–102. [CrossRef]

108. Wu, N.; Yuan, Z.; Du, K.Y.; Fang, L.; Lyu, J.; Zhang, C.; He, A.; Eshaghi, E.; Zeng, K.; Ma, J.; et al. Translation of Yes-Associated Protein (YAP) Was Antagonized by Its Circular RNA via Suppressing the Assembly of the Translation Initiation Machinery. *Cell Death Differ.* **2019**, *26*, 2758–2773. [CrossRef]

109. Zhou, B.; Yu, J.-W. A Novel Identified Circular RNA, CircRNA_010567, Promotes Myocardial Fibrosis via Suppressing MiR-141 by Targeting TGF-B1. *Biochem. Biophys. Res. Commun.* **2017**, *487*, 769–775. [CrossRef]

110. Yuan, S.X.; Wang, J.; Yang, F.; Tao, Q.F.; Zhang, J.; Wang, L.; Yang, Y.; Liu, H.; Wang, Z.G.; Xu, Q.G.; et al. Long Noncoding RNA DANCR Increases Stemness Features of Hepatocellular Carcinoma by Derepression of CTNNB1. *Hepatology* **2016**, *63*, 499–511. [CrossRef]

111. Choe, M.H.; Yoon, Y.; Kim, J.; Hwang, S.-G.; Han, Y.-H.; Kim, J.-S. MiR-550a-3-5p Acts as a Tumor Suppressor and Reverses BRAF Inhibitor Resistance through the Direct Targeting of YAP. *Cell Death Dis.* **2018**, *9*, 640. [CrossRef] [PubMed]

112. Zhao, L.; Han, S.; Hou, J.; Shi, W.; Zhao, Y.; Chen, Y. The Local Anesthetic Ropivacaine Suppresses Progression of Breast Cancer by Regulating MiR-27b-3p/YAP Axis. *Aging* **2021**, *13*, 16341–16352. [CrossRef] [PubMed]

113. Chen, R.; Qian, Z.; Xu, X.; Zhang, C.; Niu, Y.; Wang, Z.; Sun, J.; Zhang, X.; Yu, Y. Exosomes-Transmitted MiR-7 Reverses Gefitinib Resistance by Targeting YAP in Non-Small-Cell Lung Cancer. *Pharmacol. Res.* **2021**, *165*, 105442. [CrossRef] [PubMed]

114. Higashi, T.; Hayashi, H.; Ishimoto, T.; Takeyama, H.; Kaida, T.; Arima, K.; Taki, K.; Sakamoto, K.; Kuroki, H.; Okabe, H.; et al. MiR-9-3p Plays a Tumour-Suppressor Role by Targeting TAZ (WWTR1) in Hepatocellular Carcinoma Cells. *Br. J. Cancer* **2015**, *113*, 252–258. [CrossRef] [PubMed]

115. Chen, X.; Wang, A.-L.; Liu, Y.-Y.; Zhao, C.-X.; Zhou, X.; Liu, H.-L.; Lin, M.-B. MiR-429 Involves in the Pathogenesis of Colorectal Cancer via Directly Targeting LATS2. *Oxid. Med. Cell Longev.* **2020**, *2020*, 5316276. [CrossRef] [PubMed]

Article

Genetic Variations *miR-10a*A>T, *miR-30c*A>G, *miR-181a*T>C, and *miR-499b*A>G and the Risk of Recurrent Pregnancy Loss in Korean Women

Hui-Jeong An [1,2,†], Sung-Hwan Cho [1,3,†], Han-Sung Park [1], Ji-Hyang Kim [4], Young-Ran Kim [4], Woo-Sik Lee [5], Jung-Ryeol Lee [6], Seong-Soo Joo [2], Eun-Hee Ahn [4,*] and Nam-Keun Kim [1,*]

1   Department of Biomedical Science, College of Life Science, CHA University, Seongnam 13488, Korea
2   College of Life Science, Gangneung-Wonju National University, Gangneung 25457, Korea
3   Department of Otolaryngology-Head and Neck Surgery, College of Medicine, Soonchunhyang University Cheonan Hospital, Cheonan 31151, Korea
4   Department of Obstetrics and Gynecology, CHA Bundang Medical Center, School of Medicine, CHA University, Seongnam 13488, Korea
5   Fertility Center of CHA Gangnam Medical Center, CHA University, Seoul 06135, Korea
6   Department of Obstetrics and Gynecology, Seoul National University Bundang Hospital, Seongnam 13620, Korea
*   Correspondence: bestob@chamc.co.kr (E.-H.A.); nkkim@cha.ac.kr (N.-K.K.)
†   These authors contributed equally to this work.

Citation: An, H.-J.; Cho, S.-H.; Park, H.-S.; Kim, J.-H.; Kim, Y.-R.; Lee, W.-S.; Lee, J.-R.; Joo, S.-S.; Ahn, E.-H.; Kim, N.-K. Genetic Variations *miR-10a*A>T, *miR-30c*A>G, *miR-181a*T>C, and *miR-499b*A>G and the Risk of Recurrent Pregnancy Loss in Korean Women. *Biomedicines* 2022, 10, 2395. https://doi.org/10.3390/biomedicines10102395

Academic Editors: Milena Rizzo and Elena Levantini

Received: 19 May 2022
Accepted: 20 September 2022
Published: 25 September 2022

Publisher's Note: MDPI stays neutral with regard to jurisdictional claims in published maps and institutional affiliations.

Abstract: This study investigated the genetic association between recurrent pregnancy loss (RPL) and microRNA (miRNA) polymorphisms in *miR-10a*A>T, *miR-30c*A>G, *miR-181a*T>C, and *miR-499b*A>G in Korean women. Blood samples were collected from 381 RPL patients and 281 control participants, and genotyping of *miR-10a*A>T, *miR-30c*A>G, *miR-181a*T>C, and *miR-499b*A>G was carried out by TaqMan miRNA RT-Real Time polymerase chain reaction (PCR). Four polymorphisms were identified, including *miR-10a*A>T, *miR-30c*A>G, *miR-181a*T>C, and *miR-499b*A>G. *MiR-10a dominant model* (AA vs. AT + TT) and *miR-499b*GG genotypes were associated with increased RPL risk (adjusted odds ratio [AOR] = 1.520, 95% confidence interval [CI] = 1.038–2.227, p = 0.032; AOR = 2.956, 95% CI = 1.168–7.482, p = 0.022, respectively). Additionally, both *miR-499* dominant (AA vs. AG + GG) and recessive (AA + AG vs. GG) models were significantly associated with increased RPL risk (AOR = 1.465, 95% CI = 1.062–2.020, p = 0.020; AOR = 2.677, 95% CI = 1.066–6.725, p = 0.036, respectively). We further propose that *miR-10a*A>T, *miR-30c*A>G, and *miR-499b*A>G polymorphisms effects could contribute to RPL and should be considered during RPL patient evaluation.

Keywords: recurrent pregnancy loss; single-nucleotide polymorphism (SNP); microRNA

1. Introduction

Recurrent pregnancy loss (RPL) is generally defined as three or more consecutive losses of pregnancy before 20 weeks of gestation. However, the American Society for Reproductive Medicine recently redefined RPL as more than two consecutive pregnancy losses [1]. Worldwide, RPL is a serious health problem that is significantly associated with morbidity and mortality. Factors contributing to the etiology of RPL include advanced maternal age, maternal anatomic anomalies, placental anomalies, chromosomal abnormalities, endocrine dysfunction, antiphospholipid syndrome, hereditary thrombophilia, psychological trauma, and environmental factors, such as smoking, excessive alcohol consumption, and stress [2]. Additionally, women who miscarry during their first pregnancy are 5% more likely to develop RPL than healthy women [3]. Although many relevant factors have been identified, the root cause of most cases of RPL remains unknown. RPL is also associated with blood clotting angiogenesis and immune disorders.

MicroRNAs (miRNAs) are small (approximately 23 nucleotides), noncoding, single stranded RNA molecules that form base pairs with complementary target messenger RNA (mRNAs) [4]. It has been demonstrated that miRNAs modulate gene expression via destabilization or translational repression of target mRNAs [5,6]. Furthermore, miRNAs have been implicated in the regulation of several biochemical pathways in various eukaryotic organisms [7,8]. RNA polymerase II transcribes miRNAs into long precursor transcript known as primary (pri)-miRNAs, which are subsequently converted into pre-miRNAs by DROSHA, which is a ribonuclease type III enzyme that forms a functional complex with Di George syndrome critical region 8 [9,10]. The pre-miRNA is then exported to the cytoplasm by the exportin5 (XPO5)-RAS–related nuclear protein (RAN)-guanosine-5'-triphosphate (GTP) complex [11]. RAN is a small GTP-binding protein, and the RAN GTPase-XPO5 complex forms a heterotrimer with the pre-miRNA [12]. The pre-miRNA is processed by RNase III DICER to release the miRNA duplex, which is a double-stranded RNA approximately 23 nucleotides in length. DICER also initiates the formation of the RNA-induced silencing complex (RISC) [13], which is responsible for miRNA-mediated gene silencing and RNA interference. The biological function of the miRNA is initiated by binding to the 3'-untranslated region (UTR) of the target mRNA, thereby repressing its expression. A single miRNA can regulate the expression of multiple target mRNAs, thus serving as a master controller of gene expression.

Multiple studies have recently demonstrated the roles of miRNAs in the pathophysiology of several ovarian diseases, including polycystic ovary syndrome (PCOS) and primary ovarian insufficiency (POI) [14,15]. POI, which is also known as premature ovarian failure, is characterized by insufficient or premature depletion of ovarian reserves, which leads to infertility [16]. The findings of the present study suggest that miRNAs play an essential role in the normal function and regulation of reproductive organs.

The expression of a given gene may be affected or regulated by its genetic variations, and single-nucleotide polymorphisms (SNPs) are the most common genetic variation affecting DNA [17]. SNPs or mutations in genes encoding miRNAs can affect miRNA properties, resulting in their altered expression and/or maturation [18]. Sequence variations around the processing sites of miRNAs or in the mature miRNA itself, particularly in the seed sequence, can profoundly affect miRNA biogenesis and function [19]. Polymorphisms in pre-miRNAs were first reported in 2005 [20], and several studies on the associations of these polymorphisms have since been reported [21,22]. Aberrant miRNA expression has been implicated in numerous diseases; therefore, considerable research efforts are currently being made for miRNA-based therapies [23]. In the present study, we performed a database search and identified four SNPs in pre-form miRNAs: *miR-10a*A>T (rs3809783), *miR-30c*A>G (rs113749278), *miR-181a*T>C (rs16927589), and *miR-499b*A>G (rs3746444). All of these miRNAs are reportedly associated with various reproductive diseases [24–27]. Therefore, we hypothesized that the SNPs *miR-10a*, *miR-30c*, *miR-181a*, and *miR-499b* play a role in the development of RPL. The minor allele frequency of these SNPs is >5% in the Asian population; however, whether they are genetically associated with RPL or whether miRNA expression varies as a function of these pre-form polymorphisms remains unclear. We, therefore, investigated the correlation between RPL and these miRNA polymorphisms.

2. Materials and Methods

2.1. Study Participants

Blood samples were collected from 381 RPL patients (mean age ± standard deviation [SD], 33.00 ± 5.73 years) and 281 control participants (33.03 ± 4.36 years). Blood samples were collected prior to 20 weeks of gestation based on human chorionic gonadotropin (hCG) levels. The RPL patients were recruited from the Department of Obstetrics and Gynecology or the Fertility Center at the CHA Bundang Medical Center in Seongnam, South Korea between March 1999 and February 2010. Women in the control group were recruited from CHA Bundang Hospital and met the following criteria: history of at least one spontaneous pregnancy; current pregnancy; regular menstrual cycles; karyotype 46

XX; and no history of miscarriage. The study abided by the Declaration of Helsinki and was approved by the Institutional Review Board of CHA Bundang Medical Center (IRB approval no. BD2010-123D), and written informed consent was obtained from all participants. All RPL patients had suffered a minimum of two consecutive spontaneous miscarriages at an average gestational stage of 7.36 ± 1.93 weeks. Pregnancy loss was diagnosed based on the results of hCG tests, ultrasound, and/or physical examination before 20 weeks of gestation. None of the participants had a history of smoking or alcohol use. The following parameters were also measured: activated partial thromboplastin time (aPTT), body mass index (BMI), blood urea nitrogen (BUN), creatinine, estradiol (E2), follicle-stimulating hormone (FSH), luteinizing hormone (LH), platelet (PLT) count, and prothrombin time (PT), using participant blood samples.

Patients with the following conditions were excluded from the study: RPL or implantation failure due to hormonal, genetic, anatomic, infectious, autoimmune, or thrombotic causes. Anatomic causes were evaluated using hysterosalpingogram, hysteroscopy, computed tomography, and magnetic resonance imaging to detect intrauterine adhesions, septate uterus, and uterine fibroids. Hormonal causes, including hyperprolactinemia, luteal insufficiency, and thyroid disease, were evaluated by blood analyses. Infectious causes, such as the presence of Ureaplasma urealyticum or Mycoplasma hominis, were evaluated by bacterial culture. Autoimmune causes, including antiphospholipid syndrome or lupus, were evaluated using lupus anticoagulant and anticardiolipin antibodies. Thrombotic causes, such as thrombophilia, were evaluated by identification of protein C and S deficiencies and by detection of β-2-glycoprotein 1 antibodies.

2.2. Antibody Preparation

A total of 150 µL of whole blood and fluorochrome-labeled monoclonal antibodies against anti-CD3-FITC (1:100, 555339), anti-CD4-PE(1:100, 357404) anti-CD8-PE-cy5 (1:20, 344769) anti-CD19-APC (1:100, 392503), anti-CD56-PE-Cy7 (1;100, 392411) NK cells were added to each tube. All antibodies were obtained from Biolegend (San Diego, CA, USA). The tubes were vortexed and incubated in the dark at room temperature for 40 min. Next, 2 mL of Lyse solution (diluted 1:10; BD Bioscience, Sunnyvale, CA, USA) was added, and the tubes were vortexed again, incubated at room temperature for 30 min, and centrifuged at 1200 rpm for 5 min. The cells were then washed three times with 2 mL of PBS each wash, and the cells were suspended in 250 µL of PBS and analyzed by flow cytometry (BD Bioscience).

2.3. Chromosome Analysis

Chromosome analysis was conducted according to standard cytogenetic methods. Peripheral blood lymphocytes were cultured for 70 h, and then KaryoMAX Colcemid Solution (Gibco) was added when the chromosomes were at the metaphase stage. KCl (0.05 M) was added as a hypotonic agent, and the cells were fixed for harvest using a fixative formed by adding one volume of acetic acid to two volumes of methanol. Metaphase chromosome preparations obtained after cell culture were stained using the Giemsa-Trypsin-Giemsa (GTG) banding method.

2.4. Genotyping

Genomic DNA was extracted from anticoagulant-treated peripheral blood samples using a G-DEX Genomic DNA extraction kit (iNtRON Biotechnology, Seongnam, Korea) [28,29]. Briefly, Proteinase K was added to a microcentrifuge tube, followed by 30 µL of blood. Next, 300 µL of Lysis solution was added, and the samples were vortexed and incubated at 55 °C for 10 min. A total of 350 µL of ethanol was then added to each sample, and the samples were bound, washed, and eluted according to the manufacturer's protocol. Four miRNAs (SNPs) were selected using the NCBI human genome SNP database (dbSNP, http://www.ncbi.nlm.nih.gov/snp (accessed on 13 March 2019)). The SNPs *miR-10a*A>T (rs3809783), *miR-30c*A>G (rs113749278), *miR-181a*T>C (rs16927589), and *miR-499b*A>G (rs37464444) are either mature-form (rs3746444, rs-formnp8978) or pri-form (rs3809783, rs16927589). *miR-10a*A>T, *miR-30c*A>G, *miR-181a*T>C, and *miR-499b*A>G were genotyped according to TaqMan® SNP Genotyping Assays system (Applied Biosystems, Foster City, CA, USA). Based on the intensity of fluorescence signals of FAM and VIC, samples were automatically classified into one of three groups corresponding to the geno types AA, AG, or TT of *miR-10a*A>T; AA, AG, or GG of *miR-30c*A>G; TT, TC, or CC of *miR-181a*T>C; and AA, AG, or GG of *miR-499b*A>G. The basic principle of the assay is as follows: when the allele-specific probe is fully hybridized to the template DNA, Taq polymerase cleaves the reporter dye, leading to fluorescence emission. However, if a single base mismatch exists between the probe and template DNA, hybridization is inefficient, and reporter dye fluorescence is thus reduced. The sequences of the SNPs were as follows *miR-10a*A>T, CTCTT ATTTTTCCAG AAGAAAAAAA[A/T]ATATATATAT GTATATG TAG TATTT; *miR-30c*A>G, TACTTTCCACAGCTG AGAGTGTAGG[A/G]DTGTTTACAGT ATCTGTCGCT CAGTG; *miR-181a*T>C, AAAAT AGCACAAAAT TATCCAATTG[T/C] GACAGTTCTT ATCACATTTC ACTTT; and *miR-499b*A>G, ATGTTTAACT CCTCTC CACG TGAAC[A/G]TCACAGCAAG TCTGTGCTGC TTCCC. Information regarding the miRNA probes was as follows: *miR-10a*A>T, wild type homozygous AA (VIC reaction & FAM no reaction), heterozygous AT (VIC reaction & FAM reaction), mutant homozygous TT (VIC no reaction & FAM reaction); *miR-30c*A>G, wild homozygous AA (VIC reaction & FAM no reaction), heterozygous AG (VIC reaction & FAM reaction), mutant homozygous GG (VIC no reaction & FAM reaction); *miR-181a*T>C, wild homozygous TT (VIC reaction & FAM no reaction), heterozygous TC (VIC reaction & FAM reaction), mutant homozygous CC (VIC no reaction & FAM reaction); *miR-499b*A>G, wild homozygous AA (VIC reaction & FAM no reaction), heterozygous AG (VIC reaction & FAM reaction), mutant homozygous GG (VIC no reaction & FAM reaction).

2.5. Assessment of Plasminogen Activator Inhibitor (PAI-1), Homocysteine, Total Cholesterol, Uric Acid Levels, and Blood Coagulation Status

Plasma PAI-1, total cholesterol, uric acid, and homocysteine levels were measured in participant blood samples. Plasma was separated by centrifugation of whole blood at $1000 \times g$ for 15 min. PAI-1 levels were determined using a human serpin E1/PAI-1 immunoassay (R&D Systems, Minneapolis, MN, USA). Uric acid and total cholesterol levels were measured using enzymatic colorimetric tests (Roche Diagnostics, GmbH, Mannheim, Germany). Homocysteine levels were measured using a fluorescence polarization immunoassay with an Abbott IMx analyzer (Abbott Laboratories, Abbott Park, IL, USA).

2.6. Statistical Analyses

The significance of differences in the frequencies of the *miR-10a*A>T, *miR-30c*A>G, *miR-181a*T>C, and *miR-499b*A>G SNPs between the control and patient groups were assessed using Fisher's exact test and a logistic regression model. *p*-values were calculated using two-sided *t*-tests for continuous variables and chi-square tests for categorical variables. Allele frequencies were calculated to investigate the deviation from Hardy–Weinberg equilibrium. The genotype distribution of RPL patients and controls with $\geq$h or $\geq$o pregnancy loss was investigated. Odds ratios (ORs), adjusted odds ratios (AORs), and 95% confidence intervals (CIs) were used to examine the associations between various miRNA polymorphisms and RPL risk. Data are presented as the mean $\pm$ SD for continuous variables or a percentage for categorical variables. The results of the allele and genotype combination analysis were consistent with those derived from Fisher's exact test during regression analysis.

Statistical analyses were carried out using MedCalc software, version 12.1.4 (MedCalc Software bvba, Mariakerke, Belgium) or GraphPad Prism 4.0 software (GraphPad Software, Inc., San Diego, CA, USA). Logistic regression analysis was applied to data regarding baseline characteristics, genotype frequencies, genotype combinations, and allele combinations for quantitative traits shown in Table 2, Table 3, Table 4 and Table 5. The HAPSTAT program (v.3.0, www.bios.unc.edu/~lin/hapstat/ (accessed on 10 April 2018)), which exhibits a strong synergistic effect, was used to estimate the frequencies of polymorphic haplotypes. A p-value < 0.05 indicated statistical significance. HAPSTAT allows testing of haplotype (or allele combination) effects by maximizing the likelihood (from the observed data) that properly accounts for phase uncertainty and study design. False-positive discovery rate (FDR) correction was used to adjust multiple comparison tests and associations with FDR-adjusted p-values < 0.05 were considered statistically significant [30]. FDR calculation is also used for multiple hypotheses testing to correct for multiple comparisons. Multifactor dimensionality reduction (MDR) analysis was used to determine the best-model gene-gene interaction for RPL risk. The advantage of using MDR is that it overcomes the sample size limitations often encountered during logistic regression analysis in studies of high-level interactions. The MDR method consists of two main steps. First, the best combination of multi-factors is selected, and second, genotype combinations are classified into high- and low-risk groups [31]. We constructed all possible allelic combinations by MDR analysis to identify combinations with strong synergy. Allelic combinations for multiple loci were estimated using the expectation-maximization algorithm with SNPAlyze (v. 5.1; DYNA-COM Co, Ltd., Yokohama, Japan), and those having frequencies < 1% were excluded from statistical analysis. We also applied multiple regression models to further explain the results of the allelic combination analysis. Genetic interaction analyses were performed using the open-source MDR software package (v.2.0), which is available at www.epistasis.org (accessed on 15 March 2018).

2.7. Expression Vector Construction (miR-10aA>T, miR-30cA>G, and miR-181aT>C)

The pre-miRs *(miR-10a, miR-30c,* and *miR-181a)* and their flanking regions were amplified from human genomic DNA and cloned into the vector pcDNA3.1(−) (Invitrogen, Carlsbad, CA, USA). The primers used in the study included F: 5′-TGC GAA CTG GCT ACT TGA AA-3′, R: 5′-TTC CAA TAA AGC CTC CCT GA-3′ *(miR-10a)*; F: 5′-GCA CCA TGT GTC ACA CAG GT-3′, R: 5′-CAA GTG TTG GGA AGA TGC TAT-3′ *(miR-30c)*; and F: 5′-ACA TTT TCT CAG ACA TTC AT-3′, R: 5′-ATG TGA GAA AAC TGA GAC AC -3′ *(miR-181a)*. For single-point mutations, we used an Intron Muta-direct kit (Intron, Seoul, Korea). The sequences of these three vectors were confirmed by direct sequencing, and the SNPs were the only differences detected. To generate the miRNAs target gene::luciferase reporter constructs, similar to the cloning vectors, fragments of the *PAI-1* gene corresponding to the 3′-UTR region clone (OriGene, Rockville, MD, USA) were amplified and cloned into the pGL4.13-luciferase vector (Promega, Madison, WI, USA). The resulting cDNAs were PCR amplified using the following primers: forward 5′-CCC TGG GGA AAG ACG CCT T-3′ and reverse 5′-TTC GTA TTT ATT TAT TTT ATT TTT T-3′ with *Xba*I (TCTAGA)and *Fse*I (GGCCGGCC) linker (New England Biolabs, Ipswich, MA, USA), and all constructs were verified by sequencing. Cells from a human endometrial cell line (Ishikawa) were plated at 1×10^6 cells per well in 6-well plates and transfected 24 h later using JetPRIME transfection reagent (Polyplus, France). Transfection reactions for miR-10a contained 500 ng of miR10a-A (in pcDNA3.1-) or 500 ng of miR-10a-T (in pcDNA3.1-) with 500 ng of 3′-UTR-PAI-1 in pGL4.13 and 200 ng of pGL4.75 (Renilla-normalization control); for miR-30c, reactions contained 500 ng of miR-30c-A (in pcDNA3.1-) or 500 ng of miR-30c-G (in pcDNA3.1-) with 500 ng of 3′-UTR-PAI-1 in pGL4.13 and 200 ng of pGL4.75 (Renilla-normalization control), for miR-181a2, reactions contained 500 ng of miR-181a-T (in pcDNA3.1-) or 500 ng of miR-181a-G (in pcDNA3.1-) with 500 ng of 3′-UTR-PAI-1 in pGL4.13 and 200 ng of pGL4.75 (Renilla-normalization control).

2.8. Quantitative Real-Time PCR (miR-10a, miR-30c, miR-181a Pre- and Mature-Form Primers)

TRIzol reagent (Invitrogen, Waltham, Massachusetts, USA) was used to isolate total RNA from Ishikawa cells that were transfected with 2.5 µg of vector after 16 h. Total RNA was then reverse transcribed using an M-MLV reverse transcriptase PCR kit (Biofact, Co., Ltd., Daejeon, Korea) and random or oligo dT20 primers (Invitrogen, Waltham,

MA, USA) in addition to specific primers for *PAI-1* and *glyceraldehyde 3-phosphate dehydrogenase (GAPDH)*. Quantitative real-time PCR (qPCR) was performed as 20 μL reactions containing each sequence-specific primer and quantitative PCR master mix (Solgent, Co. Ltd., Daejeon, Korea), using a Rotor-Gene 6000 real-time PCR system (Qiagen, Co., Ltd. Hilden, Germany). Expression levels were calculated according to the comparative threshold cycle (Ct) method using the formula $2^{-\Delta\Delta Ct}$. Primer sequences for amplification were as follows: has-miR-10a-pre forward: 5′-CCG AAT TTG TGT AAG GAA TTT TG-3′ and reverse 5′-AAG AGC GGA GTG TTT ATG TCA A-3′; has-miR-10a-mature forward: 5′-TAC CCT GTAG ATC CGA ATT T and reverse: universal primer (Qiagen Cat# 218193); has-miR-30c-pre forward: 5′-TGT GTA AAC ATC CTA CAC TCT CAG C-3′ and reverse: 5′-CCA TGG CAG AAG GAG TAA ACA-3′; has-miR-30c-mature forward: 5′-AAA CAT CCT ACA CTC TCA GC-3′ and reverse universal primer (Qiagen Cat# 218193); has-*miR-181a*-pre forward:5′-TAT CAG CCC AGC CTT CAG AG-3′ and reverse: 5′-AAT CCC AAA CTC ACC GAC AG-3′; *miR-181a*-mature forward:5′- TTC AAC GCT GTC GGT GAG TT-3′ and reverse: universal primer (Qiagen Cat# 218193); Human RNU6B (RNU6-2) forward:5′-ACC CAA ATT CGT GAA GCG TT-3′ and reverse universal primer (Qiagen Cat# 218193).

2.9. Prediction of miRNA Binding and Luciferase Reporter Assay

An online search was conducted to identify targets for *miR-10a, miR-30c, miR-181a* and *miR-499b* using the TargetScan (http://www.targetscan.org (accessed on 21 May 2018) and miRIAD databases (http://bmi.ana.med.uni-muenchen.de/miriad/ (accessed on 16 May 2018)). We used these databases to predict miRNAs that target overlapping regions of *PAI-1* mRNA transcripts. Target mRNA sequences, particularly within the 3′-UTR, are often obtained from the National Center for Biotechnology Information (www.ncbi.nlm.nih.gov (accessed on 26 September 2018)). We found that *miR-30c, miR-10a*, and *miR-181a* were predicted to be targets of the *PAI-1* 3′-UTR. Therefore, a luciferase reporter assay was used to evaluate the roles of *miR-30c, miR-10a*, and *miR-181a* in regulating the expression of target genes, as previously described. Briefly, wild-type pGL4.13-luciferase vector (Promega, Madison, WI, USA). constructs containing the 3′-UTRs of the *PAI-1* gene were generated by amplifying the 3′-UTR region clone (OriGene, Rockville, MD, USA) and cloning the amplification products into the downstream region of the pGL4.13 vector (Promega, Madison, WI, USA) using the *Xba*I and *Fse*I endonucleases (New England BioLabs, Ipswich, MA, USA). Positive clones were selected by sequence-specific PCR, restriction enzyme digestion, and DNA sequencing. Ishikawa cells were cultured in Dulbecco's modified Eagle's medium (DMEM) (Thermo Fisher Scientific, Inc. Waltham, Massachusetts, USA). All medium was supplemented with 10% fetal bovine serum (FBS) (Thermo Fisher Scientific, Inc. Waltham, Massachusetts, USA) and 1% penicillin/streptomycin (Thermo Fisher Scientific, Inc. Waltham, Massachusetts, USA). All cell lines were maintained in a CO_2 incubator (5% CO_2) at 37 °C. The Ishikawa cells used in this study were endometrial and are commonly used in RPL studies. Next, *miR-10a, miR-30c*, and *miR-181a* mimics (50 nM) were co-transfected into Ishikawa cells with 200 ng of the 3′-UTR of *PAI-1* in pGL4.13 constructs using lipofectamine 2000 (Invitrogen, Carlsbad, CA, USA). After 16 h of incubation, the luciferase activity was measured using a dual-luciferase reporter assay system (Promega Madison, WI, USA). Each transfection was performed as triplicates.

3. Results

3.1. Baseline Characteristics of Recurrent Pregnancy Loss Patients and Control Subjects

The characteristics of RPL patients and control subjects are summarized in Table 1. The mean age was approximately 33 years for both groups, and both groups were 100% female. PLT count, aPTT, and concentrations of E2 and LH were greater in RPL patients than in controls ($p = 0.0007$, $p = 0.005$, $p = 0.001$, and $p = 0.011$, respectively). There were no significant differences in age, BMI, uric acid level, or FSH level between the two groups.

Table 1. Baseline characteristics of recurrent pregnancy loss patients and control subjects.

Characteristics	Controls (n = 281)	RPL Patients (n = 381)	p *
Age (years, mean ± SD)	33.00 ± 5.73	33.03 ± 4.36	0.94
BMI (kg/m^2, mean ± SD)	21.58 ± 3.18	21.35 ± 4.04	0.558
Previous pregnancy losses	None	3.01 ± 1.50	
Average no. of gestational weeks	39.28 ± 1.67	7.36 ± 1.93	<0.0001
CD56 NK cells (%, mean ± SD)	None	18.12 ± 7.98	
Homocysteine (μmol/L, mean ± SD)	None	6.98 ± 2.10	
Folate (nmol/L, mean ± SD)	None	14.18 ± 12.01	
Total cholesterol (mg/dL, mean ± SD)	None	187.73 ± 49.41	
Uric acid (mg/dL, mean ± SD)	4.19 ± 1.44	3.80 ± 0.83	0.172
PLT (10^3/μL, mean ± SD)	235.17 ± 63.60	255.43 ± 59.22	0.0007
aPTT (sec, mean ± SD)	30.77 ± 4.60	32.23 ± 4.32	0.005
PAI-1 (ng/mL)	None	10.53 ± 5.72	
BUN (mg/dL)	None	9.98 ± 2.76	
Creatinine (mg/dL)	None	0.72 ± 0.12	
FSH (mIU/mL)	8.11 ± 2.84	7.51 ± 10.54	0.557
LH (mIU/mL)	3.32 ± 1.74	6.32 ± 12.11	0.011
E2 (pg/mL)	26.00 ± 14.74	35.64 ± 29.53	0.001
PT (sec, mean ± SD)	11.53 ± 3.10	11.58 ± 0.85	0.84

Abbreviations: aPTT, activated partial thromboplastin time; BMI, body mass index; BUN, blood. urea nitrogen; E2, estradiol; FSH, follicle-stimulating hormone; LH, luteinizing hormone; PLT, platelets; PT, prothrombin time; RPL, recurrent pregnancy loss; SD, standard deviation. * p-values were calculated by a two-sided t-test for continuous variables and a chi-square test for categorical variables.

3.2. Genotype Frequencies of miRNA Polymorphisms According to the Number of Recurrent Pregnancy Losses

Table 2 shows the distribution of genotypes in RPL patients with ≥3 or ≥4 pregnancy losses and control subjects. Significant differences in the *miR-10a* SNP were observed between the RPL and control groups and were significantly correlated with RPL prevalence. Consistently, the absence of these miRNA polymorphisms showed a negative correlation with RPL. The associations of these polymorphisms were very interesting in RPL patients because the miRNA polymorphisms were related to decreased RPL, but they were not associated with RPL risk (Table 2). In addition, the number of RPL patients with risk factors was very small. Therefore, the associations with RPL occurrence will require further investigation. *miR-10a*A>T (chr17:48579816, rs3809783), *miR-30c*A>G (chr6:71377017, rs113749278), *miR-181a*T>C (chr9:124692981, rs16927589), and *miR-499b*A>G (chr20: 34990400, rs37464444) were all in the miRNA mature-form (rs3746444, rs-formnp8978) or pri-form (rs3809783, rs16927589). The SNPs in miRNA genes, including pri-miRNAs, pre-miRNAs, and mature miRNAs, could potentially influence the processing and/or target selection of miRNAs. Since we selected four SNPs in pri-form or mature-form, we wanted to determine whether all these miRNAs could influence the expression and regulation of target genes. Based on the intensity of FAM and VIC fluorescence, samples were automatically classified into one of three groups corresponding to genotypes AA, AT, or TT of *miR-10a*A>T; AA, AT, or GG of *miR-30c*A>G; TT, TC, or CC of *miR-181a*T>C; and AA, AG, or GG of *miR-499b*A>G.

Table 2. Genotype frequencies of miRNA gene polymorphisms in control subjects and recurrent pregnancy loss patients.

Genotype	Controls (n = 281)	RPL Patients (n = 381)	AOR (95% CI)	p	FDR-p	PL ≥ 3 (n = 201)	AOR (95% CI)	p	FDR-p	PL ≥ 4 (n = 81)	AOR (95% CI)	p	FDR-p
miR-10aA>T													
AA	230 (81.9)	285 (74.8)	1.000 (reference)			151 (75.1)	1.000 (reference)			60 (74.1)	1.000 (reference)		
AT	50 (17.8)	88 (23.1)	1.420(0.963–2.094)	0.077	0.584	44 (21.9)	1.365 (0.865–2.154)	0.181	0.362	19 (23.5)	1.470 (0.805–2.683)	0.21	0.21
TT	1 (0.4)	8 (2.1)	6.476(0.804–52.176)	0.079	0.237	6 (3.0)	9.484 (1.128–79.759)	0.038	0.274	2 (2.5)	7.931 (0.705–89.228)	0.094	0.156
Dominant (AA vs. AT+TT)			1.520(1.038–2.227)	0.032	0.709		1.524 (0.979–2.372)	0.062	0.124		1.595 (0.889–2.862)	0.117	0.156
Recessive (AA+AT vs. TT)			6.003(0.746–48.285)	0.092	0.276		8.847 (1.055–74.186)	0.045	0.631		7.206 (0.644–80.641)	0.109	0.156
HWE P	0.318	0.695				0.217				0.738			
miR-30cA>G													
AA	106 (37.7)	163 (42.8)	1.000 (reference)			130 (64.7)	1.000 (reference)			52 (64.2)	1.000 (reference)		
AG	144 (51.2)	182 (47.8)	0.821(0.591–1.139)	0.237	0.237	64 (31.8)	1.044 (0.647–1.686)	0.86	0.994	27 (33.3)	1.170 (0.640–2.139)	0.611	0.994
GG	32 (11.0)	36 (9.4)	0.742(0.432–1.276)	0.281	0.422	7 (3.5)	-	0.994	0.994	2 (2.5)	-	0.994	0.994
Dominant (AA vs. AG+GG)			0.810(0.591–1.111)	0.191	0.191		1.162 (0.724–1.864)	0.534	0.994		1.262 (0.696–2.288)	0.443	0.994
Recessive (AA+AG vs. GG)			0.842(0.506–1.400)	0.507	0.606		-	0.994	0.994		-	0.994	0.994
HWE P	0.104	0.144											
miR-181aT>C													
TT	198 (70.5)	247 (64.8)	1.000 (reference)			79 (39.3)	1.000 (reference)			32 (39.5)	1.000 (reference)		
TC	78 (27.8)	125 (32.8)	1.286(0.916–1.805)	0.147	0.221	104 (51.7)	1.376 (0.866–2.185)	0.177	0.223	42 (51.9)	1.403 (0.779–2.525)	0.259	0.345
CC	5 (1.8)	9 (2.4)	1.483(0.488–4.509)	0.487	0.487	18 (9.0)	2.094 (0.812–5.399)	0.126	0.223	7 (8.6)	2.035 (0.621–6.665)	0.241	0.345
Dominant (TT vs. TC+CC)			1.294(0.929–1.803)	0.128	0.191		1.448 (0.925–2.269)	0.106	0.223		1.462 (0.826–2.585)	0.192	0.345
Recessive (TT+TC vs. CC)			1.337(0.443–4.037)	0.606	0.606		1.764 (0.707–4.400)	0.223	0.223		1.664 (0.548–5.057)	0.369	0.369
HWE P	0.393	0.137											
miR-499bA>G													
AA	188 (66.9)	221 (58.0)	1.000 (reference)			116 (57.7)	1.000 (reference)			46 (56.8)	1.000 (reference)		
AG	87 (31.0)	139 (36.5)	1.361(0.977–1.896)	0.068	0.204	77 (38.3)	2.037 (1.240–3.347)	0.005	0.01	34 (42.0)	2.274 (1.241–4.168)	0.008	0.016
GG	6 (2.1)	21 (5.5)	2.956(1.168–7.482)	0.022	0.066	8 (4.0)	3.890 (0.767–19.730)	0.101	0.135	1 (1.2)	1.970 (0.152–25.590)	0.604	0.805
Dominant (AA vs. AG+GG)			1.465(1.062–2.020)	0.02	0.06		2.136 (1.314–3.472)	0.002	0.008		2.259 (1.240–4.114)	0.008	0.016
Recessive (AA+AG vs. GG)			2.677(1.066–6.725)	0.036	0.108		2.998 (0.605–14.857)	0.179	0.179		1.361 (0.111–16.739)	0.81	0.81
HWE P	0.263	0.888											

Abbreviations: AOR, adjusted odds ratio; CI, confidence interval; FDR-P, false-positive discovery rate-corrected; PL, pregnancy loss; RPL, recurrent pregnancy loss.

3.3. Adjusted Odds Ratios for Risk of Recurrent Pregnancy Loss Associated with miRNA Polymorphisms Combined with Clinical Factors

The *miR-30c*AG+GG genotype was associated with decreased risk of RPL for age < 33 years (odds ratio [OR] = 0.583; 95% confidence interval [CI] = 0.371–0.918; p = 0.022) (Table 3). However, the *miR-181a*TC+CC genotype was associated with increased risk of RPL for age < 33 years (OR = 1.677; 95% CI = 1.038–2.709; p = 0.035), and the *miR-499b*AG + GG genotype was associated with increased risk of RPL for age $\geq$ 33 years (OR = 1.631; 95% CI = 1.028–2.588; p = 0.038). The *miR-10a*AT+TT genotype was associated with increased risk of RPL for BMI $\geq$ 25 kg/m^2 (OR = 2.840; 95% CI = 1.544–5.223; p = 0.001). The *miR-499b*AG + GG genotype was associated with increased risk of RPL for BMI <25 kg/m^2 (OR = 1.456; 95% CI = 1.029–2.059; p = 0.034) and with increased risk of RPL for BMI $\geq$25 kg/m^2 (OR = 2.284; 95% CI = 1.377–3.789; p = 0.001). The *miR-181a*TC + CC genotype was associated with increased risk of RPL for PLT count <255.62$\times 10^3$/µL) (OR = 1.779; 95% CI = 1.038–3.048; p = 0.036). Finally, the *miR-30c*AG+GG was associated with decreased risk of RPL for aPTT < 32.83 s (OR = 0.364; 95% CI = 0.185–0.717; p = 0.004).

Table 3. Adjusted odds ratios for risk of recurrent pregnancy loss associated with miRNA polymorphisms combined with clinical factors.

Variable	*miR-10a*AT + TT		*miR-30c*AG + GG		*miR-181a*TC + CC		*miR-499b*AG + GG	
	AOR (95% CI)	p	AOR (95% CI)	p	AOR (95% CI)	p	AOR (95%CI)	p
Age (years)								
<33	1.476 (0.870–2.505)	0.149	0.583 (0.371–0.918)	0.02	1.677 (1.038–2.709)	0.035	1.329 (0.848–2.083)	0.216
$\geq$33	1.566 (0.902–2.717)	0.111	1.124 (0.721–1.754)	0.606	0.996 (0.626–1.583)	0.985	1.631 (1.028–2.588)	0.038
Homocysteine								
<6.97 µmol/L	1.186 (0.127–11.086)	0.881	0.364 (0.040–3.344)	0.372	-	-	3.063 (0.333–28.174)	0.323
$\geq$6.97 µmol/L	1.690 (0.190–15.041)	0.638	0.275 (0.032–2.399)	0.243	0.590 (0.124–2.816)	0.509	0.846 (0.177–4.050)	0.835
BMI								
<25 kg/m^2	1.399 (0.928–2.108)	0.109	0.834 (0.593–1.174)	0.298	1.401 (0.979–2.005)	0.065	1.456 (1.029–2.059)	0.034
$\geq$25 kg/m^2	2.840 (1.544–5.223)	0.001	0.949 (0.591–1.524)	0.829	1.194 (0.725–1.967)	0.485	2.284 (1.377–3.789)	0.001
Platelet								
<255.62 $\times 10^3$/µL	1.133 (0.624–2.057)	0.681	1.008 (0.606–1.678)	0.976	1.779 (1.038–3.048)	0.036	1.468 (0.878–2.455)	0.144
$\geq$255.62 $\times 10^3$/µL	2.019 (0.933–4.370)	0.075	0.539 (0.287–1.011)	0.054	0.820 (0.429–1.569)	0.55	1.256 (0.665–2.369)	0.483
PT								
$\geq$11.58 s	1.476 (0.870–2.505)	0.149	1.845 (0.468–7.277)	0.382	0.557 (0.145–2.139)	0.394	0.368 (0.090–1.514)	0.166
<11.58 s	1.566 (0.902–2.717)	0.111	0.699 (0.335–1.458)	0.339	1.031 (0.495–2.151)	0.935	1.023 (0.522–2.006)	0.947
aPTT								
<32.83 s	1.476 (0.870–2.505)	0.149	0.364 (0.185–0.717)	0.004	1.714 (0.862–3.409)	0.125	1.409 (0.763–2.604)	0.273
$\geq$32.83 s	1.566 (0.902–2.717)	0.111	0.976 (0.426–2.237)	0.954	1.069 (0.459–2.493)	0.877	0.639 (0.284–1.439)	0.279

Abbreviations: aPTT, activated partial thromboplastin time; AOR, adjusted odds ratio; BMI, body mass index; CI, confidence interval; PT, prothrombin time; RPL, recurrent pregnancy loss. The aPTT was below the 15% cut-off level in RPL patients and controls. Platelets were above the 15% cut-off level in RPL patients and controls.

3.4. Combination Analysis of miRNA Polymorphisms between Recurrent Pregnancy Loss Patients and Control Subjects

The results of combined gene-genotype analyses are shown in Table 4. The *miR-10a/miR-30c* combined genotype AT/AG was associated with increased RPL risk (OR = 2.156; 95% CI = 1.120–4.151; p = 0.022). The *miR-10a*A>T/*miR-181a*T>C combined genotype AT/TT was associated with increased RPL risk (OR = 1.974; 95% CI = 1.065–3.658; p = 0.031). The *miR-10a*A>T/*miR-499*A>G combined genotype AT/AG was associated with increased RPL risk (OR = 2.195; 95% CI = 1.156–4.169; p = 0.016). The *miR-30c*A>G/*miR-181a*T>C combined genotype AG/TT was also associated with increased RPL risk (OR = 1.839; 95% CI = 1.054–3.210; p = 0.032). Similarly, increased RPL risk was associated with the *miR-30c*A>G/*miR-499*A>G combined genotypes AA/GG (OR = 4.324; 95% CI = 1.423–13.141; p = 0.010) and AG/AG (OR = 1.921; 95% CI = 1.145–3.224; p = 0.013). The *miR-181a*T>C/*miR-499*A>G combined genotype TT/GG was also associated with increased RPL risk (OR = 8.320; 95% CI = 1.043–66.384; p = 0.046). However, after false-discovery rate (FDR)-p correction, there were no significant differences between RPL patients and controls in the ORs for the combined genotypes, except for the *miR-30c*A>G/*miR-499*A>G combined genotypes AA/GG and AG/AG.

Table 4. Combination analysis of miRNA polymorphisms between recurrent pregnancy loss patient and control subjects.

Genotype Combination	Controls (n = 281)	RPL Patients (n = 381)	AOR (95% CI)	p [a]	FDR-p [b]
*miR-10a*A>T/*miR-30c*A>G					
AA/AA	162 (57.7)	190 (49.9)	1.000 (reference)		
AT/AA	35 (12.5)	51 (13.4)	1.243 (0.770–2.008)	0.374	0.499
AT/AG	14 (5.0)	35 (9.2)	2.156 (1.120–4.151)	0.022	0.088
AT/GG	1 (0.4)	2 (0.5)	1.770 (0.159–19.773)	0.643	0.643
TT/AA	1 (0.4)	6 (1.6)	4.958 (0.589–41.748)	0.141	0.282
*miR-10a*A>T/*miR-181a*T>C					
AA/TT	88 (31.3)	113 (29.7)	1.000 (reference)		
AA/TC	121 (43.1)	142 (37.3)	0.915 (0.632–1.324)	0.637	0.812
AA/CC	21 (7.5)	30 (7.9)	1.079 (0.576–2.020)	0.812	0.812
AT/TT	18 (6.4)	45 (11.8)	1.974 (1.065–3.658)	0.031	0.124
AT/TC	22 (7.8)	37 (9.7)	1.272 (0.698–2.317)	0.433	0.812
*miR-10a*A>T/*miR-499*A>G					
AA/AA	154 (54.8)	168 (44.1)	1.000 (reference)		
AA/AG	71 (25.3)	102 (26.8)	1.317 (0.907–1.914)	0.148	0.197
AA/GG	5 (1.8)	15 (3.9)	2.719 (0.964–7.665)	0.059	0.118
AT/AA	34 (12.1)	47 (12.3)	1.264 (0.772–2.069)	0.351	0.351
AT/AG	15 (5.3)	36 (9.4)	2.195 (1.156–4.169)	0.016	0.064
AT/GG	1 (0.4)	5 (1.3)	4.508 (0.520–39.109)	0.172	0.344
TT/AG	1 (0.4)	1 (0.3)	0.922 (0.057–14.874)	0.954	0.954
*miR-30c*A>G/*miR-181a*T>C					
AA/TT	79 (28.1)	101 (26.5)	1.000 (reference)		
AA/TC	89 (31.7)	120 (31.5)	1.058 (0.707–1.583)	0.784	0.784
AA/CC	30 (10.7)	26 (6.8)	0.692 (0.377–1.268)	0.233	0.466
AG/TT	25 (8.9)	58 (15.2)	1.839 (1.054–3.210)	0.032	0.128
AG/TC	52 (18.5)	59 (15.5)	0.886 (0.551–1.425)	0.617	0.784
*miR-30c*A>G/*miR-499*A>G					
AA/AA	137 (48.8)	146 (38.3)	1.000 (reference)		
AA/AG	57 (20.3)	83 (21.8)	1.355 (0.898–2.044)	0.147	0.196
AA/GG	4 (1.4)	18 (4.7)	4.324 (1.423–13.141)	0.01	0.026
AG/AA	49 (17.4)	68 (17.8)	1.319 (0.852–2.041)	0.214	0.214
AG/AG	27 (9.6)	55 (14.4)	1.921 (1.145–3.224)	0.013	0.026
AG/GG	2 (0.7)	2 (0.5)	0.908 (0.124–6.641)	0.924	0.924
*miR-181a*T>C/*miR-499*A>G					
TT/AA	72 (25.6)	98 (25.7)	1.000 (reference)		
TT/AG	33 (11.7)	54 (14.2)	1.155 (0.677–1.970)	0.597	0.796
TT/GG	1 (0.4)	11 (2.9)	8.320 (1.043–66.384)	0.046	0.184
TC/AA	95 (33.8)	107 (28.1)	0.818 (0.542–1.236)	0.34	0.68
TC/AG	47 (16.7)	67 (17.6)	1.040 (0.642–1.685)	0.872	0.872

Abbreviations: AOR, adjusted odds ratio; CI, confidence interval; RPL, recurrent pregnancy loss. [a] Fisher's exact test. [b] False-discovery rate-adjusted p-value.

3.5. Allele Combination Analysis of miRNA Polymorphisms in Recurrent Pregnancy Loss Patients and Control Subjects

The results of allele combination analyses of miRNA polymorphisms in RPL patients and control subjects are shown in Table 5 and Supplementary Tables S2–S4. The allele combinations *miR-10a/miR-30c/miR-181a/miR-499b* A-T-G-G (OR = 1.952; 95% CI = 1.120–3.149 p = 0.006), A-C-A-G (OR = 2.343; 95% CI = 1.111–4.942; p = 0.026), A-C-G-A (OR = 2.136 95% CI = 1.095–4.165; p = 0.028), T-T-G-A (OR = 0.455; 95% CI = 0.215–0.962; p = 0.044), and T-C-G-A (OR = 13.020; 95% CI = 0.739–229.300; p = 0.017) were associated with an increased risk of RPL. However, after FDR-p correction, there were no significant differences between RPL patients and controls in the ORs of the allele combinations, except for the A-T-G-G and T-C-G-A allele combinations.

Table 5. Allele combination analysis of miRNA polymorphisms in recurrent pregnancy loss patients and control subjects.

Allele Combination	Controls	RPL Patients	OR (95% CI)	p [a]	FDR-p [b]
	(n = 281)	(n = 381)			
	miR-10aA>T/miR-181aT>C/miR-30cA>G/miR-499A>G				
A-T-A-A	236 (41.9)	271 (35.6)	1.000 (reference)		
A-T-A-G	38 (6.9)	66 (8.6)	1.468 (0.953–2.263)	0.085	0.128
A-T-G-A	128 (22.9)	139 (18.2)	0.935 (0.695–1.257)	0.705	0.705
A-T-G-G	27 (4.9)	63 (8.2)	1.952 (1.210–3.149)	0.006	0.036
A-C-A-A	44 (7.8)	59 (7.7)	1.163 (0.759–1.784)	0.516	0.619
A-C-A-G	9 (1.7)	26 (3.5)	2.343 (1.111–4.942)	0.026	0.056
A-C-G-A	12 (2.3)	31 (4.1)	2.136 (1.095–4.165)	0.028	0.056
T-T-A-G	7 (1.2)	22 (3.0)	2.851 (1.202–6.764)	0.014	0.056
T-T-G-A	20 (3.7)	10 (1.4)	0.455 (0.215–0.962)	0.044	0.088
T-T-G-G	3 (0.7)	1 (0.2)	0.434 (0.079–2.391)	0.425	0.425
T-C-A-A	9 (1.6)	18 (2.3)	1.735 (0.765–3.937)	0.235	0.313
T-C-G-A	0 (0.0)	6 (0.9)	13.020 (0.739–229.300)	0.017	0.017

Abbreviations: CI, confidence interval; FDR, false-discovery rate; OR, odds ratio; RPL, recurrent pregnancy loss. ORs and 95% CIs for each allele combination were calculated with reference to frequencies of all others using Fisher's exact test. [a] Fisher's exact test. [b] FDR-adjusted p-value.

3.6. Differential Expression of the miR-10aA>T, miR-30cA>G, and miR-499bA>G Polymorphisms

The impact of SNPs on the interaction of *miR-10aA>T, miR-30cA>G, miR-181aT>C,* and *miR-499bA>G* on their targets was investigated by constructing various expression plasmids (*pri-miR-10aA, pri-miR-10aG, pri-miR-30cA, pri-miR-30cG, pre-miR-181aT, pre-miR-181aG, pri-miR-499bA,* and *pri-miR-499bG*) under control of the cytomegalovirus (CMV) promoter with either the major or minor allele. These plasmids were used in a dual luciferase assay performed with the 3′UTR of PAI-1, one of the predicted targets of miR-10a, miR-30c and miR181a, in Ishikawa human endometrial cells. A schematic diagram of a gene with a 3′-UTR of PAI-1 containing possible *miR-10a* and *miR-30c* binding sites in a conserved region is shown in Figure 1A,B. The luciferase activity of the 3′UTR of PAI-1 was significantly lower in *pre-miR-10a* having the A allele as compare to *pre-miR-10a* having the T allele (p < 0.05) (Figure 1C). Similarly, the luciferase activity of the 3′UTR of PAI-1 was significantly lower in the pre-miR-30c with the A allele as compared to pre-miR-30c with the G allele (p < 0.05) (Figure 1D).

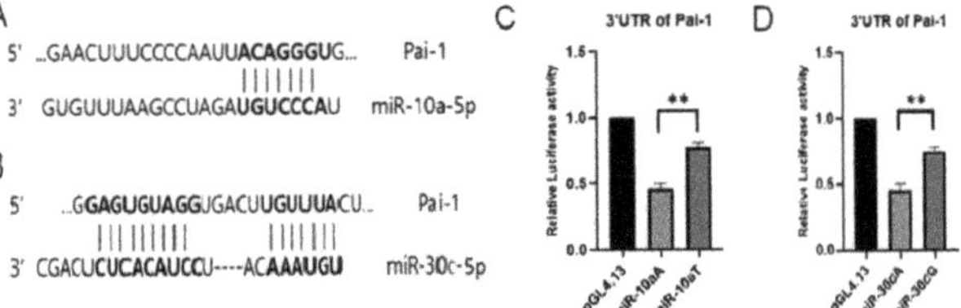

Figure 1. Expression of *miR-10aA>T, miR-30cA>G,* and the regulation of 3′-UTR of PAI-1 by *miR-10a* and *miR-30c*. (**A,B**) A schematic representation of gene with 3′-UTR of *PAI-1* that contain possible *miR-10a* and *miR-30c* binding sites in conserved regions. (**C**) Dual-luciferase reporter assays were performed to test the interaction of hsa-*miR-10aA>T* and its targeting sequence in the PAI-1 3′-UTR using constructs containing the predicted targeting sequence (pGL4.13-PAI-1 3′-UTR) cloned into the 3′-UTR of the reporter gene. Luciferase expression levels were normalized against Renilla luciferase expression. Data represent three independent exper-iments with triplicate measurements. ** p < 0.05. (**D**) Dual-luciferase reporter assays were performed to test the interaction of *miR-30cA>G* and its target sequence in the PAI-1 3′-UTR using constructs containing the predicted targeting sequence (pGL4.13-PAI-1 3′-UTR) cloned into the 3′-UTR of the reporter gene. Luciferase expression levels were normalized against Renilla luciferase expression. Data represent three independent experiments with triplicate measurements. ** p < 0.05.

3.7. Differences of Various Clinical Parameters According to miRNA Polymorphisms in RPL Patients

Associations between miRNA polymorphisms and the levels of homocysteine, folate total cholesterol, uric acid, blood urea nitrogen (BUN), estradiol (E2), thyroid-stimulating hormone (TSH), FSH, LH, prolactin, creatinine, platelets (PLT), as well as $CD3^+$, $CD4^+$ $CD8^+$, $CD19^+$, and $CD56^+$ NK cells, in addition to the PT and aPTT were assessed by ordinal logistic regression analyses. We divided the risk factors into 10 grades and performed ordinal logistic regression using a proportional odds model. We found that the genotype frequency of *miR-30cA>G* was significantly associated with aPTT (AA: 32.46 ± 4.71, GG 27.56 ± 3.59, $p = 0.001$), creatinine (AA: 1.19 ± 1.94, GG: 6.26 ± 3.71, $p = 0.001$), and E2 (AA: 1.19 ± 1.94, GG: 6.26 ± 3.71, $p = 0.001$). Levels of FSH differed significantly ($p < 0.05$) between the *miR-30cA>G* AA (mean ± SD, 32.36 ± 4.30 and 6.96 ± 4.29, respectively) and GG genotypes (30.49 ± 3.02 and 33.82 ± 55.85, respectively) (Table 6, Figure 2A,C,D) Additionally, levels of hematocrit (Hct) and total cholesterol (T. chol) differed significantly ($p < 0.05$) between the *miR-30c*AA and GG genotypes (36.65 ± 3.73 and 34.14 ± 4.49, 172.56 ± 64.85 and 38.15 ± 76.11, respectively). The *miR-181a*T>C genotype frequency was significantly associated with levels of creatinine (TT: 2.38±3.24, TC: 1.17±1.76, $p = 0.011$), Hcy (TT: 6.76 ± 2.01, CC: 9.98 ± 4.50, $p = 0.001$), LH (TT: 4.81 ± 2.74, CC: 4.20 ± 0.71, $p = 0.038$), PT (TT: 11.43 ± 1.14, CC: 10.20 ± 0.28, $p = 0.048$), and T. chol (TT: 136.96 ± 86.18, TC: 185.65 ± 76.23, $p = 0.001$). The *miR-499b*A>G genotype frequency was significantly associated with aPTT (TT: 31.20 ± 4.29, GG: 32.10 ± 4.18, $p = 0.026$) (Table 6, Figure 2B).

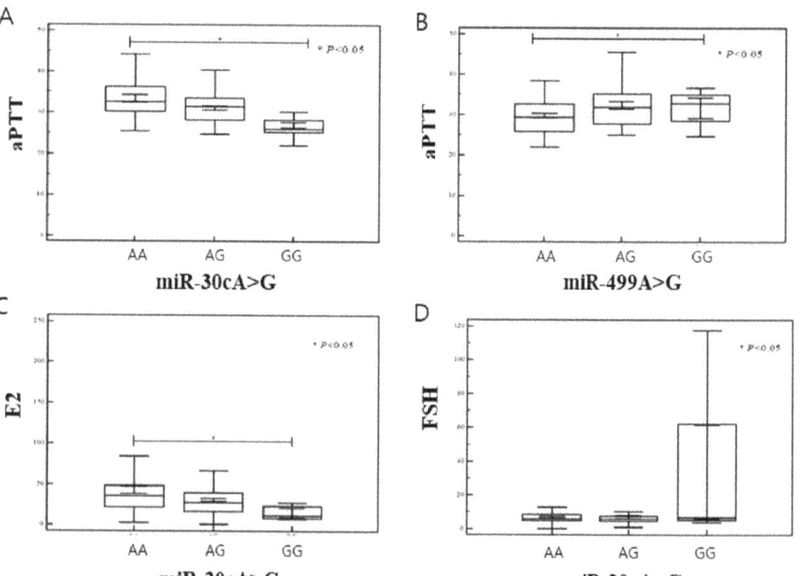

Figure 2. Analysis of variance of aPTT levels according to miRNA polymorphisms (PL ≥ 3) and analysis of variance of E2 and FSH levels according to miRNA polymorphisms. (**A**) The aPTT level was significantly different ($p = 0.001$) between miR-30cA>G AA (mean ± SD, 32.46 ± 4.71), AG (31.92 ± 4.11), and GG (27.56 ± 3.59). (**B**) The aPTT level was significantly different ($p = 0.026$) between miR-499 AA (mean ± SD, 31.20 ± 4.29), AG (32.62 ± 4.69), and GG (32.10 ± 4.18). (**C**) The E2 level was significantly different ($p < 0.05$) between miR-30cA>G AA (mean ± SD, 43.02 ± 38.58), AG (30.80 ± 19.30), and GG (15.10 ± 10.66). (**D**) The FSH level was significantly different ($p < 0.05$) between miR-30A>G AA (mean ± SD, 6.96 ± 4.29), AG (6.98 ± 8.47), and GG (33.82 ± 55.85).

Table 6. Differences of various clinical parameters according to miRNA polymorphisms in RPL patients.

Genotype	aPTT	Creatinine (mg/dL)	E2 (pg/mL)	FSH (mIU/mL)	Hct	Hcy	LH (mIU/mL)	PT	T. Chol (mg/dL)
	Mean ± SD	Mean ± SD	Mean ± SD	Mean ± SD	Mean ± SD	Mean ± SD	Mean ± SD	Mean ± SD	Mean ± SD
miR-10a A>T									
AA	31.86 ± 4.43	1.87 ± 2.75	35.07 ± 25.54	7.84 ± 11.93	36.28 ± 3.90	6.92 ± 2.00	5.72 ± 7.14	11.61 ± 1.87	154.54 ± 83.14
AT	31.39 ± 4.64	2.47 ± 3.36	38.07 ± 41.91	6.30 ± 3.79	36.36 ± 4.20	7.31 ± 2.37	8.38 ± 21.83	11.41 ± 0.89	140.93 ± 89.67
TT	31.47 ± 3.42	0.73 ± 0.15	31.20 ± 9.11	9.94 ± 4.91	36.63 ± 6.47	5.64 ± 1.10	5.52 ± 1.60	11.80 ± 0.35	251.00 ± 118.01
p	0.719	0.53	0.837	0.642	0.974	0.228	0.435	0.696	0.079
miR-30c A>G									
AA	32.46 ± 4.71	1.19 ± 1.94	43.02 ± 38.58	6.96 ± 4.29	36.65 ± 3.73	6.84 ± 2.00	5.41 ± 3.63	11.83 ± 2.21	172.56 ± 64.85
AG	31.92 ± 4.11	1.55 ± 2.35	30.80 ± 19.30	6.98 ± 8.47	36.27 ± 4.02	7.01 ± 2.16	6.46 ± 14.43	11.41 ± 1.28	167.25 ± 79.17
GG	27.56 ± 3.59	6.26 ± 3.71	15.10 ± 10.66	33.82 ± 55.85	34.14 ± 4.49	7.73 ± 1.86	20.00 ± 32.76	11.68 ± 0.83	38.15 ± 76.11
p	0.001	0.001	0.015	0.001	0.013	0.199	0.062	0.135	0.001
miR-181a T>C									
TT	31.51 ± 4.45	2.38 ± 3.24	33.91 ± 21.97	6.42 ± 3.17	36.26 ± 4.08	6.76 ± 2.01	4.81 ± 2.74	11.43 ± 1.14	136.96 ± 86.18
TC	32.23 ± 4.48	1.17 ± 1.76	39.85 ± 41.63	9.76 ± 17.78	36.45 ± 3.70	7.36 ± 1.99	9.50 ± 20.64	11.91 ± 2.48	185.65 ± 76.23
CC	36.05 ± 3.32	-	25.67 ± 5.08	6.73 ± 1.80	31.35 ± 0.64	9.98 ± 4.50	4.20 ± 0.71	10.20 ± 0.28	-
p	0.17	0.011	0.409	0.115	0.19	0.001	0.038	0.048	0.001
miR-499bA>G									
AA	31.20 ± 4.29	2.14 ± 3.00	35.86 ± 26.02	6.84 ± 7.73	36.43 ± 3.85	7.00 ± 2.06	5.37 ± 4.90	11.54 ± 1.72	144.26 ± 86.46
AG	32.62 ± 4.69	1.82 ± 2.84	34.70 ± 37.33	9.32 ± 15.27	36.25 ± 4.24	7.01 ± 2.15	8.62 ± 20.48	11.54 ± 1.25	164.37 ± 84.75
GG	32.10 ± 4.18	1.67 ± 2.54	38.39 ± 23.85	5.46 ± 3.78	34.60 ± 3.54	6.80 ± 1.99	4.55 ± 4.01	12.17 ± 3.59	166.09 ± 82.36
p	0.026	0.66	0.94	0.251	0.18	0.942	0.198	0.453	0.223

Abbreviations: aPTT, activated partial thromboplastin time; E2, estradiol; FSH, follicle-stimulating hormone; Hct, hematocrit; T. chol, total cholesterol; Hcy, homocysteine level; LH, luteinizing hormone; RPL, recurrent pregnancy loss; PT, prothrombin time.

4. Discussion

Increasing evidence suggests that miRNAs play critical roles in the pathophysiology of various reproductive disorders [14,15,32]. Here, we investigated whether four pre-miRNA SNPs (*miR-10a, miR-30c, miR-181a*, and *miR-499b*) were associated with the risk of RPL in a cohort of Korean women. Specifically, we focused on the genotypes and allele combination of the selected miRNA polymorphisms and aimed to determine how they affected the risk of RPL. Using a genotype-based analysis method, we found that the GG and dominant (AA vs. AG + GG) *miR-499b* genotypes were significantly more common in RPL patients (PL $\geq$ 3 and PL $\geq$ 4, $p < 0.05$) than control subjects. In allele combination analyses, the AA/GG and AG/AG genotypes of *miR-30c*A>G/*miR-499*A>G were significantly more common in RPL patients than in controls.

As the activities of many genes are interconnected in complex conditions such as RPL, gene-gene interactions may affect gene-disease associations. The MDR method enables the detection of gene-gene interactions, regardless of the chromosomal locations of the genes [33]. We used a novel genotype-based MDR approach to examine the effects of potential interactions between different miRNAs on RPL risk. These results of these analyses, which examined the effects of four miRNA polymorphisms associated with RPL, suggested that gene-gene interactions involving these four miRNA polymorphisms also play roles in determining the risk of RPL. Allele combination MDR analyses indicated that the two combination conferred by *the miR-10a*A>T/*miR-181a*T>C/*miR-30c*A>G/*miR-499*A>G (A-T-G-G and T-C-G-A), the two combination conferred by the *miR-10a*A>T/*miR-181a*T>C/*miR-30c*A>G (T-T-A, T-C-G), the two combination conferred by the *miR-10a*A>T/*miR-30c*A>G/*miR-499*A>G allele combination (C-A-G, C-G-A), and the genotype conferred by the *miR-10a*A>T/*miR-30c*A>G allele combination (T-A) occur more frequently in patients with RPL than control subjects, suggesting a significant association with increased risk of RPL (all $p < 0.05$). In addition, the *miR-10a*A>T/*miR-181a*T>C/*miR-30c*A>G allele combination T-T-G and the *miR-10a*A>T/*miR-30c*A>G/*miR-499*A allele combination C-G-G were found to be less frequent in RPL patients than controls, suggesting these combinations exert a protective effect (all $p < 0.05$).

SNPs that occur in miRNA genes, miRNA machinery genes, or miRNAs that target genes involved in miRNA synthesis or function could adversely affect downstream gene expression [34]. Several studies have provided evidence supporting the critical role of miRNAs in RPL [35]. A previous study demonstrated that *miR-499* was associated with the transforming growth factor (TGF)-β signaling pathway [24]. Furthermore, the 3′-UTR of the *TGF-β3* gene has been shown to contain a putative binding site for *miR-30c* (rs928508 (http://www.targetscan.org (accessed on 21 May 2018)), which targets the drug metabolism gene *SULT1A1* [25]. Several TGF-β superfamily members perform critical functions in the female reproductive system. Specifically, these proteins regulate all processes of ovarian follicle development, including granulosa and theca cell proliferation, primordial follicle recruitment, gonadotropin receptor expression, ovulation, oocyte maturation, luteinization, and corpus luteum formation [36]. Additionally, the 3′-UTR of the prostaglandin F2 receptor inhibitor gene has been shown to contain a predicted binding target for *miR-604* (http://www.targetscan.org (accessed on 21 May 2018)), and prostaglandin F2 is required for placenta retention [37]. Furthermore, the *miR-10a*A>T polymorphism has been associated with regulation of IL-6 expression [26], and a previous study reported abnormal IL-6 expression in both animal models and patients with recurrent spontaneous abortions [38].

An online search for *miR-10a, miR-30c, miR-181a*, and *miR-499b* targets using the Target Scan and miRIAD databases (http://bmi.ana.med.uni-muenchen.de/miriad/ (accessed on 21 May 2018)) returned many putative mRNA targets. Among these targets, we focused on *PAI-1* for further functional analyses of *miR-10a, miR-30c*, and *miR-181a* because this gene has been shown to play several important roles in pregnancy and infertility [27]. PAI-1 is the primary inhibitor of plasminogen activators, including tPA and uPA. In the human placenta, *PAI-1* is expressed in the extravillous interstitial and vascular trophoblasts. During implantation and placentation, PAI-1 inhibits extracellular matrix degradation which thereby inhibits trophoblast invasion. We reviewed the literature regarding various reproductive diseases in which PAI-1 plays a role. Elevated *PAI-1* levels have been detected in patients with RPL, preeclampsia, intrauterine growth restriction, gestational diabetes mellitus (GDM), endometriosis, and PCOS. Furthermore, both GDM and PCOS development have been reported to be related to the genetic role of the 4G/5G polymorphism in *PAI-1*. *In general, elevated blood levels of PAI-1* are associated with an increased risk of infertility and poor pregnancy outcomes. In contrast, deficiency of *PAI-1* results in transiently impaired placentation in mice [39], and deficiency of the *PAI-1* gene is associated with abnormal bleeding after trauma or surgery in humans [40]. PAI-1 functions as a major inhibitor of fibrinolysis, and its overexpression leads to fibrin accumulation and placental insufficiency during pregnancy. PAI-1 acts as a major inhibitor of fibrinolysis, resulting in fibrin accumulation and insufficient placental formation due to overexpression. Previous reports also suggested that elevation of PAI-1 levels is the most frequent hemostasis-related abnormality associated with unexplained RPL [41]. Thus, increased expression of PAI-1 leading to inhibition of fibrinolysis is believed to be the main cause of RPL.

To determine whether polymorphisms in *miR-10a*, *miR-30c*, and *miR-181a* affect target gene expression, we compared the expression levels of the $3'$-UTR of *PAI-1* harboring the different polymorphisms of *miRNAs* in Ishikawa human endometrial cells. Aberrant *PAI-1* expression resulting from the expression of *miR-10a* with the A allele was significantly lower ($p < 0.05$) than aberrant *PAI-1* expression resulting from the expression of *miR-10a* with the T allele. In addition, the expression of *miR-30c* with the A allele was significantly lower ($p < 0.05$) than expression of premature and mature *miR-30c* with the G allele.

Expression of genotypes of *miR-30c*G as well as those of *miR-10a*T led to reduced expression of *PAI-1* mRNA. These results suggest that SNPs in *miR-30c* and *miR-10a* regulate the expression of the *PAI-1* gene. PAI-1-mediated inhibition of fibrinolysis and fibrin accumulation is currently believed to be the principal culprits for RPL; however, further studies are required to fully elucidate the underlying mechanisms.

FSH is the primary gonadotropin responsible for regulating the progression of pregnancy [42]. Optimal levels of FSH, especially during the first few months of pregnancy, are critical for proper formation of the placenta [43]. Our clinical data indicated significant changes in FSH levels in RPL patients harboring the *miR-30c*A>G polymorphism. We, therefore, hypothesized that abnormal regulation of *PAI-1* expression mediated by mutant *miR-30c* SNP results in aberrant FSH expression or disruption of the normal response to FSH. Imbalances in homocysteine and folate levels in particular are thought to contribute to low birth weight [44]. Specifically, higher homocysteine and lower folate concentrations during early pregnancy have been reported to be associated with lower placental weight and birth weight. However, we did not observe any associations between folate and homocysteine concentrations and placental weight.

We found that the dominant *miR-499b* AG genotype (AA vs. AG + GG) was significantly more frequent in RPL patients ($p < 0.05$). Earlier studies used a global approach to identify and profile miRNA expression at important stages during the estrous cycle and found a role of miRNAs in ovulation. Additionally, one-way ANOVA analysis of variance of data from RPL patients (Table 6) revealed that in comparison with *miR-30c*AA, the *miR-30c*GG genotype was associated with significantly lower aPTT, E2 (pg/mL), Hct, and T. chol (mg/dL) and significantly higher creatinine (mg/dL) and FSH (mIU/mL). Compared with *miR-181a*TT, the *miR-181a*CC genotype was associated with significantly higher homocysteine levels, suggesting this genotype is associated with increased risk of RPL ($p < 0.05$). Compared with *miR-181a*TT, the *miR-181a*TC genotype was associated with significantly higher T. chol levels, suggesting this genotype is associated with increased risk of RPL ($p < 0.05$). However, in the case of creatinine levels, the *miR-181a*TC genotype was associated with significantly lower levels than the *miR-181a*TT genotype, indicating a protective effect, although the results were inconsistent with OR and therefore, the difference was not significant.

5. Conclusions

We investigated the relationship between various miRNA polymorphisms and the occurrence and risk of RPL. Several genotypes and allele combinations were positively correlated with RPL occurrence and unfavorable prognoses according to reproductive disease risk factors, including FSH, LH, and E2 levels. However, this study has several limitations. First, how the miRNA polymorphisms in the *PAI-1* gene affect the development of RPL remains unclear. In addition to studies of *PAI-1*, future follow-up studies of other RPL-related genes and the *miR-10a* and *miR-30c* targets are planned, particularly studies of the role of genes related to the TGF-β signaling pathway. As TGF-β regulates cell proliferation, apoptosis, and homeostasis, it plays a critical role in regulating the progression of pregnancy. Second, the control subjects in our study were not completely healthy because some of them had sought medical attention for other issues. Our experience shows that recruiting healthy participants through imaging and laboratory testing results in significantly reduced enrollment rates. However, enrollment of participants without imaging and laboratory testing can introduce another challenge to risk factor assessment. Lastly, the study population was restricted to Korean patients. Although the results of our study provide the first evidence suggesting that miRNA polymorphisms in the *PAI-1* gene may serve as diagnostic and prognostic biomarkers for RPL, a prospective study involving a larger cohort of patients is warranted to validate these findings. A genome-wide analysis (using transcriptome-seq and miRNA-seq) is needed to identify the primary target genes, particularly the common genes regulated by these miRNAs. Determining the expression of these genes in the relevant gene-miRNA networks would provide stronger evidence in support of the results of the present research.

Supplementary Materials: The following supporting information can be downloaded at: https://www.mdpi.com/article/10.3390/biomedicines10102395/s1, Table S1: Genotype frequencies of miRNA gene polymorphisms in control subjects and recurrent pregnancy loss patients; Table S2: Four allele combination analysis of miRNA polymorphisms in recurrent pregnancy loss patients and control subjects; Table S3. Three allele combination analysis of miRNA polymorphisms in recurrent pregnancy loss patients and control subjects; Table S4. Two allele combination analysis of miRNA polymorphisms in recurrent pregnancy loss patients and control subjects.

Author Contributions: Conceived and designed the experiments: S.-H.C., H.-J.A., S.-S.J. and N.-K.K.; Performed the experiments: H.-J.A., S.-H.C. and H.-S.P.; Analyzed the data and statistical analyses: J.-H.K., E.-H.A. and Y.-R.K.; Contributed reagents/material/analysis tools: J.-H.K., Y.-R.K., J.-R.L. and E.-H.A.; Wrote the main manuscript text: S.-H.C., H.-J.A. and N.-K.K.; Reference collection and data management: J.-H.K., Y.-R.K., W.-S.L. and N.-K.K. All authors have read and agreed to the published version of the manuscript.

Funding: This study was partially supported by a grant from the National Research Foundation of Korea (NRF) funded by the Ministry of Education (2021R1I1A1A01050945, NRF-2020R1F1A107452511 and NRF-2020R1I1A1A01072008), and by the Korea Health Technology R&D Project through the Korea Health Industry Development Institute (KHIDI), funded by the Ministry of Health & Welfare, Republic of Korea (grant number: HI18C19990200). The funders had no role in study design, data collection and analysis, decision to publish, or preparation of the manuscript.

Institutional Review Board Statement: The study was conducted in accordance with the Declaration of Helsinki, and approved by the Institutional Review Board of CHA Bundang Medical Center (IRB approval no. BD2010-123D).

Informed Consent Statement: Informed consent was obtained from all subjects involved in the study.

Data Availability Statement: The data that support the findings of this study are available from the corresponding author upon reasonable request.

Conflicts of Interest: The authors declare no competing financial interests.

References

Coulam, C.B.; Coulam, C.B.; Clark, D.A.; Beer, A.E.; Kutteh, W.H.; Kwak, J.; Stephenson, M. Clinical Guidelines Recommendation Committee for Diagnosis and Treatment of Recurrent Spontaneous Abortion. Current clinical options for diagnosis and treatment of recurrent spontaneous abortion. *Am. J. Reprod. Immunol.* **1997**, *38*, 57–74. [CrossRef] [PubMed]

Rai, R.; Regan, L. Recurrent miscarriage. *Lancet* **2006**, *368*, 601–611. [CrossRef]

Sierra, S.; Stephenson, M. Genetics of Recurrent Pregnancy Loss. *Semin. Reprod. Med.* **2006**, *24*, 017–024. [CrossRef] [PubMed]

Marjorie, P.P.; Perron, M.P.; Provost, P. Protein interactions and complexes in human microRNA biogenesis and function. *Front. Biosci.* **2008**, *13*, 2537–2547. [CrossRef]

Bartel, D.P. MicroRNAs: Genomics, biogenesis, mechanism, and function. *Cell* **2004**, *116*, 281–297. [CrossRef]

Ambros, V. The functions of animal microRNAs. *Nature* **2004**, *431*, 350–355. [CrossRef]

Lim, L.P.; Lau, N.; Garrett-Engele, P.; Grimson, A.; Schelter, J.M.; Castle, J.; Bartel, D.P.; Linsley, P.S.; Johnson, J.M. Microarray analysis shows that some microRNAs downregulate large numbers of target mRNAs. *Nature* **2005**, *433*, 769–773. [CrossRef]

Cuellar, T.L.; McManus, M.T. MicroRNAs and endocrine biology. *J. Endocrinol.* **2005**, *187*, 327–332. [CrossRef]

Lee, Y.; Kim, M.; Han, J.; Yeom, K.-H.; Lee, S.; Baek, S.H.; Kim, V.N. MicroRNA genes are transcribed by RNA polymerase II. *EMBO J.* **2004**, *23*, 4051–4060. [CrossRef]

10. Gregory, R.I.; Chendrimada, T.P.; Shiekhattar, R. MicroRNA Biogenesis: Isolation and Characterization of the Microprocessor Complex. *MicroRNA Protoc.* **2006**, *342*, 33–48. [CrossRef]

11. Yi, R.; Qin, Y.; Macara, I.G.; Cullen, B.R. Exportin-5 mediates the nuclear export of pre-microRNAs and short Hairpin RNAs. *Genes Dev.* **2003**, *17*, 3011–3016. [CrossRef] [PubMed]

12. Moore, M.S.; Biobel, G. The GTP-binding protein Ran/TC4 is required for protein import into the nucleus. *Nature* **1993**, *365*, 661–663. [CrossRef] [PubMed]

13. O'Toole, A.S.; Miller, S.; Haines, N.; Zink, M.C.; Serra, M.J. Comprehensive thermodynamic analysis of 3′ double-nucleotide overhangs neighboring Watson-Crick terminal base pairs. *Nucleic Acids Res.* **2006**, *34*, 3338–3344. [CrossRef] [PubMed]

14. Song, F.J.; Chen, K.X. Single-nucleotide polymorphisms among microRNA: Big effects on cancer. *Chin. J. Cancer* **2011**, *30*, 381–391. [CrossRef]

15. Imbar, T.; Eisenberg, I. Regulatory role of microRNAs in ovarian function. *Fertil. Steril.* **2014**, *101*, 1524–1530. [CrossRef]

16. Iwai, N.; Naraba, H. Polymorphisms in human pre-miRNAs. *Biochem. Biophys. Res. Commun.* **2005**, *331*, 1439–1444. [CrossRef]

17. Lal, A.; Navarro, F.; Maher, C.A.; Maliszewski, L.E.; Yan, N.; O'Day, E.; Chowdhury, D.; Dykxhoorn, D.M.; Tsai, P.; Hofmann, O.; et al. miR-24 inhibits cell proliferation by targeting E2F2, MYC, and other cell-cycle genes via binding to "seedless" 3′UTR microRNA recognition elements. *Mol. Cell* **2009**, *35*, 610–625. [CrossRef]

18. He, X.-Y.; Chen, J.-X.; Zhang, Z.; Li, C.-L.; Peng, Q.; Peng, H.-M. The let-7a microRNA protects from growth of lung carcinoma by suppression of k-Ras and c-Myc in nude mice. *J. Cancer Res. Clin. Oncol.* **2010**, *136*, 1023–1028. [CrossRef]

19. Toloubeydokhti, T.; Bukulmez, O.; Chegini, N. Potential Regulatory Functions of MicroRNAs in the Ovary. *Semin. Reprod. Med.* **2008**, *26*, 469–478. [CrossRef]

20. Medeiros, L.A.; Dennis, L.M.; Gill, M.E.; Houbaviy, H.; Markoulaki, S.; Fu, D.; White, A.C.; Kirak, O.; Sharp, P.A.; Page, D.C.; et al. *Mir-290–295* deficiency in mice results in partially penetrant embryonic lethality and germ cell defects. *Proc. Natl. Acad. Sci. USA* **2011**, *108*, 14163–14168. [CrossRef]

21. Butz, H.; Likó, I.; Czirják, S.; Igaz, P.; Korbonits, M.; Rácz, K.; Patócs, A. MicroRNA profile indicates downregulation of the TGF pathway in sporadic non-functioning pituitary adenomas. *Pituitary* **2011**, *14*, 112–124. [CrossRef] [PubMed]

22. Nguyen Dien, G.T.; Smith, R.A.; Haupt, L.M.; Griffiths, L.R.; Nguyen, H.T. Genetic polymorphisms inmiRNAs targeting the estrogen receptor and their effect on breast cancer risk. *Meta Gene* **2014**, *2*, 226–236. [CrossRef] [PubMed]

23. Santamaria, X.; Taylor, H. MicroRNA and gynecological reproductive diseases. *Fertil. Steril.* **2014**, *101*, 1545–1551. [CrossRef] [PubMed]

24. Suzuki, H.I. MicroRNA Control of TGF-β Signaling. *Int. J. Mol. Sci.* **2018**, *28*, 19. [CrossRef] [PubMed]

25. Knight, P.G.; Glister, C. TGF-beta superfamily members and ovarian follicle development. *Reproduction* **2006**, *132*, 191–206. [CrossRef]

26. Dong, K.; Xu, Y.; Yang, Q.; Shi, J.; Jiang, J.; Chen, Y.; Song, C.; Wang, K. Associations of functional microRNA binding site polymorphisms in IL23/Th17 inflammatory pathway genes with gastric cancer risk. *Mediat. Inflamm.* **2017**, 6974696. [CrossRef]

27. Li, C.; Zhu, H.; Bai, W.; Su, L.-L.; Liu, J.-Q.; Cai, W.-X.; Zhao, B.; Gao, J.-X.; Han, S.-C.; Li, J.; et al. MiR-10a and miR-181c regulate collagen type I generation in hypertrophic scars by targeting PAI-1 and uPA. *FEBS Lett.* **2015**, *589*, 380–389. [CrossRef]

28. Ryu, C.S.; Sakong, J.H.; Ahn, E.H.; Kim, J.O.; Ko, D.; Kim, J.H.; Lee, W.S.; Kim, N.K. Association study of the three functional polymorphisms (TAS2R46G>A, OR4C16G>A, and OR4X1A>T) with recurrent pregnancy loss. *Genes Genom.* **2019**, *41*, 61–70. [CrossRef]

29. Wang, H.; Wu, S.; Wu, J.; Sun, S.; Wu, S.; Bao, W. Association analysis of the SNP (rs345476947) in the FUT2 gene with the production and reproductive traits in pigs. *Genes Genom.* **2018**, *40*, 199–206. [CrossRef]

30. Hochberg, Y.; Benjamini, Y. More powerful procedures for multiple significance testing. *Stat. Med.* **1990**, *9*, 811–818. [CrossRef]

31.	Ritchie, M.D.; Hahn, L.W.; Roodi, N.; Bailey, L.R.; Dupont, W.D.; Parl, F.F.; Moore, J.H. Multifactor-Dimensionality Reduction Reveals High-Order Interactions among Estrogen-Metabolism Genes in Sporadic Breast Cancer. *Am. J. Hum. Genet.* **2001**, *69*, 138–147. [CrossRef] [PubMed]

32.	Teague, E.M.C.O.; Print, C.; Hull, M.L. The role of microRNAs in endometriosis and associated reproductive conditions. *Hum. Reprod. Updat.* **2010**, *16*, 142–165. [CrossRef] [PubMed]

33.	Hahn, L.W.; Ritchie, M.D.; Moore, J.H. Multifactor dimensionality reduction software for detecting gene-gene and gene-environment interactions. *Bioinformatics* **2003**, *19*, 376–382. [CrossRef] [PubMed]

34.	Moszyńska, A.; Gebert, M.; Collawn, J.F.; Bartoszewski, R. SNPs in microRNA target sites and their potential role in human disease. *Open Biol.* **2017**, *7*, 170019. [CrossRef]

35.	Barchitta, M.; Maugeri, A.; Quattrocchi, A.; Agrifoglio, O.; Agodi, A. The Role of miRNAs as Biomarkers for Pregnancy Outcomes: A Comprehensive Review. *J. Genom.* **2017**, *2017*, 8067972. [CrossRef]

36.	Liu, X.; Li, M.; Peng, Y.; Hu, X.; Xu, J.; Zhu, S.; Yu, Z.; Han, S. miR-30c regulates proliferation, apoptosis and differentiation via the Shh signal-ing pathway in P19 cells. *Exp. Mol. Med.* **2016**, *29*, e248. [CrossRef]

37.	Bider, D.; Dulitzky, M.; Goldenberg, M.; Lipitz, S.; Mashiach, S. Intraumbilical vein injection of prostaglandin F2α in retained placenta. *Eur. J. Obstet. Gynecol. Reprod. Biol.* **1996**, *64*, 59–61. [CrossRef]

38.	Wolff, M.V.; Thaler, C.J.; Strowitzki, T.; Broome, J.; Stolz, W.; Tabibzadeh, S. Regulated expression of cytokines in human endometrium throughout the menstrual cycle: Dysregulation in habitual abortion. *Mol. Hum. Reprod.* **2000**, *6*, 627–634. [CrossRef]

39.	Labied, S.; Blacher, S.; Carmeliet, P.; Noël, A.; Frankenne, F.; Foidart, J.-M.; Munaut, C. Transient reduction of placental angiogenesis in PAI-1-deficient mice. *Physiol. Genom.* **2011**, *43*, 188–198. [CrossRef]

40.	Fay, W.P.; Parker, A.C.; Condrey, L.R.; Shapiro, A.D. Human plasminogen activator inhibitor-1 (PAI-1) deficiency: Characterization of a large kindred with a null mutation in the PAI-1 gene. *Blood* **1997**, *90*, 204–208. [CrossRef]

41.	Gris, J.-C.; Ripart-Neveu, S.; Maugard, C.; Tailland, M.L.; Brun, S.; Courtieu, C.; Biron, C.; Hoffet, M.; Hédon, B.; Marès, P. Respective evaluation of the prevalence of haemostasis abnormalities in unexplained primary early recurrent miscarriages. The Nimes Obstetricians and Haematologists (NOHA) Study. *Thromb. Haemost.* **1997**, *77*, 1096–1103. [PubMed]

42.	Kim, Y.K.; Wasser, S.K.; Fujimoto, V.Y.; Klein, N.A.; Moore, D.E.; Soules, M.R. Utility of follicle stimulating hormone (FSH), luteinizing hormone (LH), oestradiol and FSH:LH ratio in predicting reproductive age in normal women. *Hum. Reprod.* **1997**, *12*, 1152–1155. [CrossRef] [PubMed]

43.	Stilley, J.A.W.; Segaloff, D.L. FSH Actions and Pregnancy: Looking Beyond Ovarian FSH Receptors. *Endocrinology* **2018**, *159*, 4033–4042. [CrossRef] [PubMed]

44.	Bergen, N.E.; Jaddoe, V.W.; Timmermans, S.; Hofman, A.; Lindemans, J.; Russcher, H.; Raat, H.; Steegers-Theunissen, R.P.; Steegers, E.A. Homocysteine and folate concentrations in early pregnancy and the risk of adverse pregnancy outcomes: The Generation R Study. *BJOG* **2012**, *119*, 739–751. [CrossRef]

 *biomedicines*

Review

Small Nucleolar Derived RNAs as Regulators of Human Cancer

Alexander Bishop Coley [1,†], Jeffrey David DeMeis [1,†], Neil Yash Chaudhary [1] and Glen Mark Borchert [1,2,*]

1   Department of Pharmacology, College of Medicine, University of South Alabama, Mobile, AL 36688, USA; abc1323@jagmail.southalabama.edu (A.B.C.); jdd1527@jagmail.southalabama.edu (J.D.D.); nyc1921@jagmail.southalabama.edu (N.Y.C.)
2   School of Computing, University of South Alabama, Mobile, AL 36688, USA
*   Correspondence: borchert@southalabama.edu; Tel.: +1-251-461-1367
†   These authors contributed equally to this work.

Abstract: In the past decade, RNA fragments derived from full-length small nucleolar RNAs (snoR-NAs) have been shown to be specifically excised and functional. These sno-derived RNAs (sdRNAs) have been implicated as gene regulators in a multitude of cancers, controlling a variety of genes post-transcriptionally via association with the RNA-induced silencing complex (RISC). In this review, we have summarized the literature connecting sdRNAs to cancer gene regulation. SdRNAs possess miRNA-like functions and are able to fill the role of tumor-suppressing or tumor-promoting RNAs in a tissue context-dependent manner. Indeed, there are many miRNAs that are actually derived from snoRNA transcripts, meaning that they are truly sdRNAs and as such are included in this review. As sdRNAs are frequently discarded from ncRNA analyses, we emphasize that sdRNAs are functionally relevant gene regulators and likely represent an overlooked subclass of miRNAs. Based on the evidence provided by the papers reviewed here, we propose that sdRNAs deserve more extensive study to better understand their underlying biology and to identify previously overlooked biomarkers and therapeutic targets for a multitude of human cancers.

Keywords: cancer; gene regulation; small nucleolar RNA (snoRNA); small nucleolar derived RNA (sdRNA); microRNA (miRNA); RNA; snoRNA; sdRNA; miRNA; genetics

Citation: Coley, A.B.; DeMeis, J.D.; Chaudhary, N.Y.; Borchert, G.M. Small Nucleolar Derived RNAs as Regulators of Human Cancer. *Biomedicines* **2022**, *10*, 1819. https://doi.org/10.3390/biomedicines10081819

Academic Editors: Milena Rizzo and Elena Levantini

Received: 30 May 2022
Accepted: 21 July 2022
Published: 28 July 2022

Publisher's Note: MDPI stays neutral with regard to jurisdictional claims in published maps and institutional affiliations.

1. Introduction

Until the first association of microRNA (miRNA) dysregulation and cancer in 2002, noncoding RNAs (ncRNA) in cancer were largely disregarded by many as "junk RNA" or transcriptional noise [1]. miRNAs are ~22 nucleotide (nt) long small RNAs which form part of the RNA Induced Silencing Complex (RISC) and suppress messenger RNA expression through miRNA-mRNA sequence complementarity. The pathway in which miRNA facilitates post-transcriptional gene suppression of target mRNA is formally known as RNA interference (RNAi). Elucidation of this RNAi pathway afforded the scientific community with a broader understanding of the existence of short ncRNA species and their relevance with respect to regulation of protein coding gene expression, including tumor suppressor genes in the context of human cancer. While a plethora of studies confirmed roles for dysregulated miRNAs in cancer, others wondered if additional short ncRNA species with distinct functions could similarly arise from full-length ncRNA transcripts. In 2008, Kawaji et al. performed an unbiased investigation of small RNA sequencing data obtained from HepG2 cells. In this study, miRNA-like fragments were aligned to the human genome [2] which resulted in the first ever characterization of noncoding-derived RNA (ndRNA) excised from other full length ncRNAs including transfer RNA, ribosomal RNA, and small nucleolar RNA (snoRNA). Similar to miRNAs, many of the observed ndRNA species were initially disregarded as degradation products or transcriptional noise and were assumed to lack functional purpose. However, many studies have since demonstrated that ndRNAs can be specifically excised from longer ncRNA species and can exert microRNA-like regulatory function on target genes. This distinguishes ndRNAs from the canonical roles of the longer transcripts from which they were derived. That said, annotation of these novel ndRNAs supports the belief that the human ncRNA regulatory network is vastly larger than currently appreciated. Moreover, expansion of the repertoire of tractable ncRNA targets for biomarker detection and drug discovery to include the emerging classes of ndRNAs is imperative. Clinical recognition of ndRNAs which possess comparable roles in gene regulation and disease pathogenesis to miRNAs will expand our collection of diagnostic markers. This will hopefully contribute to earlier disease detection and thereby confer improved clinical outcomes. In this review we describe the functional role of ndRNAs which arise from snoRNA transcripts, henceforth referred to as sno-derived RNAs (sdRNAs), and how dysregulation of these RNAs contributes to oncogenesis and cancer progression. To date, sdRNAs have been identified to function similar to miRNA and PIWI-associated RNA (piRNA). Given the profound role of miRNA and piRNA with respect to post-transcriptional gene regulation and the extensive literature characterizing their influence in cancer, investigation of sdRNAs and their role in cancer is of utmost importance.

2. snoRNA Structure and Function

One class of ncRNA found to give rise to functional ndRNAs is the snoRNA. SnoRNAs constitute a class of ncRNA localized to the nucleolus that can be further divided into two main subgroups: C/D box and H/ACA box snoRNA [3]. C/D box snoRNA are structurally distinguished by the presence of a 5′ box C motif (RUGAUGA, R = purine) and its C′ box partner, as well as a 3′ box D motif (CUGA) with its D′ box partner. Self-complementarity drives the formation of the C/D box snoRNA stem structure that in turn binds accessory proteins NOP56, NOP58, 15.5 K, and fibrillarin to form the functional sno-ribonucleoprotein (snoRNP) (Figures 1 and 2). Situated in the nucleolus, antisense domains located within the C/D box snoRNA serve as guides for the snoRNP to bind target ribosomal RNAs (rRNAs). The methyltransferase fibrillarin is then responsible for carrying out 2′-O-ribose-methylation of the rRNA [4–6]. H/ACA box snoRNA contain a 3′ ACA box (ACANNN, N = any nt) and two hairpins linked by an H box (ANANNA, N = any nt). The H/ACA box snoRNP complex is formed by association with NHP2, NOP10, GAR1, and dyskerin proteins (Figures 1 and 3). The H/ACA snoRNA hairpins guide the snoRNP to target rRNAs where dyskerin carries out pseudouridylation [7–9]. Erroneous expression and/or activity of either of these classes of snoRNA have been extensively linked to tumorigenesis. However, reports have implied that several proteins comprising the snoRNP are required for the biogenesis of distinct species of sdRNA [10].

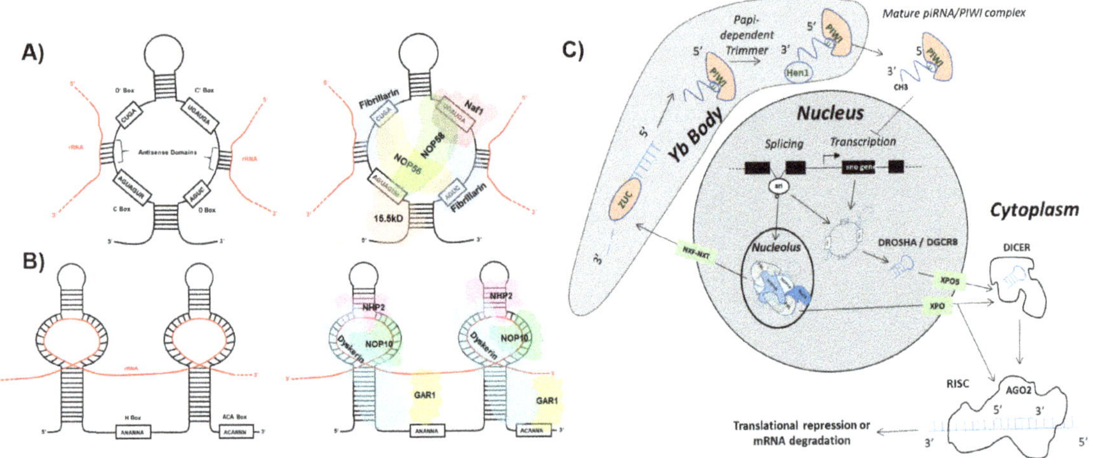

Figure 1. sdRNAs arise from full-length snoRNAs. (**A**) C/D box snoRNA structure and accessory proteins. (**Left**) The C/D box snoRNA consists of a 5′ C box (RUGAUGA) motif, a 3′ D box (CUGA) motif, and the C′ and D′ boxes located internally. Antisense domains identify target rRNAs (red) via complementarity. (**Right**) The C/D box snoRNP complex consists of NOP56, NOP58, 15.5 K, and fibrillarin proteins [4–6]. (**B**) H/ACA box snoRNA structure and accessory proteins. (**Left**) The H/ACA box snoRNA consists of a 3′ ACA box (ACANNN, N = any nt) and two hairpins that target rRNA (red) linked by an H box (ANANNA, N = any nt). (**Right**) The H/ACA box snoRNP complex consists of NHP2, NOP10, GAR1, and dyskerin proteins [7–9]. (**C**) sdRNA biogenesis and function. Biogenesis of Argonaute 2 (AGO2)-associating sno-derived RNAs (sdRNAs) Full length small-nucleolar RNAs (snoRNAs) are generated either as products of transcription or splicing [11–14]. snoRNAs produced by transcription can give rise to microRNA-like sdRNAs which are specifically excised from parent snoRNA transcripts by employment of the classical microRNA (miRNA) processing pathway. This occurs by processing of parent snoRNAs into smaller transcripts by the microprocessor complex which consists of Drosha Ribonuclease III (DROSHA) and DiGeorge syndrome critical region 8 (DGCR8). The intermediate snoRNA then undergoes cytoplasmic exportation via exportin 5 (XPO5). Following this, the smaller cytoplasmic snoRNA is processed by Dicer RNase III endonuclease (DICER) to generate the mature sdRNA which associates with AGO2, leading to the formation of the RNA-induced Silencing Complex (RISC). Similar to miRNAs, these sdRNAs function in post-transcriptional gene suppression by antisense binding to target mRNA transcripts within RISC [15]. That said, snoRNAs produced by splicing can also enter the classical miRNA processing pathway. Spliced snoRNAs, however, can bypass processing from DROSHA/DGCR8 and/or DICER as a result of trafficking to the nucleolus and subsequent processing by the fibrillarin complex followed by cytoplasmic export via a transporter belonging to the Exportin (XPO) family of proteins [10]. Biogenesis of PIWI-associated RNA (piRNA) like sdRNAs Spliced snoRNAs which arrive at the nucleolus for fibrillarin processing can be trafficked into Yb bodies via Nuclear RNA Export Factor (NXF1)/ Nuclear Transport Factor 2 Like Export Factor 1 (NXT1) where the 3′ end is cleaved by Zucchini (ZUC) and subsequently degraded. The remaining transcript is processed further within the Yb body by the papi-dependent trimmer. Following this, HEN1 double-stranded RNA binding protein binds at the 3′ end of the transcript where it adds a methyl group, generating a mature piRNA/PIWI complex which is exported to the cytoplasm. This piRNA/PIWI complex can then be shuttled back into the nucleus where it functions to inhibit transcription [16,17].

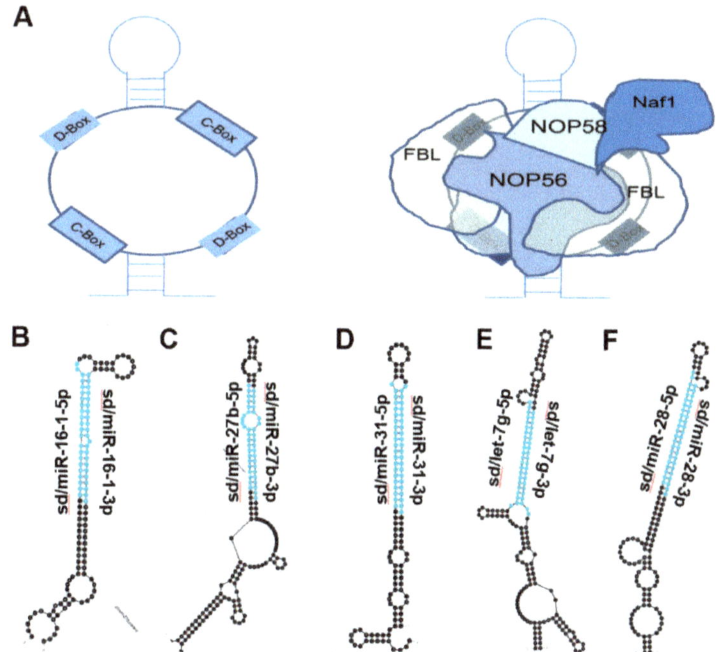

Figure 2. miRNAs derived from C/D Box snoRNA-like transcripts that bind fibrillarin. (**A**) Cartoon illustration of C/D Box snoRNA (**left**) and schematic of C/D Box snoRNA bound to fibrillarin complex, giving rise to snoRNP (**right**). Fibrillarin complex consists of NOP56, NOP58, NAF1 and Fibrillarin [18]. (**B–F**) Most thermodynamically stable secondary structures of C/D Box snoRNA-like transcripts described to bind fibrillarin complex [19] were obtained from mfold [20]. Mature miRNA sequences embedded in these transcripts are highlighted in blue. (**B**) miR16-1 (ENSG00000208006) is downregulated in chronic lymphocytic leukemia, gastric, NSCLC, Osteosarcoma, and breast cancer and is embedded within a C/D Box snoRNA-like transcript (GRCh38: chr 13:50048958-50049077:-1) known to bind fibrillarin [1,21–24]. (**C**) miR-27b (ENSG00000207864) is downregulated in prostate, lung, and bladder cancer and is located within a C/D Box snoRNA-like transcript (GRCh38: chr 9:95085436-95085592:1) known to bind fibrillarin [25–27]. (**D**) miR-31 (ENSG00000199177) is upregulated in colorectal, HNSCC, and lung cancer but is downregulated in glioblastoma, melanoma, and prostate cancer. miR31 is also located within a C/D Box snoRNA-like transcript (GRCh38: chr 9:21512102-21512221:-1) known to bind fibrillarin [28–33]. (**E**) miR-let7g (ENSG00000199150) has been shown to be downregulated in NSCLC, colorectal and ovarian cancers and is also embedded within a C/D Box snoRNA-like transcript (GRCh38: chr 3:52268239-52268408:-1) known to bind fibrillarin [34–36]. (**F**) miR-28 (ENSG00000207651) has been shown to be downregulated in B-cell lymphoma, prostate and breast cancer and is located within a C/D Box snoRNA-like transcript (GRCh38: chr 3:188688746-188688887:1) known to bind fibrillarin [37–39].

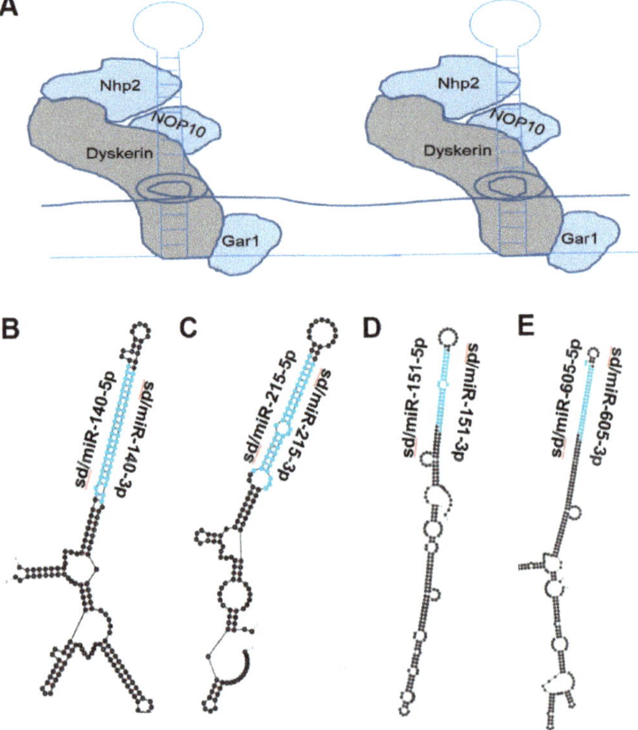

Figure 3. miRNAs derived from Box H/ACA snoRNA-like transcripts that bind dyskerin. (**A**) Schematic of box H/ACA snoRNA bound to dyskerin complex, giving rise to snoRNP. Dyskerin complex includes Dyskerin, NHP2, NOP10, and GAR1 [18]. (**B–E**) Most thermodynamically stable secondary structures of box H/ACA snoRNA-like transcripts known to bind dyskerin complex [40] were generated by mfold [20]. Mature miRNA sequences derived from these transcripts are highlighted in blue. (**B**) Box H/ACA snoRNA-like transcript (GRCh38: chr 16:69933072-69933264:1) was identified to bind dyskerin and encompasses miR-140 (ENSG00000208017) which functions as a tumor-suppressing RNA and is downregulated in prostate cancer [41]. (**C**) Box H/ACA snoRNA-like transcript (GRCh38: chr1:220117845-220118007:-1) was identified to bind dyskerin and surrounds miR-215 (ENSG00000207590) which is known to function as a tumor suppressor and is downregulated in ovarian, colorectal, prostate and lung cancer [42,43]. (**D**) Box H/ACA snoRNA-like transcript (GRCh38: chr8:140732552-140732807:-1) was determined to bind dyskerin and encapsulates miR-151 (ENSG00000254324) which is recognized as a tumor-suppressing RNA and is downregulated in prostate cancer [44–46]. (**E**) Box H/ACA snoRNA-like transcript (GRCh38: chr 10:51299393-51299708:1) was determined to bind dyskerin and surrounds miR-605 (ENSG00000207813) which is described to act as a tumor-suppressing RNA in melanoma, colorectal, breast, lung and prostate cancer [47–51].

3. sdRNA Biogenesis

3.1. miRNA-Like Biogenesis of sdRNAs

Despite requiring the existence of a parental snoRNA for sdRNA biogenesis, the transcriptional abundance of any given sdRNA and its corresponding parent snoRNA are largely unrelated [52]. SnoRNAs are approximately 100–300 nt in length and primarily arise from intronic transcripts produced by splicing mechanisms [11–13,53]. However, there is evidence supporting that correct expression and cellular localization of H/ACA snoRNA are mediated by RNA polymerase II dependent production of the snoRNA precursor, as is the case for snoRNA U64 [14]. After generation of the parent snoRNA, processing and excision of the mature sdRNA is thought to follow the classic microRNA biogenesis pathway. Consistent with this, there is evidence suggesting that H/ACA box and C/D box snoRNA are the evolutionary ancestors of a subset of miRNA precursors [19,40]. Indeed, the mature sdRNA and mature miRNA are nearly indistinguishable in structure. The exception to this is that sdRNA length varies slightly based on whether the parental snoRNA belongs to the C/D box family (~27 nt) or to the H/ACA box family (17–19 nt) while miRNAs are defined as 21–22 nt in length [2,19,40,54]. The key difference lies in biogenesis, as sdRNAs (1) arise from snoRNA transcripts and (2) can arise in a distinct manner from miRNAs (Figure 1). However, the full-length snoRNAs are believed to be processed into sdRNAs along the canonical miRNA processing pathway. In this biogenesis pathway, full length snoRNAs are processed into shorter transcripts by the microprocessor complex composed of Drosha and DGCR-8, similar to the conversion of primary microRNAs into precursor microRNAs [54,55]. Following this, the snoRNA fragment is exported from the nucleus into the cytoplasm via Exportin-5 in a RAN-GTP dependent manner. Once in the cytoplasm, the snoRNA fragment is processed further by the enzyme Dicer, yielding the mature sdRNA which can then associate with Argonaute-2 to form the RNA Induced Silencing Complex (RISC) [15,54]. At this point, the mature sdRNA functions as a molecular guide, directing RISC to a target mRNA. Upon perfect complementary binding of the sdRNA to the target mRNA, RISC then cleaves the mRNA resulting in its degradation. Alternatively, an imperfect binding of the sdRNA to the target results in a translational blockade wherein the presence of RISC impedes translation. That said, some studies on sdRNA processing may hint to the existence of disparate mechanisms for the biogenesis of miRNA-like sdRNAs which proceed independently of the microprocessor complex and DICER. Indeed, recent reports have indicated that some sdRNA species can arise from a mechanism independent of the microprocessor complex [10]. In lieu of the microprocessor complex, biogenesis of this particular sdRNA, derived from SNORD44, requires the presence of NOP58 and fibrillarin, two proteins necessary for the formation of snoRNP. Others have reported DICER independent production of miRNA-like sdRNAs [54]. Nonetheless, the remaining processing events and functional implementation are identical to the aforementioned biogenesis pathway for miRNA-like sdRNAs.

3.2. Non-miRNA-Like Biogenesis of sdRNAs

In contrast to miRNA-like biogenesis of sdRNAs, other reports have suggested the existence of sdRNA biogenesis pathways which deviate from the aforementioned pathway. For instance, the production of piRNA-like sdRNAs follows a distinct biogenesis pathway to which snoRNA transcripts are exported to Yb bodies where the 3′ ends undergo enzymatic cleavage from Zucchini (ZUC), followed by attachment of the PIWI domain to the 5′ end. The intermediate piRNA-like sdRNA is then further processed by Papi-dependent Trimmer and is trafficked to the cytoplasm, and then can re-enter the nucleus to suppress transcription (Figure 1) [16,17]. That said, there may be additional alternative pathways utilized in the biogenesis of sdRNAs which have not been realized.

4. sdRNAs in Cancer

Considering the expanding role of sdRNAs in cancer and the growing scientific interest surrounding them, we have provided a review of the existing literature highlighting specific sdRNAs involved in malignant pathology. The sdRNAs presented here represent the best studied sdRNAs to date. For convenience, Table 1 compiles these into a readily available reference with key information about each sdRNA.

Table 1. Summary of sdRNAs implicated in cancer.

sdRNA	Sequence (5′-sdRNA :: 3′-sdRNA) **	Parental snoRNA	Annotated as miR?	Cancer	Expression	Phenotypic Effect	Target	Reference
				sdRNAs Misannotated as Traditional miRNAs				
**sd/miR-664a (ENSG00000281696)	5′-ACUGGCUAGGGAAAAUGAUUGGAU-3′ :: 5′-UAUUCAUUUAUCCCCAGCCUACA-3′	SNORA36B (ENSG00000222370)	Yes	Hepatocellular carcinoma	Downregulated in tumor	Tumor-suppressing	AKT2	[56]
				Cervical	Downregulated in tumor	Tumor-suppressing	c-Kit	[57]
				Cutaneous squamous cell carcinoma	Upregulated in Tumor	Tumor-promoting	IFR2	[58]
sd/miR-1291 (ENSG00000281842)	5′-UGGCCCUGACUGAAGACCAGCAGU-3′	SNORA2C (ENSG00000221491)	Yes	Pancreatic	Downregulated in Tumor	Tumor-suppressing	FOXA2	[59]
				Pancreatic	UNDETERMINED	Tumor-suppressing	FOXA2	[60]
				Renal Cell Carcinoma	Downregulated in Tumor	Tumor-suppressing	GLUT1	[61]
				Prostate	Downregulated in Tumor	Tumor-suppressing	MED1	[62]
				Breast	Downregulated in metastases	Tumor-suppressing	UNDETERMINED	[63]
sd/miR-1248 (ENSG00000283958)	5′-ACCUUCUUGUAUAAGCACUGUGCUAAA-3′	SNORA81 (ENSG00000221420)	Yes	Prostate	Upregulated aggressive tumor	UNDETERMINED	UNDETERMINED	[64]
sd/miR-3651 (ENSG00000281156)	5′-CAUAGCCCGGUCGCUGGUACAUGA-3′	SNORA84 (ENSG00000239183)	Yes	Colorectal	Upregulated in tumor	Tumor-promoting	TBX1	[65]
				Esophageal	Downregulated in tumor	UNDETERMINED	UNDETERMINED	[66]
**sd/miR-768 (ENSG00000223224)	5′-GUUGGAGGAUGAAAGUACGGAGUGAU-3′ :: 5′-UCACAAUGCUGACACUCAAACUGCUGAC-3′	SNORD71 (ENSG00000223224)	Yes	Breast	UNDETERMINED	UNDETERMINED	YB-1	[67]
				Gastric	Downregulated in tumor	UNDETERMINED	UNDETERMINED	[68]
				Lung, Breast, Ovary, Melanoma, Liver, Parotid Gland, Thyroid Gland, Large Cell	Downregulated in tumor	Tumor-suppressing, UNDETERMINED	KRAS	[69]

Table 1. *Cont.*

sdRNA	Sequence (5′-sdRNA :: 3′-sdRNA) **	Parental snoRNA	Annotated as miR?	Cancer	Expression	Phenotypic Effect	Target	Reference
sdRNAs not Previously Annotated as miRNAs								
sd/hsa-sno-HBII-296B	NA	SNORD91B (ENSG00000275084)	No	Pancreatic ductal adenocarcinoma	Downregulated in Tumor	UNDETERMINED	UNDETERMINED	[70]
sd/hsa-sno-HBII-85-29	NA	SNORD116-29 (ENSG00000207245)	No	Pancreatic ductal adenocarcinoma	Downregulated in Tumor	UNDETERMINED	UNDETERMINED	[70]
sno-miR-28	5′-AAUAGCAUGUUAGAGUUCUGAUGG-3′	SNORD28 (ENSG00000274544)	No	Breast	Upregulated in Tumor	Tumor-promoting	TAF9B	[71]
sdRNA-93	5′-GCCAAGGAUGAGAACUCUAAUCUGAUUU-3′	SNORD93 (ENSG00000221740)	No	Breast	Upregulated in Tumor	Tumor-promoting	PIPOX	[72]
sdRNA-D19b	5′-AUUACAAGAUCCAACUCUGAU-3′	SNORD19b (ENSG00000238862)	No	Prostate	Upregulated in tumor	Tumor-promoting	CD44	[73]
sdRNA-A24	5′-CUCCAUGUAUCUUUGGGACCUGUCA-3′	SNORA24 (ENSG00000275994)	No	Prostate	Upregulated in tumor	Tumor-promoting	CDK12	[73]
miRNAs that Bind Dyskerin								
**sd/miR-140 (ENSG0000208017)	5′-CAGUGGUUUUACCCUAUGGUAG-3′ :: 5′-UACCACAGGGUAGAACCACGG-3′	Binds Dyskerin	Yes	Prostate	Downregulated in tumor	Tumor-suppressing	BIRC1	[41]
**sd/miR-151 (ENSG00000254324)	5′-UCGAGGAGCUCACAGUCUAGU-3′ :: 5′-CUAGACUGAAGCUCCUUGAGG-3′	Binds Dyskerin	Yes	Prostate	Downregulated in tumor	Tumor-suppressing	UNDETERMINED	[44]
**sd/miR-215 (ENSG00000207590)	5′-AUGACCUAUGAAUUGACAGAC-3′ :: 5′-UCUGUCAUUUCUUUAGGCCAAUA-3′	Binds Dyskerin	Yes	Ovary	Downregulated in Tumor	Tumor-suppressing	XIAP (not confirmed)	[42]
				Colorectal	Downregulated in Tumor	Tumor-suppressing	EREG, HOXB9	[43]
				Prostate	Downregulated in Tumor	Tumor-suppressing	PGK1 (not confirmed)	[45]
				Lung	Downregulated in Tumor	Tumor-suppressing	Leptin, SLC2A5	[46]
**sd/miR-605 (ENSG00000207813)	5′-UAAAUCCCAUGGUGCCUUCUCCU-3′ :: 5′-AGAAGGCACUAUGAGAUUUAGA-3′	Binds Dyskerin	Yes	Melanoma	UNDETERMINED	Tumor-suppressing	INPP4B	[49]
				Prostate	UNDETERMINED	Tumor-suppressing	EN2	[51]
				Colorectal, Breast Lung	UNDETERMINED	Tumor-suppressing, UNDETERMINED	Mdm2	[47]
				Prostate	Downregulated in tumor	UNDETERMINED	UNDETERMINED	[50]
				Prostate	UNDETERMINED	UNDETERMINED	UNDETERMINED	[48]

Table 1. *Cont.*

sdRNA	Sequence (5′-sdRNA :: 3′-sdRNA) **	Parental snoRNA	Annotated as miR?	Cancer	Expression	Phenotypic Effect	Target	Reference
				miRNAs that Bind Fibrillarin				
**sd/miR-16-1 (ENSG00000208006)	5′-UAGCAGCACGUAAAUAUUGGCG-3′ :: 5′-CCAGUAUUAACUGUGCUGCUGA-3′	Binds Fibrillarin	Yes	Chronic Lymphocytotic Leukemia	Downregulated in tumor	UNDETERMINED	Multiple (not confirmed)	[1]
				Gastric	Downregulated in tumor	Tumor-suppressing	TWIST1	[21]
				Non-small cell lung cancer	Downregulated in tumor	Tumor-suppressing	TWIST1	[22]
				Osteosarcoma	Downregulated in tumor	Tumor-suppressing	FGFR2	[23]
				Breast	Downregulated in tumor	Tumor-suppressing	PGK1	[24]
**sd/miR-27b (ENSG00000207864)	5′-AGAGCUUAGCUGAUUGGUGAAC-3′ :: 5′-UUCACAGUGGCUAAGUUCUGC-3′	Binds Fibrillarin	Yes	Prostate	Downregulated in tumor	Tumor-suppressing	UNDETERMINED	[25]
				Lung	Downregulated in tumor	Tumor-suppressing	LIMK1	[26]
				Bladder	Downregulated in tumor	Tumor-suppressing	EN2	[27]
**sd/miR-31 (ENSG00000199177)	5′-AGGCAAGAUGCUGGCAUAGCU-3′ :: 5′-UGCUAUGCCAACAUAUUGCCAU-3′	Binds Fibrillarin	Yes	Colorectal	Upregulated in tumor	Tumor-promoting	UNDETERMINED	[30]
				Head and neck squamous cell carcinoma	Upregulated in tumor	Tumor-promoting	FIH (not confirmed)	[31]
				Lung	Upregulated in tumor	Tumor-promoting	LATS2, PP2R2A	[74]
				Glioblastoma	Downregulated in tumor	Tumor-suppressing	RDX	[75]
				Melanoma	Downregulated in tumor	Tumor-suppressing	UNDETERMINED	[32]
				Prostate	Downregulated in tumor	Tumor-suppressing	UNDETERMINED	[33]
**sd/let-7g (ENSG00000199150)	5′-UGAGGUAGUAGUUUGUACAGUU-3′ :: 5′-UGAGGUAGUAGUUUGUACAGUU-3′	Binds Fibrillarin	Yes	Non-small cell lung cancer	Downregulated in tumor	Tumor-suppressing	KRAS (not confirmed)	[34]
				Colorectal	Downregulated in tumor	Tumor-suppressing	UNDETERMINED	[35]
				Ovary	Downregulated in tumor	Tumor-suppressing	UNDETERMINED	[36]

Table 1. *Cont.*

sdRNA	Sequence (5′-sdRNA :: 3′-sdRNA) **	Parental snoRNA	Annotated as miR?	Cancer	Expression	Phenotypic Effect	Target	Reference
**sd/miR-28 (ENSG00000207651)	5′-AAGGAGCUCACAGUCUAUUGAG-3′ :: 5′-CACUAGAUUGUGAGCUCCUGGA-3′	Binds Fibrillarin	Yes	B-cell Lymphoma	Downregulated in tumor	Tumor-suppressing	MAD2L1, BAG1, RAP1B, RAB23	[37]
				Prostate	Downregulated in tumor	Tumor-suppressing	SREBF2	[38]
				Breast	Downregulated in tumor	Tumor-suppressing	WSB2	[39]
Sno-Derived Piwi-interacting RNAs								
pi-sno75	5′-GGGAUUUCUGAAAUUCUAUUCUGAGGCU-3′	SNORD75	No	Breast	Downregulated in Tumor	Tumor-suppressing	WDR5	[76]
pi-sno74	5′-AGUAAUGAUGAAUGCCAACCGCUCUGAUG-3′	SNORD74	No	Breast	Downregulated in Tumor	UNDETERMINED	UNDETERMINED	[76]
pi-sno44	5′-CCUGGAUGAUGAUAAGCAAAUGCUGACU-3′	SNORD44	No	Breast	Downregulated in Tumor	UNDETERMINED	UNDETERMINED	[76]
pi-sno78 (sd78-3′)	5′-GAGCAUGUAGACAAAGGUAACACUGAAG-3′	SNORD78	No	Breast	Downregulated in Tumor	UNDETERMINED	UNDETERMINED	[76]
				Prostate	Upregulated in metastases	UNDETERMINED	UNDETERMINED	[77]
pi-sno81	5′-AUUACUUGAUGACAAUAAAAUAUCUGAUA-3′	SNORD81	No	Breast	Downregulated in Tumor	UNDETERMINED	UNDETERMINED	[76]
piR-017061 (piR-33686)	5′-CUCAGUGAUGCAAUCUCUGUGUGGUUCUGAGA-3′	SNORD91A (ENSG00000212163)	No	Pancreatic ductal adenocarcinoma	Downregulated in Tumor	UNDETERMINED	UNDETERMINED	[70]

** For sdRNAs arising from two different loci on the same precursor, the 5′ sequence precedes "::" and the 3′ sequence follows. SdRNAs arising from just one locus are given as a single sequence.

Previous reports [19,40], our own literature analysis, as well as a cursory in-house alignment of miRNAs to full-length snoRNAs collectively show that many ncRNA species currently defined as miRNAs are in fact sdRNAs. Since sdRNAs have been rapidly garnering more scientific interest, it is important to retrospectively re-categorize these miRNAs as the sdRNAs that they actually constitute. For this review, we have included several such sdRNAs that have an established role in cancer, regardless of whether the author at the time referred to the ncRNA as a miRNA or sdRNA. Previous misidentification is reported in Table 1, and each section figure shows the alignments of each of these "miRNAs" to full length snoRNAs in the human genome when relevant. For clarity, we have attached the prefix "sd/" to the beginning of the miRNA's name to indicate that the miRNA is snoRNA-derived.

4.1. sd/miR-664

In 2019 Lv et al. published a paper focused on elucidating the tumor-suppressive role of sd/miR-664 in cervical cancer. Comparing sd/miR-664 expression in cervical cancer patient samples and control tissue revealed that the sdRNA is expressed at lower levels in cancer. Using the Si-Ha cell line as a model of cervical cancer, sd/miR-664 expression was found to enhance apoptosis and reduce cell viability and migration. In a Si-Ha mouse xenograft model, treatment with an sd/miR-664 mimic drastically reduced tumor volume and weight while enhancing tumor apoptosis. The binding target was predicted to be c-Kit using online target prediction software, and this was confirmed via the Renilla luciferase assay. Sd/miR-664 expression was negatively correlated with c-Kit expression, which further solidified the proto-oncogene c-Kit as a target of this tumor-suppressive sdRNA in cervical cancer [57]. To assess the role of sd/miR-664 in cutaneous squamous cell carcinoma (cSCC), Li et al. quantified the sdRNA's expression by in situ hybridization in patient samples as well as cSCC cell lines [58]. Both analyses revealed an increase in sd/miR-664 in cSCC compared to control. Phenotypic assays showed increased cSCC viability, colony formation, invasion, and migration when sd/miR-664 levels were elevated in cell lines. Additionally, a murine cSCC xenograft model treated with sd/miR-664 mimic exhibited increased tumor volume compared to control. The 3′-UTR of IRF2 was predicted and then confirmed to be a binding target of sd/miR-664. Taken together, sd/miR-664 functions as an onco-miR in cSCC by downregulating IRF2 expression. While the role of IRF2 has not been established in cSCC, IRF2 has been identified as a tumor suppressor in lung cancer and gastric cancer and, paradoxically, as an oncogene in testicular embryonal carcinoma [78–80]. This adds a layer of complexity to sd/miR-664, showing that its overall impact on tumor progression can be tissue context-dependent.

In 2020 Li et al. measured sd/miR-664-5p expression in HCC patient cancer and normal samples as well as the HCC cell lines HepG2 and SUN-475. They found that sd/miR-664-5p is significantly downregulated in HCC patient samples as well as in the HCC cell lines compared to normal control. Sd/miR-664-5p was discovered to function as a tumor-suppressing RNA, reducing cell viability, invasion, migration, and enhancing apoptosis as measured by phenotypic assays in HepG2 and SUN-475 cells. Target engagement with the AKT2 transcript 3′-UTR was confirmed and expression of sd/miR-664-5p was negatively correlated with AKT2 expression. AKT2 (PKB) is an oncogene and a critical component of the PI3K/AKT growth-promoting pathway. Taken together, this study identifies sd/miR-664-5p as a tumor-suppressing RNA that exerts its effect by transcriptionally regulating AKT2 [56].

4.2. Sd/miR-1291

Building on their 2016 publication where they found sd/miR-1291 to be downregulated in pancreatic cancer patient tissues, Tu et al. published an article in 2020 to further investigate this mechanism of action in pancreatic cancer cells [59,60]. They found that increasing sd/miR-1291 reduced ASS1 levels in the ASS1-abundant L3.3 pancreatic cancer cell line and sensitized them to arginine deprivation in vitro. In addition, the glucose transporter GLUT1 mRNA has previously been shown to be a direct target of sd/miR-1291 in renal cell carcinoma (RCC), and Tu et al. confirmed this to be the case in pancreatic cancer cells as well with a resulting decrease in glycolytic capacity [61]. Further, sd/miR-1291 treatment sensitized pancreatic cancer cell lines to cisplatin. Taken together, sd/miR-1291 has a clear role as a modulator of cell metabolism to bring about tumor suppression in pancreatic cancer.

In 2019 Cai et al. investigated the role of sd/miR-1291 in prostate cancer. They found that sd/miR-1291 is significantly downregulated in both prostate cancer patient samples as well as prostate cancer cell lines compared to normal prostate [62]. Overexpression of sd/miR-1291 in prostate cancer cells lines inhibited cell proliferation and caused cell cycle arrest at G_0/G_1, and significantly reduced tumor weight and volume in a murine model of prostate cancer. In silico target prediction identified the 3′-UTR of MED1 a a sd/miR-1291 target, and MED1 protein levels were shown to be inversely correlated with sd/miR-1291 expression both in vitro and in vivo. MED1 has previously been shown to be a marker of poor prognosis for prostate cancer [81]. This study therefore places sd/miR-1291 as a tumor-suppressing sdRNA that regulates the MED1 oncogene to restrict prostate cancer progression.

In 2020 Escuin and colleagues investigated whether miRNAs can be used as biomarkers to identify sentinel lymph node (SLN) metastasis. Sixty breast cancer patients matched primary and SLN metastasis samples were sequenced. Differential expression analysis revealed that miR-1291 expression is significantly increased in the SLN metastasis. By separating the SLN metastases by molecular subtype, the authors found that sd/miR-1291 is upregulated in HR+ breast cancer compared to the HER2+ subtype. A bioinformatic pathway analysis predicted that sd/miR-1291 regulates WNT signaling, planar cell polarity/convergent extension (PCP/CE) pathway, β-catenin independent WNT signaling, diseases of signal transduction, and signaling by receptor tyrosine kinases (RTKs), among others. While these results strongly suggest that sd/miR-1291 is a tumor-suppressing RNA, patient survival analysis revealed that expression of this sdRNA alone could not significantly predict patient survival [63].

4.3. Sd/miR-3651

A 2020 publication aimed at identifying novel contributors to colorectal cancer (CRC) found that miR-3651, which arises from SNORA84, was overexpressed in 34/40 CRC patient tumors [65]. Model cell lines of CRC were observed to have ~3-fold overexpression of sd/miR-3651. Phenotypic assays revealed that sd/miR-3651 expression was positively correlated with CRC cell proliferation, and that reduction of the sdRNA enhanced apoptosis via deactivation of PI3K/AKT and MAPK/ERK signaling. Target prediction identified the 3′-UTR of tumor suppressor TBX1 as a sd/miR-3651 target, which was confirmed via the luciferase assay. Since TBX1 is a transcription factor that has an established role as a regulator of PI3K/AKT and MAPK/ERK pathways, these results show a coherent mechanism of action by which sd/miR-3651 inactivates TBX1 and thereby activates pro growth mitogenic pathways to function as an oncogenic sdRNA.

Wang et al. assessed the expression of sd/miR-3651 in esophageal cancer patient samples and found that sd/miR-3651 expression is significantly downregulated in esophageal cancer [66]. Kaplan–Meier analysis of a separate esophageal cancer patient cohort ($n = 108$) revealed that low sd/miR-3651 expression was a marker of poor overall survival and poor disease-free progression. In contrast with studies implicating sd/miR-3651 as an oncogenic sdRNA, Wang et al. showed that sd/miR-3651's role in esophageal cancer is tumor suppressive which indicates that the effect of this sdRNA is tissue context-dependent.

4.4. sd/miR-768

In 2012, Su et al. published a study in which they interrogated gastric cancer patient samples for miRNA expression. They found that sd/miR-768-5p, which arises from SNORD71 (Figure 4), was significantly downregulated in cancer samples [68]. A 2013 study by Blenkiron et al. identified sd/miR-768-5p as a YB-1 binding partner via CO-IP in MCF7 cells [67]. The authors proposed potential explanations for the YB-1/sdRNA interaction

Biomedicines **2022**, *10*, 1819

including YB-1 functioning as a ncRNA-sponge to prevent ncRNA-mediated transcriptional repression as well as YB-1 potentially playing a role in ncRNA biogenesis.

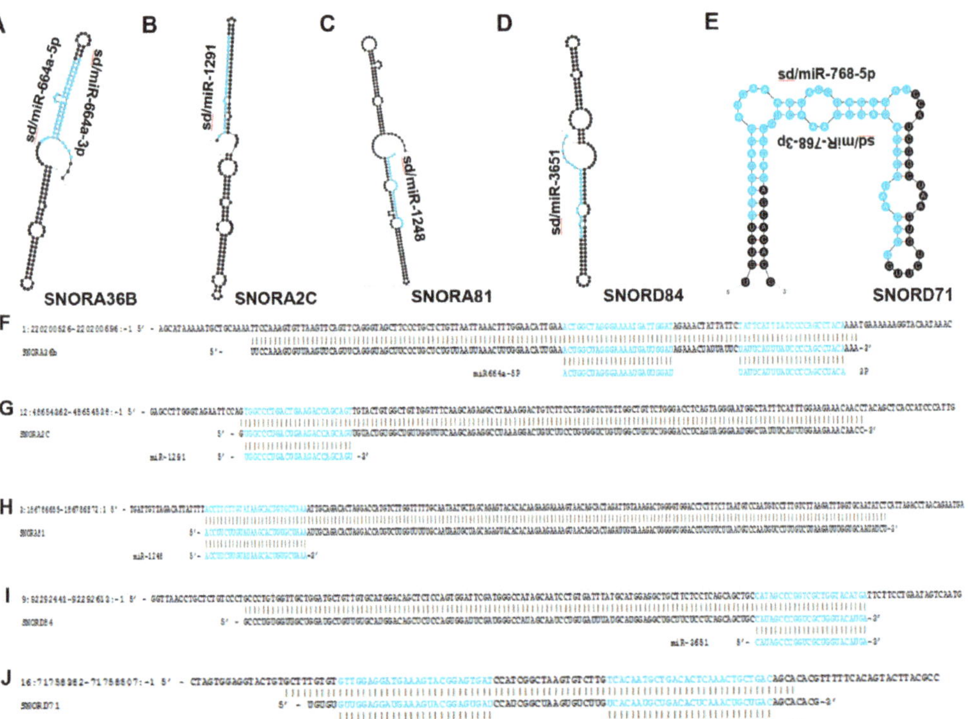

Figure 4. miRNAs that arise from snoRNA transcripts. The most thermodynamically stable secondary structures of snoRNA transcripts were generated by mfold [20]. Annotated miRNA sequences embedded in these transcripts are highlighted in blue (**A–E**). (**A**) miR-664a is embedded within the SNORA36B transcript and has been shown to function as a tumor-suppressing RNA in HCC and some types of cervical cancer while also functioning as a tumor-promoting RNA in cervical squamous cell carcinoma [56]. (**B**) miR-1291 is located within the SNORA2C transcript and is characterized as a tumor-suppressing RNA in pancreatic, prostate, breast and renal cell carcinoma [59–63]. (**C**) miR-1248 is derived from SNORA81 and has been determined to be upregulated in prostate cancer [64]. (**D**) miR-3651 is embedded within SNORA84 and is upregulated in colorectal cancer but downregulated in esophageal cancer [65,66]. (**E**) miR-768 resides within the SNORD71 transcript and has been shown to be downregulated in lung, breast, ovary, liver, parotid gland, thyroid gland cancers and melanoma [67–69]. (**F–J**) Mature miRNA sequences (highlighted in blue) were aligned to their corresponding snoRNA precursor transcripts and snoRNAs were aligned to their respective genomic regions. (**F**) Alignment between human genome (GRCh38: chr1:220200526-220200696:-1) (**top**), SNORA36B (ENSG00000222370) (**middle**), and miR664a (ENSG00000281696) (**bottom**). (**G**) Alignment between human genome (GRCh38: chr12:48654362-48654538:-1) (**top**), SNORA2C (ENSG00000221491) (**middle**) and miR-1291 (ENSG00000281842) (**bottom**). (**H**) Alignment between human genome (GRCh38: chr 3:186786655-186786872:1) (**top**), SNORA81 (ENSG00000221420) (**middle**), and miR-1248 (ENSG00000283958) (**bottom**). (**I**) Alignment between human genome (GRCh38: chr 9:92292441-92292613:-1) (**top**) SNORA84 (ENSG00000239183) (**middle**) and miR-3651 (ENSG00000281156) (**bottom**). (**J**) Alignment between human genome (GRCh38: chr 16:71758382-71758507:-1) (**top**), SNORD71 (ENSG00000223224) (**middle**) and miR-768 (ENSG00000223224) (**bottom**).

In that same year, Subramani et al. found that sd/miRNA-768-5p and its alternatively processed variant sd/miRNA-768-3p were both greatly downregulated in lung cancer and breast cancer cell lines in vitro [69]. Patient brain metastases originating from lung, breast ovary, melanoma, liver, parotid gland, thyroid gland, and large cell tumors were also found to have downregulated sd/miRNA-763-3p. Overexpression of either sd/miRNA-768-5p or sd/miRNA-768-3p increased lung and breast cancer enhanced cell viability and chemoresistance. KRAS, one of the most commonly mutated oncogenes in human cancer, was confirmed to be a sd/miRNA-768-3p target.

4.5. sd/miR-1248

There is a clear need for additional reliable biomarkers to aid care providers in their decision on which treatment to apply to prostate cancer patients following radical prostatectomy. Aimed at meeting this need, Pudova et al. published a paper in 2020 focused on identifying differentially expressed miRNAs between prostate cancer patient tumors with or without lymphatic dissemination [64]. Among the nine miRNAs found to be significantly upregulated in prostate cancer with lymphatic dissemination, sd/miR-1248 was identified which arises completely from SNORA81.

4.6. hsa-sno-HBII-296B and hsa-sno-HBII-85-29

In 2015, Müller et al. performed small RNA sequencing on six PDAC tissue samples and five normal pancreas tissue samples to assess the expression of ncRNAs not typically found via microarray analysis [70]. They considered 45 noncoding RNAs identified as significantly down-regulated in PDAC, of which there were fourteen sdRNAs and a single sno-derived piRNA. The most downregulated sdRNA in PDAC, as determined by log_2 fold change, was hsa-sno-HBII-85-29. Four sdRNAs of a total 78 ncRNAs were significantly upregulated in PDAC. The most upregulated sdRNA in PDAC, as determined by log_2 fold change, was hsa-sno-HBII-296B. Subsequent analyses were focused on a miRNA identified in the study, however this work constitutes the only publication that characterizes differentially expressed sdRNAs focused specifically in PDAC. PDAC is a particularly difficult cancer to detect and treat, with a reported 5-year survival rate of just 11% [82]. Müller et al.'s work lays the foundation for many future studies aimed at further characterizing sdRNAs in PDAC which can potentially provide tractable biomarkers/therapeutic targets to reduce the lethality of this especially morbid cancer type.

4.7. Sno-miR-28

By measuring snoRNA expression following induced P53 activation, Yu et al. found significant downregulation of all snoRNAs associated with the SNGH1 snoRNA gene [71]. One sdRNA arising from SNGH1-associated sno28, sno-miR-28, was identified in complex with AGO via HITS-CLIP data and was found to be abundantly expressed in patient breast cancer tissue. Elevating TP53 expression in vitro consequently reduced sno-mir-28, and in silico and in vitro target engagement experiments revealed the P53-stabilizing protein TAF9B as a likely sno-mir28 target. Taken together, the authors proposed that sno-mir28 directly regulates TAF9B to bring about indirect repression of P53, forming a loop as P53 overexpression decreases sno-mir28 levels. Taqman expression analysis of matched breast cancer patient tumor and normal tissues revealed that SNHG1, SNORD28 and sno-miR-28 were all significantly upregulated in tumors. Furthermore, using the MCF10A cell line as a model of undifferentiated breast epithelium, the authors found that sno-miR-28 overexpression enhanced breast cancer cells' proliferative capacity and colony formation. In sum, this study defined a role for sno-miR-28 as an oncogenic sdRNA heavily involved in suppressing the P53 pathway.

4.8. sdRNA-93

In 2017 our lab published a study in which we investigated the role of sdRNAs in the aggressive breast cancer phenotype [72]. RNAseq analysis identified 13 total full length snoRNAs differentially expressed (>7.5×) in MDA-MB-231 compared to MCF7, 10 of which consistently give rise to sdRNAs that associate with AGO in publicly available HITS-CLIP data. We elected to focus on the sdRNA arising from sno93 (sdRNA-93) (Figure 5) due to it displaying the highest differential expression in MDA-MB-231 ($\geq$75×) and previous publications implicating sdRNA-93 with miRNA-like silencing capabilities in luciferase assays [83]. SdRNA-93 silencing in MDA-MB-231 reduced cell invasion by >90% while sdRNA-93 overexpression conversely enhanced cell invasion by >100% at the same time point. By contrast, sdRNA-93 silencing in MCF7 had no significant effect on invasion while overexpression resulted in a substantial ~80% increase in cell invasion. We then employed multiple in silico miRNA target-prediction algorithms which predicted with consensus that Pipox, a gene involved in sarcosine metabolism which has been implicated in breast cancer progression, is a likely candidate for regulation by sdRNA-93. We then confirmed sdRNA-93/Pipox 3′-UTR target engagement in vitro using the Renilla luciferase assay. Taken together, these results identified sdRNA-93 as a strong regulator of breast cancer invasion, particularly in the more aggressive MDA-MB-231 cell line, with Pipox as a verified cellular target. Moving forward, sdRNA-93 could serve as a potential target for breast cancer therapeutics and biomarker studies.

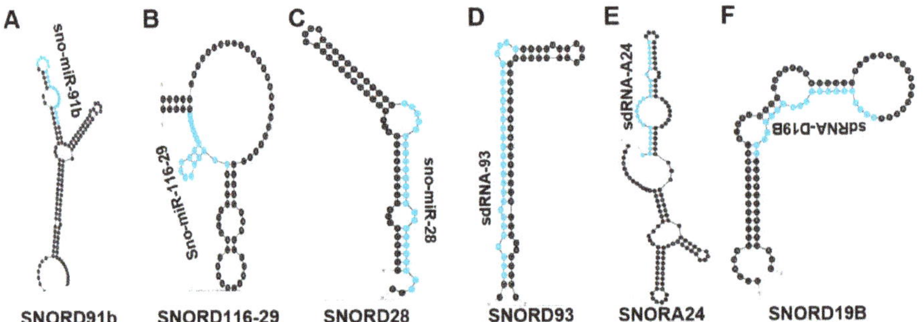

Figure 5. sdRNAs not Previously Annotated as miRNAs. (**A–F**) The most thermodynamically stable secondary structures of sdRNA producing snoRNA transcripts were generated by mfold [20]. Highlighted in blue are sdRNAs shown to function as bona fide miRNAs. (**A**) sdRNA produced by SNORD91B (ENSG00000275084) was determined to be downregulated in PDAC [70]. (**B**) sdRNA which arises from SNORD116-29 (ENSG00000207245) was determined to be downregulated in PDAC [70]. (**C**) sdRNA derived from SNORD28 (ENSG00000274544) was found to function as a tumor-promoting RNA and is upregulated in breast cancer []. (**D**) sdRNA excised from SNORD93 (ENSG00000221740) was found to function as a tumor-promoting RNA and is upregulated in breast cancer [72].(**E**) sdRNA produced from SNORA24 (ENSG00000275994) was functionally identified as a tumor-promoting RNA and is upregulated in prostate cancer [73]. (**F**) sdRNA derived from SNORD19B (ENSG00000238862) was determined to function as a tumor-promoting RNA and is upregulated in prostate cancer [73].

4.9. sdRNA-D19b and sdRNA-A24

A study published in 2022 by our lab identified 38 specifically-excised and differentially-expressed sdRNAs in prostate cancer [73]. We began by querying PCa patient TCGA datasets alongside TCGA-normal prostate using an in-house web-based search algorithm SURFR (Short Uncharacterized RNA Fragment Recognition). SURFR aligns next generation sequencing (NGS) datasets to a frequently updated database of all human ncRNAs, performs a wavelet analysis to specifically determine the location and expression of ncRNA-derived fragments (ndRNAs) and then conducts an expression analysis to identify significantly differentially expressed ndRNAs. Two sdRNAs, sdRNA-D19b and sdRNA-A24 (Figure 3), were among the most overexpressed in PCa patient tumors and were identified as AGO-associated in publicly available datasets so they were selected for further scrutiny. In vitro phenotypic assays in PC3 cells, a model cell line for castration-resistant prostate cancer (CRPC), revealed that both sdRNAs markedly increased PC3 cell proliferation and that sdRNA-D19b, in particular, greatly enhanced cell migration. When the sdRNAs were overexpressed alongside treatment with chemotherapeutic agents, both sdRNAs provided drug-specific resistances with sdRNA-D19b levels correlating with paclitaxel resistance and sdRNA-A24 conferring dasatinib resistance. Multiple in silico target-prediction algorithms provided a consensus prediction of the CD44 and CDK12 3′ UTRs as targets for sdRNA-D19b and sdRNA-A24 respectively, which we then confirmed in vitro via the Renilla luciferase assay. Taken together, this work outlines a biologically coherent mechanism by which sdRNAs downregulate tumor suppressors in CRPC to enhance proliferative/metastatic capabilities and to encourage chemotherapeutic resistance.

4.10. Sd/miR-140

Based on prior investigations linking high expression of the long-noncoding RNA MALAT1 with poor prostate cancer patient prognosis, Hao et al.'s 2020 paper focused on elucidating the mechanism of action by which MALAT1 brings about this effect [41]. In silico target prediction and in vitro confirmation via the Renilla luciferase assay identified miR-140 as a target of MALAT1, indicating that MALAT1 functions as a miRNA sponge to reduce sd/miR-140 bioavailability. This was further supported by RIP-seq analysis where direct engagement of MALAT1 and sd/miR-140 was confirmed. Target prediction indicated that sd/miR-140's target is the 3′UTR of BIRC, and this was confirmed via the Renilla luciferase assay. Knockdown of MALAT1 inhibited the growth of prostate cancer both in vitro and in vivo, an effect that the authors suggest is brought about through sd/miR-140's release from repression by MALAT1. This outlines a pathway where sd/miR-140 functions as a tumor-suppressing RNA and is tightly regulated by the lncRNA MALAT1.

4.11. Sd/miR-151

A study in 2020 by Chen et al. found that sd/miR-151 is downregulated in human prostate cancer cell lines [44]. In prostate cancer cells, Chen et al. demonstrated that sd/miR-151 overexpression inhibited cell proliferation, migration, and invasion; enhanced apoptosis; and sensitized cells to treatment with 5-FU, an antimetabolite chemotherapy. While no target prediction was performed, the authors did find that overexpression of sd/miR-151 reduces phosphorylation of PI3K/AKT. Even without a precisely mapped mechanism of action, this publication links sd/miR-151 with prostate cancer as a tumor suppressive sdRNA.

4.12. Sd/miR-215

A 2015 study by Ge et al. found that sd/miR-215 expression was reduced in both epithelial ovarian cancer cell lines and patient tissue [42]. Sd/miR-215 decreased cell proliferation, enhanced apoptosis, and enhanced sensitivity to paclitaxel treatment. In addition, increased sd/miR-205 resulted in a decrease of cellular X-linked inhibitor of apoptosis (XIAP) mRNA albeit without confirmation of target engagement. Vychytilova Faltejskova et al. found in 2017 that sd/miR-215-5p is significantly reduced in CRC patient samples and that low sd/miR-215-5p expression is associated with late clinical stages of CRC as well as poor overall survival for CRC patients [43]. Overexpression of sd/miR-215-5p reduced cell proliferation, viability, colony formation, invasion, migration, in vivo tumor volume, and weight while enhancing apoptosis. Epiregulin (EREG) and HOXB9 mRNA were confirmed to be two targets of sd/miR-215-5p. Both genes are involved in epithelial growth factor receptor (EGFR) signaling, a critical pathway exploited by CRC to promote tumor growth.

In 2020 Chen et al. found that sd/miR-215-5p is downregulated in prostate cancer patient samples and that lower sd/miR-215-5p expression predicted worse overall survival [45]. Sd/miR-215-5p expression was found to be inversely correlated with cell viability, migration, and invasion. Target prediction algorithms identified the PGK1 mRNA as a likely target, and in vitro protein quantification revealed an inverse relationship between sd/miR-215-5p expression and PGK1. A 2020 study by Jiang et al. found that the lncRNA lnc-REG3G-3-1 regulates sd/miR-215-3p availability by adsorption in lung adenocarcinoma (LAD) [46]. Overexpression of sd/miR-215-3p in LAD cells was found to reduce cell viability, invasion, and migration. Both Leptin and SLC2A5 mRNA were confirmed as targets for sd/miR-215-3p.

4.13. Sd/miR-605

With their 2011 publication, Xiao et al. found that sd/mir-605 overexpression significantly reduced cell survival and enhanced apoptosis in breast and colon cancer cell lines [47]. The MDM2 mRNA was predicted as a target of sd/mir-605, and sd/mir-605 overexpression was found to cause a reduction in MDM2 protein. MDM2 is a regulator of P53 tumor suppressor, and sd/mir-605 transfection was shown to increase P53 activity without increasing P53 protein levels, suggesting the sdRNA relieves MDM2-mediated repression of P53. In 2014 a study by Huang et al. found that single nucleotide polymorphisms (SNPs) in the sd/miR-605 precursor were significantly correlated with a shorter prostate cancer biochemical recurrence for patients [48].

Chen et al. found in 2017 that sd/miR-605 was significantly downregulated in a panel of melanoma cell lines and patient samples. Overexpression of sd/miR-605 markedly decreased cell proliferation, colony formation, and soft-agar growth in melanoma cell lines while reducing tumor volume in a murine model. INPP4B mRNA was confirmed as a sd/miR-605 target. The authors found that overexpression of sd/miR-605 decreases cellular INPP4B, leading them to conclude that sd/miR-605 suppresses melanoma growth through the inhibition of INPP4B [49].

In 2016 Alhasan et al. measured small noncoding RNA expression in prostate cancer patient serum samples and found that sd/miR-605 expression was significantly lower in low-risk prostate cancer and normal tissue compared to high-risk prostate cancer [50]. This publication established that sd/miR-605 is detectable in patient serum and is negatively correlated with the aggressive prostate cancer phenotype. A 2017 study from Zhou et al. found that sd/miR-605 is significantly downregulated in prostate cancer cell lines and patient samples [51]. EN2 mRNA was confirmed to be a sd/miR-605 target, and sd/miR-605 mimic decreased EN2 protein levels while antisense sd/miR-605 enhanced EN2 production. Prostate cancer proliferation and invasion were significantly reduced by sd/miR-605 overexpression, and this effect was rescued by transfection with EN2 cDNA. Cell-cycle analysis via flow cytometry revealed that sd/miR-605 overexpression resulted in a higher count of prostate cancer cells arrested at G_0/G_1. Taken together, the authors concluded that sd/miR-605 functions as a prostate cancer tumor-suppressing RNA by downregulating EN2.

4.14. Sd/miR-16-1

S/dmiR-16-1 is one of the miRNAs first implicated in human cancer [1]. In 2002, a paper by Calin et al. found that the sd/miR-16-1 gene was deleted in more than 65% of chronic lymphocytotic leukemia (CLL). Northern blot analysis of sd/miR-16-1 expression in CLL patient cancer and normal tissue revealed that miR-16-1 is downregulated in cancer. Sd/miR-16-1 has since been verified as a tumor suppressor in a multitude of cancers, several of which have been detailed in this 2009 review by Aqeilan et al. [84]. In 2017, the sdRNA was demonstrated to function as a tumor-suppressing RNA in gastric cancer by Wang et al. [21]. The authors found sd/miR-16-1 to be downregulated in gastric cancer patient samples. Additionally, overexpression decreased gastric cancer cell migration, invasion, and colony formation in vitro while reducing tumor volume in vivo. Target engagement with the 3′-UTR of TWIST1 was confirmed via the Renilla luciferase assay. Interestingly, a 2018 paper by a separate group, Feng et al., discovered that sd/miR-16-1 also targets TWIST1 in non-small cell lung cancer (NSCLC) [22]. Overexpression inhibited

NSCLC proliferation, migration, and invasion in vitro. Together these publications add to the numerous examples of sd/miR-16-1 functioning as a tumor-suppressing RNA.

In 2019, Maximov et al. published an article in which they found that sd/miR-16-1 was downregulated in osteosarcoma (OS) patient samples [23]. Overexpression of sd/miR-16-1 was tumor suppressive in vitro resulting in decreased colony formation and invasion while enhancing apoptosis and chemosensitivity. Increasing sd/miR-16-1 expression in a murine model of OS resulted in a reduction of tumor volume and weight. Target prediction followed by Renilla luciferase engagement confirmation identified the 3′-UTR of FGFR2 as sd/miR-16-1's target. Even more recently, Ye et al. conducted a similar study in 2020 where they found that sd/miR-16-1 targets PGK1 to suppress breast cancer malignancy [24]. In sum, sd/miR-16-1 is among the most well studied sdRNAs in cancer where its role as a tumor-suppressing RNA has been cemented.

4.15. Sd/miR-27b

In 2012, Ishteiwy et al. found that sd/miR-27b is downregulated in castration-resistant prostate tumors compared to primary prostate cancer and normal tissue [25]. Increasing sd/miR-27b expression in CRPC cell lines reduced invasion, metastasis, and colony formation. While there was no confirmation of target interactions, the authors found that modulating sd/miR-27b expression had a subsequent impact on Rac1 activity and E-cadherin expression. These findings indicate a tumor-suppressive role for sd/miR-27b in CRPC whose mechanism may involve the tumor suppressor gene E-cadherin.

Motivated by prior studies linking sd/miR-27b as a tumor-suppressing RNA, Wan et al. published a study in 2014 where they identified sd/miR-27b as downregulated in NSCLC patient tissues and cell lines [26]. Overexpression of the sdRNA reduced NSCLC proliferation, colony formation, migration, and invasion in vitro. The mRNA target was confirmed to be LIMK1, which is an oncogene that has been implicated as an enhancer of metastatic potential [85]. Regulation of LIMK1 expression by sd/miR-27b therefore suppresses the NSCLC malignant phenotype.

A 2021 publication by Li et al. measured sd/miR-27b expression in patient bladder cancer and adjacent normal tissue, finding that the sdRNA had lower expression in cancer. Overexpression of sd/miR-27b in bladder cancer cell lines caused decreases in cell proliferation, invasion, and migration while enhancing apoptosis. Target engagement with the 3′-UTR of the transcription factor EN2 was confirmed in vitro, and overexpression of sd/miR-27b was capable of reducing cellular EN2 protein as measured via western blot. Therefore, in bladder cancer the sdRNA sd/miR-27b functions as a tumor-suppressing RNA by reducing expression of the EN2 oncogene [27].

4.16. Sd/miR-31

The diverse role of miR-31 in cancer has been well studied and is the subject of two excellent reviews both in 2013 and more recently in 2018 [28,29]. Some examples from the aforementioned reviews include sd/miR-31 acting as a tumor-promoting RNA in CRC, HNSCC, and lung cancer while acting as a tumor-suppressing RNA in glioblastoma, melanoma, and prostate cancer [30–33,74,75].

4.17. Sd/let-7g

The let-7g miRNA belongs to the let-7 family of miRNAs, all of which have been well studied in the context of human cancer. Reviews published in 2010, 2012, and 2017 detail the let-7 family as well as let-7g specifically [86–88]. One example from this includes an early publication from 2008 by Kumar et al. where let-7g was found to suppress the NSCLC phenotype both in vitro and in vivo through the RAS pathway [34]. Many publications implicating sd/let-7g as a tumor-suppressing RNA in cancer have emerged since the aforementioned review, including a 2019 publication by Chang et al. in colorectal cancer and a 2019 paper by Biamonte et al. in ovarian cancer [35,36].

4.18. Sd/miR-28

Sd/miR-28 has been identified as a tumor-suppressing RNA in several cancers, including a 2014 publication by Schneider et al. focused on B-cell lymphoma (BCL) [37]. The sdRNA was downregulated in patient BCL samples (n = 25) compared to normal B cells (n = 4), and sd/miR-28 overexpression in BCL cell lines suppressed proliferation and colony formation while enhancing apoptosis. Target engagement was confirmed for sd/miR-28 with the 3′-UTRs of MAD2L1, BAG1, RAP1B, and RAB23 mRNA. Taken together, sd/miR-28 suppresses the BCL phenotype by regulating the expression of four oncogenes involved in cell cycle progression and apoptosis.

Two additional papers linking sd/miR-28-5p as a tumor-suppressing RNA were published recently in 2020. Fazio et al. found that sd/miR-28-5p binds to the 5′-UTR of SREBF2 to inhibit prostate cancer malignancy [38]. A separate publication by Ma et al. discovered that sd/miR-28-5p binds to the 3′-UTR of WSB2 mRNA to suppress breast cancer tumor migration [39]. The available literature indicates that in the majority of cases, sd/miR-28 gene products function as tumor-suppressing agents in a variety of cancers of diverse tissue origin.

4.19. pi-sno74, pi-sno75, pi-sno44, pi-sno78, and pi-sno81

Small RNA-seq (smRNAseq) of patient breast cancer tissue identified five differentially expressed sno-derived piRNAs (pi-snos) within the long noncoding RNA (lncRNA) GAS5 locus [76]. These pi-sno's included pi-sno74, pi-sno75, pi-sno44, pi-sno78, and pi-sno81 (Figure 6). Each of these were found to be downregulated in breast cancer compared to adjacent normal tissue. Microarray and PCR revealed that pi-sno75 upregulates expression of tumor necrosis factor (TNF)-related apoptosis-inducing ligand (TRAIL), a proapoptotic protein. In silico target prediction identified a locus of 169 basepairs (bp) upstream of the TRAIL transcription start site that was highly complementary to pi-sno75 with predicted thermodynamically stable binding. To further investigate this mechanism of action, knockout (KO) and Co-IPs were performed to identify PIWIL1/4 binding partners during pi-sno75 overexpression. This revealed that pi-sno75/PIWIL can specifically interact with the methyltransferase complex protein WDR5, thereby increasing TRAIL expression. In MCF7 cells, pi-sno75 overexpression greatly increased TRAIL expression and, in combination with doxorubicin treatment, enhanced apoptosis. Treatment with pi-sno75 alone resulted in marked reductions of tumor volume in MCF7 and MDA-MB-231 mouse xenografts. In summary, the sno-derived piRNA pi-sno75 recruits epigenetic machinery to specifically upregulate TRAIL, and thus functions as a tumor-suppressive sdRNA.

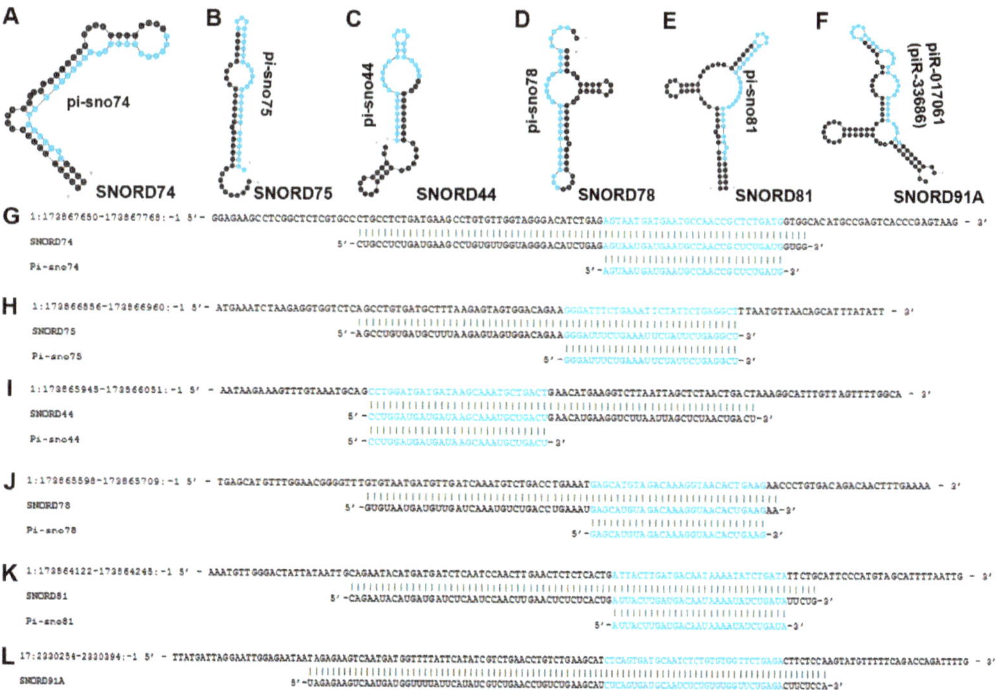

Figure 6. Sno-Derived Piwi-Interacting RNAs. (**A–F**) The most thermodynamically stable secondary structures of snoRNAs containing Piwi-interacting RNAs were obtained from mfold in which the piRNA sequence is highlighted in blue [20]. (**A**) pi-sno74 is embedded within SNORD74 (GRCh38: chr 1:173867674-173867745:-1) and is downregulated in breast cancer [76]. (**B**) Pi-sno75 is located within SNORD75 (GRCh38: chr 1:173866879-173866938:-1) and has been shown to be downregulated in breast cancer [76]. (**C**) pi-sno44 resides within SNORD44 (GRCh38: chr 1:173865968-173866028:-1) and is downregulated in breast cancer [76]. (**D**) pi-sno78 is embedded in SNORD78 (GRCh38: chr 1:173865622-173865686:-1) and is upregulated in prostate cancer metastasis but downregulated in breast cancer [76,77]. (**E**) pi-sno81 resides within SNORD81 (GRCh38: chr 1:173864146-173864222:-1) and is downregulated in breast cancer [76]. (**F**) piR-017061 (piR-33686) is embedded within SNORD91A (HBII-296A) (GRCh38: chr 17:2330279-2330370:-1) and was determined to be downregulated in PDAC [70]. (**G–L**) sno-derived piRNA sequences (highlighted in blue) were aligned to their respective snoRNA precursor sequences, and the snoRNA transcripts were aligned to their corresponding genomic loci. (**G**) Alignment between human genome (GRCh38: chr 1:173867650-173867768:-1) (**top**), SNORD74 (GRCh38: chr 1:173867674-173867745:-1) (**middle**) and pi-sno74 (**bottom**). (**H**) Alignment between human genome (GRCh38: chr1:173866856-173866960:-1) (**top**), SNORD75 (GRCh38: chr 1:173866879-173866938:-1) (**middle**), and pi-sno75 (**bottom**). (**I**) Alignment between human genome (GRCh38: chr1:173865945-173866051:-1) (**top**), SNORD44 (GRCh38: chr 1:173865968-173866028:-1) (**middle**), and pi-sno44 (**bottom**). (**J**) Alignment between human genome (GRCh38: chr1:173865598-173865709:-1) (**top**), SNORD78 (GRCh38: chr 1:173865622-173865686:-1) (**middle**) and pi-sno78 (**bottom**). (**K**) Alignment between human genome (GRCh38: chr1:173864122-173864245:-1) (**top**), SNORD81 (GRCh38: chr 1:173864146-173864222:-1) (**middle**) and pi-sno81 (**bottom**). (**L**) Alignment between human genome (GRCh38: chr17:2330254-2330394:-1) (**top**), SNORD91A (HBII-296A) (GRCh38: chr 17:2330279-2330370:-1) (**middle**) and piR-017061 (piR-33686) (**bottom**). (**A–E,G–K**) SNORD74, SNORD75, SNORD44, SNORD78, and SNORD81 arise from the GAS5 locus (ENSG00000234741) which is also annotated as a small nucleolar host gene (SNHG2).

4.20. pi-sno78 (Sd78-3′)

In 2011, a study by Martens-Uzunova et al. that focused on identifying a miRNA expression signature of prostate cancer also identified significantly differentially expressed sdRNAs [89]. While the majority of reads aligned to defined microRNAs, sdRNAs were found to be predominantly upregulated in the metastatic LN-PCa compared to local prostate cancer. The authors then focused on canonical miRNAs, leaving these enticing putative sdRNA drivers of metastatic LN-PCa uncharacterized until their follow-up publication in 2015. In their 2015 publication, Martens-Uzunova et al. specifically focused on investigating the role of sdRNAs in prostate cancer progression [77]. Consistent with their 2011 study, they identified an sdRNA from the 3′ end of snoRNA78 "sd78-3′" that was upregulated in the more aggressive LN-PCa patient samples. By examining expression at multiple stages of PCa, the authors concluded that globally overexpressed sdRNAs, including sd78-3′, are already present at early stages of cancer but exhibit striking overexpression concordantly with malignant transformation. While the authors did not investigate the mechanism of action of sd78-3′, this sdRNA is actually the pi-sno78 characterized in the above study by He et al. [76].

4.21. piR-017061

Revisiting the 2015 publication by Muller et al. which queried small RNA expression in PDAC patient samples and normal pancreas controls, a total of 123 ncRNAs were found to be significantly differentially expressed [70]. This approach considered 45 noncoding RNAs identified as significantly down-regulated in PDAC, of which there were fourteen sdRNAs and a single sno-derived piRNA, piR-017061.

5. SdRNAomes

In a large-scale pan-cancer analysis of TCGA patient small RNA sequencing data and normal tissue controls, Chow et al. created a pan-cancer "sdRNAome" which was used to investigate the role of sdRNAs in the context of tumor immunity and clinical outcomes [52]. Small RNAseq data from a total of 10,262 tumor samples with 675 adjacent normal samples across 32 cancer types were aligned to full-length human snoRNAs to generate the pan-cancer "sdRNAome". To determine whether a relationship exists between sdRNA expression and tumor immunity, the sdRNAome was queried alongside expression of PD-L1, CD8$^+$ T cell infiltration, GZMA (serine protease granzyme A) expression, and other factors such as tumor vascularization (EndothelialScore) and overall survival (OS). Taking together correlation with PD-L1 expression, CD8$^+$ T cell abundance, GZMA expression, and patient survival to calculate an ImmuneSurv score, the authors identified 267 sdRNAs that had a score of at least two (fulfilling at least two of the ImmuneSurv categories) in at least one cancer type. After including copy number variation as a category, 133 sdRNAs were found to be significant in all five tested categories. In sum, this far-reaching study provides a detailed examination of the human sdRNAome in patient tumor samples. Critically, sdRNA expression is linked to tumor immunity and patient survival in distinct cancer types, providing evidence that sdRNAs could function as useful biomarkers to assess cancer response to immunotherapies and to predict patient survival. Since a large number of sdRNAs were associated with cancers in this study, several representative sdRNAs highlighted in the paper have been compiled into Table 1 for the sake of clarity.

In addition to this, other publications reviewed above have reported comprehensive sdRNAomes comprised of differentially expressed sdRNAs in their respective cancers. In 2015 Martens-Uzunova et al. identified 657 unique sdRNAs from C/D-box snoRNAs and 244 unique sdRNAs from H/ACA-box snoRNAs in their analysis of patient prostate cancer samples. Of these, they identified at least 78 to be significantly differentially expressed in prostate cancer [77] (Table 2). Additionally, in 2017 our group identified 10 unique sdRNAs in MDA-MB-231 breast cancer cells that form complexes with AGO (Table 2). Further, five of these were significantly overexpressed in MDA-MB-237 compared to MCF7 breast cancer [72]. Most recently, in 2022, our lab similarly described 38 unique sdRNAs differentially expressed in prostate cancer patient samples [73] (Table 2).

Table 2. Cancer "sdRNAomes".

Cancer	# of sdRNAs Identified	Experimental Validation	Reference
32 TCGA Cancer Types	>300	133 sdRNAs correlate with PD-L1 expression, CD8+ T cell abundance, GZMA expression, patient survival, and copy number variation	[52]
Prostate	78	sd78-3′ was found to be overexpressed in aggressive patient tumors	[77,89]
Breast	10	sdRNA-93 was confirmed to correlate with malignant invasion in vitro and cancer type in vivo	[72]
Prostate	38	sdRNA-D19b and sdRNA-A24 were confirmed to correlate with the malignant phenotype in vitro and cancer type in vivo	[73]

6. Discussion/Conclusions

The past decade has clearly demonstrated the existence and functional relevance of small RNAs arising from full length snoRNAs. This new class of ncRNA has a similar form and function to that of miRNAs, and sdRNAs differ from traditional miRNAs primarily in terms of transcriptional origin. Additionally, alternative processing pathways to DICER/DROSHA have been proposed for sdRNA genesis, potentially distinguishing sdRNAs further from their miRNA cousins. In this review we have collected and detailed the growing catalogue of sdRNAs implicated in human cancer, including both tumor-promoting and tumor-suppressing sdRNAs. Like miRNAs, sdRNAs post-transcriptionally regulate gene expression in a myriad of ways that can result in highly variable effects on oncogenesis and malignant pathology. Despite differences in genetic origin and some possible bioprocessing divergence, the majority of sdRNAs appear to be functionally identical to canonical miRNAs. What then, is the purpose of pointing out sdRNAs? The answer to this question is two-fold.

First, and of extreme importance, is the fact that sdRNAs are routinely discarded from miRNA databases and consequently omitted from any miRNA-focused studies that follow. Even with mounting evidence to the contrary, sdRNAs are still considered by many to be random degradation products without relevance to gene regulation. As shown in this study, many such misannotated sdRNAs clearly contribute to a variety of cancers. Omission of sdRNAs unnecessarily hinders efforts to discover valuable new ncRNA biomarkers and therapeutic targets. We stress that, instead of discarding sdRNAs, they deserve inclusion in miRNA databases. By bringing the wealth of cancer sdRNA literature to the light, we hope that researchers who aim to search for ncRNAs regulating cancer will include sdRNAs in their search moving forward.

Second, we simply do not know precisely how distinct sdRNAs and miRNAs are from one another. There is ample evidence that sdRNAs are frequently processed independently of the DROSHA/DICER pathway [90]. Still, the preponderance of evidence indicates that they are extremely similar, especially regarding their function. Indeed, we have identified a 200 kb locus on human chromosome 14 that produces about 50 snoRNAs and 50 miRNAs (Supplementary Figure S1), indicating that these ncRNAs are often co-regulated and functionally similar.

As evidenced by the sdRNAs outlined in this review, we suggest that including and emphasizing sdRNAs in exploratory ncRNA cancer regulation studies as well as closing the knowledge gaps regarding what makes sdRNAs distinct, including but not limited to their alternative biogenesis and mechanism of action pathways, will be of significant value in enhancing our understanding of cancer gene regulation. Notably, sdRNAs may be thought of as the lowest hanging fruit in cancer genomics research as, to date, the principal miRNA discovery algorithms have simply, erroneously, disregarded any and all small RNAs aligning to snoRNAs (example text: "Aligned sequences with the following annotations were eliminated (as potential microRNAs): tRNA, snoRNA, … " [91]). A more complete understanding and more extensive cataloguing of sdRNAs in cancer will likely result in novel biomarkers and treatment targets that now lay only just out of reach. By appreciating sdRNAs as functionally relevant ncRNA molecules in cancer, a great deal of readily attainable information regarding cancer gene regulation and the molecules responsible for pathological phenotypes may soon be realized.

Supplementary Materials: The following supporting information can be downloaded at: https://www.mdpi.com/article/10.3390/biomedicines10081819/s1, Figure S1: Human Chromosome 14 microRNA and snoRNA cluster (C14MSC).

Author Contributions: A.B.C. and J.D.D. wrote the manuscript and produced these figures and tables. N.Y.C. conducted extensive and detailed literature searches and checked and corrected the language of the manuscript for publication. G.M.B. assisted at all stages. All authors have read and agreed to the published version of the manuscript.

Funding: Funding was provided in part by NSF RAPID grant NSF2030080 (G.M.B.) and NSF CAREER grant 1350064 (with co-funding provided by the NSF EPSCoR program) (G.M.B.) both awarded by the Division of Molecular and Cellular Biosciences. This work was also funded in part by the USA COM IGP award 1828 (G.M.B.). Graduate funding was also provided in part by Alabama Commission on Higher Education ALEPSCoR grant 210471 (J.D.D.). The project used an instrument funded, in part, by the National Science Foundation MRI Grant No. CNS-1726069. Research reported in this publication was also supported by the National Center for Advancing Translational Research of the National Institutes of Health under award number UL1TR001417.

Institutional Review Board Statement: Not applicable.

Informed Consent Statement: Not applicable.

Data Availability Statement: Not applicable.

Acknowledgments: We thank the University of South Alabama College of Medicine Department of Pharmacology for their ongoing support.

Conflicts of Interest: The authors declare no conflict of interest.

References

1. Calin, G.A.; Dumitru, C.D.; Shimizu, M.; Bichi, R.; Zupo, S.; Noch, E.; Alder, H.; Rattan, S.; Keating, M.; Rai, K.; et al. Frequent deletions and down-regulation of micro- RNA genes miR15 and miR16 at 13q14 in chronic lymphocytic leukemia. *Proc. Natl. Acad. Sci. USA* **2002**, *99*, 15524. [CrossRef] [PubMed]
2. Kawaji, H.; Nakamura, M.; Takahashi, Y.; Sandelin, A.; Katayama, S.; Fukuda, S.; O Daub, C.; Kai, C.; Kawai, J.; Yasuda, J.; et al. Hidden layers of human small RNAs. *BMC Genom.* **2008**, *9*, 157. [CrossRef] [PubMed]
3. Balakin, A.G.; Smith, L.; Fournier, M.J. The RNA world of the nucleolus: Two major families of small RNAs defined by different box elements with related functions. *Cell* **1996**, *86*, 823–834. [CrossRef]
4. Kiss-László, Z.; Henry, Y.; Kiss, T. Sequence and structural elements of methylation guide snoRNAs essential for site-specific ribose methylation of pre-rRNA. *EMBO J.* **1998**, *17*, 797–807. [CrossRef]
5. Cavaille, J.; Nicoloso, M.; Bachellerie, J.P. Targeted ribose methylation of RNA in vivo directed by tailored antisense RNA guides. *Nature* **1996**, *383*, 732–735. [CrossRef]
6. Kiss-László, Z.; Henry, Y.; Bachellerie, J.P.; Caizergues-Ferrer, M.; Kiss, T. Site-specific ribose methylation of preribosomal RNA: A novel function for small nucleolar RNAs. *Cell* **1996**, *85*, 1077–1088. [CrossRef]
7. Ganot, P.; Bortolin, M.L.; Kiss, T. Site-specific pseudouridine formation in preribosomal RNA is guided by small nucleolar RNAs. *Cell* **1997**, *89*, 799–809. [CrossRef]
8. Ni, J.; Tien, A.L.; Fournier, M.J. Small nucleolar RNAs direct site-specific synthesis of pseudouridine in ribosomal RNA. *Cell* **1997**, *89*, 565–573. [CrossRef]
9. Bortolin, M.L.; Ganot, P.; Kiss, T. Elements essential for accumulation and function of small nucleolar RNAs directing site-specific pseudouridylation of ribosomal RNAs. *EMBO J.* **1999**, *18*, 457–469. [CrossRef]
10. Bai, B.; Yegnasubramanian, S.; Wheelan, S.J.; Laiho, M. RNA-Seq of the Nucleolus Reveals Abundant SNORD44-Derived Small RNAs. *PLoS ONE* **2014**, *9*, e107519. [CrossRef]
11. Leverette, R.D.; Andrews, M.T.; Maxwell, E.S. Mouse U14 snRNA is a processed intron of the cognate hsc70 heat shock pre-messenger RNA. *Cell* **1992**, *71*, 1215–1221. [CrossRef]

12. Fragapane, P.; Prislei, S.; Michienzi, A.; Caffarelli, E.; Bozzoni, I. A novel small nucleolar RNA (U16) is encoded inside a ribosoma protein intron and originates by processing of the pre-mRNA. *EMBO J.* **1993**, *12*, 2921. [CrossRef]

13. Tycowski, K.T.; Shu MDi Steitz, J.A. A mammalian gene with introns instead of exons generating stable RNA products. *Natur* **1996**, *379*, 464–466. [CrossRef] [PubMed]

14. Richard, P.; Kiss, A.M.; Darzacq, X.; Kiss, T. Cotranscriptional recognition of human intronic box H/ACA snoRNAs occurs in splicing-independent manner. *Mol. Cell Biol.* **2006**, *26*, 2540–2549. [CrossRef] [PubMed]

15. Ender, C.; Krek, A.; Friedländer, M.R.; Beitzinger, M.; Weinmann, L.; Chen, W.; Pfeffer, S.; Rajewsky, N.; Meister, G. A huma snoRNA with MicroRNA-like functions. *Mol. Cell* **2008**, *32*, 519–528. [CrossRef]

16. Chen, S.; Ben, S.; Xin, J.; Li, S.; Zheng, R.; Wang, H.; Fan, L.; Du, M.; Zhang, Z.; Wang, M. The biogenesis and biological functio of PIWI-interacting RNA in cancer. *J. Hematol. Oncol.* **2021**, *14*, 93. [CrossRef] [PubMed]

17. Nishimasu, H.; Ishizu, H.; Saito, K.; Fukuhara, S.; Kamatani, M.K.; Bonnefond, L.; Matsumoto, N.; Nishizawa, T.; Nakanaga, K Aoki, J.; et al. Structure and function of Zucchini endoribonuclease in piRNA biogenesis. *Nature* **2012**, *491*, 284–287. [CrossRef [PubMed]

18. Wajahat, M.; Bracken, C.P.; Orang, A. Emerging functions for snoRNAs and snoRNA-derived fragments. *Int. J. Mol. Sci.* **2021**, *22*, 2 [CrossRef]

19. Ono, M.; Scott, M.S.; Yamada, K.; Avolio, F.; Barton, G.J.; Lamond, A.I. Identification of human miRNA precursors that resembl box C/D snoRNAs. *Nucleic Acids Res.* **2011**, *39*, 3879. [CrossRef]

20. Zuker, M. Mfold web server for nucleic acid folding and hybridization prediction. *Nucleic Acids Res.* **2003**, *31*, 3406–341! [CrossRef]

21. Wang, T.; Hou, J.; Li, Z.; Zheng, Z.; Wei, J.; Song, D.; Hu, T.; Wu, Q.; Yang, J.Y.; Cai, J.C. miR-15a-3p and miR-16-1-3p Negativel Regulate Twist1 to Repress Gastric Cancer Cell Invasion and Metastasis. *Int. J. Biol. Sci.* **2017**, *13*, 122. [CrossRef] [PubMed]

22. Qq, F.; Zq, D.; Zhou, Y.; Zhang, H.; Long, C. miR-16-1-3p targets TWIST1 to inhibit cell proliferation and invasion in NSCL(*Bratisl. Lek. Listy* **2018**, *119*, 60–65. [CrossRef]

23. Maximov, V.V.; Akkawi, R.; Khawaled, S.; Salah, Z.; Jaber, L.; Barhoum, A.; Or, O.; Galasso, M.; Kurek, K.C.; Yavin, E.; et a MiR-16-1-3p and miR-16-2-3p possess strong tumor suppressive and antimetastatic properties in osteosarcoma. *Int. J. Cance* **2019**, *145*, 3052–3063. [CrossRef] [PubMed]

24. Ye, T.; Liang, Y.; Zhang, D.; Zhang, X. MicroRNA-16-1-3p Represses Breast Tumor Growth and Metastasis by Inhibiting PGK1 Mediated Warburg Effect. *Front. Cell Dev. Biol.* **2020**, *8*, 615154. [CrossRef] [PubMed]

25. Ishteiwy, R.A.; Ward, T.M.; Dykxhoorn, D.M.; Burnstein, K.L. The microRNA -23b/-27b cluster suppresses the metastati phenotype of castration-resistant prostate cancer cells. *PLoS ONE* **2012**, *7*, e52106. [CrossRef] [PubMed]

26. Wan, L.; Zhang, L.; Fan, K.; Wang, J. MiR-27b targets LIMK1 to inhibit growth and invasion of NSCLC cells. *Mol. Cell Biochem* **2014**, *390*, 85–91. [CrossRef]

27. Li, Y.; Duan, Q.; Gan, L.; Li, W.; Yang, J.; Huang, G. microRNA-27b inhibits cell proliferation and invasion in bladder cancer b targeting engrailed-2. *Biosci. Rep.* **2021**, *41*, BSR20201000. [CrossRef]

28. Laurila, E.M.; Kallioniemi, A. The diverse role of miR-31 in regulating cancer associated phenotypes. *Genes Chromosom. Cance* **2013**, *52*, 1103–1113. [CrossRef]

29. Yu, T.; Ma, P.; Wu, D.; Shu, Y.; Gao, W. Functions and mechanisms of microRNA-31 in human cancers. *Biomed. Pharmacother.* **2018** *108*, 1162–1169. [CrossRef]

30. Bandrés, E.; Cubedo, E.; Agirre, X.; Malumbres, R.; Zárate, R.; Ramirez, N.; Abajo, A.; Navarro, A.; Moreno, I.; Monzó, M.; et a Identification by Real-time PCR of 13 mature microRNAs differentially expressed in colorectal cancer and non-tumoral tissue *Mol. Cancer.* **2006**, *5*, 29. [CrossRef]

31. Liu, C.J.; Tsai, M.M.; Hung, P.S.; Kao, S.Y.; Liu, T.Y.; Wu, K.J.; Chiou, S.H.; Lin, S.C.; Chang, K.W. miR-31 ablates expression of th HIF regulatory factor FIH to activate the HIF pathway in head and neck carcinoma. *Cancer Res.* **2010**, *70*, 1635–1644. [CrossRef [PubMed]

32. Greenberg, E.; Hershkovitz, L.; Itzhaki, O.; Hajdu, S.; Nemlich, Y.; Ortenberg, R.; Gefen, N.; Edry, L.; Modai, S.; Keisari, Y.; et a Regulation of cancer aggressive features in melanoma cells by MicroRNAs. *PLoS ONE* **2011**, *6*, e18936. [CrossRef] [PubMed]

33. Fuse, M.; Kojima, S.; Enokida, H.; Chiyomaru, T.; Yoshino, H.; Nohata, N.; Kinoshita, T.; Sakamoto, S.; Naya, Y. Nakagawa, M.; et al. Tumor suppressive microRNAs (miR-222 and miR-31) regulate molecular pathways based on mi croRNA expression signature in prostate cancer. *J. Hum. Genet.* **2012**, *57*, 691–699. [CrossRef] [PubMed]

34. Kumar, M.S.; Erkeland, S.J.; Pester, R.E.; Chen, C.Y.; Ebert, M.S.; Sharp, P.A.; Jacks, T. Suppression of non-small cell lung tumo development by the let-7 microRNA family. *Proc. Natl. Acad. Sci. USA* **2008**, *105*, 3903–3908. [CrossRef] [PubMed]

35. Chang, C.M.; Wong, H.S.C.; Huang, C.Y.; Hsu, W.L.; Maio, Z.F.; Chiu, S.J.; Tsai, Y.T.; Chen, B.K.; Wan, Y.J.Y.; Wang, J.Y.; et a Functional effects of let-7g expression in colon cancer metastasis. *Cancers* **2019**, *11*, 489. [CrossRef] [PubMed]

36. Biamonte, F.; Santamaria, G.; Sacco, A.; Perrone, F.M.; Di Cello, A.; Battaglia, A.M.; Salatino, A.; Di Vito, A.; Aversa, I. Venturella, R.; et al. MicroRNA let-7g acts as tumor suppressor and predictive biomarker for chemoresistance in human epithelia ovarian cancer. *Sci. Rep.* **2019**, *9*, 5668. [CrossRef] [PubMed]

37. Schneider, C.; Setty, M.; Holmes, A.B.; Maute, R.L.; Leslie, C.S.; Mussolin, L.; Rosolen, A.; Dalla-Favera, R.; Bassoa, K. microRN/ 28 controls cell proliferation and is down-regulated in B-cell lymphomas. *Proc. Natl. Acad. Sci. USA* **2014**, *111*, 8185–819C [CrossRef] [PubMed]

38. Fazio, S.; Berti, G.; Russo, F.; Evangelista, M.; D'Aurizio, R.; Mercatanti, A.; Pellegrini, M.; Rizzo, M. The miR-28-5p targetome discovery identified SREBF2 as one of the mediators of the miR-28-5p tumor suppressor activity in prostate cancer cells. *Cells* **2020**, *9*, 354. [CrossRef]
39. Ma, L.; Zhang, Y.; Hu, F. miR-28-5p inhibits the migration of breast cancer by regulating WSB2. *Int. J. Mol. Med.* **2020**, *46*, 1562–1570. [CrossRef]
40. Scott, M.S.; Avolio, F.; Ono, M.; Lamond, A.I.; Barton, G.J. Human miRNA precursors with Box H/ACA snoRNA features. *PLOS Comput. Biol.* **2009**, *5*, e1000507. [CrossRef] [PubMed]
41. Hao, T.; Wang, Z.; Yang, J.; Zhang, Y.; Shang, Y.; Sun, J. MALAT1 knockdown inhibits prostate cancer progression by regulating miR-140/BIRC6 axis. *Biomed. Pharmacother.* **2020**, *123*, 109666. [CrossRef] [PubMed]
42. Ge, G.; Zhang, W.; Niu, L.; Yan, Y.; REn, Y.; Zou, Y. MiR-215 functions as a tumor suppressor in epithelial ovarian cancer through regulation of the X-chromosome-linked inhibitor of apoptosis. *Oncol. Rep.* **2016**, *35*, 1816–1822. [CrossRef] [PubMed]
43. Vychytilova-Faltejskova, P.; Merhautova, J.; Machackova, T.; Gutierrez-Garcia, I.; Garcia-Solano, J.; Radova, L.; Brchnelova, D.; Slaba, K.; Svoboda, M.; Halamkova, J.; et al. MiR-215-5p is a tumor suppressor in colorectal cancer targeting EGFR ligand epiregulin and its transcriptional inducer HOXB9. *Oncogene* **2017**, *6*, 399. [CrossRef] [PubMed]
44. Liao, B.; Chen, S.; Ke, S.; Cheng, S.; Wu, T.; Yang, Y. MicroRNA-151 regulates the growth, chemosensitivity and metastasis of human prostate cancer cells by targeting PI3K/AKT. *J. BUON* **2020**, *25*, 2045–2050.
45. Chen, J.-Y.; Xu, L.-F.; Hu, H.-L.; Wen, Y.-Q.; Chen, D.; Liu, W.-H. MiRNA-215-5p alleviates the metastasis of prostate cancer by targeting PGK1. *Gene* **2017**, *626*, 344–353.
46. Jiang, C.; Zhao, H.; Yang, B.; Sun, Z.; Li, X.; Hu, X. lnc-REG3G-3-1/miR-215-3p Promotes Brain Metastasis of Lung Adenocarcinoma by Regulating Leptin and SLC2A5. *Front. Oncol.* **2020**, *10*, 1344. [CrossRef] [PubMed]
47. Xiao, J.; Lin, H.; Luo, X.; Luo, X.; Wang, Z. miR-605 joins p53 network to form a p53:miR-605:Mdm2 positive feedback loop in response to stress. *EMBO J.* **2011**, *30*, 5021. [CrossRef]
48. Huang, S.P.; Levesque, E.; Guillemette, C.; Yu, C.C.; Huang, C.Y.; Lin, V.C.; Chung, I.C.; Chen, L.C.; Laverdière, I.; Lacombe, L.; et al. Genetic variants in microRNAs and microRNA target sites predict biochemical recurrence after radical prostatectomy in localized prostate cancer. *Int. J. Cancer.* **2014**, *135*, 2661–2667. [CrossRef] [PubMed]
49. Chen, L.; Cao, Y.; Rong, D.; Wang, Y.; Cao, Y. MicroRNA-605 functions as a tumor suppressor by targeting INPP4B in melanoma. *Oncol. Rep.* **2017**, *38*, 1276–1286. [CrossRef]
50. Alhasan, A.H.; Scott, A.W.; Wu, J.J.; Feng, G.; Meeks, J.J.; Thaxton, C.S.; Mirkin, C.A. Circulating microRNA signature for the diagnosis of very high-risk prostate cancer. *Proc. Natl. Acad. Sci. USA* **2016**, *113*, 10655–10660. [CrossRef]
51. Zhou, Y.; Yang, H.; Xia, W.; Cui, L.; Xu, R.; Lu, H.; Xue, Z.; Zhang, B.; Tian, Z.; Cao, Y.; et al. Down-regulation of miR-605 promotes the proliferation and invasion of prostate cancer cells by up-regulating EN2. *Life Sci.* **2017**, *190*, 7–14. [CrossRef] [PubMed]
52. Chow, R.D.; Chen, S. Sno-derived RNAs are prevalent molecular markers of cancer immunity. *Oncogene* **2018**, *37*, 6442–6462. [CrossRef] [PubMed]
53. Edoh, D.; Kiss, T.; Filipowicz, W. Activity of U-snRNA genes with modified placement of promoter elements in transfected protoplasts and stably transformed tobacco. *Nucleic Acids Res.* **1993**, *21*, 1533–1540. [CrossRef] [PubMed]
54. Taft, R.J.; Glazov, E.A.; Lassmann, T.; Hayashizaki, Y.; Carninci, P.; Mattick, J.S. Small RNAs derived from snoRNAs. *RNA* **2009**, *15*, 1233–1240. [CrossRef] [PubMed]
55. Gregory, R.I.; Yan, K.P.; Amuthan, G.; Chendrimada, T.; Doratotaj, B.; Cooch, N.; Shiekhattar, R. The Microprocessor complex mediates the genesis of microRNAs. *Nature* **2004**, *432*, 235–240. [CrossRef] [PubMed]
56. Li, H.; Guo, D.; Zhang, Y.; Yang, S.; Zhang, R. miR-664b-5p inhibits hepatocellular cancer cell proliferation through targeting oncogene AKT2. *Cancer Biother. Radiopharm.* **2020**, *35*, 605–614. [CrossRef] [PubMed]
57. Lv, M.; Ou, R.; Zhang, Q.; Lin, F.; Li, X.; Wang, K.; Xu, Y. MicroRNA-664 suppresses the growth of cervical cancer cells via targeting c-Kit. *Drug Des. Dev. Ther.* **2019**, *13*, 2371. [CrossRef]
58. Li, X.; Zhou, C.; Zhang, C.; Xie, X.; Zhou, Z.; Zhou, M.; Chen, L.; Ding, Z. MicroRNA-664 functions as an oncogene in cutaneous squamous cell carcinomas (cSCC) via suppressing interferon regulatory factor 2. *J. Dermatol. Sci.* **2019**, *94*, 330–338. [CrossRef]
59. Tu, M.J.; Pan, Y.Z.; Qiu, J.X.; Kim, E.J.; Yu, A.M. MicroRNA-1291 targets the FOXA2-AGR2 pathway to suppress pancreatic cancer cell proliferation and tumorigenesis. *Oncotarget* **2016**, *7*, 45547. [CrossRef]
60. Tu, M.J.; Duan, Z.; Liu, Z.; Zhang, C.; Bold, R.J.; Gonzalez, F.J.; Kim, E.J.; Yu, A. MicroRNA-1291-5p sensitizes pancreatic carcinoma cells to arginine deprivation and chemotherapy through the regulation of arginolysis and glycolysis. *Mol. Pharmacol.* **2020**, *98*, 686. [CrossRef]
61. Yamasaki, T.; Seki, N.; Yoshino, H.; Itesako, T.; Yamada, Y.; Tatarano, S.; Hidaka, H.; Yonezawa, T.; Nakagawa, M.; Enokida, H. Tumor-suppressive microRNA-1291 directly regulates glucose transporter 1 in renal cell carcinoma. *Cancer Sci.* **2013**, *104*, 1411. [CrossRef] [PubMed]
62. Cai, Q.; Zhao, A.; Ren, L.; Chen, J.; Liao, K.; Wang, Z.; Zhang, W. MicroRNA-1291 mediates cell proliferation and tumorigenesis by downregulating MED1 in prostate cancer. *Oncol. Lett.* **2019**, *17*, 3253. [CrossRef] [PubMed]
63. Escuin, D.; López-Vilaró, L.; Bell, O.; Mora, J.; Moral, A.; Pérez, J.I.; Arqueros, C.; Ramón y Cajal, T.; Lerma, E.; Barnadas, A. MicroRNA-1291 is associated with locoregional metastases in patients with early-stage breast cancer. *Front. Genet.* **2020**, *11*, 562114. [CrossRef] [PubMed]

64. Pudova, E.A.; Krasnov, G.S.; Nyushko, K.M.; Kobelyatskaya, A.A.; Savvateeva, M.V.; Poloznikov, A.A.; Dolotkazin, D.R.; Klimina, K.M.; Guvatova, Z.G.; Simanovsky, S.A.; et al. miRNAs expression signature potentially associated with lymphatic dissemination in locally advanced prostate cancer. *BMC Med. Genom.* **2020**, *13* (Suppl. S8), 129. [CrossRef]

65. Li, C.; Ding, D.; Gao, Y.; Li, Y. MicroRNA-3651 promotes colorectal cancer cell proliferation through directly repressing T-box transcription factor 1. *Int. J. Mol. Med.* **2020**, *45*, 956–966. [CrossRef]

66. Wang, C.; Guan, S.; Chen, X.; Liu, B.; Liu, F.; Han, L.; Un Nesa, E.; Song, Q.; Bao, C.; Wang, X.; et al. Clinical potential of miR-365 as a novel prognostic biomarker for esophageal squamous cell cancer. *Biochem. Biophys. Res. Commun.* **2015**, *465*, 30–34. [CrossRef] [PubMed]

67. Blenkiron, C.; Hurley, D.G.; Fitzgerald, S.; Print, C.G.; Lasham, A. Links between the oncoprotein YB-1 and small non-coding RNAs in breast cancer. *PLoS ONE* **2013**, *8*, e80171. [CrossRef] [PubMed]

68. Su, Y.; Ni, Z.; Wang, G.; Cui, J.; Wei, C.; Wang, J.; Yang, Q.; Xu, Y.; Li, F. Aberrant expression of microRNAs in gastric cancer and biological significance of miR-574-3p. *Int. Immunopharmacol.* **2012**, *13*, 468. [CrossRef] [PubMed]

69. Subramani, A.; Alsidawi, S.; Jagannathan, S.; Sumita, K.; Sasaki, A.T.; Aronow, B.; Warnick, R.E.; Lawler, S.; Driscoll, J.J. The brain microenvironment negatively regulates miRNA-768-3p to promote K-ras expression and lung cancer metastasis. *Sci. Rep.* **2013**, *3*, 2392. [CrossRef] [PubMed]

70. Müller, S.; Raulefs, S.; Bruns, P.; Afonso-Grunz, F.; Plötner, A.; Thermann, R.; Jäger, C.; Schlitter, A.M.; Kong, B.; Regel, I.; et al. Next-generation sequencing reveals novel differentially regulated mRNAs, lncRNAs, miRNAs, sdRNAs and a piRNA in pancreatic cancer. *Mol. Cancer* **2015**, *14*, 94. [CrossRef] [PubMed]

71. Yu, F.; Bracken, C.P.; Pillman, K.A.; Lawrence, D.M.; Goodall, G.J.; Callen, D.F.; Neilsen, P.M. p53 Represses the oncogenic Sno-MiR-28 derived from a SnoRNA. *PLoS ONE* **2015**, *10*, e0129190. [CrossRef]

72. Patterson, D.G.; Roberts, J.T.; King, V.M.; Houserova, D.; Barnhill, E.C.; Crucello, A.; Polska, C.J.; Brantley, L.W.; Kaufman, G.C.; Nguyen, M.; et al. Human snoRNA-93 is processed into a microRNA-like RNA that promotes breast cancer cell invasion. *NPJ Breast Cancer* **2017**, *3*, 25. [CrossRef] [PubMed]

73. Coley, A.B.; Stahly, A.N.; Kasukurthi, M.V.; Barchie, A.A.; Hutcheson, S.B.; Houserova, D.; Huang, Y.; Watters, B.C.; King, V.M.; Dean, M.A.; et al. MicroRNA-like snoRNA-Derived RNAs (sdRNAs) promote castration-resistant prostate cancer. *Cells* **2022**, *11*, 1302. [CrossRef]

74. Liu, X.; Sempere, L.F.; Ouyang, H.; Memoli, V.A.; Andrew, A.S.; Luo, Y.; Demidenko, E.; Korc, M.; Shi, W.; Preis, M.; et al. MicroRNA-31 functions as an oncogenic microRNA in mouse and human lung cancer cells by repressing specific tumor suppressors. *J. Clin. Investig.* **2010**, *120*, 1298. [CrossRef]

75. Hua, D.; Ding, D.; Han, X.; Zhang, W.; Zhao, N.; Foltz, G.; Lan, Q.; Huang, Q.; Lin, B. Human miR-31 targets radixin and inhibits migration and invasion of glioma cells. *Oncol. Rep.* **2012**, *27*, 700–706. [CrossRef] [PubMed]

76. He, X.; Chen, X.; Zhang, X.; Duan, X.; Pan, T.; Hu, Q.; Zhang, Y.; Zhong, F.; Liu, J.; Zhang, H.; et al. An Lnc RNA (GAS5)/SnoRNA-derived piRNA induces activation of TRAIL gene by site-specifically recruiting MLL/COMPASS-like complexes. *Nucleic Acids Res.* **2015**, *43*, 3712. [CrossRef] [PubMed]

77. Martens-Uzunova, E.S.; Hoogstrate, Y.; Kalsbeek, A.; Pigmans, B.; Vredenbregt-van den Berg, M.; Dits, N.; Nielsen, S.J.; Baker, A.; Visakorpi, T.; Bangma, C.; et al. C/D-box snoRNA-derived RNA production is associated with malignant transformation and metastatic progression in prostate cancer. *Oncotarget* **2015**, *6*, 17430. [CrossRef] [PubMed]

78. Liang, C.; Zhang, X.; Wang, H.M.; Liu, X.M.; Zhang, X.J.; Zheng, B.; Qian, G.R.; Ma, Z.L. MicroRNA-18a-5p functions as an oncogene by directly targeting IRF2 in lung cancer. *Cell Death Dis.* **2017**, *8*, e2764. [CrossRef] [PubMed]

79. Chen, Y.J.; Luo, S.N.; Wu, H.; Zhang, N.P.; Dong, L.; Liu, T.T.; Liang, L.; Shen, X.Z. IRF-2 inhibits cancer proliferation by promoting AMER-1 transcription in human gastric cancer. *J. Transl Med.* **2022**, *20*, 68. [CrossRef]

80. Wang, X.; Zhang, X.; Wang, G.; Wang, L.; Lin, Y.; Sun, F. Hsa-miR-513b-5p suppresses cell proliferation and promotes P53 expression by targeting IRF2 in testicular embryonal carcinoma cells. *Gene* **2017**, *626*, 344–353. [CrossRef]

81. Hulf, T.; Sibbritt, T.; Wiklund, E.D.; Patterson, K.; Song, J.Z.; Stirzaker, C.; Qu, W.; Nair, S.; Horvath, L.G.; Armstrong, N.J.; et al. Epigenetic-induced repression of microRNA-205 is associated with MED1 activation and a poorer prognosis in localized prostate cancer. *Oncogene* **2012**, *32*, 2891–2899. [CrossRef] [PubMed]

82. Siegel, R.L.; Miller, K.D.; Fuchs, H.E.; Jemal, A. Cancer statistics, 2022. *CA Cancer J. Clin.* **2022**, *72*, 7–33. [CrossRef]

83. Brameier, M.; Herwig, A.; Reinhardt, R.; Walter, L.; Gruber, J. Human box C/D snoRNAs with miRNA like functions: Expanding the range of regulatory RNAs. *Nucleic Acids Res.* **2011**, *39*, 675–686. [CrossRef] [PubMed]

84. Aqeilan, R.I.; Calin, G.A.; Croce, C.M. miR-15a and miR-16-1 in cancer: Discovery, function and future perspectives. *Cell Death Differ.* **2009**, *17*, 215–220. [CrossRef] [PubMed]

85. Chen, Q.; Jiao, D.; Hu, H.; Song, J.; Yan, J.; Wu, L.; Xu, L.Q. Downregulation of LIMK1 level inhibits migration of lung cancer cells and enhances sensitivity to chemotherapy drugs. *Oncol. Res.* **2013**, *20*, 491–498. [CrossRef] [PubMed]

86. Boyerinas, B.; Park, S.M.; Hau, A.; Murmann, A.E.; Peter, M.E. The role of let-7 in cell differentiation and cancer. *Endocr. Relat. Cancer* **2010**, *17*, F19–F36. [CrossRef] [PubMed]

87. Wang, X.; Cao, L.; Wang, Y.; Wang, X.; Liu, N.; You, Y. Regulation of let-7 and its target oncogenes (Review). *Oncol Lett.* **2012**, *3*, 955–960. [CrossRef]

88. Balzeau, J.; Menezes, M.R.; Cao, S.; Hagan, J.P. The LIN28/let-7 pathway in cancer. *Front. Genet.* **2017**, *8*, 31. [CrossRef] [PubMed]

). Martens-Uzunova, E.S.; Jalava, S.E.; Dits, N.F.; Van Leenders, G.J.L.H.; Møller, S.; Trapman, J.; Bangma, C.H.; Litman, T.; Visakorpi, T.; Jenster, G. Diagnostic and prognostic signatures from the small non-coding RNA transcriptome in prostate cancer. *Oncogene* **2011**, *31*, 978–991. [CrossRef]

). Saraiya, A.A.; Wang, C.C. snoRNA, a Novel precursor of microRNA in giardia lamblia. *PLoS Pathog.* **2008**, *4*, e1000224. [CrossRef]

1. Lai, E.C.; Tomancak, P.; Williams, R.W.; Rubin, G.M. Computational identification of drosophila microRNA genes. *Genome Biol.* **2003**, *4*, R42. [CrossRef] [PubMed]

biomedicines

Article

Transposons Acting as Competitive Endogenous RNAs: In-Silico Evidence from Datasets Characterised by L1 Overexpression

Mauro Esposito [1], Nicolò Gualandi [1], Giovanni Spirito [1,2], Federico Ansaloni [1,3], Stefano Gustincich [2,3] and Remo Sanges [1,3,*]

1   Computational Genomics Laboratory, Area of Neuroscience, Scuola Internazionale Superiore di Studi Avanzati (SISSA), 34136 Trieste, Italy
2   CMP3vda, via Lavoratori Vittime del Col Du Mont 28, 11100 Aosta, Italy
3   Central RNA Laboratory, Istituto Italiano di Tecnologia, 16132 Genova, Italy
*   Correspondence: remo.sanges@gmail.com

Abstract: LINE L1 are transposable elements that can replicate within the genome by passing through RNA intermediates. The vast majority of these element copies in the human genome are inactive and just between 100 and 150 copies are still able to mobilize. During evolution, they could have been positively selected for beneficial cellular functions. Nonetheless, L1 deregulation can be detrimental to the cell, causing diseases such as cancer. The activity of miRNAs represents a fundamental mechanism for controlling transcript levels in somatic cells. These are a class of small non-coding RNAs that cause degradation or translational inhibition of their target transcripts. Beyond this competitive endogenous RNAs (ceRNAs), mostly made by circular and non-coding RNAs, have been seen to compete for the binding of the same set of miRNAs targeting protein coding genes. In this study, we have investigated whether autonomously transcribed L1s may act as ceRNAs by analyzing public dataset in-silico. We observed that genes sharing miRNA target sites with L1 have a tendency to be upregulated when L1 are overexpressed, suggesting the possibility that L1 might act as ceRNAs. This finding will help in the interpretation of transcriptomic responses in contexts characterized by the specific activation of transposons.

Keywords: transposons; LINE1; miRNA; ceRNA; p53

Citation: Esposito, M.; Gualandi, N.; Spirito, G.; Ansaloni, F.; Gustincich, S.; Sanges, R. Transposons Acting as Competitive Endogenous RNAs: In-Silico Evidence from Datasets Characterised by L1 Overexpression. *Biomedicines* **2022**, *10*, 3279. https://doi.org/10.3390/biomedicines10123279

Academic Editors: Milena Rizzo and Elena Levantini

Received: 30 October 2022
Accepted: 11 December 2022
Published: 17 December 2022

Publisher's Note: MDPI stays neutral with regard to jurisdictional claims in published maps and institutional affiliations.

1. Introduction

Transposable elements (TE) are interspersed and repeated DNA elements that can change their position within the genome. Based on the mechanism used for the transposition, they can be divided into DNA transposons and retrotransposons. The first ones are excised from the original genomic locus and inserted into a new position with the activity of a transposase enzyme. Retrotransposons instead can generate new copies by reverse transcribing their RNA molecules [1]. Among the several TE families, LINE L1 is the only one that contains autonomous elements considered to be still active in the human genome. L1 retrotransposons account for 17% of the human genome [2]. Despite the human genome being composed of more than 500,000 L1 fragments, the vast majority of these are inactive since they underwent 5′ truncations and accumulation of mutations [3]. A full-length L1 retrotransposon is 6 kb long and it is composed of a 5′UTR with an internal promoter, two open reading frames (ORF1 and ORF2) that provide the retrotransposition proteins, and a 3′UTR region. The human genome contains more than 5000 full-length elements and 100 out of 150 copies that are still potentially capable to mobilize [4] since they have intact and functional ORFs. It has been observed that L1 novel insertions are not randomly distributed in the genome [5]. This has suggested an underlying evolutionary force that

guided the co-option of these elements for beneficial cellular functions [6]. The positive selection of L1 elements in specific genomic regions might be linked to their capability to influence the expression of nearby genes. L1 insertions close to a given gene can provide promoters and other cis-regulatory elements [7] resulting in changes of transcriptional levels in the flanking *loci*. Conversely, insertions that induce alternative splicing events [8] or introduce miRNA target sites in 3′UTR regions [9] are examples of L1 contribution to post-transcriptional regulation. Despite selected beneficial functions, deregulated L1 activity can be detrimental to the cell and is involved in diseases, e.g., cancer and neurodegenerative diseases [10]. As a result, L1s activity is repressed at all stages of the retrotransposition process through several cellular defense mechanisms. At the genomic level, L1 loci are generally found to be inaccessible because DNA methylation and repressive epigenetic marks decrease the accessibility to the transcriptional machinery [11,12]. In addition, especially in germ cells, the Piwi-interacting RNA (piRNA) pathway acts at post-transcriptional levels to degrade L1s RNA or to inhibit its translation. The interfering activity is mediated by a complex of the small non-coding RNA piRNA and the Argonaute (Ago) proteins of the PIWI subfamily [13]. In the final stages of retrotransposition, new L1 integration sites are continuously targeted by DNA repair mechanisms for preventing the insertion of new elements [14]. Despite all these mechanisms, we do not yet have a comprehensive understanding of all the regulatory layers affecting L1 expression and how their deregulation can impact cells' homeostasis.

One of the main post-transcriptional regulatory mechanisms acting in somatic cells is represented by miRNAs, a class of small non-coding RNAs that have a fundamental role in gene expression control [15]. In the canonical pathway, miRNA genes are transcribed into primary miRNA (pri-miRNA) and processed into precursor miRNA (pre-miRNA) through the Microprocessor complex [16]. The pre-miRNA is then exported to the cytoplasm by Exportin 5 and finally processed in the mature miRNA duplex by the endonuclease Dicer [17]. The loading of the miRNA guide strand into the Argonaute protein allows the formation of the miRNA-induced silencing complex (RISC) [18]. This complex allows the binding of the specific target transcript with the loaded miRNA and then proceeds to the following steps of the process. In detail, the fate of target transcripts depends mostly on the binding between specific sequences located in the 3′ UTR regions and the miRNA. Although still not completely understood, it is believed that the degree of miRNA-target complementarity might determine whether the target transcript undergoes translational inhibition or degradation [19]. The activity of the Agonaute protein is essential for the choice of the fate. Indeed, a full pairing between the target site and the 5′ seed region (nucleotides 2–8) of the miRNA activates the endonuclease function of Argonaute, allowing mRNA slicing [20]. On the other hand, a limited miRNA-target base pairing triggers a slicer-independent pathway, leading to the recruitment of factors involved in translational inhibition or deadenylation followed by transcript decay [21,22].

In the last few years, knowledge became available on the existence of miRNA-based mechanisms used by the cell to inhibit the translation of L1 elements. In 2015, Hamdorf et al. hypothesized that miRNA can protect non-germ cells from L1 retrotransposition [23]. Particularly, they demonstrated that miR-128 can bind the ORF2 region of L1 RNAs, inducing the degradation of the transcripts. In a more recent publication, Tristàn-Ramos et al. [24] demonstrated that let-7 miRNAs bind to the ORF2 region of L1 RNAs, inhibiting their translation and altering the protein amount needed for the retrotransposition to happen.

RNA molecules that regulate transcript levels of other genes through the competition for the binding of the same pool of miRNAs have also been discovered [25]. This system formed by competitive endogenous RNAs (ceRNAs) creates crosstalk between different coding and non-coding transcripts. The first evidence of this mechanism was observed in the Poliseno study [26] which demonstrated that overexpressed 3′UTR of PTENP1 pseudogene was able to act as ceRNA for miRNAs. Indeed, the sequestering of miRNAs by PTENP1 pseudogene transcript resulted in the increase of the levels of the functional tumor suppressor PTEN transcript. The involvement of an apparent non-functional gene in this mechanism drew attention to long non-coding RNAs (lncRNAs). These are ideal candidate molecules for a widespread ceRNA mechanism given their activity as regulatory RNAs since they lack protein-encoding open reading frames [25]. In this context, the RIDL hypothesis [27] proposed that the enrichment of TEs embedded in lncRNA transcripts provides functional domains, allowing interactions with proteins and/or other nucleic acids. Accordingly, the functional properties of lncRNAs may strictly depend on their different domain combinations. As a representative example, the antisense lncRNAs *AS Uchl1* showed the ability to increase translation of its sense target gene through the activity of an effector domain made by an embedded TE and an antisense sequence as binding domain [28]. The antisense BACE1-AS is a clear example of a lncRNA in which embedded TEs provide miRNA target sites to sequester and therefore subtract miRNAs from their canonical targets [29].

Here, we aim at performing a first evaluation of the possibility that TEs might generally sequester miRNAs impacting on regulatory mechanisms in human cells. To this end, we carried out transcriptomic analyses of different cellular conditions known to undergo L1s overexpression [30–33]. We observed that genes sharing a high number of miRNA target sites with overexpressed L1s are more upregulated with respect to genes sharing a lower number of miRNA target sites with L1s. Our results are consistent with a possible ceRNAs activity mediated by L1 retrotransposon transcripts. These findings might help in the interpretation of transcriptional responses to the deregulation of TEs expression.

2. Materials and Methods

2.1. Data Collection and Pre-Processing

To explore how regulatory mechanisms control L1 transcript levels in human cells, we took advantage of different publicly available datasets overexpressing L1s. The dataset from Jonsson et al. [30] is composed of poly-A RNA-seq data generated from embryo-derived human neural epithelial-like stem cell line Sai2. In this study, the authors produced 3 controls and 3 *DNMT1* samples in which *LacZ* and *DNMT1* genes were respectively knocked out with CRISPR-Cas9 technology. We retrieved paired-end raw FASTQ files from the ENA-EBI database (PRJNA420729 accession code). For validating the reliability of our method to analyze non-autonomous L1 transcription, we used the Marasca et al. [31] dataset. It is composed of total-RNA derived from quiescent naive CD4$^+$ T cells and naive CD4$^+$ T cells activated with anti-CD3-antiCD28 beads (3 individuals, 24 total sequencing) The download of paired-end raw FASTQ files was made from the ENA-EBI database (PRJEB41930 accession code). For testing the miRNA-L1 expression levels association, we used RNA-seq and short RNA-seq data produced by the Geuvadis Project [34]. Quantification of miRNAs was retrieved from the ArrayExpress [35] repository while raw reads were retrieved for the TE quantification from ENA-EBI database (PRJEB3366 accession code). Aiming to investigate a cellular context into which L1 was artificially overexpressed, we took advantage of the publicly available RNA-seq data derived from the Ardeljan et al. study [32]. In this work, the authors performed RNA sequencing of human *Retinal Pigment Epithelial Cells* (RPE) encoding a doxycycline-inducible (Tet-On) codon-optimized L1 (ORFeus) [36] or luciferase as control. The two groups of cultures were sequenced in triplicates and the paired-end raw files were made publicly available in the ENA-EBI database (PRJNA491205 accession code). For exploring other cellular contexts in which L1s should be deregulated, we analyzed the Deneault et al. [33] polyA RNA-seq data in which, the *ATRX* gene was knocked-out in reprogrammed human induced pluripotent stem cells (iPSC) and iPSC differentiated in neuronal cells. The entire dataset composed of 12 controls and 8 treated samples was retrieved from the ENA-EBI database (PRJNA422099 accession code). The quality assessment of all the retrieved reads was performed with FastQC [37].

2.2. Analysis of Locus-Specific TE Expression

To analyze the TE transcriptome, we used the *SQuIRE* [38] software for quantifying the locus-specific TE expression starting from RNA-seq reads. The customized parameter

–build hg19 was used to allow the download of the needed files referring to the human reference genome *hg19*. Once we obtained the quantifications, we performed the differential expression analysis of TE using the *R* package *DESeq2* [39], classifying as differentially expressed TE the elements showing p-adjusted < 0.05 and |fold change| > 1.5.

2.3. Detection of L1 Autonomous Transcription

To understand if an L1 up-regulation is caused by the autonomous transcription of L1 loci, we developed a specific *R* pipeline that analyzes paired-end reads of an RNA-seq experiment. In our method, reads are aligned on the L1 consensus sequence [40] with *BWA* [41] (*mem* command in default parameters). Then, not primary and supplementary alignments are discarded using *Samtools* [42] (*-F 2304*). The remaining fragments (referred to as a pair of reads) are filtered out if at least one read of the pair has more than 20% of nucleotides that are not perfectly matched on the L1 consensus sequence. At this point, we label the fragments that are completely aligned inside the L1 elements as *"Inside"* fragments and the fragments with exclusively one read aligned inside the L1 as *"Outside"* fragments. These sets of fragments are then aligned on the human reference genome *hg19* with *BWA* [41] (*mem* command in default parameters). On the resulting BAM file, we apply the same previously described filters. In the final step, we use *Bedtools* [43] (*intersect* command) to intersect the mapping coordinates of the reads with a BED file containing the L1 elements annotated on the human reference genome (*RepeatMasker* [44] track retrieved from UCSC Table Browser [45]). Fragments with at least one read aligned on an annotated L1 element are kept into account for the final calculation. For each sample, the ratio between the number of *Inside* and *Outside* fragments is used as an indicator for the L1 autonomous transcription level in all the group comparisons except for the *ORFeus* model in which the same ratio was calculated before the genome mapping step.

2.4. Gene Expression Analysis

To investigate the gene expression levels, we aligned the FASTQ RNA-seq reads to the human reference genome *hg19* (FASTA file of primary assembly retrieved from *Ensembl* [46]) using *STAR* [47]. The indexing step was performed giving *–sjdbGTFfile* argument the GRCh37.87 GTF file retrieved from *Ensembl*. The *–sjdbOverhang* argument was instead set to *max(ReadLength)-1* for the different datasets. Then, the mapping was performed with the parameters *–quantMode GeneCounts* and *–twopassMode Basic*. After the generation of BAM files, the counting of mapped reads for each gene was obtained with *HTSeq* [48] (*htseq-count* command with the arguments *-t exon -i gene_id*). Finally, the differential gene expression analysis was performed using *R* package *DESeq2* [39] classifying, as differentially expressed, the genes showing p-adjusted < 0.05 and |fold change| > 1.5.

2.5. Functional Enrichment Analysis

To explore the transcriptomic changes in the *DNMT1* model, the differentially expressed genes were divided into upregulated and downregulated according to previous thresholds. The two separated groups were then used as input to perform the enrichment analysis with the *R* package *gProfiler2* [49] (*exclude_iea = T, user_threshold = 0.1* and *correction_method = "fdr"*). The background (*custom_bg* argument) used for the analysis was composed of genes with at least 5 mapped reads in at least 50% of the samples. Gene sets composed of more than 1000 genes were filtered out since they described too many general cellular processes. For the final exploration, gene sets with at least 10 enriched query genes and FDR $\leq$ 0.1 were considered significant.

2.6. Overlap Analysis

To study how L1 overexpression can impact regulatory gene networks, we performed an overlap analysis between genomic coordinates of L1s and protein-coding genes. For this analysis, we used the GRCh37.87 GTF annotation file retrieved from *Ensembl* [46]. From this, we selected all the exonic and UTR annotations of protein-coding genes. Then, we used

Bedtools merge [43] for collapsing together the features belonging to different transcript of the same gene. *Bedtools subtract* was used to exclude both the UTR annotations from the exons and to generate the intron coordinates for each gene. As a result, we created a BED file containing a single gene model for each protein-coding gene. The model were thus composed of the coordinates of genic structures 5′UTR, exons, introns, and 3′UTR. These coordinates were intersected with L1 coordinates (retrieved from *SQuIRE* [38] output) by using *Bedtools intersect* with "-s" argument to force the strandedness. Each genic structure was then classified based on two pieces of information: the up/down-regulation of the belonging gene and the overlap with at least one up/down-regulated L1s. At this point, for each type of genic structure, we calculated the frequency of each possible pair of classifications. For the final calculation of the Z-score, the genic classifications were randomized 1000 times recomputing the calculation of the frequencies each time, which represents the random distribution.

2.7. Analysis of miRNA Target Sites Sharing

To determine whether the set of upregulated extragenic L1 elements are sharing miRNA target sites with the 3′UTR of protein-coding genes, we used the BED file of genic structures created in the overlap analysis reported above. From this BED, we selected the 3′UTR regions longer than 30 nucleotides belonging to expressed protein-coding genes. A gene was considered as expressed if it was quantified with at least one mapped read in at least one sample. The upregulated L1 elements were identified with *SQuIRE* [38] as reported above, while the extragenic positioning of these elements was determined by intersecting the genomic coordinates of L1 and features annotated in the GRCh37.87 GTF file. All L1s that did not overlap with any concordant/discordant feature (*Bedtools intersect* with -s and -S arguments [43]) were considered extragenic. Having selected the 3′UTRs and L1 genomic coordinates, we used *Bedtools getfasta* to retrieve the FASTA file of each element and a custom *Perl* script to identify target sites by looking only for 8-mer seed-matched sites [19] of the entire set of human miRNAs retrieved from *miRBase* database (Release 22.1) [50]. Hence, for each 3′UTR of protein-coding genes, the number of miRNA target sites that were also found in the pool of extragenic upregulated L1s was calculated. T-test was finally used for comparing the number of L1-shared miRNA target sites between up and down-regulated genes.

2.8. Identification of miRNAs Sequestered by L1s

To search for miRNAs that are possibly sequestered by L1s, we analyzed miRNAs that were seen to potentially target at least one upregulated extragenic L1. For each of the miRNAs we identified the targeted protein-coding genes by searching for 8-mer seed-matched sites. The targeted genes were then classified in up or down-regulated based on the previous differential gene expression analysis. At this point, for each miRNA, we performed a proportion test (*prop.test R* function) to compare the proportion of up and down-regulated targeted genes to the proportion of all up and down-regulated genes. miRNAs in which the proportion of targeted upregulated genes was significantly (FDR < 0.1) higher than the proportion of downregulated ones were selected for further analyses.

2.9. miRNA-Gene Networks Identification

To explore miRNA-genes interactions, we used the MIENTURNET [51] web-based tool. We provided the miRBase ID list of 117 miRNAs potentially decoyed by L1s. The miRNA-target enrichment analysis was performed with the minimum number of miRNA target interactions set to 2 and an adjusted p-value (FDR) threshold of 0.1. The interactions categorized by *miRTarBase* [52] as strongly and weakly validated were taken into account. The following enrichment analysis of 57 genes upregulated in the *DNMT1* model was performed with *EnrichR* [53] considering significant the gene set enriched with an FDR < 0.1.

2.10. Analysis of miRNA-L1 Expression

To investigate the association between the expression levels of miRNAs and LINE1 elements in Geuvadis dataset [34], we used the quantifications retrieved from ArrayExpress [35] or obtained with SQuIRE [38] respectively. For each sample, the L1 expression level was obtained by summing the DESeq2 [39] normalized counts of elements belonging to the L1HS/L1PA subfamilies and longer than 5000 bp. The sample "NA18861" was discarded since it represented an outlier. The miRNA-L1 expression levels association was analyzed by performing Pearson's correlation tests.

2.11. Analysis of TE Expression at the Consensus Level

In order to investigate the TE transcriptome of the *ORFeus-OE* model, we used the *TEspeX* [54] software. For the analysis, we provided a modified version of TE consensus sequences from the *Dfam* database [55] that includes the L1-ORFeus sequence (https://www.addgene.org/browse/article/28204003/ accessed on 24 August 2021). Furthermore, the annotation files of coding and non-coding transcripts were retrieved from *Ensembl* [46] and referred to the *hg19* version of the human genome. After obtaining the quantifications, we used the *R* package *DESeq2* for differential expression analysis, classifying as differentially expressed TE the subfamilies showing p-adjusted < 0.05 and fold change > |1.5|.

3. Results

3.1. L1s Are Autonomously Transcribed upon DNMT1-KO

The first set of analyses was carried out to explore the transcriptomic changes of human neural progenitor cells (hNPCs) produced in the Jonsson's study [30]. This dataset is composed of RNA-seq data derived from three controls and three samples in which the *DNMT1* gene was knocked-out with CRISPR-Cas9 technology. As Jonsson and colleagues demonstrated, the abrogated activity of the most important maintenance DNA methyltransferase [56] erases a fundamental epigenetic repressive layer from the L1 defense mechanisms. Aiming to confirm whether L1 elements were autonomously transcribed [57,58] in this model, we used *SQuIRE* [38] to quantify the locus-specific expression of TEs. Our differential expression analysis highlighted that the KO of *DNMT1* leads to a strong upregulation of TEs: 3015 TEs result upregulated and 277 downregulated. Interestingly, 1660 L1s were upregulated, resulting the most upregulated TE family in this experiment, as shown in Figure 1A.

Considering that ~99% of L1 RNA sequences in human arise from L1s embedded in other transcripts rather than from L1 promoters [59], we aimed to confirm the autonomous transcription of L1 elements in this model. In the method we developed (Figure 1B), paired-end reads are first aligned on the L1 consensus sequence and then on the human reference genome. The goal is to identify the number of fragments that are completely aligned inside the L1 elements (*Inside* fragments) and the number of fragments with a read aligned inside an L1 and the other one mapping outside the L1 on the reference genome (*Outside* fragments). We reasoned that the sequencing of L1 autonomous transcripts should produce mostly paired-end reads mapping within the internal part of L1 RNAs, hence mainly producing *Inside* fragments. Accordingly, by computing the *Inside/Outside* ratio, high ratio levels should result when L1 elements are autonomously transcribed. Applying our methodology to the *DNMT1* experiment, we were able to calculate a mean *Inside/Outside* ratio of 7.53 in the control group and a mean ratio of 9.33 in the KO group. A significantly (*t*-test *p*-value = 0.018) higher *Inside/Outside* ratio was observed in the *DNMT1* KO samples with respect to controls (Figure 1C), confirming that the observed upregulation of L1s is due to autonomous and independent L1 transcription and that the KO of *DNMT1* causes the transcriptional activation of genomic L1 loci.

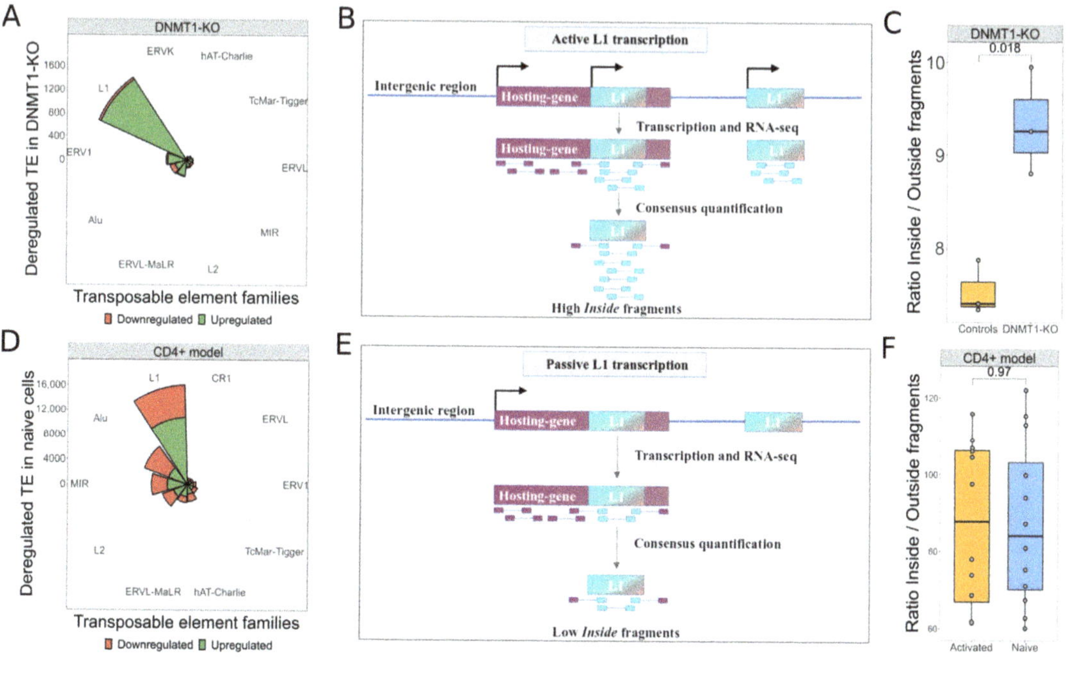

Figure 1. The KO of *DNMT1* gene leads to autonomous L1 transcription. (**A**). Top-10 deregulated TE families in *DNMT1* model. L1 is the most upregulated TE family with 1660 elements. (**B**). Rationale of our method for detecting autonomous transcription of L1 elements. Autonomous transcription of L1s will produce a higher amount of *Inside* fragments with both reads mapped inside the L1 consensus resulting in a higher *Inside/Outside* ratio. (**C**). Analysis to detect autonomous transcription of L1 elements. Upon the KO of *DNMT*, the upregulation of L1s derives by autonomous transcription of elements. (**D**). Top-10 deregulated TE families in naive T *CD4+* cells. L1 is the most upregulated TE family with 10,794 elements. (**E**). Rationale of our method for detecting non-autonomous transcription of L1 elements. Transcription of L1s embedded in other transcriptional units will produce a low amount of Inside fragments. (**F**). Analysis to detect autonomous transcription of L1 elements. In naive T CD4+ cells, the upregulation of L1s derives by a non-independent transcription of elements. Given that the two distributions are similar the overexpression of L1s is due to their transcription as part of other transcriptional units.

In order to verify the capability of our method to identify situations in which L1 transcriptional upregulation is determined by non-autonomous and non-independent L1 transcription (i.e., L1s transcription mainly results from the transcription of their fragments as part of canonical genes), we took advantage of the Marasca et al. [31] dataset. In this study, the authors observed that naive CD4+ T cells transcribe L1-containing transcripts as non-canonical splicing variants. Upon cell activation, modifications in the splicing pattern induce the downregulation of these alternative transcripts promoting the canonical ones. We started quantifying the locus-specific expression of TE with *SQuIRE* [38]. Our differential expression analysis confirmed that naive cells, with respect to activated ones, are characterized by overexpression of L1 elements: 10,794 L1s are up-regulated and 5289 are down-regulated (Figure 1D). To assess if L1 overexpression is the result of L1s transcribed as part of other transcriptional units as Marasca et al. demonstrated, we applied our methodology to calculate the *Inside/Outside* ratio (Figure 1E). Applying our methodology to the *CD4+* experiment, we were able to calculate a mean *Inside/Outside* ratio of 87.43 in the naive group and a mean ratio of 87.12 in the active group. Since no statistically significant differences between the two groups were observed (Figure 1F), the *CD4+* model allowed us to confirm the lack of autonomous L1 transcription in naive T cells and therefore the reliability of our method to discriminate between the autonomous and non-autonomous transcription of L1 elements when upregulation of L1s is observed with standard procedures.

3.2. L1 Transcripts Could Act as ceRNA

In order to inspect the transcriptional changes characterizing *DNMT1 KO* in the experiment by Jonsson and colleagues, we carried out differential gene expression analysis: 2188 genes resulted upregulated and 627 downregulated. Functional enrichment analyses revealed that genes belonging to the piRNA pathway (GO:0034587) [13] and to p53 transcriptional gene network (WP:WP4963) [60] were significantly enriched among the upregulated genes, while several proliferation-related genesets (e.g., GO:0042127) [61] were globally enriched among downregulated genes. These enrichments suggest that cells might be responding to transcriptional activation of L1s and possibly to a genotoxic stimulus [62]. Once the cell response was established, we performed an overlap analysis of genomic coordinates between L1 and the genic structures of protein-coding genes. The results revealed that upregulated genes are significantly enriched (Z-score > 3) to be in overlap with upregulated L1 elements in all the considered genic structures, as indicated in Figure 2A. The most intriguing finding was the observation that the 3′UTRs of upregulated genes were also significantly enriched to host upregulated L1 elements.

We then wondered about the existence of a possible ceRNA activity mediated by L1 transcripts. In our hypothesis, overexpressed L1 transcripts share miRNA target sites with the 3′UTRs of specific gene target sets. In this way, L1s might be able to recruit miRNAs from the target transcripts. As a result of this mechanism, transcripts from which miRNAs are sequestered should result upregulated. In order to verify this hypothesis, we selected the 799 upregulated extragenic L1s to represent the group of potentially autonomously transcribed L1 elements (referred to as *active* L1s). This set was chosen since they should less likely be involved in passive transcription as part of canonical transcripts [54], therefore reducing possible background noise from exonized L1s. We then identified the miRNA target sites both in the *active* L1s and in the 3′UTRs of protein-coding genes. In this pilot study we exclusively searched for exact matches to 8-mer seed-matched sites because it is believed to be the most effective single canonical site [19]. From this analysis, upregulated genes resulted to share on average 147 miRNA target sites with *active* L1s while the downregulated ones showed an average of about 114 miRNA target sites in common with the *active* L1s. The difference between the two groups is significant and indicates that upregulated genes, with respect to downregulated ones, have a significantly higher number of miRNA target sites in common with *active* L1s (Figure 2B, *p*-value = 2.35^{-7}; Supplementary Table S1). This result suggests that the group of upregulated genes, sharing more miRNA target sites with autonomously transcribed L1s, might be subjected to a possible ceRNA activity mediated by upregulated L1s, adding support to our hypothesis.

In an attempt to identify the miRNAs that could bind *active* L1s, we hypothesized that genes usually targeted by these should result mostly upregulated. With this in mind, we analyzed 2563 miRNAs that were targeting at least one *active* L1s. For each miRNA, the protein-coding genes containing target sites were classified as up or downregulated. Then, we searched for miRNAs in which the proportion of targeted upregulated genes was significantly (FDR < 0.1) higher than the proportion of downregulated ones. With this procedure, we identified 117 miRNAs that represent the most suitable pool of elements undergoing the putative ceRNA activity mediated by L1s (Figure 2C, Supplementary Table S2). Then, to understand which genes are putatively upregulated upon the L1 ceRNA activity, we used the MIENTURNET [51] tool. In this analysis, we provided the list of 117 identified miRNAs for discovering miRNA-genes interactions categorized in the miRTarBase database [52] as experimentally validated. The tool found 3878 miRNA-genes interactions involving 108 miRNAs and 543 genes. Of them, 57 were upregulated in the *DNMT1* model. The enrichment analysis of these 57 genes revealed that they are involved in processes that modulate transcription (GO:0006355) and particularly the p53 transcriptional program (WP4963). These results suggest that a putative L1-miRNA-transcript network might be involved in the regulation of genes that are part of a defense mechanism, specifically against L1 overexpression. Interestingly, four members of the let-7 family (let-7a-3p, let-7b-3p, let-7f-1-3p and miR-98-3p) [63], already known to target L1 transcripts, were found to be among the top-10 miRNAs with the highest number of connections to the upregulated genes (nine connections with respect to an average of 3.58 per miRNA, Figure 2D) which corroborates our hypothesis.

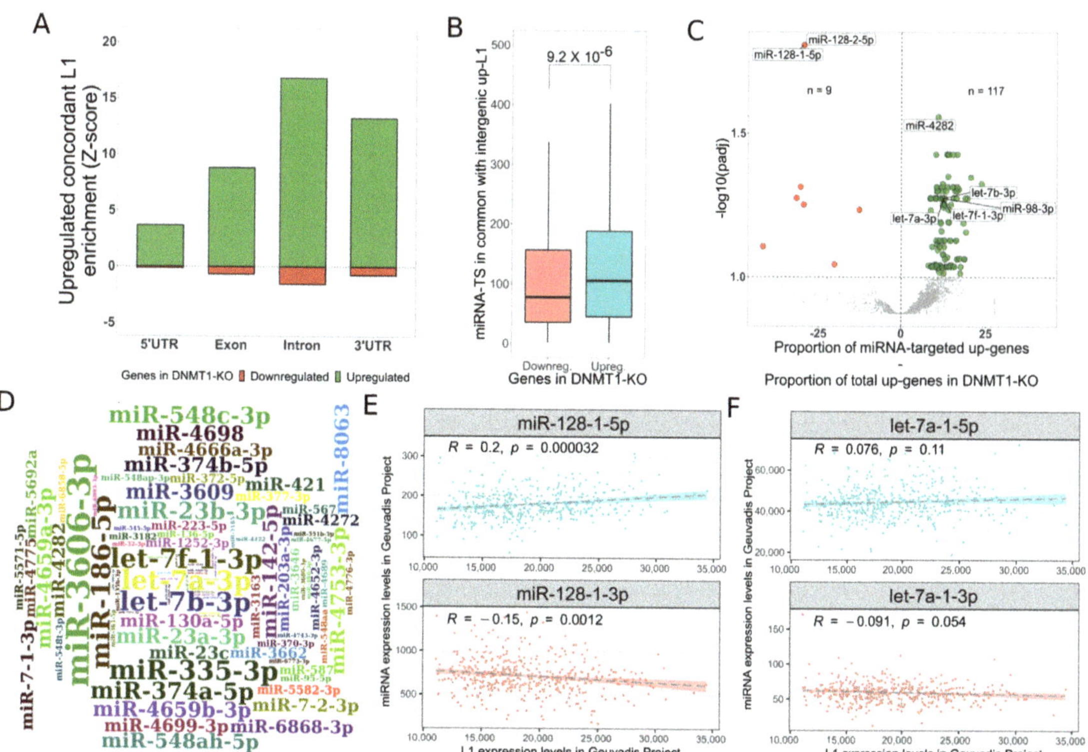

Figure 2. L1 transcripts might act as ceRNA. (**A**). Overlap analysis of L1s and protein-coding genes. Regions of upregulated genes are significantly enriched to contain upregulated L1 fragments. We found particularly interesting the strong enrichment in the 3′UTRs. (**B**). Analysis of miRNA target sites sharing between autonomously transcribed L1s and the 3′UTRs of protein-coding genes in *DNMT1* model. The upregulated genes share a significantly higher number of miRNA target sites with *active* L1s, adding support to a possible ceRNA activity of L1 transcripts. (**C**). Analysis to identify miRNAs sequestered by L1s. Each analyzed miRNA is represented by a point with the X-axis indicating the delta between the proportion of targeted upregulated genes and the total upregulated genes proportion. The Y-axis represents the -log10(FDR) of the proportional statistical test applied. The 117 miRNA in green represent the most probable pool of miRNAs that are undergoing the ceRNA activity from L1s. (**D**). In this wordcloud are shown the 117 miRNAs whose size font is proportional to the absolute number of experimentally validated target genes upregulated in *DNMT1* model. The let-7 family is among the top-10 miRNAs with the higher number of connections to these upregulated genes. (**E**). Correlation analysis between miR-128-1 (Y-axis) and L1 (X-axis) expression levels in Geuvadis dataset. Upon the L1 overexpression, the miR-128-1-5p levels concordantly increases probably as part of defense cellular mechanisms. (**F**). Correlation analysis between let-7a-1 (Y-axis) and L1 (X-axis) expression levels in Geuvadis dataset. No significant associations were found.

On the other hand, the strong enrichment for downregulated genes to share miR-128 targets with L1s seems to go against our hypothesis. MiR-128 has indeed been demonstrated to also target L1 transcripts [23], and therefore we were expecting to observe an enrichment in upregulated genes such as the one observed for members of the let-7 family. Instead, we observed the opposite with miR-128 resulting the most significant miRNA associated to downregulated genes (Figure 2C). In order to explain this difference, we took advantage of the Geuvadis dataset [34] which contains data from small RNAseq and RNAseq of lymphoblastoid cell lines from ~500 individual. These data allowed us to correlate the expression of all miRNAs (short RNAseq) with the expression of any other cellular transcript (RNAseq). We exploited them to evaluate the correlation between the expression of miR-128 and L1s and between let-7 and L1s. As shown in Figure 2E, the expression of the miR-128-1-5p results positively correlated (p-value = 0.000032) to L1 transcript levels. The putative passenger 3p strand results negatively correlated (p-value = 0.0012), reflecting its possible decay [64]. Conversely, as already observed [24], the expression levels of let-7-a-1 do not correlate with L1 transcript abundance (Figure 2F), suggesting that the expression of this miRNA can be interpreted as rather constant and independent by L1 expression levels. Overall, these results suggest that miR-128 levels increase when L1 transcript levels increase possibly as part of a response mechanism against TEs. The increased amount of miR-128 following L1 upregulation would then target not only L1 transcripts [65], but also, at least in part, canonical transcript targets of miR-128, causing their downregulation. This would not happen in the case of let-7, because its transcription does not change following L1 upregulation.

3.3. Support for L1 ceRNA Activity Using an Independent Experiment in Which a Specific L1 Is Artificially Overexpressed

To add support to our model, we analyzed the Ardeljan et al. [32] dataset composed of RNA-seq data derived from retinal pigment epithelial cells (RPE). In this cellular model, three samples overexpressing a codon-optimized L1 ORFeus construct were compared to three control samples. The aim of our analyses was to validate the activity of L1 as a ceRNA exploiting a cellular context perturbed exclusively by the overexpression of a single L1 without affecting DNA methylation. To assess the capability of *ORFeus-OE* model to overexpress the artificial L1 construct, we used the consensus-specific tool *TEspeX* [54], quantifying TE transcription levels. From the differential expression analysis, seven TE subfamilies were upregulated and two downregulated. The strongest (log2FC = 10.92) upregulated TE was the L1 construct overexpressed in the experiment, as visible in Figure 3A. To confirm the autonomous transcription of the L1 construct, our methodology was applied by using the L1 ORFeus sequence as consensus and without considering the genome mapping step in the pipeline (see Methods) because the expression construct is of exogenous origin. With this procedure, we calculated a mean *Inside/Outside* ratio of 2.94 in the control group and a mean of 10.11 in the ORFeus-OE samples. Hence, we observed a significantly (*t*-test, p-value = 0.00036) higher *Inside/Outside* ratio for the ORFeus-OE samples with respect to controls (Figure 3B) confirming the autonomous increase in the amount of L1 transcripts in this experimental setting.

To understand if the overexpressed L1 construct could be acting as ceRNA, we performed the differential gene expression analysis, identifying 3352 upregulated genes and 2890 downregulated ones. Then, we identified miRNA target sites in the L1 construct and in the 3′UTRs of protein-coding genes. Upregulated genes resulted to share an average of 10.38 miRNA target sites with the L1 construct. This is significantly higher with respect to the downregulated ones, showing an average of 9.38 (Figure 3C, p-value = 0.014). Moreover, in this case, our observations support the idea that the group of genes sharing a higher number of miRNA target sites with the L1 overexpressed construct could be undergoing ceRNA activity, resulting in their upregulation.

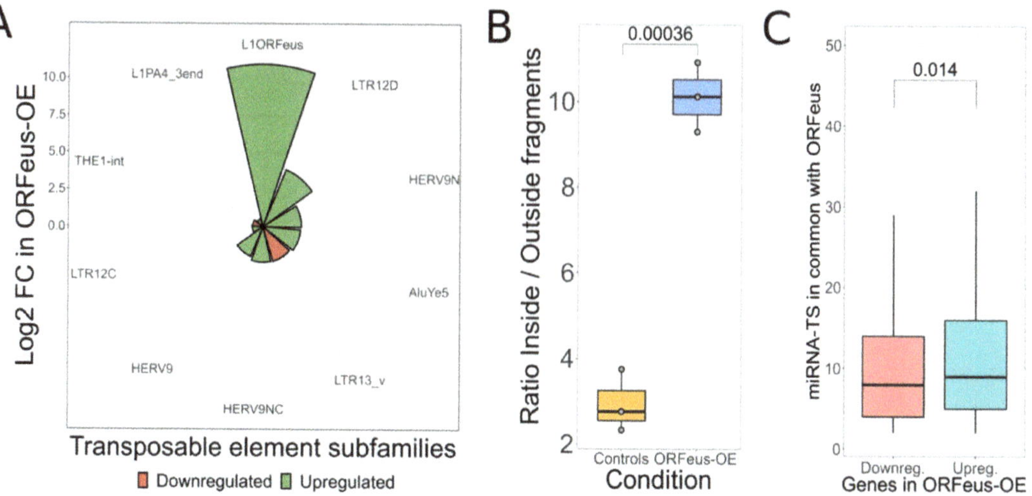

Figure 3. L1 could act as ceRNA when artificially overexpressed. (**A**). TE subfamilies significantly deregulated in the ORFeus-OE model. In this experiment, the L1 construct is the most upregulated TE with a log2 fold-change of 10.92. (**B**). Analysis to detect autonomous transcription of L1. In ORFeus-OE cells, the upregulation of L1 is deriving from an autonomous transcription of the ORFeus element. (**C**). Analysis of miRNA target sites sharing between artificial L1 construct and the 3′UTRs of protein-coding genes in ORFeus-OE model. The upregulated genes share a significantly high number of miRNA target sites with the L1 construct, reflecting a possible ceRNA activity of the artificial construct transcript.

3.4. Putative L1 ceRNA Activity Might Depend on Autonomous L1 Transcription and AGO2

Having collected evidence that overexpressed L1 transcripts might work as ceRNA, we decided to investigate other conditions leading to L1 deregulation. To this end, we analyzed a cellular model in which Deneault et al. [33] sequenced RNA of cells characterized by the KO of *ATRX* gene in reprogrammed human induced pluripotent stem cells (iPSC) and in iPSC differentiated in neuronal cells. *ATRX* is a chromatin remodeler gene that causes epigenetic modifications in retrotransposons *loci* [66]. To explore TE transcriptomic levels in the *ATRX* model, we used *SQuIRE* [38] to quantify the *locus*-specific expression of TEs. The differential expression analysis revealed a clear deregulation of TEs. Particularly, in iPSC cells the deregulation was more evident with 4490 up and 1217 down-regulated TEs, while the neuronal counterpart had 1820 up and 540 down-regulated elements. A common feature of both cell lines was that ~50% of upregulated TEs was represented by L1 elements: 2438 were the upregulated TEs in iPSC cells and 965 in neurons (Figure 4A). As observed in the *DNMT1* and *ORFeus-OE* models, in this experimental setting, the L1 is again the most upregulated TE family.

To investigate if the upregulation of L1s derived from a general active or passive TEs transcription, we applied our previously described method (Figure 4B). In iPSC cells, there was a clear difference in the *Inside/Outside* ratio between the control and the *ATRX*-KO groups (mean controls = 28.1 and mean *ATRX*-KO = 47.0). The neuronal cells, instead, were characterized by similar ratio between the two different groups (mean controls = 24.3 and mean *ATRX*-KO = 27.0). Therefore, while autonomous L1 transcription can be confirmed for iPSC cells (t-test, *p*-value = 0.0013), this is not the case for the neuronal cells (t-test, *p*-value = 0.77). These results suggest that neuronal cells carrying the *ATRX* KO, differently from the iPSC, might not undergo autonomous L1s upregulation. To explore the effects of L1 transcripts on the putative ceRNA dynamics, we identified the deregulated protein-coding genes. In iPSC, we found 818 up- and 1441 down-regulated genes while in neurons, they were respectively 319 and 369. We then compared miRNA target sites between the *active* L1s (1181 in iPSC and 479 in neurons) and the 3′UTRs of protein-coding genes. In contrast to *DNMT1* and *ORFeus-OE* models, the downregulated genes shared a significantly (p-value < 0.5) higher number of miRNA target sites with the *active* L1s in both experiments (Figure 4C). These results go against our hypothesis. While this could be explained for neuronal cells, since L1 transcripts could be passively transcribed as part of hosting genes which would decrease their capability to sequester miRNAs, we need a different explanation for the iPSC experiment.

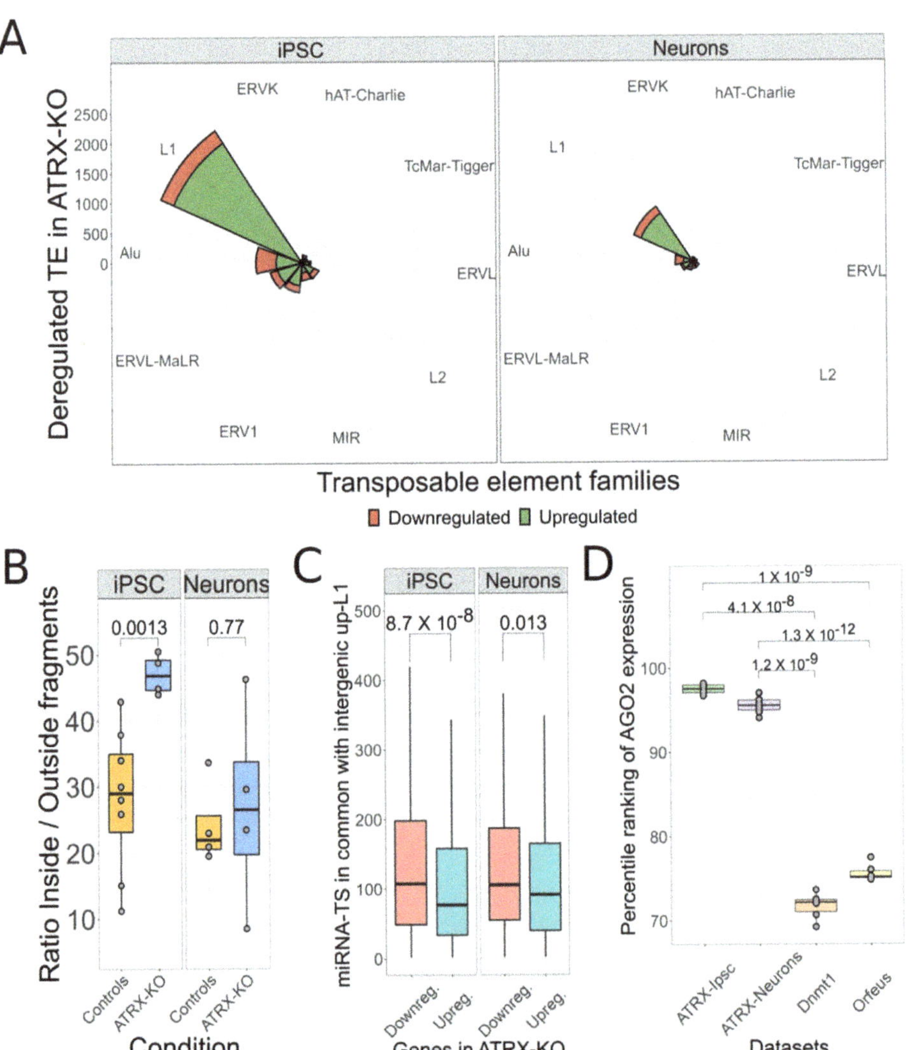

Figure 4. L1 ceRNA activity potentially relies on autonomous L1 transcription and Ago2 levels. (**A**). Top-10 deregulated TE families in *ATRX* model. Both in iPSC cells and in neurons, L1 family is the most upregulated TE family. (**B**). Analysis to detect autonomous transcription of L1 elements in *ATRX* model. Upon the KO of ATRX, the upregulation of L1 elements seems not to derive from an autonomous transcription of elements in neurons. (**C**). Analysis of miRNA target sites sharing between *active* L1s and the 3′UTRs of protein-coding genes in *ATRX* model. Downregulated genes show a higher number of miRNA target sites in common with overexpressed L1s. (**D**). Comparative analysis of Ago2 expression levels in all analyzed datasets.

The current understanding of the ceRNAs pathway indicates that the limiting factor for the ceRNA effect is the availability of Ago2 protein and not the amount of miRNA molecule. Indeed, it has been shown that a low amount of Ago2 is crucial for the ceRNA effect to take place while a high amount of it does not give raise to any target sites competition because there is no limitation of molecules for which to compete [67]. We therefore decided to investigate the expression levels of *AGO2* in a comparative manner between all the considered experiments. To make the expression levels of *AGO2* comparable among different experiments, we transformed the normalized counts related to *AGO2* expression in the corresponding percentile, as calculated with respect to the distribution of counts for all the transcripts relatively to each analyzed sample. From this analysis we observed that *AGO2* is expressed at a much higher level in the *ATRX* experiment (~95 percentile) with respect to the *DNMT1* and ORFeus (~73 percentile) ones (Figure 4D). These observations suggest that the reason why the *ATRX* experiment does not support a general L1s ceRNA effect could be due to the lack of a limiting dose of the Ago2 protein.

4. Discussion

L1 is the unique TE known to be still autonomously mobile in the human genome [1]. Although some elements are co-opted for cellular beneficial functions [6], deregulated L1 activity can be detrimental to cells [10]. As a result, cells use several mechanisms to regulate L1 transcript levels like the interfering activity of miRNAs [23]. Beyond this, TEs embedded in lncRNA are supposed to provide miRNA target sites competing with target genes for the same pool of miRNAs [29]. Despite the increasing interest in this field, a full comprehension of the regulatory layers in which L1 transcripts are involved is still lacking. Furthermore, it remains unclear how their expression levels are regulated and how their deregulation might impact transcriptional programs. In this work, we reanalyzed RNA-seq data of cells carrying mutations in genes affecting L1 transcript levels or overexpressing a specific L1 construct.

Analyzing the *DNMT1*-KO model [30], we confirmed a strong transcriptional activation of L1 loci following global demethylation. The upregulated protein-coding genes were enriched to contain L1 element fragments especially in their introns and 3′UTRs. This made us speculate that L1 transcripts might have ceRNA activity, competing with miRNAs normally targeting genes, resulting in upregulation. To support this hypothesis, we demonstrated that the upregulated genes, with respect to the downregulated ones, share a higher number of miRNA target sites with the autonomously transcribed L1 elements. This feature is in support of a possible L1 ceRNA activity. In our model, overexpressed L1 transcripts sharing more miRNA target sites with the 3′UTR of a given set of transcripts are prone to recruit miRNAs from them. As a consequence, transcripts from which more miRNAs are sequestered result upregulated. Identifying the putatively sequestered miRNAs, we found a pool of 117 miRNAs including different miRNAs belonging to the let-7 family. This family is experimentally validated to bind the ORF2 of L1 transcripts [24], supporting the reliability of our experimental procedure for identifying miRNAs sequestered by L1s. Analyzing the genes that should be upregulated upon L1 ceRNA activity, we found an enrichment for genes involved in the p53 transcriptional gene network. p53 is a tumor suppressor gene that induces transcriptional programs for responding to a variety of stress signals. Among these, direct [60] or indirect [68] functions are shown to control L1 activity. The L1 ceRNA activity might be a mechanism that induces a cellular response, like p53 targets transcription, which in turn act against L1 when these elements are overexpressed. The competition between L1 transcripts and canonical transcripts for the binding of a selected pool of miRNAs, might cause the transcriptional activation of genes that coherently work to repress L1 activities, protecting the well-being of cells. In this experiment, we also found an enrichment for downregulated genes to share miR-128 targets with L1s, which goes against our hypothesis since also this miRNA has been demonstrated to target L1s. Nevertheless, our model still holds if we take into account that the expression of miR-128 might increase or be induced in *DNMT1*-KO cells, which is reasonable because the overexpression of miR-128 can also, at least in part, justify the increase we have observed in the expression of p53 target-genes [69]. In addition, the analysis of Geuvadis dataset clearly shows a positive correlation between miR-128 expression and L1 transcript levels in lymphoblastoid cell lines, suggesting that, conversely to let-7 miRNAs whose transcription does not change, an induction in the expression of mir-128 happens in response to L1 overexpression.

We then analyzed an artificial condition with L1 RNA overexpression driven from a synthetic plasmid [32]. Further, in this case, we confirmed that upregulated genes are enriched for genes sharing a higher number of miRNA targets with the overexpressed L1 adding further support to a possible L1 ceRNA activity.

As a third study with overexpression of L1s, we analyzed a dataset derived from knocking out the *ATRX* gene in cultured neurons and iPSC cell lines as a model of autism [33]. Initially, an enrichment was found for shared miRNA targets among genes downregulated upon *ATRX*-KO. While these data were in contrast with our hypothesis, we took then into account that the ceRNA effect requires that Ago2 protein is expressed in limiting amounts. When Ago2 is in high quantity, the competing effect does not manifest because other molecules such as miRNAs are generally present in non-limiting abundance [67,70]. Reassuringly, the amount of *AGO2* mRNA in the *ATRX*-KO cell model was significantly higher with respect to those in the *DNMT1*-KO and in the Orfeus construct experiments, thus further supporting our model.

In conclusion, we have shown that cellular conditions, with strong autonomous L1 transcription, are characterized by the sharing of miRNA target sites between overexpressed L1s and upregulated genes when Ago2 availability is limited. This sharing might be at the basis of a competition for miRNAs targeting. The sequestering of miRNAs by L1 transcripts could then result in the upregulation of a given set of transcripts. Thus, our study provides initial evidence in support of the hypothesis of transposon acting as ceRNAs. In this model, the ceRNA activity might even result as a way for the cell to trigger defense mechanisms, such as the p53 transcriptional program, when the L1 transcript levels overcome a certain threshold.

Our results and evidence have been so far produced exclusively in-silico, and therefore specific validation of the model herein proposed is needed to ascertain the pervasive activity of TEs at the ceRNAs level. Once experimentally validated, we believe that our hypothesis will help future studies for dissecting cellular responses in both developmental [71] and pathological [10] conditions characterized by the transcriptional activation of L1 elements.

Supplementary Materials: The following supporting information can be downloaded at: https://www.mdpi.com/article/10.3390/biomedicines10123279/s1, Table S1: Information about miRNA target sites in common between protein-coding genes and intergenic upregulated L1 in *DNMT1* experiment; Table S2: Number of deregulated target genes for each analyzed miRNA in *DNMT1* experiment.

Author Contributions: Conceptualization, R.S. and M.E.; formal analysis, M.E.; methodology, M.E., N.G., G.S. and R.S.; software, M.E. and R.S.; supervision, R.S. and F.A.; visualization, M.E.; writing—original draft, M.E.; writing—review & editing, R.S., M.E., N.G., G.S., F.A. and S.G. All authors have read and agreed to the published version of the manuscript.

Funding: This research received no external funding.

Institutional Review Board Statement: Not applicable.

Informed Consent Statement: Not applicable.

Data Availability Statement: RNA-seq data from *DNMT1*-KO used in this study are available in the ENA-EBI repository (Accession code: PRJNA420729) https://www.ebi.ac.uk/ena/browser/view/PRJNA420729 (accessed on 19 November 2020); RNA-seq data from CD4$^+$ used in this study are available in the ENA-EBI repository (Accession code: PRJEB41930) https://www.ebi.ac.uk/ena/browser/view/PRJEB41930 (accessed on 8 March 2022); RNA-seq data from Geuvadis project used in this study are available in the ENA-EBI repository (Accession code: PRJEB3366) https://www.ebi.ac.uk/ena/browser/view/PRJEB3366 (accessed on 19 November 2020); Short RNA-seq data from Geuvadis project used in this study are available in the ArrayExpress repository https://www.ebi.ac.uk/

arrayexpress/files/E-GEUV-3/analysis_results/GD452.MirnaQuantCount.1.2N.50FN.samplename resk10.txt (accessed on 27 October 2022); RNA-seq data from ORFeus used in this study are available in the ENA-EBI repository (Accession code: PRJNA491205) https://www.ebi.ac.uk/ena/browser view/PRJNA491205 (accessed on 19 July 2021); RNA-seq data from *ATRX*-KO used in this study are available in the ENA-EBI repository (Accession code: PRJNA422099) https://www.ebi.ac.uk/ena browser/view/PRJNA422099 (accessed on 18 September 2021).

Conflicts of Interest: The authors declare no conflict of interest.

References

1. Wicker, T.; Sabot, F.; Hua-Van, A.; Bennetzen, J.L.; Capy, P.; Chalhoub, B.; Flavell, A.; Leroy, P.; Morgante, M.; Panaud, O.; et al. A Unified Classification System for Eukaryotic Transposable Elements. *Nat. Rev. Genet.* **2007**, *8*, 973–982. [CrossRef] [PubMed]
2. Lander, E.S.; Linton, L.M.; Birren, B.; Nusbaum, C.; Zody, M.C.; Baldwin, J.; Devon, K.; Dewar, K.; Doyle, M.; FitzHugh, W.; et al. Initial Sequencing and Analysis of the Human Genome. *Nature* **2001**, *409*, 860–921. [CrossRef] [PubMed]
3. Morrish, T.A.; Gilbert, N.; Myers, J.S.; Vincent, B.J.; Stamato, T.D.; Taccioli, G.E.; Batzer, M.A.; Moran, J.V. DNA Repair Mediated by Endonuclease-Independent LINE-1 Retrotransposition. *Nat. Genet.* **2002**, *31*, 159–165. [CrossRef] [PubMed]
4. Mills, R.E.; Bennett, E.A.; Iskow, R.C.; Luttig, C.T.; Tsui, C.; Pittard, W.S.; Devine, S.E. Recently Mobilized Transposons in the Human and Chimpanzee Genomes. *Am. J. Hum. Genet.* **2006**, *78*, 671–679. [CrossRef]
5. Sultana, T.; van Essen, D.; Siol, O.; Bailly-Bechet, M.; Philippe, C.; Zine El Aabidine, A.; Pioger, L.; Nigumann, P.; Saccani, S.; Andrau, J.-C.; et al. The Landscape of L1 Retrotransposons in the Human Genome Is Shaped by Pre-Insertion Sequence Biases and Post-Insertion Selection. *Mol. Cell* **2019**, *74*, 555–570.e7. [CrossRef]
6. Kazazian, H.H. Mobile Elements: Drivers of Genome Evolution. *Science* **2004**, *303*, 1626–1632. [CrossRef]
7. Jordan, I.K.; Rogozin, I.B.; Glazko, G.V.; Koonin, E.V. Origin of a Substantial Fraction of Human Regulatory Sequences from Transposable Elements. *Trends Genet. TIG* **2003**, *19*, 68–72. [CrossRef]
8. Ni, J.Z.; Grate, L.; Donohue, J.P.; Preston, C.; Nobida, N.; O'Brien, G.; Shiue, L.; Clark, T.A.; Blume, J.E.; Ares, M. Ultraconserved Elements Are Associated with Homeostatic Control of Splicing Regulators by Alternative Splicing and Nonsense-Mediated Decay. *Genes Dev.* **2007**, *21*, 708–718. [CrossRef]
9. Spengler, R.M.; Oakley, C.K.; Davidson, B.L. Functional MicroRNAs and Target Sites Are Created by Lineage-Specific Transposition. *Hum. Mol. Genet.* **2014**, *23*, 1783–1793. [CrossRef]
10. Rangasamy, D.; Lenka, N.; Ohms, S.; Dahlstrom, J.E.; Blackburn, A.C.; Board, P.G. Activation of LINE-1 Retrotransposon Increases the Risk of Epithelial-Mesenchymal Transition and Metastasis in Epithelial Cancer. *Curr. Mol. Med.* **2015**, *15*, 588–597. [CrossRef]
11. Sanchez-Luque, F.J.; Kempen, M.-J.H.C.; Gerdes, P.; Vargas-Landin, D.B.; Richardson, S.R.; Troskie, R.-L.; Jesuadian, J.S.; Cheetham, S.W.; Carreira, P.E.; Salvador-Palomeque, C.; et al. LINE-1 Evasion of Epigenetic Repression in Humans. *Mol. Cell* **2019**, *75*, 590–604.e12. [CrossRef] [PubMed]
12. Hatanaka, Y.; Inoue, K.; Oikawa, M.; Kamimura, S.; Ogonuki, N.; Kodama, E.N.; Ohkawa, Y.; Tsukada, Y.; Ogura, A. Histone Chaperone CAF-1 Mediates Repressive Histone Modifications to Protect Preimplantation Mouse Embryos from Endogenous Retrotransposons. *Proc. Natl. Acad. Sci. USA* **2015**, *112*, 14641–14646. [CrossRef] [PubMed]
13. De Fazio, S.; Bartonicek, N.; Di Giacomo, M.; Abreu-Goodger, C.; Sankar, A.; Funaya, C.; Antony, C.; Moreira, P.N.; Enright, A.J.; O'Carroll, D. The Endonuclease Activity of Mili Fuels PiRNA Amplification That Silences LINE1 Elements. *Nature* **2011**, *480*, 259–263. [CrossRef] [PubMed]
14. Servant, G.; Streva, V.A.; Derbes, R.S.; Wijetunge, M.I.; Neeland, M.; White, T.B.; Belancio, V.P.; Roy-Engel, A.M.; Deininger, P.L. The Nucleotide Excision Repair Pathway Limits L1 Retrotransposition. *Genetics* **2017**, *205*, 139–153. [CrossRef]
15. Lee, R.C.; Ambros, V. An Extensive Class of Small RNAs in Caenorhabditis Elegans. *Science* **2001**, *294*, 862–864. [CrossRef]
16. Nguyen, T.A.; Jo, M.H.; Choi, Y.-G.; Park, J.; Kwon, S.C.; Hohng, S.; Kim, V.N.; Woo, J.-S. Functional Anatomy of the Human Microprocessor. *Cell* **2015**, *161*, 1374–1387. [CrossRef]
17. Grishok, A.; Pasquinelli, A.E.; Conte, D.; Li, N.; Parrish, S.; Ha, I.; Baillie, D.L.; Fire, A.; Ruvkun, G.; Mello, C.C. Genes and Mechanisms Related to RNA Interference Regulate Expression of the Small Temporal RNAs That Control C. Elegans Developmental Timing. *Cell* **2001**, *106*, 23–34. [CrossRef]
18. Rivas, F.V.; Tolia, N.H.; Song, J.-J.; Aragon, J.P.; Liu, J.; Hannon, G.J.; Joshua-Tor, L. Purified Argonaute2 and an SiRNA Form Recombinant Human RISC. *Nat. Struct. Mol. Biol.* **2005**, *12*, 340–349. [CrossRef]
19. Bartel, D.P. MicroRNAs: Target Recognition and Regulatory Functions. *Cell* **2009**, *136*, 215–233. [CrossRef]
20. Jo, M.H.; Shin, S.; Jung, S.-R.; Kim, E.; Song, J.-J.; Hohng, S. Human Argonaute 2 Has Diverse Reaction Pathways on Target RNAs. *Mol. Cell* **2015**, *59*, 117–124. [CrossRef]
21. Behm-Ansmant, I.; Rehwinkel, J.; Doerks, T.; Stark, A.; Bork, P.; Izaurralde, E. MRNA Degradation by MiRNAs and GW182 Requires Both CCR4:NOT Deadenylase and DCP1:DCP2 Decapping Complexes. *Genes Dev.* **2006**, *20*, 1885–1898. [CrossRef] [PubMed]
22. Braun, J.E.; Truffault, V.; Boland, A.; Huntzinger, E.; Chang, C.-T.; Haas, G.; Weichenrieder, O.; Coles, M.; Izaurralde, E. A Direct Interaction between DCP1 and XRN1 Couples MRNA Decapping to 5′ Exonucleolytic Degradation. *Nat. Struct. Mol. Biol.* **2012**, *19*, 1324–1331. [CrossRef] [PubMed]

3. Hamdorf, M.; Idica, A.; Zisoulis, D.G.; Gamelin, L.; Martin, C.; Sanders, K.J.; Pedersen, I.M. MiR-128 Represses L1 Retrotransposition by Binding Directly to L1 RNA. *Nat. Struct. Mol. Biol.* **2015**, *22*, 824–831. [CrossRef] [PubMed]

4. Tristán-Ramos, P.; Rubio-Roldan, A.; Peris, G.; Sánchez, L.; Amador-Cubero, S.; Viollet, S.; Cristofari, G.; Heras, S.R. The Tumor Suppressor MicroRNA Let-7 Inhibits Human LINE-1 Retrotransposition. *Nat. Commun.* **2020**, *11*, 5712. [CrossRef] [PubMed]

5. Salmena, L.; Poliseno, L.; Tay, Y.; Kats, L.; Pandolfi, P.P. A CeRNA Hypothesis: The Rosetta Stone of a Hidden RNA Language? *Cell* **2011**, *146*, 353–358. [CrossRef] [PubMed]

6. Poliseno, L.; Salmena, L.; Zhang, J.; Carver, B.; Haveman, W.J.; Pandolfi, P.P. A Coding-Independent Function of Gene and Pseudogene MRNAs Regulates Tumour Biology. *Nature* **2010**, *465*, 1033–1038. [CrossRef] [PubMed]

7. Johnson, R.; Guigó, R. The RIDL Hypothesis: Transposable Elements as Functional Domains of Long Noncoding RNAs. *RNA* **2014**, *20*, 959–976. [CrossRef]

8. Carrieri, C.; Cimatti, L.; Biagioli, M.; Beugnet, A.; Zucchelli, S.; Fedele, S.; Pesce, E.; Ferrer, I.; Collavin, L.; Santoro, C.; et al. Long Non-Coding Antisense RNA Controls Uchl1 Translation through an Embedded SINEB2 Repeat. *Nature* **2012**, *491*, 454–457. [CrossRef]

9. Fu, Z.; Li, G.; Li, Z.; Wang, Y.; Zhao, Y.; Zheng, S.; Ye, H.; Luo, Y.; Zhao, X.; Wei, L.; et al. Endogenous MiRNA Sponge LincRNA-ROR Promotes Proliferation, Invasion and Stem Cell-like Phenotype of Pancreatic Cancer Cells. *Cell Death Discov.* **2017**, *3*, 1–10. [CrossRef]

0. Jönsson, M.E.; Ludvik Brattås, P.; Gustafsson, C.; Petri, R.; Yudovich, D.; Pircs, K.; Verschuere, S.; Madsen, S.; Hansson, J.; Larsson, J.; et al. Activation of Neuronal Genes via LINE-1 Elements upon Global DNA Demethylation in Human Neural Progenitors. *Nat. Commun.* **2019**, *10*, 3182. [CrossRef]

1. Marasca, F.; Sinha, S.; Vadalà, R.; Polimeni, B.; Ranzani, V.; Paraboschi, E.M.; Burattin, F.V.; Ghilotti, M.; Crosti, M.; Negri, M.L.; et al. LINE1 Are Spliced in Non-Canonical Transcript Variants to Regulate T Cell Quiescence and Exhaustion. *Nat. Genet.* **2022**, *54*, 180–193. [CrossRef]

2. Ardeljan, D.; Steranka, J.P.; Liu, C.; Li, Z.; Taylor, M.S.; Payer, L.M.; Gorbounov, M.; Sarnecki, J.S.; Deshpande, V.; Hruban, R.H.; et al. Cell Fitness Screens Reveal a Conflict between LINE-1 Retrotransposition and DNA Replication. *Nat. Struct. Mol. Biol.* **2020**, *27*, 168–178. [CrossRef] [PubMed]

3. Deneault, E.; White, S.H.; Rodrigues, D.C.; Ross, P.J.; Faheem, M.; Zaslavsky, K.; Wang, Z.; Alexandrova, R.; Pellecchia, G.; Wei, W.; et al. Complete Disruption of Autism-Susceptibility Genes by Gene Editing Predominantly Reduces Functional Connectivity of Isogenic Human Neurons. *Stem Cell Rep.* **2018**, *11*, 1211–1225. [CrossRef] [PubMed]

4. Lappalainen, T.; Sammeth, M.; Friedländer, M.R.; 't Hoen, P.A.C.; Monlong, J.; Rivas, M.A.; Gonzàlez-Porta, M.; Kurbatova, N.; Griebel, T.; Ferreira, P.G.; et al. Transcriptome and Genome Sequencing Uncovers Functional Variation in Humans. *Nature* **2013**, *501*, 506–511. [CrossRef] [PubMed]

5. Athar, A.; Füllgrabe, A.; George, N.; Iqbal, H.; Huerta, L.; Ali, A.; Snow, C.; Fonseca, N.A.; Petryszak, R.; Papatheodorou, I.; et al. ArrayExpress Update—From Bulk to Single-Cell Expression Data. *Nucleic Acids Res.* **2019**, *47*, D711–D715. [CrossRef] [PubMed]

6. An, W.; Dai, L.; Niewiadomska, A.M.; Yetil, A.; O'Donnell, K.A.; Han, J.S.; Boeke, J.D. Characterization of a Synthetic Human LINE-1 Retrotransposon ORFeus-Hs. *Mob. DNA* **2011**, *2*, 2. [CrossRef] [PubMed]

7. Andrews, S. FastQC: A Quality Control Tool for High Throughput Sequence Data [Online]. 2010. Available online: http://www.bioinformatics.babraham.ac.uk/projects/fastqc/ (accessed on 20 November 2020).

8. Yang, W.R.; Ardeljan, D.; Pacyna, C.N.; Payer, L.M.; Burns, K.H. SQuIRE Reveals Locus-Specific Regulation of Interspersed Repeat Expression. *Nucleic Acids Res.* **2019**, *47*, e27. [CrossRef]

9. Love, M.I.; Huber, W.; Anders, S. Moderated Estimation of Fold Change and Dispersion for RNA-Seq Data with DESeq2. *Genome Biol.* **2014**, *15*, 550. [CrossRef]

0. Dombroski, B.A.; Scott, A.F.; Kazazian, H.H. Two Additional Potential Retrotransposons Isolated from a Human L1 Subfamily That Contains an Active Retrotransposable Element. *Proc. Natl. Acad. Sci. USA* **1993**, *90*, 6513–6517. [CrossRef]

1. Li, H.; Durbin, R. Fast and Accurate Short Read Alignment with Burrows-Wheeler Transform. *Bioinforma. Oxf. Engl.* **2009**, *25*, 1754–1760. [CrossRef]

2. Li, H.; Handsaker, B.; Wysoker, A.; Fennell, T.; Ruan, J.; Homer, N.; Marth, G.; Abecasis, G.; Durbin, R. 1000 Genome Project Data Processing Subgroup The Sequence Alignment/Map Format and SAMtools. *Bioinformatics* **2009**, *25*, 2078–2079. [CrossRef] [PubMed]

3. Quinlan, A.R.; Hall, I.M. BEDTools: A Flexible Suite of Utilities for Comparing Genomic Features. *Bioinformatics* **2010**, *26*, 841–842. [CrossRef] [PubMed]

4. Tarailo-Graovac, M.; Chen, N. Using RepeatMasker to Identify Repetitive Elements in Genomic Sequences. *Curr. Protoc. Bioinforma.* **2009**, *4*, 4–10. [CrossRef] [PubMed]

5. Karolchik, D.; Hinrichs, A.S.; Furey, T.S.; Roskin, K.M.; Sugnet, C.W.; Haussler, D.; Kent, W.J. The UCSC Table Browser Data Retrieval Tool. *Nucleic Acids Res.* **2004**, *32*, D493–D496. [CrossRef]

6. Hubbard, T.; Barker, D.; Birney, E.; Cameron, G.; Chen, Y.; Clark, L.; Cox, T.; Cuff, J.; Curwen, V.; Down, T.; et al. The Ensembl Genome Database Project. *Nucleic Acids Res.* **2002**, *30*, 38–41. [CrossRef]

7. Dobin, A.; Davis, C.A.; Schlesinger, F.; Drenkow, J.; Zaleski, C.; Jha, S.; Batut, P.; Chaisson, M.; Gingeras, T.R. STAR: Ultrafast Universal RNA-Seq Aligner. *Bioinformatics* **2013**, *29*, 15–21. [CrossRef]

48. Anders, S.; Pyl, P.T.; Huber, W. HTSeq–a Python Framework to Work with High-Throughput Sequencing Data. *Bioinforma. Ox Engl.* **2015**, *31*, 166–169. [CrossRef]
49. Kolberg, L.; Raudvere, U.; Kuzmin, I.; Vilo, J.; Peterson, H. Gprofiler2—An R Package for Gene List Functional Enrichmen Analysis and Namespace Conversion Toolset g:Profiler. *F1000Research* **2020**, *9*, ELIXIR-709. [CrossRef]
50. Griffiths-Jones, S.; Grocock, R.J.; van Dongen, S.; Bateman, A.; Enright, A.J. MiRBase: MicroRNA Sequences, Targets and Gen Nomenclature. *Nucleic Acids Res.* **2006**, *34*, D140–D144. [CrossRef]
51. Licursi, V.; Conte, F.; Fiscon, G.; Paci, P. MIENTURNET: An Interactive Web Tool for MicroRNA-Target Enrichment an Network-Based Analysis. *BMC Bioinformatics* **2019**, *20*, 545. [CrossRef]
52. Huang, H.-Y.; Lin, Y.-C.-D.; Li, J.; Huang, K.-Y.; Shrestha, S.; Hong, H.-C.; Tang, Y.; Chen, Y.-G.; Jin, C.-N.; Yu, Y.; et al. MiRTarBas 2020: Updates to the Experimentally Validated MicroRNA–Target Interaction Database. *Nucleic Acids Res.* **2020**, *48*, D148–D15 [CrossRef] [PubMed]
53. Chen, E.Y.; Tan, C.M.; Kou, Y.; Duan, Q.; Wang, Z.; Meirelles, G.V.; Clark, N.R.; Ma'ayan, A. Enrichr: Interactive and Collaborativ HTML5 Gene List Enrichment Analysis Tool. *BMC Bioinform.* **2013**, *14*, 128. [CrossRef] [PubMed]
54. Ansaloni, F.; Gualandi, N.; Esposito, M.; Gustincich, S.; Sanges, R. TEspeX: Consensus-Specific Quantification of Transposabl Element Expression Preventing Biases from Exonized Fragments. *Bioinformatics* **2022**, *38*, 4430–4433. [CrossRef] [PubMed]
55. Hubley, R.; Finn, R.D.; Clements, J.; Eddy, S.R.; Jones, T.A.; Bao, W.; Smit, A.F.A.; Wheeler, T.J. The Dfam Database of Repetitiv DNA Families. *Nucleic Acids Res.* **2016**, *44*, D81–D89. [CrossRef] [PubMed]
56. Hermann, A.; Goyal, R.; Jeltsch, A. The Dnmt1 DNA-(Cytosine-C5)-Methyltransferase Methylates DNA Processively with Hig Preference for Hemimethylated Target Sites. *J. Biol. Chem.* **2004**, *279*, 48350–48359. [CrossRef]
57. Lanciano, S.; Cristofari, G. Measuring and Interpreting Transposable Element Expression. *Nat. Rev. Genet.* **2020**, *21*, 721–73 [CrossRef]
58. Gualandi, N.; Iperi, C.; Esposito, M.; Ansaloni, F.; Gustincich, S.; Sanges, R. Meta-Analysis Suggests That Intron Retention Ca Affect Quantification of Transposable Elements from RNA-Seq Data. *Biology* **2022**, *11*, 826. [CrossRef]
59. Deininger, P.; Morales, M.E.; White, T.B.; Baddoo, M.; Hedges, D.J.; Servant, G.; Srivastav, S.; Smither, M.E.; Concha, M DeHaro, D.L.; et al. A Comprehensive Approach to Expression of L1 Loci. *Nucleic Acids Res.* **2017**, *45*, e31. [CrossRef]
60. Tiwari, B.; Jones, A.E.; Caillet, C.J.; Das, S.; Royer, S.K.; Abrams, J.M. P53 Directly Represses Human LINE1 Transposons. *Gene Dev.* **2020**, *34*, 1439–1451. [CrossRef]
61. Belgnaoui, S.M.; Gosden, R.G.; Semmes, O.J.; Haoudi, A. Human LINE-1 Retrotransposon Induces DNA Damage and Apoptosi in Cancer Cells. *Cancer Cell Int.* **2006**, *6*, 13. [CrossRef]
62. De Cecco, M.; Ito, T.; Petrashen, A.P.; Elias, A.E.; Skvir, N.J.; Criscione, S.W.; Caligiana, A.; Brocculi, G.; Adney, E.M Boeke, J.D.; et al. L1 Drives IFN in Senescent Cells and Promotes Age-Associated Inflammation. *Nature* **2019**, *566*, 73–7 [CrossRef] [PubMed]
63. Roush, S.; Slack, F.J. The Let-7 Family of MicroRNAs. *Trends Cell Biol.* **2008**, *18*, 505–516. [CrossRef] [PubMed]
64. Ha, M.; Kim, V.N. Regulation of MicroRNA Biogenesis. *Nat. Rev. Mol. Cell Biol.* **2014**, *15*, 509–524. [CrossRef] [PubMed]
65. Wee, L.M.; Flores-Jasso, C.F.; Salomon, W.E.; Zamore, P.D. Argonaute Divides Its RNA Guide into Domains with Distinc Functions and RNA-Binding Properties. *Cell* **2012**, *151*, 1055–1067. [CrossRef]
66. Sadic, D.; Schmidt, K.; Groh, S.; Kondofersky, I.; Ellwart, J.; Fuchs, C.; Theis, F.J.; Schotta, G. Atrx Promotes Heterochromatir Formation at Retrotransposons. *EMBO Rep.* **2015**, *16*, 836–850. [CrossRef]
67. Loinger, A.; Shemla, Y.; Simon, I.; Margalit, H.; Biham, O. Competition between Small RNAs: A Quantitative View. *Biophys. J* **2012**, *102*, 1712–1721. [CrossRef]
68. Wylie, A.; Jones, A.E.; D'Brot, A.; Lu, W.-J.; Kurtz, P.; Moran, J.V.; Rakheja, D.; Chen, K.S.; Hammer, R.E.; Comerford, S.A.; et a P53 Genes Function to Restrain Mobile Elements. *Genes Dev.* **2016**, *30*, 64–77. [CrossRef]
69. Adlakha, Y.K.; Saini, N. MicroRNA-128 Downregulates Bax and Induces Apoptosis in Human Embryonic Kidney Cells. *Cell Mo Life Sci.* **2011**, *68*, 1415–1428. [CrossRef]
70. Vickers, T.A.; Lima, W.F.; Nichols, J.G.; Crooke, S.T. Reduced Levels of Ago2 Expression Result in Increased SiRNA Competitior in Mammalian Cells. *Nucleic Acids Res.* **2007**, *35*, 6598–6610. [CrossRef]
71. Jachowicz, J.W.; Bing, X.; Pontabry, J.; Bošković, A.; Rando, O.J.; Torres-Padilla, M.-E. LINE-1 Activation after Fertilizatior Regulates Global Chromatin Accessibility in the Early Mouse Embryo. *Nat. Genet.* **2017**, *49*, 1502–1510. [CrossRef]

MDPI

St. Alban-Anlage 66

4052 Basel

Switzerland

www.mdpi.com

Biomedicines Editorial Office

E-mail: biomedicines@mdpi.com

www.mdpi.com/journal/biomedicines